COMBINATORIAL NUMBER THEORY

A Treatise on Growth, Based on the GOODSTEIN-SKOLEM Hierarchy,

Including a Critique of Non-Constructive or 1st Order Logic

COMBINATORIAL NUMBER THEORY

A Treatise on Growth, Based on the GOODSTEIN-SKOLEM Hierarchy,

Including a Critique of Non-Constructive or 1st Order Logic

Al-Haj Muhammad Ali McBETH,
Ph. D. (Lond.), C. Math.

Formerly Lecturer in Mathematics,
University of Khartoum, D. R. Sudan

The Edwin Mellen Press
Lewiston / Queenston / Lampeter

Library of Congress Cataloging-in-Publication Data

McBeth, Muhammad Ali.
Combinatorial number theory : a treatise on growth, based on the Goodstein-Skolem hierarchy, including a critique on non-constructive or first-order logic / Muhammad Ali Mc Beth.
p. cm.
Includes bibliographical references and index.
ISBN 0-7734-9085-X
1. Hierarchies. 2. First-order logic.
QA9.62.M38 1994
511.3'3--dc20

94-13953
CIP

A CIP catalog record for this book
is available from the British Library.

The Edwin Mellen Press
Box 450
Lewiston, New York
USA 14092-0450

The Edwin Mellen Press
Box 67
Queenston, Ontario
CANADA L0S 1L0

The Edwin Mellen Press, Ltd.
Lampeter, Dyfed, Wales
UNITED KINGDOM SA48 7DY
Tel. 01570 423356

Second printing, with corrections, January, 1996

Printed in Great Britain by
Antony Rowe Ltd, Chippenham, Wiltshire

PREFACE

This monograph in hierarchy arithmetic contains work done by myself and others, during the period 1970-1993. The content is unified and presented from the strictly finitist viewpoint.

There is a long list of acknowledgements owing, and at the very minimum I must include the following:

I thank Professor Angus MacINTYRE for encouragement in the writing of this book, from the time of a meeting at Greg-y-Nog, in September, 1986. It is also a pleasure to acknowledge the interest and support of Professor Crispin NASH-WILLIAMS.

In drafting Ch. 1, my debt to Michael MOSS for many long conversations became clear; these continued up until a few years before I received the sad news of his death. In particular the discussions centred on the pivotal importance of diagonalization; and the relation between **EP**, and GENTZEN's consistency proof for first order arithmetic. I discussed the ideas of Ch. II with Dr. John LAKE in 1977; and one of the reasons the paper [94] [1] was written, on which this book is based, was the unequivocal encouragement he gave me then.

Dr. David SINGMASTER has repeatedly fired my interest in pure mathematics; it is much hoped that this is evident in a work whose area belongs essentially to mathematical logic. I am particularly indebted for his detailed checking of the main arguments of Ch. IV.

Many early drafts of manuscripts of the main chapters were read by my friend and teacher, Dr. Geoffrey KNEEBONE. In the time since our meeting in 1970, I came to take his interest in the work for granted. His understanding and style are to be found in all the root ideas of the text.

Sincere thanks are due to Dr. John CLEAVE for reading the whole text and passing on many helpful comments.

I am indebted to Professor Wilfrid HODGES for helping with the proof-reading of Ch. II, and hardly less so for many past conversations. Also to Professor N. G. de BRUIJN, for supplying a comprehensive document on trees; and for general encouragement of ideas occurring in Ch. IV.

As a research student, Mr. Rob STACEY introduced himself after I had given a seminar at Bedford College, Regent's Park, in July, 1980, on the subject-matter of Ch. VII. From that moment I was impressed by a mathematician of appalling excellence, and of the highest level of independence. No idea on which I had worked gave him the least difficulty; on the other hand he generously shared his knowledge of classical mathematics with me. His effortless grasp of combinatorics was truly miraculous. He affirmed much of what is in Chs. II, III, IV, VI, VII, VIII, about half of which I had already published before we met. Rob gave me proofs of a

[1] Drafted, 1976; published, 1980.

surprising and intricate identity of CAYLEY, which links the sequence whose terms are the numbers of normal trees on respectively $1, 2, 3, \ldots$ vertices, to a certain infinite product; and of several identities determined by other classes of finite trees. Therefore all of the main results of Ch. XII were given to me by him. Rob died on 2nd March, 1982, aged 26.

I have to say that during a period of great personal hardship, the kind and courageous intervention of Prof. Maurice JASWON prevented my research activities from slipping out of my control.

More recently, the greatest kindness has been shown to me by Dr. Robin WHITTY. Robin encouraged me to give seminars at Goldsmith's College, on the content of Ch. IV. It is to him that I owe the most timely reminder of a gap (easily plugged) in an essential definition, namely that of $\mathfrak{T}$. Separate discussions with Robin and with David SINGMASTER lead me to seek the linear realization of **EP** set out in Ch. XI. On top of this, Robin introduced me to TeX computer typing.

I would like to mention my unrepayable debt to Dr. John HOGARTH: it was he who proved to me, in our student days, that mathematics is the root of science.

Finally, I offer my sincere apologies to the many others whose names I have omitted, whose selflessness towards me has made the work possible.

M. A. McB.

Goldsmiths' College,
University of London,
March, 1994.

SUMMARY OF CONTENTS

Contents

INTRODUCTION

Preamble

The Introduction contains a short account of the aims and content of each Chapter (Appendices included), and of some of their interrelations. It concludes with an indication of the place of the subject-matter in the technical literature; and of fixed notation conventions.

Aims and Content

This monograph is intended to provide a reliable and comprehensive source of reference in growth theory for graduate students and research workers in mathematics. The goal of the theory is to replace the GOEDEL scorpion, which goes on stinging itself for ever, by a hierarchy of paroxysmal upward acceleration. The text is also aimed at a wider readership, including all who feel inclined to re-examine their mathematical background, from a more exacting, finitist viewpoint.

The book is the outcome of work spanning 2 decades, in the area of fundamental sequences. From the outset, the finitist scruple centres attention on a realization of the first stage of CANTOR's $\varepsilon-$hierarchy: the system **EP**. The text includes published and unpublished original material by myself, besides standard results and constructions of W. ACKERMANN, H. BACHMANN, N. G. de BRUIJN, E. A. CICHON, A. EHRENFEUCHT, R. L. GOODSTEIN, G. H. HARDY, M. H. LÖB, G. MILLS, R. PÉTER, G. POLYA, Th. SKOLEM, R. STACEY, W. SIERPIŃSKI, S. S. WAINER and others.

The overall aim of the book is to lay a secure foundation, for a finitary attack on one of the great unsolved problems of 20th century pure mathematics: that of finding a complete system of fundamental sequences. The problem reveals, once again, the reflex nature of the field: the more successful are the partial solutions, the clearer it becomes how difficult is the global problem. A complementary aim is to establish that, besides being the first step of a larger construction, **EP** is an object of study of unique interest.

There is no call for an elaborate introduction, as the Chapters are in order of dependence. With the exception Ch. 1, forward references have been suppressed. Each Chapter has a brief introduction.

Since the arising of interest in the codification of mathematical reasoning, definite axiomatizations have become established through widespread usage. One of the most common is $\underline{N}$, the first-order version of the FREGE-DEDEKIND-PEANO formalization of arithmetic. [2] So-called finitizations, which comprise reformulation of classical theorems so that they become expressible in $\underline{N}$, such as FRIEDMAN's work on the KRUSKAL tree theorem, and MILL's reformulation of KÖNIG's lemma, have aims quite different from those of the present work. FRIEDMAN's reformulation of the homeomorphic embedding theorem is expressible in $\underline{N}$; but not provable, even in $\underline{N}+$ a 1st order version of {all GOODSTEIN sequences are finite}. It therefore

[2] See WANG 1957, 'The axiomatization of arithmetic', *J. Symbolic Logic*, **22**, 145-158.

seemed imperative to begin with a Chapter outlining what might be called 'elementary finitism'. This includes a critique of the *non-constructive* (viz. 1st or higher order) logics.

The definition and immediate dissection of the GOODSTEIN-SKOLEM hierarchy **EP** are contained in the (sub-)Title Chapter. **EP** is remarkable in 2 ways:

1. in the ε−correspondence mentioned above;

2. in the applications, or developments to which it gives rise.

Numerous definitions and results are accumulated for use in later Chs; the material is separated with intervals of discussion, e.g. of a higher-type approach to **EP**. Similarly with Ch. III: this introduces the most direct and elementary link between fundamental sequences and diagonalization. The main section treats the canonical assignment for **EP**; later sections deal with properties of the assignment and its applications, in particular to reduction of exponentials of arbitrary height. Ch. IV contains an exhaustive examination of a fundamental natural form inherent in the polynomials of **EP**, the normal tree. This is the bed-rock on which rests the proof of the descending chain condition, [3] DCC. All studies of rapidly growing arithmetic functions ultimately depend on this finiteness principle. Besides two main sections, respectively treating the recurrent lexicographic order, and the one-to-one correspondence between trees and exponentials, there are several subsidiaries. These introduce and reinforce the main sections, and pursue further details of the theory, particularly domination and the frequently-occurring concept of post-order. The reader who has progressed thus far will have detected the fairly tight interpenetration of arithmetic and combinatoric ideas. This is exploited in Ch. V. Given an arbitrary $e \in$ **EP**, suitable iteration of the canonical reduction of Ch. III uniquely characterizes e. The procedure is finite, by the DCC. This second normal form interlocks with that of Ch. II, via explicit coefficients in the post order. A further Ch., VI, is required in order to found rapidly-growing hierarchies on **EP**. The key device which it contains is the diagonalization of induction on the natural numbers. This arises naturally from the ideas of Chs. II, III. The theorems of generalized induction (GEN IND) and of the DCC are all but equivalent in power. Different routes, leading from one to a proof of the other theorem are treated in some detail, and a GOODSTEIN-type 'cumulative' proof of the GEN IND, free of dependence on the DCC, is given: GEN IND may be viewed as 'useful software'. The various forms of downward induction are examined, besides other relations between foundation and induction. What might be regarded as the goal of the book is reached in Ch. VII: a diagonalization of ACKERMANN's generalized exponential, based on Chs. II, III, IV and VI. The finiteness of the calculation of the generalized form based on $e \in$ **EP** at value y follows at once from the apparition of the (finite) plane-structure

[3] Garrett BIRKHOFF, *Lattice Theory*, Amer. Math. Soc. Publ. **25** (1940).

binary tree $T = T(e, y)$ described by the calculation. In §4 of Ch. VII it is proved that the calculation traverses T in pre-order.

Each one of the techniques of Chs. I-VII is applied with the intention of establishing that certain finitary constructions are possible. As detailed in the introduction to Ch. VIII, there are many interesting, if complex, techniques which contain the potential for carrying the construction further. However, only those which arise as a direct result of earlier material are taken up in Ch. VIII; the object being to probe the possibilities of advancement, using type-1 generalized polynomials. The first phase is the enlargement of **EP**; the second, diagonalization based on the various enlargements. The feather diagram is introduced and elaborated in later sections. This fundamental reduction of the structure of any infinite hierarchy promises clarification of many concepts in Mathematical Logic.

Ch. IX is a return to the subject-matter of Ch. II, with particular emphasis on the relation of *immediate domination*, and including a recently-published application of complete trees (Mc BETH, 1991); Chs. X-XII contain further elaborations of **EP** and its substructure $\mathfrak{T}$ of normal trees. The treatment of GOODSTEIN sequences of Ch. X is wholly based on **EP**, leading to new proofs of the finiteness of the GOODSTEIN reduction, and of the KIRBY-PARIS result that this is slower than canonical reduction. Applications to formal systems, usually independence results, e.g. BECKMAN and McALOON (1982); DELON, LASCAR and PARIGOT (1986) lie outside the confines of this volume, and the reader is referred to these sources, besides others cited at the end of Ch. X. It is an easy matter to 'lay out' **EP** on the real unit interval, using continued fractions. In Ch. XI this technique is applied to some of the ideas of earlier Chs. The final chapter seeks to establish the combinatorial foundation of **EP**.

It is hoped that the Appendices (A to E), itemized in the Table of Contents, will be of use to readers seeking background material in contemporary logic and elementary combinatorics. These are needed in order to assess the location and importance of the theory of hierarchies. About a dozen entries in the Bibliography were obtained by on-line search. The General Index, included to give fast access to key definitions, etc., is reasonably long and contains a number of redundancies.

It is not possible in the present volume, which concentrates on the prerequisites for, and the constructions of Chs. VII and VIII, to even survey the wider territory of the 'analysis of **EP**'. However the several diversions in this direction, throughout the text, are hopefully sufficient to show how interesting it is. Other topics occur together with lists of open problems in the End-of-Chapter Notes.

The headings covered are summarized before the table of contents. It should be mentioned that very few books in print contain substantial parts of the above material. HINMAN's *Recursion-Theoretic Hierarchies* (Springer, 1978) does not even contain a citation to WAINER's important 1970 paper on the classification of 'ordinal recursive functions'. Harvey ROSE' monograph *Subrecursion* (Oxford, 1984) is introduced as mostly '...ultraconstructivist or finitist'. Yet the basic issues of the acceptability of general recursiveness (ref. Ch. I, §3.1) and of (segments of) CANTOR's 'second number class' [4] (Ch. II, §2.18) are not confronted. Despite the importance of the order-type of **EP** under eventual domination (viz. that generally agreed to be required to obtain a consistency proof for $\underline{\mathbf{N}}$), even recent texts on mathematical logic do not cite SKOLEM's 1956 paper.

Finite trees arise in different places in the theory e.g. as the inherent normal tree stucture of an exponential, in calculations arising from the application of **EP** and in structures obtained by canonical reduction of the feather diagram. This creates objects of discourse, in many cases amenable to treatment by the techniques of classical combinatorics.

Some Abbreviations, Notations and Conventions

Aside from applications in the text, numbers enclosed in square brackets are citations to the bibliography. The abbreviations '$\neg$' ('not'), '&' ('and'), '$\vee$' ('or'), '$\Rightarrow$' ('implies'), '$\Leftrightarrow$', '$\equiv$' ('is equivalent to'), '$\forall$' ('for all') '$\exists$' ('there exists') and '$\exists!$' ('there exists exactly one') are used throughout. '$\Rightarrow$' is used in the non-strict sense, so that '$A \Rightarrow B$' allows the possibility '$A \equiv B$'. Otherwise, symbols such as '$\subset$,' '$<$'and '$\prec$' are used in the *strict* sense. References to articles, viz. theorems, lemmas, etc., are to those within the current chapter, unless prefixed by a chapter number. Roman numerals have been used for these prefixes. In most cases articles are delimited by the end-sign

$$\square.$$

The word 'section' and the sign '§', for reference purposes, are used interchangeably. Upper-case roman letters or letter-strings, in some cases bold face (e.g. **EP**) are used to denote predicates whose meaning does not vary throughout the text; calligraphic letters (e.g. $\mathcal{X}$) are used to denote predicates whose interpretation may vary from section to section. Some notation and frequently-used abbreviations are contained in Ch. II, §1.3. Generally letters only are used for abbreviations (e.g. 'si', 'ct' respectively for 'strictly increasing', and 'canonical term'). The abbreviation for 'induction hypothesis', as an aid to readability, is distinguished by: 'i.h.'.

[4] Ref. the central construction of this book, Ch. II, §2.2, defn. 3.

SPECIAL ABBREVIATIONS

With Page of Definition or of First Use

The list does not include all function or predicate symbols which take the form of abbreviations.

Main Abbreviations

1. **AC:** Axiom of Choice, p22
2. **LM:** Law of the Excluded Middle, p23
3. **cst:** constant; **si:** strictly increasing; **nd:** non-decreasing, p45
4. **ec:** explicit coefficient, p62; **ei:** explicit index, p103
5. **fm:** finitely many, p63
6. **rp:** register polynomial, p66
7. **td:** **TAKEUTI** diagram, p68
8. **cd:** compound diagonal, p107
9. **rs:** regular sequence, p95
10. **cr:** canonical reduction, p109
11. **ct:** canonical term, p97
12. **pv:** principal vertex, p124
13. **DCC:** descending chain condition, p22; p151
14. **GEN IND:** generalized induction, p215

Other Abbreviations

1. **dee:** dominating explicit exponent, p160
2. **ls:** local substitution, p181
3. **DS:** density/ starting, p307
4. **fbp:** finite basis property, p339

Chapter I

FINITISM

Section Headings

1.1 Introduction

The linear system **EP** of GOODSTEIN-SKOLEM exponential polynomials is a highly explicit infinite structure. The polynomials are unglamorous objects, like bricks, but with only modest demands on time and effort, they can be built into profound and indestructible arithmetic, combinatoric, algebraic and logic hierarchies.

This introductory chapter, which is intended to provide a setting for the construction of **EP** and its ramifications, is in 2 main parts. Section 2 outlines the finitist approach taken in all later chapters. Section 3 contains a short critique of modern logic (Mathematical Logic). This Chapter alone contains forward references.

1.2 Finitism in Mathematics and Logic

The purpose of this research monograph is to expose to the mathematical, and to the wider scientific community, uncultivated territory. The approach is finitary, and the territory comprises the hierarchy of growth of arithmetic functions. The text takes the forms of presentation, construction, interpretation and application. It is a great responsibility to be writing on so interesting a subject.

C. F. GAUSS, widely regarded as the greatest mathematician since antiquity, asserted in 1836 (see C. W. DUNNINGTON [31]) that '...the use of an infinite magnitude as something completed ...is never permissible'. Taking 'infinite sets' to mean 'completed infinite totalities', there appears to be little short of unanimity among titans of mathematics, prepared to re-affirm this injunction: POINCARÉ, BIRKHOFF, ZYGMUND, HILBERT, HARDY to name only some.[1] Some more quotations are contained in appendix E. So, what is the position? Consider the fundamental sequence

$$\mathrm{N} \equiv 1, 2, 3, 4, 5, 6, 7, ...$$

of natural numbers. Is this sequence finite or infinite? Of course it is infinite, endless. The jump from this assertion, obvious to anyone from an early age, to the intimation that these numbers can all be collected up, so as to form a set, is one that has trapped generations of thinkers.

Without further discussion, some will say: there is a largest natural number, N. There is no evidence for this in mathematics, and no use will be made here of this solution to the problem of infinite sets.

Thus, 'finitary' means, in the present context, 'without any reliance at all on the notion of infinite set'. Consider a single example. In the axiomatized CANTORIAN theory, in which the individual natural numbers are regarded as sets, every one of $1, 2, 3, \ldots$ is surpassed by a greater, infinite, set. None of the definitions, constructions, proofs or concepts of this monograph entertain this notion. However, consider the constant powers $x, x^2, x^3, \ldots$ of the continuous variable x. Of course, for $x > 1$, x^{n+1} dominates x^n :

$$x \prec x^2 \prec x^3 \prec \ldots .$$

What is more, for every n, for x suitably large, the nth term is dominated by the exponential function:

$$x \prec x^2 \prec x^3 \prec \ldots \prec e^x .$$

This example suggests that, far from being a suicidal renunciation of CANTOR's ideas, the uncompromising rejection [2] of infinite sets leaves open the possibility of re-creation of the work, using tried and tested classical notions.

---

[1] There is the most explicit denunciation of 'infinite sets' in an early paper of SKOLEM [154].

[2] Note that the sets 'constructible' in the sense of GÖDEL are *not* constructible in the present, finitist sense; since GÖDEL's construction assumes (vide. T. J. JECH, *Lectures in Set Theory* , §11, p32; *Springer Lecture Notes in Mathematics* **217**, New York (1971))the entire CANTOR hierarchy of 'ordinals' at the first step.

1.2.1 Transforms and Functions

How is a function to be conceived? The oft-quoted conception of an infinite set of ordered pairs, on the left the argument value, on the right the corresponding function value, actually misses the essentially verb-like character of functions. A function does something, changes something. Considerations of time generally do not apply, although many functions which arise in the physical world are interpreted dynamically.

In any case, in this presentation of the GOODSTEIN-SKOLEM hierarchy, infinite sets are disqualified, and a function is taken to be a 'black box' that is, a transform with one (or, of course, many) inputs, and an output. The transform accepts an input, inflicts upon it a change, then discharges it as the output. The means effecting the change are generally unseen, but for example the general notion of the arithmetic function, understood as a black box, may be replaced by the more explicit 'glass box' where, as in the case of a polynomial function, the mechanism of action upon an argument value may be observed in detail.

The black box concept encompasses all functions, whether arising in a physical context, such as the oscillation of elastic membranes, as in the case of BESSEL functions; or in pure mathematics, as in for example, the number of finite graphs satisfying a fixed condition. For present purposes, very often the black box will be such as to accept only a single natural number, and with certainty to return either the same or another number.

1.2.2 Predicates

One of the most fundamental ideas of symbolic reasoning is that of predicate. E.g. let 'Quad' mean 'square-free' as applied to factorizations in **N**. Then e.g.'Quad(6)' may be written '6 $\in$ Quad' which is to be read '6 *is* square-free'. This, after all was the intention of the inventor of the notation, Giuseppe PEANO, (1858-1932) who chose the first letter of the Greek $\epsilon\sigma\tau\iota$ ('is') (Ref. KNEEBONE [75], II, Ch. 5, Section 5, p146). Similarly, 98 $\notin$Quad, etc. Just as there can be sufficient agreement on the property of being a natural number, so can there be on that of Quad. Conversely, the infinitely many square-free numbers can no more be collected up to form a completed totality than can all the numbers $1, 2, 3, \ldots$. 'is' is the *exclusive* meaning of $\in$ in this monograph, so that 'let n be a number of **N**', 'let n be an element of **N**', etc. are abbreviations for 'let n be a number of the kind **N**'. A predicate P, false for all argument values, is the *trivial* predicate. Otherwise, P is a *non-trivial* predicate.

The notation used throughout is based on the above convention. E.g. '$x \in$ P$-$Q' means 'P(x) & $\neg$Q(x)'. In no case does use of the PEANO $\in$-notation imply the acceptance of infinite sets.

1.2.3 The Axiom of Choice

The principle contained in this axiom (AC), which is rejected outright in the present volume, has no intrinsic connection with sets. [3] Therefore the mere rejection of infinite sets still leaves the problem of the validity of the AC.

The axiom (introduced, in 1908 by Ernst ZERMELO, and subsequently promulgated by A. A. FRAENKEL) as understood here, is essentially [4] the following principle: *given any non-trivial predicate* P, *an* x *is given for which* P(x). It has been widely rejected [5].

First of all, if P is non-trivial, how can this be known, without actually having an x for which P(x) to hand? Consider the case of the number $\pi(x)$ of prime numbers not exceeding the natural number x. RIEMANN (1859) conjectured that, for $x > 2$,

$$\pi(x) < \mathrm{li}\, x = \int_0^x \frac{dr}{\log r}.$$

J.E. LITTLEWOOD (1911; see [88], 1914) disproved this; using the PHRAGMÉN-LINDELÖF maximum-modulus theorem; by showing that $d(x) = \pi(x) - \mathrm{li}(x)$ changes sign infinitely often as $x \to \infty$. But it was 2 decades before the famous SKEWES bound vindicated LITTLEWOOD's non-constructive argument. The predicate P(x) iff $x > 2$ & $d(x) > 0$ was therefore, known during this time to have been non-trivial; yet no x for which P(x) could have (nor has ever) been given. Some references to work on $d(x)$ are given in Appx. C.

Turning to an example, suppose it is known that the non-trivial predicate $\mathcal{X}$ and the linear relation $\prec$ satisfy the descending chain condition (DCC), viz. there are no endlessly descending sequences

$$\ldots \prec b \prec a$$

of objects with property $\mathcal{X}$. Now let P be a non-trivial predicate for which P(x) $\Rightarrow$ $\mathcal{X}(x)$. Suppose there is no $\prec$ −smallest x such that P(x). Since P is non-trivial, by the AC let a be such that P(a). Define $\mathrm{P}_a(x) \Leftrightarrow \mathrm{P}(x)$ & $x \prec a$. Since a is not the $\prec$ −least object with property P, P_a is non-trivial. Thus let b be such that $\mathrm{P}_a(b)$. Define $\mathrm{P}_b(x) \Leftrightarrow \mathrm{P}(x)$ & $x \prec b$. Similarly for subsequent predicates. Thus $\ldots \prec b \prec a$, contrary to the DCC. Therefore it is proved that a $\prec$ −minimal object has the property P. Without the AC, this theorem cannot be proved. [6]

---

[3] Although a formulation of the AC is usually included in first-order axiomatizations of the CANTOR set theory. See in particular the remark of JECH, loc. cit., §12, head of p12.

[4] See particularly the detailed discussion in ROSSER [133]. The AC for sets is usually given in the stronger form, AC_∞: *on every non-empty collection* X *of non-empty sets* x, *there is defined a 'choice function'* f, *for which* $f(x) \in x$. If the x are pairwise disjoint, the subset $\{f(x) \mid x \in X\}$ of $\bigcup X$ is the 'choice set' of X, and the AC takes the equivalent form: *every* X *has a choice set.* Thus, the form of the AC prohibited in later Chs. is AC_1 ('one-point'), the weakest in the the scale of implications $\mathrm{AC}_\infty \Rightarrow \mathrm{AC}_{\mathrm{denum.}} \Rightarrow \mathrm{AC}_{\mathrm{finite}} \Rightarrow \mathrm{AC}_1$. The subscripts refer to restrictions on the cardinal $\mid X \mid$.

[5] See SIERPIŃSKI's monograph [135], Ch. VI, p92. See also the comments of B. L. van der WAERDEN [148] in the preface of his introductory treatise *Modern Algebra* .

[6] Ref. SIERPIŃSKI [138], Ch. XII, §7, p262.

A last but important point should be made: the AC is not needed in the construction and development of the GOODSTEIN-SKOLEM hierarchy of exponential polynomials. Returning to the above example, the existence of the $\prec$ −minimal object (ref. Ch. VI, §6, note 3) is the root of the notion of well-order. Constructions in subsequent Chapters require no more than the logically weaker notion of *sequential order*, viz. $\prec$ is linear, and satisfies the DCC. There is thus no question of omitting any interesting material, now requiring the AC while possibly being presented otherwise, after some later advance.

1.2.4 Logic and Mathematics

This monograph does not contain a new foundation, or system of foundations for mathematics, or for logic. It is first of all concerned with a hierarchy, feather-like in structure, yet harder than the fundamental particles of matter and radiation.

However, it is in the foundations of logic that the question of the validity of the Law of the Excluded Middle [7] (LM) arises. Logic is independent of mathematics. Its occurrence in fields such as natural justice establish this. Once routine applications are struck out, mathematics plays no part in this important branch of knowledge. On the other hand, mathematics is beyond logic. Mathematical reasoning is only a concession to the world, which takes form after the event. Finally, mathematics and logic have a most intimate relationship. Nothing places the demands on logic which mathematics places, [8] and this is why Mathematical Logic is the most interesting branch of logic. The conclusion which has to be drawn is that logic and mathematics are mutually independent. Therefore the question of the infinite in logic demands still further rigour. On the one hand, infinite sentences occur in nature, just as do infinite sequences. For example consider EUCLID's proof of the infinitude of the primes $2, 3, 5, \ldots = p_1, p_2, p_3, \ldots$. For any natural number N, given $p_1, p_2, \ldots, p_N$, another prime $p \leq 1 + p_1 p_2 \cdots p_N$ is found. Therefore the theorem is

$$(\forall x)(\exists y)\, x < p_y, \tag{I}$$

where quantification is over the natural numbers, **N**. Let $\mathrm{P}(x) \equiv (\exists y)\, x < p_y$. (I) is therefore an infinite sentence:

$$\mathrm{P}(1)\ \&\ \mathrm{P}(2)\ \&\ \mathrm{P}(3)\ \&\ \ldots,$$

which has to be accepted as a natural phenomenon. On the other hand, consider EUCLID's proof by contradiction that $\sqrt{2} \neq \frac{a}{b}$, no matter how the integers a, b are chosen. Suppose $\sqrt{2} = \frac{a}{b}$. Then $2b^2 = a^2$. Therefore 2 exactly divides the LHS an odd number of times, while for the RHS the number is even. From this contradiction,

---

[7]Viz. for any finite or infinite proposition A, either A itself is true, or the negation $\neg$ A of A is true. It may be noted that the Law of Contradiction states that, for all A, the conjunction A $\& \neg$ A is false.

[8]This point is taken up by GOODSTEIN in the introduction to [49].

it is to be concluded that $\sqrt{2} \neq \frac{a}{b}$. The proof is so compelling; but one is left wanting to know what $\sqrt{2}$ *is*, having learned what it is not. Thus, a constructive proof, viz. one which does not rely on the LM applied to infinite sentences (ref. Appx. A, §2·1), might be:

1. the continued fraction expression for any rational $\frac{a}{b}$ is found, by a terminating algorithm, to be *finite*.

2. $\sqrt{2} = \frac{2+\sqrt{2}}{1+\sqrt{2}}$ gives $1 + \frac{1}{2+\frac{1}{2+\frac{1}{2+\dots}}}$.

The LM, although a considerable weakening of the axiom of choice, is plainly unacceptable for the purposes of construction. It should be regarded as useful, but a vestige of everyday thinking, which raises the alarm, thereby initiating the work of evading danger.

Observing the rule: 'look after the technicalities, and the logic will look after itself', no attempt will be made to classify the very many possibilities which arise from the applications of different combinations of rigours (e.g. the intuitionist rejection of the law of the excluded middle, but acceptance of the 'reductio' [9] rule of inference).

1.3 Critique of Mathematical Logic

The following summary outline is intended only to serve as a basis for discussion of questions of finitism. Besides predicates informally referred to as 'classes', in line with common usage, attention is mainly focussed on arithmetic predicates and functions of up to 3 arguments.

1.3.1 The Class $\mathcal{R}$ of General Recursive Functions

Consider the space $\mathcal{N}$ of arithmetic functions, viz. of $f : \mathrm{N} \to \mathrm{N}$. Suppose these could be exhaustively listed: $f_1, f_2, f_3, \dots$. Now CANTOR's diagonal F defined by $F(x) = f_x(x) + 1$ is in $\mathcal{N}$, therefore $F = f_y$ for some fixed $y \in \mathrm{N}$. But in that case

$$F(y) = f_y(y) + 1 > f_y(y),$$

a contradiction, therefore F was missed out in the list. It follows [10] therefore that $\mathcal{N}$ cannot be listed, or enumerated. On the other hand, consider an explicitly given finite subclass $\mathcal{X}_1 \subset \mathcal{N}$. Suppose a procedure P is at hand, whereby, given such a subclass, P may be applied algorithmically so as to obtain, after a finite number of

[9] Viz. if A⇒B and A⇒ ¬B, then ¬ A.

[10] This argument depends on the LM: the 'missing F' is constructed from an enumeration which is proved impossible.

operations, an enlarged class $\mathcal{X}_2$ from $\mathcal{X}_1$. P may now be applied to $\mathcal{X}_2$ to give $\mathcal{X}_3$, and so on. The closure

$$\mathcal{X} = \cup \mathcal{X}_i$$

of this process is itself an enumerable class. According to the thesis [11] of A. CHURCH, the largest possible $\mathcal{X}$ obtainable in this way, coincides with the (functions of a single variable of the) class $\mathcal{R}$ of *general recursive functions*, to be formulated more precisely below.

Characterizations of $\mathcal{R}$

For those who accept the concept $\mathcal{R}$ as sound, it has many different but equivalent formulations. 3 of these have received particular attention: $f \in \mathcal{R}$ iff f

1. is the solution of a recursive system of equations; or
2. can be computed by a TURING machine; or
3. is numeralwise representable in 1st order arithmetic.

μ–Recursive Functions

The characterization 1. is due to KLEENE, and gives its name to the system $\mathcal{R}$. It is very often itself reformulated in terms of the unbounded 'least-number' operator, so that $\mathcal{R}$ is obtained [12] from initial functions by closure under the 3 operations of substitution, recursion and:

$$\lambda x, y \cdot f(x, y) \in \mathcal{R} \Rightarrow F \in \mathcal{R},$$

where

$$F(x) = (\mu y)[f(x, y) = 1].$$

The sign '(μy)' should be read as 'the least y such that ...'. Given x, there need not be a number y such that $f(x, y) = 1$, as is the case with the recursive function $x + y$. Thus a function of $\mathcal{R}$ is **not** *necessarily defined at any given argument value.*

Remark

An operation $^{-1}$ of *inverse* is readily defined, using the μ–operator: given a function f on the natural numbers, f^{-1} is

$$f^{-1}(y) = (\mu x)[y \leq f(x)].$$

If f is unbounded, then it is guaranteed that after a finite number of examinations $x = 1, 2, 3, \ldots$, an x will be found which satisfies the condition.

[11] [20], *Amer. J. Math.* **58** (1936), pp345-363. See also S. C. KLEENE [74], §60, p. 300. §§54-55 contain a discussion of the arising of the class $\mathcal{R}$, in particular of ACKERMANN's 1928 solution to a problem of HILBERT, 1926.

[12] See R. PÉTER, [122], §§17.2-11, p199 et seq.

Subclasses of $\mathcal{R}$

The subclass $\mathcal{P}$ of *primitive recursive* functions of $\mathcal{R}$ has been investigated from the time of POINCARÉ. $\mathcal{P}$ is the smallest class which contains the common arithmetic functions; and is closed under the operations of substitution, and of recursion with respect to a single variable. Thus, writing $\underline{x}$ for the **N**-vector $x_1, x_2, \ldots, x_k$, if $f, g \in \mathcal{P}$, and

$$e(\underline{x}, 1) = f(\underline{x}) \text{ and for } n = 1, 2, 3, \ldots,$$
$$e(\underline{x}, n+1) = g\{\underline{x}, n, e(\underline{x}, n)\},$$

then $e \in \mathcal{P}$.

It is known that $\mathcal{P}$ is a proper subclass of $\mathcal{R}$. Indeed, the only occurrence of $\mathcal{P}$ in the remainder of the text will be to prove this fact (Ch VII, §1). Proper subclasses of $\mathcal{P}$ include the CSILLAG-KALMÁR class $\mathcal{E}$; and the GRZEGORCZYK hierarchy

$$\mathcal{E}^0 \subset \mathcal{E}^1 \subset \mathcal{E}^2 \subset \mathcal{E}^3 = \mathcal{E} \subset \mathcal{E}^4 \subset \ldots,$$

for which $\mathcal{P} = \cup_n \mathcal{E}^n$. The classes $\mathcal{E}^n$ are obtained by restricting the rate of growth of new functions obtained by application of the recursion schema of $\mathcal{P}$. The smallest, $\mathcal{E}^0$, contains none which grows faster than $\lambda x \cdot 2x$. This, however, does not imply that the $\mathcal{E}^n$ are small classes: the theorem 5.3 of GRZEGORCZYK [53] p53 on $\mathcal{E}^0$ is certain evidence to the contrary.

TURING Computable Functions

Consider now the above characterization (2) of $\mathcal{R}$. The conceptual computing machine, introduced in 1936 [13] before the computer era by Alan TURING, may be replaced by the equivalent, but in some ways more applicable, SHEPHERDSON-STURGIS [14] machine, for which a tape of scanned squares is replaced by a set of accumulators. Consider 2 simple programs.

Firstly, given (as a 'rooted' di-graph) a test for zero:

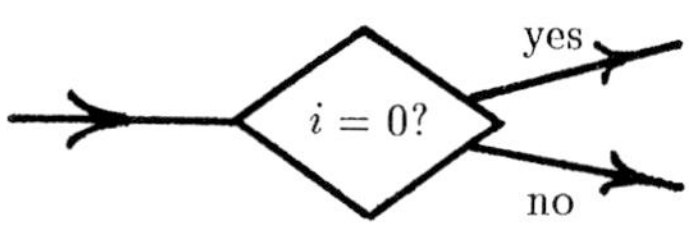

it is an easy matter to construct a comparison of natural numbers r and R, say : [15]

[13] TURING, 'On computable numbers, with an application to the Entscheidungsproblem', *Proc. Lond. Math. Soc.*, **42**, 230-265 (1936-37); ibid, **43**. 544-46 (1937). See KLEENE [74], p. 356.

[14] J. C. SHEPHERDSON and H. E. STURGIS, 1963, 'Computability of recursive functions', *J. Assoc. Computing Machinery*, 10, pp217-255. See also the survey of work of MINSKY, MELZAK, LAMBEK and BORGER in: J. P. JONES and Y. V. MATIJASEVIČ, 'Register machine proof of the theorem on exponential diophantine representation of enumerable sets', *J. Symb. Logic*, **49**, 3 (1984).

[15] ':=' may be read 'becomes equal to'.

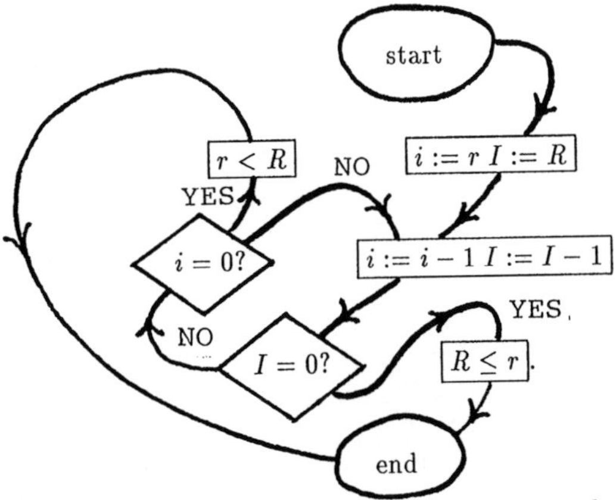

Now consider the 2 programs, which, via the above subroutine, will be built up from $+1$ and -1 only. The square boxes represent accumulators, which neeed not be numbered on account of the arrows.

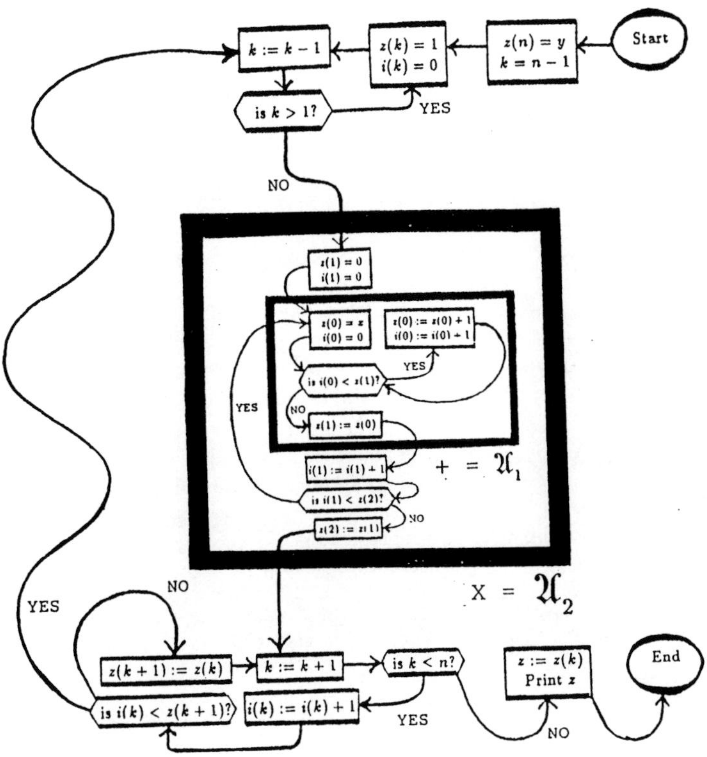

Given natural numbers x, y, n, this algorithm computes ACKERMANN's generalized exponential $\mathfrak{A}_n(x, y)$ (ref. Ch. VII, §2). One may see, in following through the algorithm, that a notional control moves initially from the start box, and generally from one box to another, as specified by an arrow. Each move brings about a change of *state*, S. It is immediately clear that such an algorithm or *TURING machine* can be found to compute any of the well-known arithmetic functions. It is also clear, that, as (finite directed) graphs, the class of all such algorithms is enumerable.

B.

Consider the rather simpler algorithm

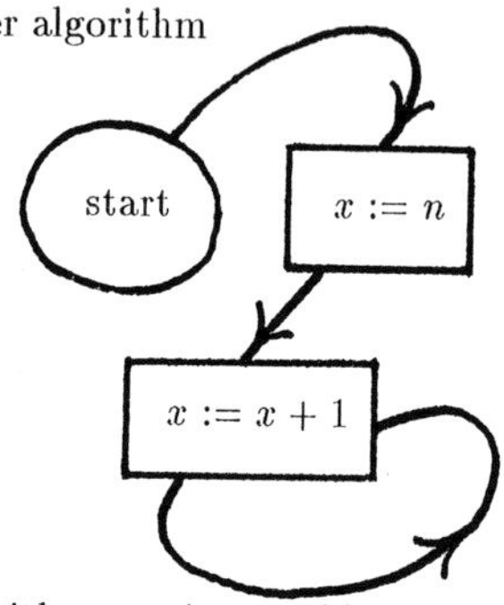

Starting at n, this algorithm continues adding one, endlessly, without ever producing a result. An easy modification e.g. yields $n/2$ if the number n is even, but produces no result if n is odd. Once again, it is clear that the class $\mathcal{R}$ of algorithms, or computable functions *includes functions which are* **not** *defined on certain argument values.*

Enumeration of $\mathcal{R}$

There is no practical difficulty in constructing actual enumerations of the computable functions, and of their computations at given values. Indeed, this may be done using primitive recursive functions. Thus, let $\phi_1, \phi_2, \ldots$ be an exhaustive enumeration of (the functions [16] of a single variable of) $\mathcal{R}$. The diagonalization $\Phi(x) = \phi_x(x) + 1$ does not lead out of $\mathcal{R}$, because it is not assured that $\Phi = \phi_y$ is defined at $x = y$. Nonetheless, the function $\phi(x) = \phi_x(x)$ is the cornerstone of the theory of $\mathcal{R}$. For suppose the following is accepted as a valid definition of a function:

$$\varphi(x) = \begin{cases} \text{undefined} & \text{if } \phi(x) \text{ is defined} \\ 1 & \text{otherwise} \end{cases} \qquad \text{(I)}$$

[16] Generally, the subclass $\mathcal{R}_n$ of functions of $\mathcal{R}$ of n variables, $n = 1, 2, 3, \ldots$, has a primitive recursive enumeration $\psi_1, \psi_2, \psi_3, \ldots$. Therefore there is a $\theta \in \mathcal{R}_{n+1}$, the *universal function* for $\mathcal{R}_n$, viz. such that for each $\underline{x}$, and $k = 1, 2, 3, \ldots$, $\theta(k; x_1, x_2, \ldots, x_n) = \psi_k(x_1, x_2, \ldots x_n)$, whenever either side is defined.

If $\varphi \in \mathcal{R}$, then $\varphi = \phi_k$, for some $k \in \mathbf{N}$. Then

$$\varphi(k) = \begin{cases} \text{undefined} & \text{if } \phi(k) \text{ is defined, i.e.} \\ & \text{if } \phi_k(k) \text{ is defined, i.e.} \\ & \text{if } \varphi(k) \text{ is defined;} \\ 1 & \text{if } \varphi(k) \text{ is undefined.} \end{cases}$$

From this contradiction, it is concluded that $\varphi \notin \mathcal{R}$. Thus,

$$\varphi(x) \text{ defined } \Leftrightarrow (\forall y)\, \neg\phi(x) = y,$$

so accepting the above definition implies that definition by prefixing a quantifier to a predicate built up from equality and functions of $\mathcal{R}$ leads to new functions, not in $\mathcal{R}$.

Firstly, the $\varphi \notin \mathcal{R}$ result was proved by obtaining 2 contradictions, one for each of the 2 horns of the LM: either $\varphi(k)$ is defined; or $\varphi(k)$ is undefined.

There is, from the finitist viewpoint, a far more serious, in fact insuperable objection to the equation (I): the value of φ (at the arbitrary value x) depends on whether or not $\phi(x)$ is defined. If $\phi(x)$ is not defined, then by definition it cannot be known that it is not defined. On these grounds, subsequent chapters will not build on the notion $\mathcal{R}$.

1.3.2 The First Order Theory $\underline{\mathbf{N}}$ of Natural Numbers

As already mentioned, the class $\mathcal{R}$ is identical with the numeralwise representable functions. Appendix A contains a short account of

1. the 1st order (quantification over individual variables) formula calculus, $\mathfrak{L}$;

2. the sequent calculus, or system of rules for deriving valid formulae, either from the axioms of logic, or from formulae already derived.

Appendix B contains a brief outline of the formal theory of natural numbers, afforded by the 1st order logic.

Numerals

The numerals $1, 1', 1'', \ldots$ of the language of $\underline{\mathrm{N}}$ of appendix B (and notation relating predicates and first-order formulae) will be tacitly assumed. Thus, e.g. the '2–place predicate A evaluated at (a, b) is true is a formal theorem of $\underline{\mathrm{N}}$' will be written '$\vdash$ A(a, b)' rather than e.g. '$\vdash \underline{\mathrm{A}}(\underline{a}, \underline{b})$'. A similar convention, as to which side of '$\vdash$' quantifiers $\mathcal{Q} = \exists, \forall$ occur, is adopted: '$\ldots \mathcal{Q} \vdash \ldots$' in the empirical sense; '$\ldots \vdash \mathcal{Q} \ldots$' in the formal sense.

Meta-theorems

In his consistency proof for N, [17] GENTZEN demonstrated the expressibility therein of the LeGENDRE symbol; and chose as an example of a theorem provable in $\underline{\mathrm{N}}$, GAUSS' law A of quadratic reciprocity. [18] Many of the results of contemporary logic are sentences describing the expressbility or otherwise; or the provability or otherwise of theorems such as A, in the system $\underline{\mathrm{N}}$ (or in other formal systems of first or higher order logic). Such results are not theorems, either in the empirical or formal senses that A itself is. They are, therefore, usually termed *meta-theorems.*

1.3.3 Undecidability

Numbering of Symbols of $\underline{\mathrm{N}}$

The establishment of an $\underline{\mathrm{N}}$−expressible sentence, evidently true from other considerations, but not provable by formal inference from the axioms of $\underline{\mathrm{N}}$, was the work of K. GÖDEL, [44], 1931. The following summary of the proof is due to J. HERBRAND [60], 1931. It depends on the procedure of *numbering the logical and mathematical symbols* of $\underline{\mathrm{N}}$. Thus, e.g.

$$1, 2, 3, \ldots, 9, 0, (,), +, \cdot, \text{ etc.}$$

may be assigned prime numbers $p_1, p_2, p_3, \ldots, p_{11}, p_{12}$, etc. Using the fundamental thm. of arithmetic, it is then a straightforward matter to assign unique numbers (indices) to strings of symbols (formulae), and to strings of strings (proofs), using only primitive recursive functions.

The Diagonalization of $\underline{\mathrm{N}}$

Now let 'formula' mean 'formula of $\underline{\mathrm{N}}$ with a single free variable'. By the foregoing, the predicate

$$\mathrm{P}(w, n, x),$$

which holds iff w is the index of the ($\underline{\mathrm{N}}$−)proof of the formula of index n at argument value x, may be constructed from equality, the propositional connectives and primitive recursive functions only. Since $\mathrm{P} \in \mathcal{P}$, if P is true at (w, n, x), then $\mathrm{P}(w, n, x)$ has a proof (in $\underline{\mathrm{N}}$):

$$\vdash \mathrm{P}(w, n, x).$$

Now let b be the index of the formula $(\forall w)\neg\mathrm{P}(w, n, n)$.

METATHEOREM 1.3.1 *Suppose* $\underline{\mathrm{N}}$ *is* ω−*consistent.* [19] *Let* A *denote the sentence* $(\forall w)\ \neg\mathrm{P}(w, b, b)$. *Then neither* $\vdash$ A, *nor* $\vdash \neg$ A.

[17] Viz. for no formula C are there derivations within $\underline{\mathrm{N}}$ of both C and ¬C.

[18] See DAVENPORT [25] Ch. III, which also contains an historical account of this theorem.

[19] That is $\underline{\mathrm{N}}$ is simply consistent, in the sense of fn. 17; and furthermore, if $\mathrm{Q}(x)$ is any formula, then

$$\{\vdash \mathrm{Q}(1) \text{ and} \vdash \mathrm{Q}(2) \text{ and} \ldots\} \text{ implies } \nvdash \neg(\forall x)\ \mathrm{Q}(x).$$

Proof. Suppose A to be a sentence which has a proof in $\underline{\mathrm{N}}$. Then A has a proof index, a say; so a is the index of the proof of (the formula) $(\forall w)\ \neg\mathrm{P}(w, n, n)$ at b, i.e.

$$\mathrm{P}(a, b, b)$$

is true, by defn. of P. So by the remark on the primitive recursiveness of P,

$$\vdash \mathrm{P}(a, b, b).$$

But on the other hand, by the inference in $\underline{\mathrm{N}}$ of $\forall$−elimination, for each natural number a, $\vdash$ A implies

$$\vdash \neg\mathrm{P}(a, b, b),$$

contrary to the assumption of the (simple) consistency of $\underline{\mathrm{N}}$. Thus, $\nvdash$ A.

Conversely, $\mathrm{P}(a, b, b)$ implies $\vdash$ A, therefore by the preceeding argument, for *every* natural number a, $\mathrm{P}(a, b, b)$ is false; so again by the remark on the primitive recursiveness of P, it follows $\vdash \neg\mathrm{P}(a, b, b)$. Thus it is proved that

$$\vdash \neg\mathrm{P}(1, b, b) \text{ and } \vdash \neg\mathrm{P}(2, b, b) \text{ and} \ldots;$$

therefore by the assumption of ω−consistency,

$$\nvdash \neg(\forall w)\neg\mathrm{P}(w, b, b),$$

viz. $\nvdash \neg$ A.

□

The sentence $(\forall w)\neg\mathrm{P}(w, b, b)$ is said to be *undecidable* in $\underline{\mathrm{N}}$; and that there is a sentence undecidable in $\underline{\mathrm{N}}$ is also expressed: $\underline{\mathrm{N}}$ is *incomplete.* [20]

Remarks

1. Often $\underline{\mathrm{N}}$ is referred to as *PEANO arithmetic.* The system actually presented by G. PEANO in 1889 (see van HEIJENOORT [154] p89, 94; also KNEEBONE [75] p146, 232-3) contains a 2nd order axiom of induction, and therefore, as is well known, is complete.

2. A quick proof of the above result was given by George BOOLOS ([15], Notices of the Amer. Math. Soc., April, 1989). This version is similar in form to RICHARD's paradox (ref. KNEEBONE, [75] pp112,127).

In J. B. ROSSER's proof ([134], 'Extension of some theorems of GÖDEL and CHURCH', *J. Symb. Logic*, **1** (1936) pp87-91; see also the same author's exposition [135], ibid. **4** (1939), pp 53-60, reprinted in DAVIS [26] *The Undecidable* , pp 223-230) of the present meta-theorem, the need for the assumption of ω−consistency is eliminated.

[20] Note that arithmetic is not incomplete. It is axiomatizations of arithmetic (such as $\underline{\mathrm{N}}$) that are incomplete.

3. If the sentence $(\forall w)\ \neg P(w, b, b)$ is added to the axioms of $\underline{N}$, then the decidability of (the new) predicate $P^{+}(w, n, x)$ in the enlarged system $\underline{N}^{+}$ is not, of course, altered. The same argument therefore establishes: $\underline{N}^{+}$ is incomplete.

4. The diagonalization above may be compared with RUSSELL's antinomy (1903): the village barber shaves just those who do not shave themselves. Who shaves the barber?

The ASCII Code

In the context of the previous paragraph, the American Standard Code for Information Interchange provides an unexpected contrast. About 250 signs

$$1, 2, 3, \ldots; a, b, \ldots, z; A, B, \ldots, Z, (,), \$, \text{ etc.}$$

are sufficient for most purposes, so e.g. a 64–bit word of computer memory has room for 8 characters of natural language. This fundamental expedient, of assigning numbers to characters, is the single means by which computers, ranging from machines that are no more than toys, to huge mainframe systems of government, business, science and education, communicate the results of their activity with society. It is more than remarkable that, an application now so important for everyone, originated as a research tool used to expose a weakness of formal logic.

'Non-Standard Models'

Let $A = (\forall w)\ \neg B(w)$ be an undecidable sentence in $\underline{N}$, such as has been exhibited. Then inclusion of the axiom $\neg\ A \equiv (\exists w)\ B(w)$ will not flaw the consistency of $\underline{N}$. Now writing $\underline{1}$, $\underline{2}$, $\underline{3}$, ... for the numerals of $\underline{N}$, evidently

$$\nvdash B(\underline{1}),\ B(\underline{2}),\ B(\underline{3}),\ \ldots,$$

otherwise the falsehood [21] of A would be a theorem of $\underline{N}$. On the other hand, the sentence $(\exists w)\ B(w)$ can only be interpreted as asserting the existence of an object ξ, not corresponding to a numeral, for which $\vdash B(\xi)$. Since the axioms of $\underline{N}$ give, for each natural number n,

$$n \leq \xi \leq n + 1 \Rightarrow \xi = n\ \vee \xi = n + 1,$$

it follows that

$$1 < 2 < 3 < 4 < \ldots < \xi$$

(as in the scale of domination considered in §2). Thus SKOLEM proved that the incompleteness of $\underline{N}$ implies $\underline{N}$ describes other entities, each greater than all the natural numbers.

[21] Note that the equivalence of $\neg A$ and $(\exists w)\ B(w)$ may not be inferred in the Intuitionistic part (ref. Appx. B, §1) of $\underline{N}$.

The above establishes that, far from being a property of the natural numbers, the incompleteness of $\underline{\mathbf{N}}$ is proof that first-order arithmetic fails to even characterize successfully the objects of study of arithmetic.

1.3.4 The Relationship Between $\mathcal{R}$ and $\underline{\mathbf{N}}$

Numeralwise Representability

Given the arithmetic function f of k variables, suppose there is a first-order formula A of $k+1$ free variables such that for all $x_1, x_2, \ldots, x_k, y$,

$$f(x_1, x_2, \ldots, x_k) = y \text{ iff } \vdash \mathrm{A}(x_1, x_2, \ldots, x_k, y).$$

and

$$\vdash (\exists! u)\, \mathrm{A}(x_1, x_2, \ldots, x_k, u).$$

Such an f is said to be *numeralwise representable* (in $\underline{\mathbf{N}}$). It is a standard metatheorem of logic (ref. KLEENE [74], thm. 32, §59, p295; or MENDELSON [108], proposition 3.23, p131) that the numeralwise representable functions *coincide* with the class $\mathcal{R}$ of general recursive functions. From this follows another proof of the incompleteness of $\underline{\mathbf{N}}$: let $\phi_1, \phi_2, \ldots, \phi_n, \ldots$ be the enumeration of $\mathcal{R}$, and φ the function outside $\mathcal{R}$, of §3.1. Suppose the recursive functions ϕ_n are represented by the formulae A_n in $\underline{\mathbf{N}}$, $n = 1, 2, 3, \ldots$. Since the enumeration is primitive recursive, the numbering of formulae of $\underline{\mathbf{N}}$ applied to the expression

$$\neg(\exists y)\, \mathrm{A}_x(x, y)$$

determines a formula of one free variable, $\mathrm{A}(x)$, say. Then for all x, $\mathrm{A}(x)$ iff $\varphi(x) = 1$. It is not true however, that for all x, $\varphi(x) = 1$ iff $\vdash \mathrm{A}(x)$, since otherwise φ would be representable, contrary to $\varphi \notin \mathcal{R}$. Therefore for some natural number a, $\varphi(a) = 1$, so that $\mathrm{A}(a)$ is true, but $\nvdash \mathrm{A}(a)$. Therefore

METATHEOREM 1.3.2 *The sentence* $\mathrm{A}(a)$ *is undecidable in* $\underline{\mathbf{N}}$, *if the theory is consistent.*

□

Remark

The above-cited representation result follows from the correspondence between

1. initial functions and basic mathematical formulae, and
2. rules, viz.
   (a) substitution, etc. of formulae and substitutions of functions,
   (b) induction schemata and recursion schemata.

Thus, as seen in §3.3,4 just as it may be impossible to compute the value of a function $f \notin \mathcal{R}$ at the fixed argument value x in a finite number of steps, so there may exist no finite derivation of the truth value of a formula A at the given numeral x in $\underline{\mathrm{N}}$. The remarks at the final section of §3.1 have equal applicability therefore to $\underline{\mathrm{N}}$, as to $\mathcal{R}$.

KLEENE's Fixed-Point Result for $\mathcal{R}$

This may be obtained as follows: since ϕ (viz. $\phi(x) = \phi_x(x)$) is in $\mathcal{R}$, for any $f \in \mathcal{R}$ the composition $f\phi \in \mathcal{R}$, so $f\phi = \phi_k$ for some k. Therefore assuming ϕ is defined at k, $\phi(k) = n$, say

$$f(n) = f\phi(k) = \phi_k(k) = n$$

if f is defined at n. By introducing an auxillary function θ, it is then straightforward to obtain: $(\forall f \in \mathcal{R})(\exists n)\ \phi_{f(n)} = \phi_n$.

Enumeration of $\underline{\mathrm{N}}$; Provably Defined Arithmetic Functions

Let $\mathrm{T}(w, n, x)$ be KLEENE's T−predicate ([74], para 56, p281). Thus, in the enumeration of $\mathcal{R}$,

$$\mathrm{T}(w, n, x) \Leftrightarrow \phi_n(x) = y,$$

where y is the result determined (i.e. $\phi_n(x)$ is defined) by the computation of number u. Since the enumeration is primitive recursive, T can be expressed in the first-order language of $\underline{\mathrm{N}}$. Now consider the class $\mathcal{K}$ of general recursive functions f, for which there is an n such that $f = \phi_n$ and

$$\vdash (\forall x)(\exists w)\ \mathrm{T}(w, n, x).$$

KREISEL [80] (1952) proved that $\mathcal{K}$ is a proper subclass of $\mathcal{R}$, in fact containing exactly the functions primitive recursive in the finitary diagonalization of ACKERMANN's function, to be considered in Ch. VII, §3. [22]

1.3.5 The Consistency of Formalized Arithmetic

First order arithmetic, $\underline{\mathrm{N}}$, is incomplete. Worse still, $\underline{\mathrm{N}}$ may contain a derivation of a contradiction:

$$\vdash \mathrm{C} \text{ and } \vdash \neg\mathrm{C}$$

for some formula C. [23]

[22] See also Ch. VIII, §4.12.

[23] Arithmetic is really the intelligent application of our knowledge of numbers. It makes no sense to speak of the 'consistency of arithmetic', because a contradiction can only arise in a logical system. Arithmetic itself is not a logical system; such systems, with only partial success, attempt to formalize ratiocinations, which often follow intuitive 'flashes', necessary for theorem-proving.

Let $N(n, \overline{n})$ be the primitive recursive predicate, built up from the formula numbering of $\underline{N}$, which holds iff n is the index of some well-formed formula C, and $\overline{n}$ is that of $\neg$ C. Also, let $P(w, n)$ be the predicate which holds iff w is the proof index of the sentence of index n. Then the sentence '$\underline{N}$ is consistent', i.e. '$\underline{N}$ does not contain a contradiction' is:

$$B \equiv (\forall w, \overline{w}, n, \overline{n}) \neg\{N(n, \overline{n}) \,\&\, P(w, n) \,\&\, P(\overline{w}, \overline{n})\},$$

which is expressible in $\underline{N}$. This sentence B is in fact undecidable. The proof is long (see, e.g. HILBERT and BERNAYS [64] 1939, *Grundlagen der Mathematik* , Vol II, Berlin). But the result is hardly more than a curiosity; if $\underline{N}$ contained a contradiction, every sentence would be provable, including B. A formal proof of B therefore, would carry no weight at all.

GENTZEN's 1936 Consistency Proof

The task at hand is to prove that $\underline{N}$ contains no contradictions. Such a proof was obtained by GENTZEN in 1936. The proof contains several well-defined parts, all but one of which are generally accepted [24] as flawless. It happens that the present monograph dwells long and in detail on the one link, in the part of the proof which might be suspected of containing a flaw, viz. the proof of the DCC.

From the finitist point of view, $\mathcal{R}$ and $\underline{N}$ are not concepts which deserve the interest which has been accorded them in the 20th century. It may be pointed out that $\underline{N}$ has not stimulated interest among great arithmeticians of modern times. *'The Higher Arithmetic'* by Harold DAVENPORT contains not a single reference to the incompleteness of $\underline{N}$, indeed no reference at all to $\underline{N}$. The same may be said of standard works by G. H. HARDY & E. M. WRIGHT, G. HALBERSTAM & K. F. ROTH, and S. RAMANUJAN. On the other hand, the chance of any strengthening of the proof of consistency seems worth taking. Therefore a short exposition of the outline of GENTZEN's proof follows. This contains nothing not to be found in many similar accounts.

The well-defined parts of the proof are

1. The reduction of sequents.

---

[24] The proof may be regarded as yielding a metatheorem about $\underline{N}$ as a universal formal system: meta-theorem 3.1 asserts that there is a sentence A for which $(\forall n)$ $\{\nvdash_{\text{via } n\text{th proof}} A \text{ and } \nvdash_n \neg A\}$. This may be re-expressed (introducing a redundant bound variable) as:

$$(\exists m, n)\ \{\vdash_m A \text{ or } \vdash_n \neg A\}$$

is impossible. The GENTZEN result establishes that for every sentence A,

$$(\exists m, n)\ \{\vdash_m A \text{ and } \vdash_n \neg A\}$$

is impossible. In the light of the remark of §3.4, any such proof could hardly be regarded as constructive.

(a) Initial sequents.

i. Logical.

ii. Arithmetic.

(b) The reduction of sequents, obtained from given sequents, by rules of inference. This is accomplished by reductions of derivations.

2. The correlation of 1st order derivations with DCC-objects.

3. Suppose α, β are respectively correlated with the derivations P, Q. Then if the reduction of Q is P,

$$\alpha \prec \beta.$$

4. Proof that the DCC objects do indeed satisfy the DCC.

These ideas are brought to bear on the question of consistency by GENTZEN as follows ([40], section 13.90): Suppose the sequent →A & ¬A were derivable in N.

Then any arbitrary sequent, e.g. $\rightarrow 1 = 2$, may be derived:
→A and → ¬A ('&-elimination'),
and also
$\neg 1 = 2 \rightarrow$A and $\neg 1 = 2 \rightarrow \neg$A, ('adjunction to antecedent')
therefore
$\rightarrow \neg\neg 1 = 2$ by the 'reductio', therefore
$\rightarrow 1 = 2$ (by '¬−elimination').
But by (1.) this is impossible, since the sequent $\rightarrow 1 = 2$ is neither in endform (Appx. B, §3), nor is it capable of reduction.

It is plain that initial sequents can be reduced to endform in a finite number of steps. But a derivation reduction, 1(b), may leave the endsequent unchanged. However, using the correlation of 2., by 3. every reduction entails a descent, therefore by 4., only finitely many reductions are possible. But a reduction step on a derivation proceeds, unless the endsequent is in endform. It therefore follows that, after finitely many reductions, every sequent is reduced to endform.

The investigation of **EP**, the centre of this monograph, was not first of all undertaken with the solution to the consistency problem for N as a goal. It is a remarkable fact though, that the only part of GENTZEN's proof not generally considered unquestionably sound is part 4., the DCC.

In 1970 Prof. P. BERNAYS[25] published an earlier version of the above proof, which GENTZEN had withdrawn at the galley stage, on BERNAYS' pointing out that this version required the 'fan' theorem. [26] This much shorter version does not use the DCC. It is surely so much easier to follow than the 1936 version (published, Math. Annalen), that one wonders whether the DCC really is necessary to achieve GENTZEN's aims.

[25]'On the original GENTZEN consistency proof for number theory' *Proc. Conf. Buffalo, N.Y.*, 1968, Publ. North-Holland.

[26]Viz. KÖNIG's lemma, Ch. IV·15. Ref. DUMMETT [160], *Elements of Intuitionism* , *Ch. I.*

Subsystems of $\underline{\mathbf{N}}$

Corresponding to the proper subclass $\mathcal{P}$ of $\mathcal{R}$ is the formal subsystem of *primitive recursive arithmetic* introduced by SKOLEM in 1923[27] and re-presented by HILBERT and BERNAYS[28] in 1934. [29] See also GOODSTEIN, 1957, *Recursive Number Theory* , North-Holland, Amsterdam, and GOODSTEIN, 1954, 'Logic-free formalizations of recursive arithmetic', *Math. Scand*, **2**.

CLEAVE and ROSE [23] have characterized the formal subsystems of $\underline{\mathbf{N}}$ corresponding to the KALMÁR-GRZEGORCZYK classes $\mathcal{E}^n$, $n = 1, 2, 3, \ldots$. One of the main results is that the formal theorem asserting the consistency of $\mathcal{E}^n$–arithmetic is provable in $\mathcal{E}^{n+1}$–arithmetic. Similar meta-theorems, corresponding to a decomposition of the class $\mathcal{K}$ of §3.4, in place of $\mathcal{P}$, are contained in ROSE [131], Ch. VI.

In another direction, an early (1929) result of Mojzesz PRESBURGER [125] establishes that the assertion of self consistency is provable in the fragment of $\underline{\mathbf{N}}$ obtained by deleting arithmetic product.

1.4 Notes on Ch. I

1. §2·3 A consequence of the AC is the BANACH-TARSKI paradox: a finite solid S may be decomposed into finitely many parts, such that these may be re-assembled without deformation, so as to give 2 solids congruent to S.

2. §3·1 No proper decomposition $\mathcal{R} = \bigcup \mathcal{R}_n$ may be obtained, by restricting to $n = 1, 2, 3, \ldots$ the number of applications of the μ–operator; since (Ref. PÉTER, loc. cit., p200) the whole of $\mathcal{R}$ may be obtained from $\mathcal{P}$, given only a single application of 'μ'.

3. §3·1. Let $\psi_1, \psi_2, \psi_3, \ldots$ be a $\mathcal{P}$-enumeration of functions of $\mathcal{R}$ of 2 variables (ref. fn. 16). Pairing functions $\pi(x, y) = z$ (viz. which define a $1-1$ onto correspondence from $\mathbf{N}^2$ to $\mathbf{N}$, as explicitly given in e.g. Ch. VI, §5·1, # 2) and projection functions $\sigma(z) = x$; $\rho(z) = y$; $\pi, \sigma, \rho \in \mathcal{P}$; may be used to describe ψ_k in terms of a function of a single variable:

$$\psi_k(x, y) = \psi_k\{\sigma(z), \rho(z)\} = f(z), \text{ say; } f \in \mathcal{R}.$$

Therefore, using the $\mathcal{P}$-enumeration of §3·1, this relationship defines a correspondence

$$\psi_k \leftrightarrow \phi_{r(k)},$$

for some primitive recursive sequence $r(k)$, $k = 1, 2, 3, \ldots$.

[27] In English translation in Van HEIJENOORT[154],1967, pp302-333

[28] *Grundlagen der Mathematik* , Vol I, pp307-346, Springer, Berlin.

[29] Viz. the *free variable calculus* obtained by taking the initial equations

$$f(\underline{x}) = g(\underline{x})$$

with f, g in $\mathcal{P}$ rather than in $\mathcal{R}$.

4. §3·1 The non-constructive argument which leads to $\varphi \notin \mathcal{R}$, leads from this to many similar conclusions. In particular, if

$$f(x) = \begin{cases} 1 & \text{if } \phi_x(x) \text{ is defined} \\ 0 & \text{otherwise,} \end{cases}$$

then $f \notin \mathcal{R}$, since otherwise f would afford a recursive definition of φ. Furthermore, if

$$\psi(x) = \begin{cases} 1 & \text{if } \phi_x(1) \text{ is defined} \\ 0 & \text{otherwise,} \end{cases}$$

Then $\psi \notin \mathcal{R}$. For, suppose otherwise. Then for each $x = 1, 2, 3, \ldots$, an $N = N(x)$ may be obtained from x by a primitive recursive procedure, such that either

$$\phi_x(x) = \phi_N(1),$$

or the expression on each side is undefined. But then $\psi \circ N = f$, contrary to $f \notin \mathcal{R}$.

5. **1.4.1 Higher Recursion Theory**

The (meta-)theory of $\mathcal{R}$ is amenable to generalizations, which should not be confused with (restricted) 'transfinite' recursion, nor with finitary forms of this recursion, such as the treatment contained in Ch. VI.

Arithmetic and Analytic Hierarchies

As noted in §3·1, prefix by quantifier leads out of $\mathcal{R}$. Prefix by several *like* quantifiers, however, causes no further enlargement. Thus, e.g using the pairing functions of note 3,

$$F(x) = (\mu y)\ [(\forall x_1)(\forall x_2)\ f(x, x_1, x_2) = y] \qquad \text{(I)}$$

has the equivalent recursive formulation

$$F(x) = (\mu y)\ [(\forall u)\ f\{x, \sigma(u), \rho(u)\} = y].$$

Alternation of quantifiers $\exists, \forall$ leads to the proper inclusions

$$\begin{array}{ccccccc} & & & \Sigma_1^0 & & & \\ & & \subset & & \subset & & \subset \\ \mathcal{R} = \Delta_1^0 & & & & & \Delta_2^0 & \ldots, \\ & & \subset & & \subset & & \subset \\ & & & \Pi_1^0 & & & \end{array}$$

where Σ_n^0, Π_n^0 denote n alternations, starting respectively with $\exists, \forall$; $\Delta_n^0 = \Sigma_n^0 \cap \Pi_n^0$. The '0' is a reference to the type of bound variable occurring in the recursive predicate, as in e.g. (I). If (recursive) predicates of function variables are allowed, then much enlarged classes Σ_n^1, Π_n^1 are obtained.

Hyper-Arithmetic Hierarchy

Let ${}_nf_1(x, \underline{x})$, ${}_nf_2(x, \underline{x}), \ldots$ be a ($\mathcal{P}$-)enumeration of the functions of $\mathcal{R}_n$ (ref. fn. 16). Let the class $\mathcal{HA}$ be the smallest obtained by taking the general recursive closure of the operation $f \in \mathcal{HA} \Rightarrow F \in \mathcal{HA}$, where

$$F(x) = (\mu y)\ [(Qx_1)\ldots(Qx_{\sigma\{f(x)\}})\ {}_{1+\sigma}f_{\rho\{f(x)\}}(x, x_1, \ldots, x_\sigma) = y],$$

where furthermore each Q denotes a quantifier alternation. This defines the 'hyper-arithmetic' functions (of a single variable). A feature of interest in this construction is quantification of a 'variable number of variables'. It should be noted that no more are the above hierarchies acceptable to finitism, than is their keystone, $\mathcal{R}$.

Chapter II

THE GOODSTEIN-SKOLEM HIERARCHY EP

Section Headings

Section Headings (Cont'd)

2.1 Introduction

2.1.1 Early Presentations of EP

The earliest published discussion of the class **EP** of arithmetic polynomial functions, with exponentiation, of the writer's finding, occurs in the June, 1944 paper [45] by R. L. GOODSTEIN. Th. SKOLEM [141] in 1956, without citing GOODSTEIN, defined **EP** inductively; and using the axiom of choice, showed that **EP** is well-ordered under the relation $\prec$ of eventual domination. At the end of [141], SKOLEM introduced some extensions of the concept **EP**, together with various conjectures. In 1986, during a visit to Oslo, I compiled a bibliography of the published work of SKOLEM (in arithmetic and logic); it appears he did not build further on the 1956 paper.

The present library of the University of Khartoum, D. R. Sudan, pleasantly situated on the banks of the Blue Nile, was the site of the old University College, Khartoum. In this place I had the good fortune to discover, without knowledge of the above 2 works, the system $(\mathbf{EP}, \prec)$. In December, 1977, while reading in the stacks of the D. M. S. WATSON library, University College, London, I chanced to open the volume of Det Kongelige Norske Videnskabers Selskabs Forhandlinger which contains SKOLEM's paper. Only after that, with the arising of general interest in GOODSTEIN sequences, did I learn of the earlier presentation.

2.1.2 Definition and Fundamental Properties

Section 2, the longest in the book, defines **EP** and establishes key properties taken for granted in the text, explicitly those relating to eventual domination and the CANTOR-type normal form. More specialized features, particularly immediate domination, the EULER-MAURER exponential, the summation hierarchy and the register polynomial are simultaneously introduced. The subsequent section is devoted to the **EP**–property which lies at the root of the GOODSTEIN sequences finiteness phenomenon. §4 re-introduces **EP** via finite types; the closing section, 'Variants of **EP**', is not at all a survey. It is intended to show how much inessential territory has to be left untouched, in pushing ahead into the pure hierarchy.

2.1.3 The GOODSTEIN Representation

GOODSTEIN's idea may be introduced informally by considering any reasonably large number. So, e.g., take the binomial coefficient $\binom{88}{11} = 3200\ 60083\ 61808$ [1] It may be noted that, relative to the usual base 10,

$$N = 3 \cdot 10^{13} + 2 \cdot 10^{12} + \ldots + 10^3 + 8 \cdot 10^2 + 8,$$

is so large that the leading exponent, 13, exceeds the base: $13 = 10^1 + 3$. Generally, GOODSTEIN considered the representation

$$n = f(m),$$

where $m, n \in \mathbf{N}$, $m \geq 2$, are given numbers and $f(m)$ is the exponential expression in m obtained by taking the 'decimal' expression of n to base m; and then by repeatedly analyzing the exponents into base–m expressions, and so on, until only an exponential cluster of occurrences of m, and of coefficients $< m$, remains. If now the occurrences of m are replaced by the sign 'x' denoting a variable over $\mathbf{N}$, a recurrent exponential function $e(x)$ is obtained. **The representing functions e obtained as m, n take the above values, constitute the class EP.** This is the characterization given by GOODSTEIN on p2 of [45]. The linearly ordered system $(\mathbf{EP}, \prec)$ is obtained by taking $e \prec f$ iff $e(m) < f(m)$, where every coefficient in e, f is less than m.

GOODSTEIN made the following application of his exponential forms: given successive increases of base, with a unit decrement at each step, viz.

$$\begin{aligned} n &= f(m) \\ N &= f(M) - 1, \text{ where } 1 < m < M \\ &= e(M), \end{aligned}$$

the finiteness of the sequences

$$\ldots \prec e \prec f$$

of representing functions in **EP** obtained, is equivalent to the form of the DCC required to support the GENTZEN consistency proof of Ch.I, §3·5. Aside from a further publication [46] in 1947, in which the above representations are generalized using 'majorant variables', GOODSTEIN did not return to the research of **EP**.

My own 1976 formulation of **EP**, a minimal inductive definition, turned out to be one and the same [2] as SKOLEM's 1956 presentation. SKOLEM did not consider fundamental sequences; nor normal trees, although a proof in the 1956 paper (thm. 2, p57) does use 'ramified systems'.

---

[1] MILLER [109], 1954.

[2] Aside from discussions of the AC.

The approach of GOODSTEIN, sketched above, will now be discussed in a little more detail. The special notations used in the remainder of §1.3 will not be used again. Let $m, n \in \mathbf{N}$, $m \geq 2$. Then $(\exists! a_k, \ldots, a_1; c_k, \ldots, c_1 \in \mathbf{N})_{a_i < m}$

$$n = a_k m^{c_k} + a_{k-1} m^{c_{k-1}} + \ldots + a_2 m^{c_2} + a_1 m^{c_1} + a_0, \qquad \text{(I)}$$

where a_0 is an integer in the range $0 \leq a_0 < m$; $c_1 < c_2 < \ldots c_k$; and where the c_i themselves assume such a form. Let $\phi_m(n)$ be an abbreviation for the above expression of n with *coefficients* $< m$ and *scale symbol* m. Then ϕ may be defined recursively by

$$\phi_m(n) \text{ is } am^{\phi_m(c)} + \phi_m(n - am^c),$$

where m^c is the greatest power of m not exceeding n; and am^c is the greatest multiple of m^c not exceeding n. Now let Sb_x^m operate on an expression of the form (I) by substituting 'x' for each occurrence of 'm'. Let $\mathrm{T}_x^m(n) = \mathrm{Sb}_x^m \phi_m(n)$. If x is now interpreted as a variable over $\mathbf{N}$, $\mathrm{T}_x^m(n)$ determines an arithmetic function. To take an example of [45], $\phi_3(106)$ is the expression $3^{3+1} + 2.3^2 + 2.3 + 1$; and $\mathrm{T}_x^3(106) = x^{x+1} + 2x^2 + 2x + 1$. GOODSTEIN introduced the separate notation

$$\mathrm{S}_x^m(n)$$

for the value determined by the expression $\mathrm{T}_x^m(n)$, when 'x' itself denotes a particular natural number. Note that $\mathrm{Sb}_x^m \phi_m(n) = \phi_x\{\mathrm{S}_x^m(n)\}$.

The expressions $\mathrm{T}_x^m(n)$ are not all distinct: Suppose $m < m'$, and $n' = \mathrm{S}_{m'}^m(n)$. Then

$$\begin{aligned} \mathrm{T}_x^{m'}(n') &= \mathrm{Sb}_x^{m'} \phi_{m'}(n') = \phi_x\{\mathrm{S}_x^{m'}(n')\} \\ &= \phi_x[\mathrm{S}_x^{m'}\{\mathrm{S}_{m'}^m(n)\}] \\ &= \phi_x\{\mathrm{S}_x^m(n)\} = \mathrm{Sb}_x^m \phi_m(n) = \mathrm{T}_x^m(n). \end{aligned}$$

Some of the lemmas assumed by GOODSTEIN in respect of the properties of the $\mathrm{T}_x^m(n)$ are proved in §3, in particular that, under the usual interpretation, $\mathrm{T}_x^{m'}(n')$ is less than $\mathrm{T}_x^m(n)$ iff $n' < \mathrm{S}_{m'}^m(n)$.

2.1.4 Notation For the Present and Subsequent Chapters

The fixed notation $\mathcal{N}$ is used for the space of arithmetic functions $f : \mathrm{N} \to \mathrm{N}$.

Given a fixed x, ambiguity can arise, leading to the confounding of a function f, and its value $f(x)$ at x. So, where practicable, fixed notations are assigned to standard functions. In particular,

ω is used *throughout* to denote the identity $\lambda x \cdot x$ on $\mathbf{N}$;

and constant functions (on $\mathbf{N}$) are denoted by the corresponding bold-face numeral. $\mathbf{0}$ is $\lambda x \cdot 0$.

Standard operations on (arithmetic) functions $\alpha, \beta, \ldots$ are denoted as usual by:

$$\alpha + \mathbf{1} = \lambda x \cdot \alpha(x) + 1; \quad \alpha + \omega = \lambda x \cdot \alpha(x) + x, \text{ etc.;}$$

$$\omega^\alpha \cdot \omega^\beta = \lambda x \cdot \{x^{\alpha(x)}\}\{x^{\beta(x)}\} = \omega^{\alpha+\beta}.$$

The predecessor $x - 1$ of a natural number x is occasionally written x^-. Lower-case Latin letters, particularly the letter h, may in some cases be used to denote either an arithmetic function, *or* a natural number.

'$X \subset Y$' always denotes *strict* containment (i.e. $X \subset Y \Leftrightarrow \{X \subseteq Y \; \& \; X \neq Y\}$.). Square brackets are occasionally used to denote integral part.

Following the notation of de BRUIJN [28], '$\mathbf{N}_0$' denotes the natural numbers augmented by number zero, viz.

$$\mathbf{N}_0 \equiv 0, 1, 2, 3, 4, 5, \ldots\ldots .$$

Abbreviations

For f a function:

1. A constant f is designated *cst*, otherwise *nc*.

2. A strictly increasing f, viz. for which

$$x < y \Rightarrow f(x) < f(y),$$

   is designated *si*. It may be noted that the si f are closed under sum, product, exponentiation and composition.

3. A non-decreasing f, viz. for which $x < y \Rightarrow f(x) \leq f(y)$ is designated *nd*.

4. '$y \in \mathrm{im} f$' means: 'for some x, $y = f(x)$'.

2.2 Exponential Polynomials

2.2.1 Natural Polynomials

The most important exponential polynomials are *natural* polynomials, viz. functions of the positive integral variable x of the form

$$e = \lambda x \cdot \sum_{i=n}^{0} a_i x^i = \sum_{i=n}^{0} a_i \omega^i, \qquad \text{(I)}$$

where $n \in \mathbf{N}_0$, and for which the coefficients a_i satisfy

1. $a_i \in \mathbf{N}_0$ if $i > 0$.
2. $a_0 \in \mathbf{N}$ if all other coefficients are zero, $a_0 \in \mathbf{N}_0$ otherwise. The class of natural polynomials is denoted by **XP**. [3]

Thus **XP** forms a proper subclass of the ring $\mathbf{Z}[x]$ of *integral polynomials* (ref. HARDY and WRIGHT, *Introduction to the Theory of Numbers* , p82). Since any integral polynomial P may be expressed as the difference $\alpha - \beta$ of natural polynomials, it follows that the congruence theory of the roots $P \equiv 0 (\text{mod } m)$, $m \in \mathbf{N}$, reduces to the study of the congruences $\alpha \equiv \beta (\text{mod } m)$. E.g. the above source contains the following example: $x^4 \equiv 1 (\text{mod } 16)$ has 8 incongruent roots. For each $m = 1, 2, 3, \ldots$, the natural polynomials whose coefficients a_i satisfy $0 \leq a_i < m$, viz. the integral polynomials (mod m), determine an important proper subclass of **XP**.

Inductive Definition of Natural Polynomials

1. The functions $\lambda x \cdot \mathbf{n}(x) = n$, $n = 1, 2, 3 \ldots$ belong to **XP**.
2. The identity $\omega = \lambda x \cdot x$ belongs to **XP**.
3. If $e_1, e_2 \in \mathbf{XP}$, then $e_1 + e_2 = \lambda x \cdot \{e_1(x) + e_2(x)\} \in \mathbf{XP}$.
4. If $e_1, e_2 \in \mathbf{XP}$, then $e_1 \cdot e_2 = \lambda x \cdot \{e_1(x) \cdot e_2(x)\} \in \mathbf{XP}$.

THEOREM 2.1 (Standard Form Theorem for Polynomials) *1. If e is of the form* (I) *above, then $e \in$* **XP**.

2. If $e \in$ **XP**, *then e can be put in the form* (I).

Proof.

1. By application of the rules (1)-(4).

[3] Or by $\mathbf{N}[x]$.

2. By induction, using the associative, commutative, distributive etc. laws of arithmetic,

□

giving the simplification

COROLLARY 2.2 *The class* **XP** *is the least for which*

Rule **XP**.*1 The function* **1** *taking the constant value* 1 *belongs to* **XP**.

Rule **XP**.*2* $e \in \mathbf{XP} \Rightarrow \lambda x \cdot \{e(x) + 1\} \in \mathbf{XP}$.

Rule **XP**.*3* $e \in \mathbf{XP} \Rightarrow \lambda x \cdot \{e(x) \cdot x\} \in \mathbf{XP}$.

□

Note that by eliminating terms with zero coefficients, and re-naming coefficients of the remaining terms, the standard form becomes the normal form:

$$e = \sum_{i=r}^{0} a_i \omega^{n_i}.$$

Examples.

1. Consider the base$-m$ representation (I) of n, of §1.2. Ignoring the representations of the exponents, this determines an $e \in \mathbf{XP}$ for which $n = e(m)$.

2. When 'x' is interpreted as a continuous variable, LEGENDRE, HERMITE, LAGUERRE, CHEBYCHEV and many other polynomials of special function theory may be expressed (after the possible introduction of a constant rational factor) as differences of polynomials in **XP**. It may be noted that equalities of the form $\int e\ dx = \int f\ dx$, for $e, f \in \mathbf{XP}, e \neq f$, between definite limits, thus follow for each orthogonality of e.g. special polynomials. It is perhaps unfortunate that the natural polynomials, building blocks in many areas of pure mathematics, have not been accorded an earlier characterization, and closer study.

3. Let $e = \sum_{i=r}^{0} a_i \omega^{n_i}$ be an arbitrary natural polynomial. Let $E = a\omega^n$, where $n = \max_i n_i$, $a = (r+1) \cdot \max_i a_i$. Then $(\forall x)\ e(x) \leq E(x)$. But for $x \geq 1 + \max\{a, n\}$, $E(x) < \psi(x) = x^x$, i.e. for large x, $e(x) < \psi(x)$. Note that there are very much slower growing elementary functions than ψ with this property, e.g. $\omega^{[\sqrt{\omega}]}$, and each term in the descending scale

$$\lambda x \cdot x^{l(x)}, x^{l\{l(x)\}}, \ldots,$$

where $l(x) = [1 + \ln x]$ (square brackets denote integral part).

Notation

Throughout this and following chapters, the notation ψ $(= \omega^\omega)$ of the above example 3 will be adopted for the distinguished function x^x of $x = 1, 2, 3, \ldots$.

The Natural Linear Forms

A subclass $\mathbf{NL} \subset \mathbf{EP}$ of some importance is that which contains exactly the expressions

$$e(x) = ax + b;\ a, b \in \mathrm{N}_0;\ b > 0 \text{ if } a = 0;$$

defining a linear function e of the positive integral variable x.

Example.

The RHS's $qx + r$ occurring in the computation of EUCLID's algorithm belong to **NL**.

THE GOODSTEIN-SKOLEM HIERARCHY EP

2.2.2 Exponential Polynomials

The following occurs in McBETH, 1980 [4], where the predicate **EP** is informally called a 'class':

DEFINITION 2.3 *1.* **EP** *is the smallest class closed under the following rules:*

Rule **EP**.*1* $\lambda x \cdot \mathbf{1}(x) \in \mathbf{EP}$.

Rule **EP**.*2* $e, f \in \mathbf{EP} \Rightarrow e + f = \lambda x \cdot \{e(x) + f(x)\} \in \mathbf{EP}$.

Rule **EP**.*3* $e \in \mathbf{EP} \Rightarrow \omega^e = \lambda x \cdot x^{e(x)} \in \mathbf{EP}$.

2. Polynomials of the class **EP** *are* **exponential polynomials**.

□

Note that $\omega = \lambda x . x^{\mathbf{1}(x)} \in \mathbf{EP}$. Also, **EP** is closed under the rules of **XP**, and $\psi \notin \mathbf{XP}$ belongs to **EP**, so

$$\mathbf{XP} \subset \mathbf{EP}.$$

Since the property of being homogeneous does not arise, $e \in \mathbf{EP}$ may conveniently be called an (exponential) *form*.

[4] [96], §2, 'Fundamental sequences for exponential polynomials', *Zeitschr. f. math. Logik Grundl. Math.*, Bd. 26, S. 115-122.

NOTATION

1. The class $\mathbf{EP} \cup \{\mathbf{0}\}$ is denoted by $\mathbf{EP_0}$.

2. Lower case Latin, $e, f, g, \ldots$ and Greek $\alpha, \beta, \ldots, \theta, \phi, \xi, \ldots$ letters are used to denote functions of **EP**. (The occasional occurrences in the text of the natural base

$$e = \lim_{n \to \infty}(1 + \frac{1}{n})^n = 1 + \frac{1 + \frac{1 + \cdots}{3}}{2} = 2.71828\ldots$$

should not be confounded with $e \in \mathbf{EP}$).

3. The *successor* $e + \mathbf{1}$ of the polynomial e, in **EP**, is denoted by e^+.

Example. The function e arising by GOODSTEIN's procedure from the binomial coefficient used in §1.2, viz.

$$e(x) = 3x^{x+3} + 2x^{x+2} + 6x^9 + 8x^6 + 3x^5 + 6x^4 + x^3 + 8x^2 + 8$$

belongs to **EP**.

Note

EP is not closed under the operation of translation of a variable: given $e \in \mathbf{EP}$, $f = \lambda x \cdot e(x+1)$, it is not in general true that $f \in \mathbf{EP}$. For example, the rules of defn. 3 cannot give the function $(x+1)^{(x+1)}$.

Remark

SKOLEM [141], (1956) included **0** in his class $\mathbf{S} = \mathbf{EP_0}$. [5]

Now suppose $e \in \mathbf{EP}$.

THEOREM 2.4 *e is either cst or si.*

Proof. $\mathbf{1} \in \mathbf{EP}$ satisfies the condition; and if $e, f \in \mathbf{EP}$ satisfy it, then so do $e + f$ and ω^e. So the result follows by induction on order-of-definition.

□

Thus, if $e \in \mathbf{EP}$ is nc, $(\forall x)\ x \le e(x)$.

DEFINITION 2.5 *1. The x- height* [6] $\mathrm{ht}(e)$ *of* $e \in \mathbf{EP_0}$ *is defined inductively:*

(a) *If e is cst,* $\mathrm{ht}(e) = 0$.

(b) *If* $\mathrm{ht}(e_i) = h_i$, $e_i \in \mathbf{EP}$, $i = 1, 2$, *then* $\mathrm{ht}(e_1 + e_2) = \max_{i=1,2} h_i$.

[5] If the domain of ψ is accepted as extensible to the positive reals, put $y = \frac{1}{x}$. Then $\ln \psi = -\frac{1}{y}\ln y$, $\to 0$ as $y \to \infty$. Therefore $\lim_{x \to 0} \psi = 1$ gives, by continuity, $\psi(0) = 0^0 = 1$.

[6] [96] defn. 2.4, p116.

(c) *If* $\mathrm{ht}(e) = h$, $e \in \mathbf{EP}$, *then* $\mathrm{ht}(\omega^e) = h + 1$.

2. $\mathrm{ht}(\mathbf{0}) = 0$.

3. *For* $h = 0, 1, 2, 3, \ldots$ *the predicates (subclasses)* $\mathbf{EP}^{(h)}$ *are defined by:*

$$e \in \mathbf{EP}^{(h)} \Leftrightarrow \{e \in \mathbf{EP} \ \& \ \mathrm{ht}(e) \leq h\}.$$

Thus, by part 1. of the defn., the x−height of the previous ex. is 2. And by part 3.:

$$\mathbf{EP}^{(0)} \subset \mathbf{EP}^{(1)} = \mathbf{XP} \subset \mathbf{EP}^{(2)} \subset \ldots \subset \mathbf{EP}^{(h)} \subset \mathbf{EP}^{(h+1)} \subset \ldots,$$

the inclusions (using the notation convention of §1.3) being strict; and $\mathbf{EP} = \bigcup_h \mathbf{EP}^{(h)}$.

2.2.3 Other Forms Designated Exponential Polynomials

The infinite system **EP**, the object of study of the present and other chapters, is only one of at least 3 designations occurring in the literature.

The BELL Exponential Polynomials

Let $r \in \mathbf{N}$, let x, t denote independent (real) variables. Writing $D_t = \frac{d}{dt}$, the polynomials $\xi_n = \xi_n(x, t; r)$, as $n = 1, 2, 3, \ldots$ of E. T. BELL [12], p258, are defined (in a paper entitled 'Exponential polynomials') by

$$\xi_n = \lambda x, t \cdot \exp(-xt^r) D_t^n \{\exp(xt^r)\}.$$

Thus, although ξ_n is obtained by operating on an exponential form by differentiation, ξ_n does not contain an exponential term. BELL gives an explicit form ([12], 1.6, p260) of ξ_n. In the case $r = 2$ this gives

$$D_t^2 \xi_n(x, t; 2) + 2xt D_t \xi_n(x, t; 2) - 2nx \xi_n(x, t; 2) = 0.$$

Putting $x = -1$ gives

$$\frac{d^2\xi}{dt^2} - 2t\frac{d\xi}{dt} + 2n\xi = 0 \qquad \text{(I)}$$

viz. HERMITE's equation. The ξ_n thus include the HERMITE polynomials as a special case. [7] BELL considered arithmetical properties of generalizations ϕ_n, Y_n of the the ξ_n. These polynomials do not contain an exponential term.[8] [12] also follows many other generalizations of, and relations between standard classes of polynomials.

[7] Alternatively, the defining equation of the ξ_n may be viewed as a generalization of RODRIGUEZ's formula

$$H_n(t) = (-1)^n e^{t^2} \frac{d^n}{dt^n}(e^{-t^2}),$$

for the solutions of (I).

[8] However it is shown that the functions $\zeta_n(x, t; 2r) = e^{-xt^{\frac{nr}{2}}} \xi_n(-x, t; 2r)$, $x > 0$ form an orthogonal set:

$$\int_{\pm\infty} \zeta_m \zeta_n dt = 0, \quad m \neq n.$$

Remark

The *exponential* or BELL [9] numbers are by definition those generated by

$$e^{e^x-1} = \sum_{n=0}^{\infty} B(n)\frac{x^n}{n!},$$

viz. a repeated exponential. For many references, see GOULD [51], (1971); in terms of STIRLING numbers (Appx. C, §3), $B(n) = \sum_{m=0}^{n}\left\{ \begin{matrix} n \\ m \end{matrix} \right\}$.

Exponential Polynomials in Function Theory

Another class of functions designated exponential polynomials consists of finite linear combinations of terms

$$h = b_1^{q_1} b_2^{q_2} \ldots b_m^{q_m} g,$$

where the functions b are additive: $b(x+y) = b(x) + b(y)$ and g is an exponential or is multiplicative: $g(x+y) = g(x)g(y)$. This definition may be found in LAIRD [83], 1979. [10] Sources of references to the earlier literature are given at the end of SZÉKELYHIDI, [145], 1982, 'A note on exponential polynomials'. A basic theorem in the study of these forms is given in ANSELONE and KOREVAR [7], 1964: every translation-invariant subspace of the continuous functions on the real line consists of exponential polynomials.

Notes

1. Exponential polynomials in the sense of the immediately present context *do* contain exponential terms. However, unlike the generating function of the BELL numbers, or the general forms of the class **EP**, the exponentiation is restricted to a single level.

2. It is to be emphasised that the exponential polynomials, briefly discussed in §2·3, will not be treated further in this volume.

2.2.4 Specialized Notation of ROBINSON

The following inductively defined repeated or *iterated* exponentiation is not an operation of closure of **EP**. But it is useful for applications and in introducing examples. The notation adopted is that of ROBINSON, 1952 [129], who used repeated exponentiation as a bound in studies in existential definability. The operation is often mentioned in publications of GOODSTEIN.

DEFINITION 2.6 $(\forall x, y \in \mathrm{N})$

$$x \star 1 = x \quad \text{and} \quad x \star (y+1) = x^{x \star y}.$$

[9] Some values are tabulated in Appx. C.

[10] A common form of exponential polynomial conforming to this definition is

$$t_1^{\mu_1 - 1} \ldots t_p^{\mu_p - 1} e^{\lambda_1 t_1 + \ldots + \lambda_p t_p}, \ \mu_i \in \mathrm{N}.$$

Examples

1. $3\star 3 \approx 7\cdot 6\times 10^{12}$; $2\star 5\approx 2\cdot 0\times 10^{19728}$.

2. For each $n\in\mathbf{N}$, $\omega\star n = \lambda x\cdot x\star n \in \mathbf{EP}$ (viz. $\omega\star n$ is obtained by n applications of the rule **EP**.3.)

3. Note that $\mathrm{ht}(\omega\star n) = n$, with $\omega\star(n+1)\in \mathbf{EP}^{(n+1)} - \mathbf{EP}^{(n)}$.

Defn. 6 is an appropriate context in which to introduce, for later applications, a function outside, but closely associated with **EP**.

NOTATION 2.2.1 *1. The function $x\star x$ of x is denoted by ε.*

2. The class $\overline{\mathbf{EP}} =_{df} \mathbf{EP}\cup\{\varepsilon\}$.

$\varepsilon = \omega\star\omega$ is a somewhat fast-growing function. The usual base-10 decimal expression of $\varepsilon(4)$ has more digits than could be inscribed on a smooth orb, the size of the Earth, using the latest X- ray engraving technologies.

The EULER-MAURER Exponential Function.

There is a wide literature of repeated exponentiation,[11] often in the limiting form

$$y = x^{x^{x\cdots}} = \lim_{n\to\infty} x\star n. \qquad (\mathrm{I})$$

KNOEBEL [75], 1981, cites over a hundred references, from the 18th century onwards. Many of these are in the context of the search for solutions of the equation

$$x^y = y^x.$$

In a very short note, MACINTYRE [89], 1966, proves the convergence of (I) in the case $x = i = e^{\frac{i\pi}{2}} = \sqrt{-1}$. A computer plot shows the sequence converges to $\sigma + i\rho$ in a spiral pattern. [12] It is well-known (KNOEBEL [76]) that for $e^{-e} < x < e^{\frac{1}{e}} = 1.44466\ldots$, (I) converges to a limit y. Thus, taking $x = \sqrt{2} < \sqrt[e]{e}$, $(\sqrt{2})^y = y$ gives the solutions $y = 2, 4$. [13] The notation ${}^n x = x\star n$ introduced by MAURER [94], 1901, occurs in GOODSTEIN [46], p129; this interesting notation is ambiguous in such expressions as $x^n y$, although a priority convention could easily enough be devised.

It is emphasised that the EULER-MAURER iterated exponential function (I) will not be considered again in this volume.

[11] Variously called 'hyperpowers', 'power sequences', 'super-exponentiation' etc.

[12] It may be noted that $i^i = e^{-\frac{\pi}{2}}$ is real. This leads to the curious formula ${}^{\sqrt{-1}}\sqrt{-1}\cdot\sqrt{-1}^{\sqrt{-1}} = 1$.

[13] The multiplicity is explained by the minimum of the function of a real variable $\psi(x)$ at $x = \frac{1}{e}$ (ref. ANDREWS and LACHER [6], prop. 3, p140).

2.2.5 Failure of Association

As is well known, the exponential operations are not associative. Thus

$$2^{(3^4)} = 2^{81} > 4096 = (2^3)^4;$$

$$2 \star (3 \star 4) = 2 \star (3^{3^{3^3}}) > 2 \star 7 > 16 \star 4 = (2 \star 3) \star 4.$$

Therefore the *right to left* convention for bracketing is adopted for $\star$, as for the usual exponentiation. E.g. $x \star y \star z = x \star (y \star z)$. It may be remarked that exponentials do not appear to have played a large role in non-associative algebra. See, e.g., SCHAFER's monograph [137][14]

A further definition is useful in technical applications:

DEFINITION 2.7 $(\forall x, y, z \in \mathbf{N})$

$$x \star (1; z) = x^z, \quad \text{and} \quad x \star (y+1; z) = x^{x \star (y;z)}.$$

As with defn. 6, the notation is intended to include functions. Thus $\lambda x \cdot x \star \{f(x); g(x)\}$ is denoted by the symbol $\omega \star (f; g)$. Also, the defns may be extended by: $x \star 0 = 1$; $x \star (0; z) = z$.

Examples.

1. $2 \star (3; 4) = 2^{2^{2^4}} = 2^{65536}$. $\omega \star (1; f) = \omega^f$.

2. For $n \in \mathbf{N}$, $\omega \star (n; \omega) = \omega \star (n+1)$.

3. The definition is often useful in the denotation of functions and their values in **EP**, such as $\omega \star (3; \psi^+)(2) = 2 \star (3; 5)$; $\omega \star (\psi; \varepsilon)(2) = 2 \star 6$.

4. According [15] to a theorem of WRIGHT[161](1951), there exists a real number ξ, such that for every $n = 1, 2, 3, \ldots$, the integer

$$[2 \star (n; \xi)]$$

is prime.

Note that, for $e \in \mathbf{EP}$, $\omega \star (n; e) \in \mathbf{EP}$.

---

[14] An explanation surely is that exponentiation is essentially non-associative, so that weak rules of association, such as $(xy)x^2 = x(yx^2)$ as in JORDAN algebras (e.g. the space of $n \times n$ real or complex symmetric matrices), or $xy^2 = (xy)y$ as in CAYLEY-DICKSON algebras, do not hold.

[15] 'A prime-representing function', *Amer. Math. Monthly* **58**, 616-618 (1951).

Note on the Laws of Exponentiation

The 4 fundamental laws of exponentiation enumerated by MACINTYRE in the end paragraph of [90], viz.

1. $x^1 = x$,
2. $(x \cdot y)^n = x^n \cdot y^n$,
3. $x^m \cdot x^n = x^{m+n}$,
4. $(x^m)^n = x^{m \cdot n}$. [16]

have no analogue (save $x \star 1 = x$) for the higher exponentiation $\star$ of §2.4. This is because (3) and (4) depend on the associativity of multiplication; while (2) depends on the commutativity of this operation. It is a remarkable fact that although exponentiation does not satisfy these laws, it may enter into combination with lower operations to yield (1) - (4). The identity

$$x \star (m; x \star n) \equiv x \star (m + n) \equiv x \star (n; x \star m)$$

has a superficial resemblance to (3); but the expressions $x \star m$, $x \star n$ are not operands of standard operations. Certain 'laws' can, however be found. E. g. for $x, y \geq 2, n \geq 1$,

$$(xy) \star (n + 1)$$

expands into a pattern which contains every possible exponential ladder of height $n + 1$, containing x's or y's. Since the expansion of $(x^y) \star n$ contains only some of these ladders,

$$(x^y) \star n < (xy) \star (n + 1).$$

2.2.6 Eventual Domination

For $e \in \mathbf{EP}$, the rate of growth of $e(x)$, as x becomes large, can be estimated most precisely by comparison with known functions, usually in **EP**. This uses the order of eventual domination, which applies generally to arithmetic and other functions:

DEFINITION 2.8 *The relation $\prec$ on $\mathbf{EP} \times \mathbf{EP}$ is defined by:*

$$(\forall e, f \in \mathbf{EP})\, e \prec f \Leftrightarrow \{(\exists N)(\forall x)\, N \leq x \Rightarrow e(x) < f(x)\}.$$

This is expressed in words: 'e is eventually dominated by f', 'for x sufficiently large, f exceeds e'; and '$e \preceq f$' means '$e \prec f \vee e = f$'.

Examples

1. $e \prec e^+ = e + 1$ in **EP**.
2. By ex. 3 foll. corol. 2, $e \prec \psi$, $e \in \mathbf{XP}$.
3. $\omega \star m \prec \omega \star n$, $(m < n)$; and $\omega \star (n; m) \prec \omega \star (n + 1)$.

[16] Laws 2, 3 and 4 are given here in the dependence order of the derivations in H. A. THURSTON, *The Number System* , Ch. H, p90.

4. Eventual domination arises in the comparison of divergent series. E.g. let

$$f_N = \sum_{n=1}^{N} \frac{1}{n^{0\cdot 1}}; \quad g_N = \sum_{n=1}^{N} \frac{1}{\ln(1+n)}.$$

As $n \to \infty$, $n^{-0\cdot 1} \to 0$ faster than $\{\ln(1+n)\}^{-1}$, yet $> 10^{14}$ terms are required, before $f_N < g_N$. (Ref. R. P. BOAS, 'Convergence, Divergence and the Computer', in HONSBERGER, Ross [32], pp151-159).

NOTATION

Occasionally $e \prec\!\prec f$ is used, to indicate that the rate of growth of f is, in the current context, very much greater than that of e.
Eventual domination is a relation between functions, of outstanding importance. In order to indicate the scope of the above definition, there follows a short digression on rates of growth. But note that the specialization to **EP** produces an extra-ordinary research tool: a rational linear hierarchy capable of supporting GENTZEN's consistency argument.

2.2.7 Rates of Growth

1. Appendix C contains tables of discrete-valued functions of classical number theory, and of algebra and combinatorics.

2. There is the general problem of assessing the rate of growth of (or 'order of infinity') of a function of a continuous variable, using a suitable apparatus of reference functions. This has proved too complicated (see HARDY [59], Ch. IV), and has been done only in special cases. What *is* possible, at least in the initial stages, is to provide a reference system which will contain a function of least growth not less than that of a given function. For the given function, points on the real axis not corresponding to an integer need play no part in the assessment, and the hierarchy **EP** provides a most delicate system of reference.

3. **Subordinacy**

   Domination arises in the theory of functions (as *subordinacy*). Thus, if

$$f = \sum_{i=0}^{\infty} a_i z^i, \quad F = \sum_{i=0}^{\infty} A_i z^i$$

are analytic functions, F univalent [17] on the unit disc D, and $f(D), F(D)$ are the respective conformal mappings of D under f, F, then by definition (in the context of §2·7)

$$f \prec F$$

[17] Viz. injective: $z_1 \neq z_2 \Rightarrow F(z_1) \neq F(z_2)$ on D.

iff $f(D) \subset F(D)$. For further details, see HILLE, *Analytic Function Theory*, Ch. 18, §5. This relation is a generalization of arithmetic domination, in the sense of the following existence theorem:

THEOREM 2.9 *There is a function γ, analytic on D, for which $|\gamma(z)| < 1$, $z \in D$; and for which*

$$f(z) = F\{\gamma(z)\},\ z \in D.$$

□

From this follows:

THEOREM 2.10 *If*

$$\| F \|_2 = \lim_{r\to\infty} \sup\{\int_{\pi}^{-\pi} | F(re^{i\theta}) |^2 \ d\theta\}^{\frac{1}{2}}$$

is finite, and if $f \prec F$, then for every n,

$$\sum_{i=0}^{n} | a_i |^2 \leq \sum_{i=0}^{n} | A_i |^2 .$$

□

2.2.8 Spanning and Cofinality

Suppose $(\mathcal{X}, <)$ is an ordered system, and $\mathcal{Y} \subseteq \mathcal{X}$.

DEFINITION 2.11 *$\mathcal{Y}$ is said to* span $\mathcal{X} = (\mathcal{X}, <)$, *or to be* cofinal in $\mathcal{X}$, *iff*

$$(\forall x \in \mathcal{X})(\exists y \in \mathcal{Y})\ x < y.$$

□

If $e = \sum_{i=r}^{0} a_i\omega^{n_i} \in \mathbf{XP}$, then $e \preceq \sum a_i \cdot \omega^{n_r} \prec \omega^{n_r+1}$. Thus, e.g., $(\mathbf{XP}, \prec)$ is spanned by the sequence $\omega, \omega^2, \omega^3, \ldots$. More generally, $\mathbf{EP}^{(h)}$ is spanned by the sequence

$$\{\omega \star (h; \mathbf{n})\}_{n=1,2,3,\ldots},$$

for every height h.

2.2.9 Properties of $(\mathbf{EP}, \prec)$

From defn. 8 follow the irreflexivity, asymmetry; and the transitivity of $\prec$:

$$e \prec f \ \&\ f \prec g \Rightarrow (\exists N)(\forall x)\ N \leq x \Rightarrow e(x) < g(x).$$

Note that from $e \prec f, \alpha \prec \beta$ in **EP**, follows $e + \alpha \prec f + \beta$. Also by thm. 4,

COROLLARY 2.12 $(\forall e \in \mathbf{EP})\, \mathbf{1} \preceq e.$

The relation o (ref. KNUTH [78], p21) between orders of greatness of functions is stronger than $\prec$, or $\prec$ is finer than o :

$$e = o(f) \Rightarrow e \prec f.$$

Now let $\mathcal{X}$ be the subclass of functions ω^e, $e \in \mathbf{EP}$. Let Φ be the $\prec$ −conserving transform defined by $\Phi(e) = \omega^e$. Then

$$e \prec f \Leftrightarrow \Phi(e) = o\{\Phi(f)\}.$$

On account of this transform, the relation $\prec$ is in effect that between orders of greatness of functions in **EP**. [18]

The relation $e \prec f$ in **EP** entails that for *some* $N = N(e, f)$, f must take larger values at all $x \geq N$. Let M be the *least* such N.

DEFINITION 2.13 *For e, f and M as above, f is said to* pass e, at M, *written* $e \prec f@M$.

Note that M may be 1. Thus e.g. $e \prec e + \mathbf{1}@1$; $\omega^3 + \omega^2 + \omega \prec \psi@4$. Note also that if f passes e at $M > 1$, then $f(M^-) \leq e(M^-)$. Indeed, there are 2 kinds of pass at M :

1. where the inequality is strict, as at $M^- = 3$ in the previous example, and
2. where equality occurs, as for $n = k \star 2$, $\omega^n \prec \omega \star 3$, $M^- = k$.

Remark. Suppose $\alpha \prec \beta@M_1$ and $\beta \prec \gamma@M_2$ in **EP**. Then for $M = max\{M_1, M_2\}$,

$$\alpha(x) < \beta(x) < \gamma(x), \text{ for } x \geq M.$$

Therefore $\alpha \prec \gamma@M' \leq M$.

Noting ex. 3 after defn. 6:

THEOREM 2.14 $(\forall h)\, e \in \mathbf{EP}^{(h)} \Rightarrow e \prec \omega \star (h + 1).$

Proof, by induction on h. For $h = 0$, the constants are eventually dominated by $\omega = \omega \star 1$. Now suppose the result is known for polynomials of $\mathbf{EP}^{(h)}$, $0 \leq h$, and let $e \in \mathbf{EP}^{(h+1)}$. Then by defn. 3,

$$e = \sum_{i=0}^{R} \omega^{e_i}$$

for some $R \in \mathbf{N}_0$, and $e_1, e_2, \ldots, e_R \in \mathbf{EP}^{(h)} \cup \{\mathbf{0}\}$. Suppose $e_i \prec \omega \star (h+1)@M_i$, $i = 0, 1, 2, \ldots, R$; and let $M = \max_i M_i$. Then

$$e_i(x) < x \star (h + 1), \quad \text{for } x = M, M + 1, M + 2, \ldots;$$

[18] HARDY [59] uses '$\prec$' to denote $= o()$.

and

$$\begin{aligned} e(x) &= \sum_{i=0}^{R} x^{e_i(x)} \leq (R+1) \cdot x^{-1+x\star(h+1)} \text{ for } x \geq M \\ &< x^{x\star(h+1)} \text{ provided } x > R \\ &= x \star (h+2). \end{aligned}$$

Therefore

$$e \prec \omega \star (h+2)@P,$$

for some $P \leq \max\{M, R\}$, and the result follows by induction.

□

Now for all h, and all $x > h$, $\quad x \star h < x \star x = \varepsilon(x)$, therefore

THEOREM 2.15 $\qquad (\forall h)\, \omega \star h \prec \varepsilon.$

□

Combining the last 2 results,

COROLLARY 2.16 $\quad \varepsilon \notin \mathbf{EP}.$

□

Note also, by thm. 14,

COROLLARY 2.17 *The sequence* $(\omega \star h)_{h=1,2,3,\ldots}$ *spans* $(\mathbf{EP}, \prec)$.

□

The relation $\prec$ has a number of important properties. The proof of the following theorem of SKOLEM is along the lines of that given in [96], 1980, theorem 2.6.

THEOREM 2.18 *The relation* $\prec$ *defined on* $\mathbf{EP} \times \mathbf{EP}$ *is connected: for any polynomials* e, f *of* $\mathbf{EP}$,

$$e \prec f \vee e = f \vee f \prec e.$$

Proof. By the remarks following defn. 5, there is a least $h \geq 0$ such that

$$e, f \in \mathbf{EP}^{(h)}.$$

The proof is by induction on h. In the initial case, $\mathbf{m} \prec \mathbf{n}$ in $\mathbf{EP}^{(0)}$ iff $m < n$ in the connected system $\mathbf{N}$. Suppose for that $\{\mathbf{EP}^{(h)}, \prec\}$, $0 \leq h$, has been proved connected, and consider $e, f \in \mathbf{EP}^{(h+1)}$. By the rules $\mathbf{EP}$.1-.3 there are

$$\alpha_i,\ i = 1, 2, \ldots, R; \quad \beta_i,\ i = 1, 2, \ldots, S$$

in $\mathbf{EP}_{\mathbf{0}}^{(h)}$ for which

$$e = \sum_{i=R}^{0} \omega^{\alpha_i}, \quad f = \sum_{i=S}^{0} \omega^{\beta_i}.$$

By the i.h., it may be assumed

$$\alpha_0 \preceq \alpha_1 \preceq \ldots \preceq \alpha_R; \quad \beta_0 \preceq \beta_1 \preceq \ldots \preceq \beta_S.$$

Either every term of one polynomial is a term of the other; or for some least j, $\alpha_{R-j} \neq \beta_{S-j}$. Therefore by the i.h., the only possibilities are

1. $R = S$, $\alpha_i = \beta_i$, $0 \leq i \leq R$; so that $e = f$.
2. (a) $R < S$, $\alpha_i = \beta_i$, $0 \leq i \leq R$; so that $e \prec f$.
   (b) $S < R$, $\alpha_i = \beta_i$, $0 \leq i \leq S$; so that $f \prec e$.
3. For some least j, $0 \leq j \leq \min(R, S)$,

   (a)
   $$\alpha_{R-j} \prec \beta_{S-j}.$$
   Then by the argument of thm. 14,
   $$\sum_{i=R-j}^{0} \omega^{\alpha_i} \preceq (R-j+1) \cdot \omega^{\alpha_{R-j}} \prec \omega^{\beta_{S-j}},$$
   since $\alpha_{R-j} \prec \beta_{S-j}$ implies there is an N such that
   $$x \geq N \Rightarrow \alpha_{R-j}^{+}(x) \leq \beta_{S-j}(x).$$
   Therefore, since the remaining terms of e are equal to the corresponding terms of f, it follows $e \prec f$. Or

   (b)
   $$\beta_{S-j} \prec \alpha_{R-j}.$$
   In this, the dual of case 3(a), $f \prec e$.

Therefore the functions of $\mathbf{EP}^{(h+1)}$ are pairwise $\prec$ – comparable. So the result for arbitrary $x-$height now follows by induction.

□

Thus, by the remarks before corol. 12

COROLLARY 2.19 *The system* $(\mathbf{EP}, \prec)$ *is linear.*

□

SKOLEM (1956, [141], p56 after thm. 1) notes that the last result implies that there are no $\alpha, \beta \in \mathbf{EP}$ which 'pass one another infinitely often' (cf. the difference $d(x)$, I, §2·3). Also

COROLLARY 2.20 *For given* $\alpha, \beta \in \mathbf{EP}$, *suppose* $(\exists N)(\forall x)\ x \geq N \Rightarrow \alpha(x) = \beta(x)$. *Then* $\alpha = \beta$.

□

Given any fixed $e \in \mathbf{EP}$, corol. 19 leads to

DEFINITION 2.21

$$(\mathbf{EP})_e = (\mathbf{EP}, \prec)_e = \{\alpha \in \mathbf{EP} : \alpha \prec e\}.$$

Examples

1. $(\mathbf{EP})_\omega$, $(\mathbf{EP})_\psi$ respectively denote the constant polynomials and the class **XP**, and generally $\mathbf{EP}^{(h)} = (\mathbf{EP})_{\omega\star(h+1)}$.
2. Let $\alpha = \omega \star (2, \mathbf{2})$, and let $\mathcal{X} \subset \mathbf{EP}$ be the class of polynomials with linear exponents. Then $\mathcal{X} = (\mathbf{EP})_\alpha$.

Remark

Definition 8 may be extended to the class $\mathcal{X}$ of integral polynomials (ref §2.1). The argument of thm. 18 then establishes $(\mathcal{X}, \prec)$ as an ordered ring. Note that both the DCC (Ch. I, §2.3) and the dual *ascending chain condition* (ACC) are violated.

2.2.10 Gradual Ascent in EP

Suppose now $e \prec \alpha$ in **EP**. Then by defn. 8,

$$(\exists N)\ x \geq N \Rightarrow e^+(x) \leq \alpha(x).$$

Therefore by corol. 20,

$$e^+ \preceq \alpha.$$

So in **EP**, there is a 'next' polynomial, e^+, following after e. e^+ is scarcely faster growing than its predecessor. Indeed, to pass along the linear ordering to reach the function $e + e$, itself a function of only modestly faster growth than that of e, it is necessary to traverse all the functions $e + \phi$, where $\phi \prec e$ in **EP**.

Remarks.

1. Repetition of the above argument gives $\mathbf{1} \prec \mathbf{2} \prec \mathbf{3} \prec \ldots$ as an initial segment of $(\mathbf{EP}, \prec)$. Rules **EP**.1-.3 give $\omega = \lambda x \cdot x$ as the $\prec$ –least term which follows these, followed by $\omega^+ = \omega + \mathbf{1} = \lambda x \cdot (x+1) \prec \omega + \mathbf{2} \prec \ldots$ etc. It is not difficult to see that $(\mathbf{XP}, \prec)$ is an initial segment of **EP**; and that the natural polynomials under eventual domination have the order-type of the right-adjusted finite words on the infinite alphabet $0, 1, 2, 3, \ldots$ under the standard dictionary order.

2. The fundamental sequence $1, 2, 3, 4, 5, \ldots$ corresponds to the initial segment of cst functions of **EP**. This is in contradistinction to $\mathbf{S} = \mathbf{EP_0}$, (for which the sequence $0, 1, 2, 3, 4, 5, \ldots$ corresponds). It is undoubtedly better to start counting at 1 (rather than at 0), where this is possible.

2.2.11 CANTOR-Type Normal Form Theorem

The connexity of $\prec$ permits $e \in \mathbf{EP}$ to be uniquely expressed in a recurrent normal form. The added terms are built up by collecting together alike terms which occur in the expressions used in thm. 18. This is essentially the CANTOR normal form, although, regrettably, CANTOR did not apply his discovery to incontravertibly mathematical objects.

THEOREM 2.22 *If e is an exponential polynomial, then e has the unique representation*

$$e = \sum_{i=r}^{1} a_i \omega^{e_i} + \mathbf{a}$$

where $r \in \mathrm{N_0}$; $\sum =_{df} \mathbf{0}$ if $r = 0$; $e_1 \prec e_2 \prec \ldots \prec e_r$ in **EP**; *$a_1, a_2, \ldots, a_r \in \mathrm{N}$; and where*

$$a \in \begin{cases} \mathrm{N_0} & \text{if } r > 0 \\ \mathrm{N} & \text{if } r = 0. \end{cases}$$

Proof. Every polynomial of the given form can be obtained by the rules of formation of **EP**. On the other hand, for rule **EP**.1, **1** is of the given (unique) form. Now suppose, for rule **EP**.2, that e and f are of the given form, and $g = e + f$. Then by thm. 18, the terms of e and f can be taken together and ordered and added so as to give an expression for g in the given form. For rule **EP**.3, given e, ω^e is of the form of the theorem. This is unique: $\omega^e = \omega^\alpha + \omega^\beta$ cannot occur in **EP**; for otherwise taking $\alpha \preceq \beta$, as in the proof of thm. 18 $\beta^+ \preceq e$ implies $2 \cdot \omega^\beta \prec \omega^e$.

□

NOTATION.

The constant function **a** of thm. 22 may be expressed $\mathbf{a} = a\omega^{\mathbf{0}}$, so that

$$e = \sum_{i=r}^{0} a_i \omega^{e_i},$$

$$\text{where } e_0 = \mathbf{0}, \text{ and } a_0 \in \begin{cases} \mathrm{N_0} & \text{if } r > 0 \\ \mathrm{N} & \text{if } r = 0. \end{cases}$$

Remarks

1. As x becomes large, the leading term $a_r\omega^{e_r}$ in the expression for e, of thm. 22 becomes dominant. Therefore, if $f \in \mathbf{EP}$ has leading term $b_s\omega^{f_s}$,

$$\lim_{x\to\infty}\frac{e(x)}{f(x)} = \begin{cases} \infty & \text{if } f_s \prec e_r, \\ \text{the rational number } \frac{a_r}{b_s} & \text{if } e_r = f_s, \\ 0 & \text{if } e_r \prec f_s. \end{cases}$$

2. Besides the rule **EP**.2, **EP** is closed under the operation of multiplication. For, taking $e = \sum\omega^{e_i}, f = \sum\omega^{f_i}$, $e\cdot f = \sum_{i,j}\omega^{e_i}.\omega^{f_j} = \sum\omega^{e_i+f_j} \in \mathbf{EP}$, as the $(e_i + f_j) \in \mathbf{EP_0}$, giving the rule of closure $e \in \mathbf{EP} \Rightarrow \omega\cdot e = \lambda x\cdot xe(x) \in \mathbf{EP}$. However, if $e \in \mathbf{EP}$, it is not in general true that $e^\omega = \lambda x\cdot\{e(x)\}^x \in \mathbf{EP}$. E.g. $\omega + \mathbf{1} \in \mathbf{EP}$, but writing $\beta = \lambda x\cdot(x+1)^x$, $\lim_{x\to\infty}\frac{\beta}{\psi} = 2.71828\ldots$, so by rem. 1, $\beta \notin \mathbf{EP}$. Alternatively, $3\psi \succ \beta \succ (2.7)\psi \succ 2\psi + \omega^n$, for every $n = 1,2,3,\ldots$. But thm. 22 gives $e = 3\psi$ as the $\prec$ –least **EP**–function with this property.

The inductively defined normal form is recurrent, in the sense that thm. 22 may be applied to the functions e_i which occur in the statement of the theorem. This gives rise to various definitions for classes of coefficients and terms contained in the expression.

DEFINITION 2.23 *In the expression of thm. 22,*

1. *The numbers $a_r, a_{r-1}, \ldots, a_1, a$ are the* basic *coefficients of e.*

2. (a) *The functions $e_i \in \mathbf{EP}, i = r, r-1, \ldots, 1$ are the* (basic) exponents *of e.*
   (b) *The leading exponent e_r is the* degree *of e,* $\deg e$.
   (c) *The function e_1 is the* minimal *or* rightmost *exponent of e.*

3. *The basic coefficients of e, of the e_i, of the exponents of the $e_i, \ldots$ taken together comprise the* explicit coefficients *(ec's) of e.*

4. *The sum* $\mathrm{ht}(e) + \sum_{a=ec\text{ of } e} a$ *is the* length, $\mathrm{lh}(e)$, *of e.*

Thus, if $\mathbf{a} \in \mathbf{EP}$ is cst, then a is the only ec. Note that if an exponent of e is a constant $\mathbf{n}$, then n is an ec of e.

THEOREM 2.24 *The explicit coefficients of e are finite in number.*

Proof, by induction on order of definition. For **EP**.1, the result is certainly true in the case $e = \mathbf{1}$. For **EP**.2, suppose $e, f \in \mathbf{EP}$ are known to have not more than m, n ec's respectively. Then $e + f$ has not more than $m + n$ ec's. For **EP**.3, if e has n ec's then ω^e has $1 + n$ ec's.

□

Note that the exponents of e occur in $e + f$ in **EP**; and the explicit coefficients of e occur in ω^e; also that $\text{lh}(e + f) \leq \text{lh}(e) + \text{lh}(f)$.

The normal form of thm. 22 evidently takes both the *recurrent* form, in which the e_i are assumed to be given; and the *explicit* form, in which the theorem is repeatedly applied, so all explicit coefficients appear. Thus, e.g. if $\alpha = 2\omega + \mathbf{3}$, $f = 3\omega^\alpha + 5\omega + \mathbf{6}$, $e = 2\omega^f + \omega^\omega + \mathbf{1}$, then $2\omega^{3\omega^{2\omega+\mathbf{3}}+5\omega+\mathbf{6}} + \omega^\omega + \mathbf{1}$ is the explicit form of e.

If $f = e_1 + e_2$ in **EP**, then $(\forall x)\ e_i(x) < f(x)$; so by defn. 8, $e_i \prec f; i = 1, 2$. Similarly, $1^{\alpha(1)} = 1$, and for $x > 1$, $x^{\alpha(x)} \geq 2^{\alpha(x)} > \alpha(x)$, so again, $\alpha \prec \lambda x \cdot x^{\alpha(x)}$. Therefore

THEOREM 2.25 *In the representation of thm. 22*

$$e_i \prec e,$$

for $i = r, r - 1, \ldots, 1$.

□

The expression of thm. 22 contains finite sequences in **EP** which may be extracted as follows. Let $e_{(1)} = e$. If e is nc, let $e_{(2)}$ be one of the exponents of $e_{(1)}$. If $e_{(2)}$ is nc, let $e_{(3)}$ be one of the exponents of $e(2)$. This choice of arbitrary exponents may be repeated, to give a sequence

$$\mathcal{S}: \qquad e_{(1)}, e_{(2)}, e_{(3)}, \ldots .$$

THEOREM 2.26 *1. $\mathcal{S}$ is finite.*

2. $e_{(1)} \succ e_{(2)} \succ e_{(3)} \succ \ldots .$

Proof.

1. For $i = 1, 2, 3, \ldots$, the passage from $e_{(i)}$ to $e_{(i+1)}$ represents a single application of rule **EP**.3 of defn. 3. Since $e = e_{(1)}$ is obtained by only finitely many (fm) applications of the rules **EP**.1-3, $\mathcal{S}$ is finite.

2. Is a re-statement of thm. 25

□

Note. Let $\mathcal{S}$ be as above. Then by the theorem, $\mathcal{S}$ has a last term, $e^{(T)}$, say. By the rules **EP**.1-.3, it follows $e^{(T)}$ is one of the constants $\mathbf{1}, \mathbf{2}, \mathbf{3}, \ldots$ (ref. rem. 1, §2·10).

DEFINITION 2.27 *1. The* constant exponents *of e are exactly the constant polynomials obtainable by taking the last term of an $\mathcal{S}-$sequence in e.*

2. *$\alpha(\in \mathbf{EP})$ is an* explicit exponent (ee) *of e iff $(\exists i)\ \alpha = e_{(i)}$, for some sequence $\mathcal{S}$, as in thm. 26.*

□

Example. For $e = \omega^3$, 3 is an explicit coefficient, and (as **3**) an explicit exponent and a constant exponent of e. [19]

Remarks

1. The result 26·1 is in no way dependent on 26·2; in other words, the finiteness of $\mathcal{S}$ does not depend on properties of $\prec$.

2. Thms. 24, 26 are proved from much the same basic properties. 24 follows from 26 by an application of KÖNIG's lemma [79]. But it should be noted that 24 has been obtained without recourse to this result.

It is clear that there are only fm **XP**−polynomials whose ec's are less than a given bound. Now let $h \in \mathbf{N}, h \geq 1$. If a similar result is assured for $\mathbf{EP}^{(h)}$−polynomials, then the finite choice of exponents and coefficients gives a similar result for $\mathbf{EP}^{(h+1)}$−polynomials. So by induction on height,

THEOREM 2.28 EP *contains only finitely many polynomials whose heights and ec's satisfy given bounds.*

□

Thus, let $\mathcal{X} \subset \mathbf{EP}$ be the least class for which

1. $\omega \in \mathcal{X}$,
2. If $e, f \in \mathcal{X}$, and for no $\xi \in \mathcal{X}$; $\alpha, \beta \in \mathcal{X}_0$, $e = \xi + \alpha$; $f = \xi + \beta$, (so that in particular, $e \neq f$), then $e + f \in \mathcal{X}$.
3. If $e \in \mathcal{X}$, then $\omega^e \in \mathcal{X}$.

Then by thm. 28, $\mathcal{X}$ contains only fm polynomials of given height.

Note that the classes $\mathbf{EP}^{(h)}$ may be characterized by a generalization of corol. 2:

THEOREM 2.29 *For $h = 1, 2, 3, \ldots$, $\mathbf{EP}^{(h)} = \mathcal{X}$, say, is the smallest class containing* **1**, *and closed under*

1. $e \in \mathcal{X} \Rightarrow e^+ \in \mathcal{X}$;
2. $(\forall i \leq h)\ \omega \star h \in \mathcal{X}$:
3. *If $e \in \mathcal{X}$, then the polynomial obtained from e by multiplying e or any nc explicit exponent of e, by ω, belongs to $\mathcal{X}$.*

□

[19] In strict conformity to the expression of thm. 22, 'ω^3' should be '$\omega^{\mathbf{3}}$;' but this rule can generally be waived.

2.2.12 Classification of Polynomials in EP.

In the expression $e = \sum_{i=r}^{1} a_i\omega^{e_i} + \mathbf{a}$ of thm. 22, it is possible that $\mathbf{a} = \mathbf{0}$, in which case $e \neq \mathbf{1}$, yet $e \in \mathbf{EP}$ has no predecessor. The exponential polynomials may be classified precisely along these lines, as in

DEFINITION 2.30 *1. The constant function* $\lambda x.1 = \mathbf{1}$ *is the* initial polynomial *of* **EP**. *Remaining polynomials are* non-initial.

2. *Any e for which there is an α such that $e = \alpha + \mathbf{1}$ in* **EP** *is a* successor polynomial, *denoted by* Suc(e). *α is the* predecessor *of e, denoted by $\alpha = e^-$, or by $e = \alpha^+$.*

3. *Any $e \in$* **EP** *for which neither $e = \mathbf{1}$ nor* Suc(e) *is a* diagonal (polynomial), *denoted by* Diag(e).

Examples. In **EP** :

1. The $\prec$ – least successor is **2**; the $\prec$ –least diagonal is ω.

2. For any e, Diag(ω^e).

3. If e, α are diagonal, then Diag($e + \alpha$).

THEOREM 2.31 (CANTOR) $(\forall e \in \mathbf{EP})(\exists \delta \in \mathbf{EP_0})(\exists a \in \mathrm{N_0})$

$$\delta = \mathbf{0} \vee \{\mathrm{Diag}(\delta) \;\&\; e = \delta + \mathbf{a}\}.$$

Proof. In the expression $e = \sum_{i=r}^{1} a_i\omega^{e_i} + \mathbf{a}$ of thm. 22, the term $\delta = \sum_{i=r}^{1} \omega^{e_i}$ is diagonal, by the above remarks, or $\delta = \mathbf{0}$. Thus δ is the required function in $\mathbf{EP_0}$, and $\mathbf{a}$ as above may be taken as the constant occurring in the expression.

□

Thus, if e is non-constant and non-diagonal, after a finite number of reductions, of the form e to e^-, a diagonal polynomial is reached.

As already noted, if in thm. 22 $\mathbf{a} = \mathbf{0}$, then $e = \sum_{i=r}^{1} a_i\omega^{e_i}$ is *diagonal.* The form of the RHS of this equation will, in the sequel, always denote a diagonal of **EP**.

A property of the initial value may be noted:

LEMMA 2.32 $(\forall e)\; e(1) = 1 \Leftrightarrow e = \mathbf{1}$ *or e is of the form $e = \omega^f$, in* **EP**.

Proof. If $e = \sum_{i=R}^{1} \omega^{e_i}$ and $R > 1$, then $e(1) > 1$.

□

Finite Classifications

Theorem 28 gives the decomposition $\mathbf{EP} = \cup_N \mathcal{X}_N$, where the finite classes $\mathcal{X}_N$ contain exactly the polynomials for which the height and the ec's are $\leq N$ (or for which the length $\leq N$).

2.2.13 Register Polynomial

Given $e \in \mathbf{EP}$, consider the formation of the sequence $\mathcal{S}$ of thm. 26. A fixed initial or successor polynomial is determined by e. [20] This is either e itself, or is reached by always choosing the minimal exponent of a diagonal $e_{(i)}$:

DEFINITION 2.33 *1. If* $\text{ht}(e) = 0$, *then* $e_1^{(T)} = e =$ register polynomial (rp) *of e. Now suppose the* rp *has been defined for all polynomials of* $\mathbf{EP}^{(h)}$, $0 \leq h$. *Let* $e = \sum \omega^{e_i} + \mathbf{a}$ *be of height* $h+1$. *Then the* rp

$$e_1^{(T)} \text{ of } e \text{ is } \begin{cases} e & \textit{if } \mathbf{a} \neq \mathbf{0}, \\ \text{rp of } e_1 & \textit{otherwise.} \end{cases}$$

2. Any e for which $\text{rp}(e) = \mathbf{1}$ is *perfect*, otherwise e is *imperfect*. If the diagonal $e = \Sigma + \omega^{\Sigma'+\mathbf{1}}$, $\Sigma' \in \mathbf{EP}$, then e is *basically imperfect*.

As remarked, the register polynomial arises by considering the $\prec$ –lowest exponent in an application of thm. 22. This is iterated until an exponent equal to $\mathbf{1}$ or to a successor is obtained. The rp is thus never diagonal. Note that $e_1^{(T)}$ occurs in the explicit normal form of e as

$$e = \Sigma_1 + \omega^{\Sigma_2 + \omega^{\cdot^{\cdot^{\cdot}}}} = {}_{\cdot^{\cdot^{\cdot}}}\Sigma_{T-1} + \omega^{e_1^{(T)}},$$

where $e_1^{(T)-} = \Sigma_T \in \mathbf{EP_0}$.

Examples

1. rp of $\mathbf{1}$ is $\mathbf{1}$; also the rp$=\mathbf{1}$ for each term in the sequence $\omega, \psi = \omega\star 2, \omega\star 3, \ldots$.

2. Every simple diagonal is perfect.

3. $e = \omega\star 5 + \omega^{\alpha}$, where $\alpha = \omega\star 3 + \omega^{\omega^2+\mathbf{1}}$. $\text{rp}(e) = \text{rp}(\alpha) = \text{rp}(\omega^2 + \mathbf{1}) = \omega^2 + \mathbf{1}$.

The (perfect) e determining the class $\mathcal{X}$ of polynomials for which every ee is perfect may be called $x-$*cluster perfect:* $\omega \in \mathcal{X}$; if $e, f \in \mathcal{X}$, then $e + f \in \mathcal{X}$; and $e \in \mathcal{X} \Rightarrow \omega^e \in \mathcal{X}$. Since any $e \in \mathbf{EP} - \mathcal{X}$ requires the introduction of $\mathbf{1}$ or a successor, the

[20] C.f. McBETH [98], 1984 'A second normal form for functions of the system **EP**,' *Zeitschr. f. math. Logik Grundl. Math.*, Bd. 30, S. 393-400, §2.

characterization $\mathcal{X}$ is complete. Notice the single detail by which $\mathcal{X}$ differs from **EP** in its inductive definition: the initial polynomial **1** is replaced by ω.

Now suppose $\alpha, \beta \in$ **EP** are diagonal. The register polynomial of $\alpha = \sum a_i \omega^{\alpha_i}$ is contained in the rightmost exponent α_1 of α. Similarly for β. Therefore $\mathrm{rp}(\alpha\beta)$ is contained in the rightmost exponent $\alpha_1\beta_1$ of $\alpha\beta$. Thus, by an induction on $x-$height, for α, β as above,

THEOREM 2.34 $\mathrm{rp}(\alpha\beta) = \mathrm{rp}(\alpha)\mathrm{rp}(\beta)$.

2.2.14 Immediate Domination in EP

The relation $\prec$ of eventual domination of §2·6 may be replaced by the stronger immediate domination of e by f (in **EP**), in which at *every* value x, f exceeds e. Because unity determines a special case, the following formulation is preferable:

DEFINITION 2.35 [21] $(\forall e, f)$

1. $e \ll f \Leftrightarrow e(1) \leq f(1)$ & $(\forall x > 1)\ e(x) < f(x)$.

2. f *is said to* immediately dominate e *iff* $e \ll f$.

3. $e \underline{\ll} f \Leftrightarrow e \ll f \vee e = f$.

□

Examples

1. Evidently, $\mathbf{1} \prec e \Rightarrow \mathbf{1} \ll e$. And if e is a successor, $e^- \ll e$.
2. Thm. 25 may be strengthened to: $e_i \ll e,\ i \leq r$. Furthermore, let a be an ec of e; and let f be obtained from e by replacing a by $a+1$ (other ec's of e of value a remaining unchanged). Then $e \ll f$.
3. Under the conditions of case 2(a) of thm. 18, $e \ll f$.

The relation $\ll$ is irreflexive, asymmetric and transitive, but not connected, as may be seen by taking the minimal incomparable pair $e = \mathbf{2},\ f = \omega$.

Remark

The notation $e \ll f$ is a relation between *arithmetic* functions. The relation $F \ll G$ of the standard analytical usage, viz. $F = O(G) : (\exists k)(\forall x > 0) \mid F(x) \mid < kG(x)$, is weaker: the present definition is uniform, in the sense that $k = 1$.

An unending sequence $\alpha \ll \beta \ll \ldots$ in **EP** is provided e.g. by the fundamental sequence of constant functions. Since the relations entail

$$\alpha(2) < \beta(2) < \ldots,$$

[21] [103] defn. 4.1, following EHRENFEUCHT [33], 1973.

no polynomial $e \in \mathbf{EP}$ can satisfy

$$\alpha \ll e, \beta \ll e, \ldots$$

for every term, therefore there is a drastic difference in length between the systems $(\mathbf{EP}, \prec)$ and $(\mathbf{EP}, \ll)$. On the other hand, $\ll$ and $\prec$ 'differ only by a constant': suppose $e, f \in \mathbf{EP}$, with $e \prec f@M$ (ref. defn. 13). Then as f is cst or si, if f is cst, $M = 1$; if f is si, the difference $f - e$, for $x = 1, 2, 3, \ldots, M - 1$ is always less than the value of f at M. So

$$e \ll F \in \mathbf{EP},$$

where $F = f + \lambda x \cdot f(M)$.

A corollary of this is that, given any ascending sequence $\alpha_1 \prec \alpha_2 \prec \alpha_3 \prec \ldots$ in $\mathbf{EP}$, for which $\alpha_{i-1} \prec \alpha_i @ M_i$, $1 < i$, by the remark following defn. 13 an immediate domination sequence

$$e_1 \ll e_2 \ll e_3 \ll \ldots$$

in $\mathbf{EP}$ is obtained by taking $e_1 = \alpha_1$; and for $i > 1$, $e_i = \alpha_i + \mathbf{a}_i$, where $a_i = \sum_{j=2}^{i} \alpha_j(M_j)$.

Note that $e = \sum_{i=r}^{1} \omega^{e_i} \ll \sum_{i=s}^{1} \omega^{f_i} = f$ can only arise, in $\mathbf{EP}$, if $(\exists i \leq s)\ e_i \ll f_i$ [22] (otherwise $f(1) < e(1)$, or $f(x) \leq e(x)$ for some $x > 1$.) On the other hand, it is not necessary that $e_i \ll f_i$ for all $i = 1, 2, 3, \ldots, r$; as with $\psi + \omega \ll \omega^{\psi} + \mathbf{1}$.

2.2.15 TAKEUTI Diagrams

Various proof-theoretic systems of notations have been introduced (e.g. GENTZEN [40] and [41]). The system of TAKEUTI, [149], 1957 and [147], 1975 will be briefly considered here. The class of TAKEUTI diagrams (td's) ([149], §1, p389) of *order* $n \in \mathbf{N}$ is the least class $\mathcal{T}$ closed under

Rule T.1 $a \in \mathbf{N} \Rightarrow a \in \mathcal{T}$

Rule T.2 $\alpha, \beta \in \mathcal{T} \Rightarrow \alpha \# \beta \in \mathcal{T}$

Rule T.3 $\alpha \in \mathcal{T}; i, a \in \mathbf{N}, i \leq n; \Rightarrow (i; a, \alpha) \in \mathcal{T}$.

The operations $\#$ of 'natural' sum, and $(i; a, \alpha)$ of exponentiation of ACKERMANN's construction [3], 1951, endow $\mathcal{T}$ with the following properties:

1. $\mathcal{T}$ posesses a linear ordering $<$, for which $(\mathcal{T}, <)$ satisfies the DCC.

2. Every formal derivation F in $\underline{\mathbf{N}}$ (ref. Appendix B) may be assigned a td.

3. If F is reduced to E, (I, §3) then

$$\text{td}(E) < \text{td}(F).$$

[22] Extending defn. 35 by writing '$e_i \ll f_i$' for $r < i \leq s$ in case $r < s$.

Given Rules T.1-.3, a glance at defn. 3 is sufficient to clearly establish the **EP** outline of a td. Indeed a td of order n may be interpreted as a $\lambda x \cdot e(x) \in \mathbf{EP}$, in which every occurrence of the symbol 'x' has inscribed within, say the left crescent the numeral for $a \in \mathbf{N}$, and within the right: that for $i \leq n$. Thus e is an *i–section* of f in **EP**,

$$e \subset_i f$$

in the sense of TAKEUTI ([146], para. 1·2, p389), iff

1. e is an explicit exponent (defn. 27·2) of f, and
2. the unique x–path (which is void if e is basic) from the term of f containing e, to the term x^e contains no right index $\geq i$; and that the 'x' of x^e has right index $= i$. (The left index may be arbitrary).

Note that, by thm. 25, $e \subset_i f \Rightarrow \omega^e \preceq f$. If $g = h + f$, and $e \subset_i f$, then by the above definition, $e \subset_i g$, as in [149], §1.2.3. This is an instance of a deep superiority of the GOODSTEIN-SKOLEM ordering by eventual domination, over the CANTOR 'well-ordering of the ordinals $< \varepsilon_0$,' which has an addition giving '$\alpha < \beta$' $\Rightarrow$ '$\omega^\alpha + \omega^\beta = \omega^\beta$', necessitating the 'natural' sum '#' (HESSENBERG (1906); see SIERPIŃSKI [138], Ch. XIV, §28). This is defined to emulate the polynomial sum, for which: $\omega^f \prec \omega^e + \omega^f$ (in **EP**).

2.2.16 EISENSTEIN's Criterion

EP does not admit negative coefficients; and therefore has no theory of divisibility. However, e is either irreducible, or has a non-trivial factorization

$$e = \prod_i e_i^{n_i}$$

in **EP**, where the $n_i \in \mathbf{N}$.

Before discussing the analogue of EISENSTEIN's criterion, it may be noted that factorization in **EP** is recurrent in the sense that exponents occuring in the factors may themselves be factorized, as in: $\alpha = \omega^{\omega+\mathbf{1}} + \omega^\omega + \omega^3 + \omega^2$ and $e = 2\omega^{\alpha+\omega^3} + 14\omega^\alpha + \omega^{\omega^3+\mathbf{2}} + 7\omega^2 = (2\omega^\alpha + \omega^2)(\omega^{\omega^3} + 7) = (2\omega^{(\omega^\omega+\omega^2)(\omega+\mathbf{1})} + \omega^2)(\omega^{\omega^3} + \mathbf{7})$. Note that it is possible for a recurrent factorization to proceed, while at each step prior to factorization all of the exponents are irreducible.

If **EP** is extended to a larger class **EP**$'$ by exchanging, for the rule **EP**.3, the rule of closure

$$e, \alpha \in \mathbf{EP}' \Rightarrow e^\alpha \in \mathbf{EP}',$$

then factorization, recurrent in an extended sense, is possible. Thus, the basic factorization

$$F = \sum (e_i)^{f_i} = \prod_i \{\sum_j (e_{ij})^{\beta_{ij}}\}^{n_i}$$

of F in **EP**$'$, besides factorization of the β_{ij}, admits factorization of the e_{ij}.

EISENSTEIN's irreducibility criterion for nc $e = \sum_{i=r}^{1} c_i\omega^{e_i} + c$ in **EP** is: there exists a prime p for which

1. $p^2 \not| c$.

2. $c_{r-1}, c_{r-2}, \ldots, c_1, c$ are each divisible by p.

3. $p \not| c_r$.

Example

$$e = \omega^{2\omega+\mathbf{3}} + 24\psi + 12\omega + \mathbf{6},$$

thus p may be taken as 2 (or 3); and e satisfies the criterion. The following special case will be considered. Suppose, as in the example, [23] that the exponents e_i of e are *linear*:

$$e_i = h_i\omega + k_i;\ h_i, k_i \in \mathbf{N}_0, \text{ not both zero.}$$

THEOREM 2.36 *Provided the basic coefficients of e satisfy the above criterion, then, division by a constant factor excepted, e has no proper decomposition as a product of factors in* **EP**.

Proof, by contradiction. This is much as in the classical proof (for integral polynomials), since the only non-basic coefficients it uses are those which determine the introduction of zero basic coefficients. Suppose the basic coefficients of e as above satisfy the criterion, and that

$$e = \alpha\beta = (\sum_{i=A}^{1} a_i\omega^{\alpha_i} + \mathbf{m}) \cdot (\sum_{i=B}^{1} b_i\omega^{\beta_i} + \mathbf{n}), \quad A, B \geq 1$$

in **EP**. Now for the purposes of the proof, it is necesssary to modify the normal form of α and β. Let

$$\alpha_i = u_i\omega + v_i, \quad \beta_i = x_i\omega + y_i;$$

$$u = \max_{i\leq A} u_i, \quad v = \max_{i\leq A} v_i; \quad x = \max_{i\leq B} x_i; \quad y = \max_{i\leq B} y_i; \quad w = \max(v, y).$$

Then α, β may be expressed

$$\alpha = \sum_{i=0}^{u}\sum_{j=0}^{w} m_{iw+j}\omega^{i\omega+j}, \quad \beta = \sum_{i=0}^{x}\sum_{j=0}^{w} n_{iw+j}\omega^{i\omega+j},$$

allowing zero coefficients. The finite double sums can be written out in lexicographic order, so that their product is an $(s+1)\times(t+1)$ array, where $(s+1) = (u+1)(w+1)$; $(t+1) = (x+1)(w+1)$; $r = s+t$. The array can now be summed along the anti-diagonals, as in the usual proof: [24] Changing to single-suffix notation,

$$e = (\sum_{i=0}^{s} m_i\omega^{\mu_i}) \cdot (\sum_{i=0}^{t} n_i\omega^{\nu_i})$$

---

[23] And as in ex. 2 after defn. 21.

[24] Ref. J. W. ARCHBOLD, *Algebra*, Ch. VIII, §4 (publ. Pitman, 1970).

say, where $m = m_0$, $n = n_0$. Now $c = mn$, so for some prime p, by part 2. of the criterion, $p \mid m$ or $p \mid n$. On the other hand, by part 1., $p \not\mid m$ or $p \not\mid n$. Therefore $p \mid m$, iff $p \not\mid n$. Thus, suppose $p \mid m$. Now for $i = 0, 1, 2, \ldots, s$, a sum of terms selected from those of

$$m_i n + m_{i-1} n_1 + \ldots + m n_i,$$ [25]

may be made, so as to include the first term, and so as to give a (basic) coefficient c_k, $k < r$, of e. The rule determining the selection is indicated in a simple case by the figure. If $m_i = 0$ or $p \mid m_i$, for $i = 1, 2, 3, \ldots, s$, then p divides the leading

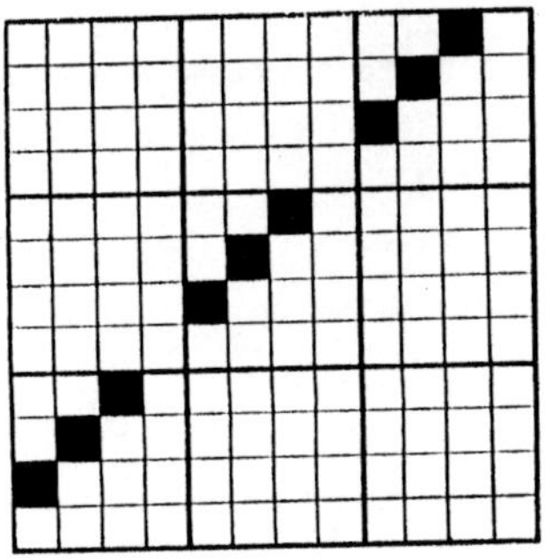

coefficient of e, contradicting 3. There is thus a least $i \leq s$, $1 \leq i$, for which $m_i > 0$ and $p \not\mid m_i$. Every m_j, $j = 0, 1, \ldots, i-1$ is zero, or is divisible by p. But $k < r$, so by 2. $p \mid c_k$. Therefore, by the above expression, $p \mid m_i$ or $p \mid n$. The first is ruled out by supposition; the second by the assumption $p \mid m$. In the same way, the assumption $p \mid n$ leads to a contradiction. Therefore $p \not\mid c$, contrary to (2). Therefore e has no proper divisors in **EP**.

□

Remark. A recurrent form of the rule indicated in the figure, would give a proof for the cases that the exponents of e are e.g. quadratic, cubic, ...

2.2.17 EP as a Hierarchy

Sub-Hierarchies

It is evident that, e.g. $\mathbf{XP} = \{\text{constants}\} \cup \mathbf{XP}_1 \cup \mathbf{XP}_2 \cup \mathbf{XP}_3 \cup \ldots$; $\mathbf{XP}_n$ = natural polynomials of degree n. **EP** and its subclasses readily telescope into hierarchies filled with towers $\mathcal{X}_1 \subset \mathcal{X}_2 \subset \mathcal{X}_3 \subset \ldots$ and their recurrences.

[25] Writing $n_i = 0$ in case $t < s$ and $t < i \leq s$.

DEFINITION 2.37

$$\mathcal{X}_\alpha = \{e \in \mathbf{EP} : e = \sum_{i=r}^{1} a_i \omega^{e_i} + \mathbf{a} \ \&\ e_r \prec \alpha\}.$$

□

Therefore $\lambda \prec \mu$ in **EP** $\Rightarrow \mathcal{X}_\lambda \subset \mathcal{X}_\mu$, giving, in particular, $\mathcal{X}_{\alpha+}$ 'much larger' than $\mathcal{X}_\alpha$. This makes an interesting comparion with the 'gradually ascending' property (§2.10) of **EP**. However, the increase in $\mathcal{X}_\lambda$ is in general greater in the case Diag(μ); and greater still in the case ht λ <ht μ. Given any ascending hierarchy $\mathcal{H} = (\mathcal{X}_\mu)_{\mu\prec\alpha}$, a new hierarchy $\mathcal{G} = (\mathcal{Y}_\mu)_{\mu\prec\alpha}$, for which $\mathcal{G} \simeq \mathcal{H}$ under inclusion, may be obtained by the operation $\mathcal{Y}_\mu = \Phi\{\mathcal{X}_\mu\}$, of §2.9. Indeed, Φ may be applied k times, so that the result becomes more 'rarefied' as k increases.

Classification of Basic Coefficients

Suppose $e \prec \omega \star (\omega; 2)$ in **EP**. Then $e \in \mathcal{X}$ has the normal form

$$e = \sum a_i \omega^{m_i\omega + n_i} + \mathbf{a}, \quad m_i, n_i \in \mathbf{N}.$$

But as in the previous section, e could be considered as a polynomial in general 'longer' than any in **XP**, viz.

$$e = \sum_{i=R}^{1} \sum_{j=S}^{1} a_{ij} \omega^{m_i\omega + n_j}, \qquad \text{(I)}$$

$m_0, n_0 = 0$. This modification to the normal form of §2.11 may require the admission of zero coefficients; not necessarily so many, however, as that used in §16. Similarly, if linear exponents are replaced by quadratic, e takes a 3−dimensional form

$$e = \sum_i \sum_j \sum_k a_{ijk} \omega^{q_{ijk}},$$

and for $\mathbf{XP_0}$−exponents, p say,

$$e = \sum_{i(1)} \cdots \sum_{i(n)} a_{\mathbf{i}} \omega^{p_{\mathbf{i}}} = \prod a\omega^p, \text{ say}, \qquad \text{(II)}$$

(where, as a special case, $\prod$ is product). In all cases, the exponents are ordered by eventual domination $\prec$ according to the rule $p_{\mathbf{i}} \prec p_{\mathbf{j}}$ iff $\mathbf{i} = \{i(1), \ldots, i(m)\}$, $\mathbf{j} = \{j(1), \ldots, j(n)\}$; and $\mathbf{i} < \mathbf{j}$, where $<$, in this immediate context only, means the usual (right-adjusted) left-to-right lexicographic order. Every $\mathbf{EP}^{(2)}$−polynomial is of the form (II).

This has gone quite far into **EP**, yet it is necessary to go considerably further in order to glimpse how vast it is. Suppose $e = \sum a_i \omega^{\alpha_i}$ now has $\alpha_i \prec \omega \star (2, \mathbf{2})$. In other words, α_i is $\mathbf{0}$, or is of the form of the RHS of (I). Then for each successive value $m_1, m_2, m_3, \ldots, m_R$, an expression of the form (II) occurs, so that

$$e = \prod \cdots \prod a\omega^{\alpha}.$$

This comprehension of segments of **EP** can be made more precise, by writing

$$\sum = \sum^{(1)}; \quad \prod = \sum \cdots \sum = \sum^{(2)};$$

and so on for $\sum^{(3)}, \sum^{(4)}$, etc. Since: for all the forms e so represented every coefficient in the exponents can vary independently, the subscript of each basic coefficient must contain an entry for every coefficient in the exponent: it follows that the e with $\mathbf{EP}^{(2)}$-exponents have

$$e = \sum_{\mathbf{i}} a_{\mathbf{i}} \omega^{\alpha_{\mathbf{i}}}, \text{ where } \alpha_{\mathbf{i}} = \sum_{\mathbf{j}}^{(2)} a_{\mathbf{ij}} \omega^{p_{\mathbf{ij}}},$$

and where $p_{\mathbf{ij}} = p_{\mathbf{i}j(1)\cdots j(n)} \in \mathbf{XP}$, and $\mathbf{i}$ is an n-dimensional array of integer indices. They constitute the class $\mathbf{EP}^{(3)}$.

2.2.18 Similarity

DEFINITION 2.38 *The ordered systems* $(X, <)$, $(Y, \prec)$ *are* similar, $(X, <) \simeq (Y, \prec)$, *if there is a* $1-1$ *correspondence* $\Phi : X \to Y$ *so that* $x < y$ *in* $X \Rightarrow \Phi x \prec \Phi y$ *in* Y.

□

Examples.

1. $(\mathbf{N}, <) \simeq (\mathbf{EP})_\omega \simeq \mathcal{X}$, [26] the class discussed after thm. 28.

2. Let $\alpha \in \mathbf{EP}$, and let $\mathcal{X}$ be the class of diagonals of $(\mathbf{EP})_\alpha$. Then for $\alpha = \sum a_i \omega^{\alpha_i} + \mathbf{a}$, $a = 0 \;\&\; \alpha_1 \succeq \omega \Rightarrow (\mathcal{X}, \prec) \simeq \{(\mathbf{EP})_\alpha, \prec\}$.

3. When a structure $\mathcal{X}$ is described by a (1st order) theory $\underline{\mathbf{T}}$, this is denoted by $\mathcal{X} \models \underline{\mathbf{T}}$. Thus, $\mathrm{N} \models \underline{\mathrm{N}}$. The meta-theorem of SKOLEM quoted in Ch. I, §3.3, establishes that there is an infinite structure $(\mathbf{M}, \prec) = \mathbf{M}$ for which $\mathbf{M} \models \underline{\mathrm{N}}$, and yet
$$\mathbf{M} \not\simeq \mathrm{N}.$$

[26] Note that the observations of Ch. I, §2.3 apply to the 'least number principle': *to prove that for an arbitrary non-trivial predicate* P, *there is a least* $N \in \mathrm{N}$ *such that* $\mathrm{P}(N)$, *i.e. that* $(\mathbf{N}, <)$ *is a well-ordering, the* AC *(as* AC_1*) is required.*

CANTOR Arithmetic

CANTOR's arithmetic of (types of) well-ordered sets can be introduced, for **EP**, by defining special operations $(+), (\cdot)$ of sum and product. In applications, the approach through **EP** is obviously more direct than via similarity: e.g. in §2·15 the point was made that for the purposes of proving the consistency of $\underline{\mathbf{N}}$, the so-called *natural sum* has to be constructed from the similarity-sum. This 'natural' sum (product) is precisely the arithmetic sum (product) of polynomials in **EP**.

For the purposes of exposition, a brief account is given of how the *reverse process* may be effected: viz. how from sum and product in **EP**, the operations $(+), (\cdot)$ may be recovered.

Similarity Sum

Suppose

$$\alpha = \sum_{i=r}^{1} a_i \omega^{\alpha_i} + \mathbf{a}, \quad \beta = \sum_{i=s}^{1} b_i \omega^{\beta_i} + \mathbf{b}$$

are normal forms in **EP**. α and β, in the given order, may be 'pushed together', until either α disappears, every term of α being eventually dominated by the leading term of β, or until a new normal form is obtained, in either case giving the similarity sum $\alpha(+)\beta$ of α, β.

For α, β as above,

DEFINITION 2.39

$$\alpha(+)\beta = \begin{cases} \beta & \textit{if } \alpha_r \prec \beta_s, \textit{ or} \\ \sum_{i=r}^{k} a_i\omega^{\alpha_i} + \beta & \textit{otherwise,} \end{cases}$$

where k is the least index for which $\beta_s \preceq \alpha_k$.

□

Examples

1. For $m, n \in \mathbf{N}$, $\mathbf{m}(+)\mathbf{n} = \mathbf{m} + \mathbf{n}$.

2. $\mathbf{1}(+)\omega = \omega$; $\omega(+)\mathbf{1} = \omega^+$, so $(+)$ is not in general commutative.

3. $\omega(+)2\omega = 3\omega = 2\omega + \omega = 2\omega(+)\omega$.

4. $\omega(+)\omega^2 = \omega^2$.

5. $\omega^k(+)\psi = \psi$, $k = 1, 2, 3, \ldots$.

6. For $\alpha = 5\omega^\psi + 3\psi + 2\omega^3 + 7$, $\beta = \psi + \omega^4$, $\gamma = \omega^{\omega^+} + \omega^5$,

$$\alpha(+)\beta = 5\omega^\psi + 4\psi + \omega^4; \ \alpha(+)\gamma = 5\omega^\psi + \omega^{\omega^+} + \omega^5.$$

From the definition, if $\alpha \prec \beta$ in **EP**; and $a, b \in \mathbf{N}$, then $b\omega^\beta(+)a\omega^\alpha = b\omega^\beta + a\omega^\alpha$. Therefore, more generally,

THEOREM 2.40 *If* $e = \sum a_i\omega^{e_i}$ *is in the normal form of thm. 22, then* $e = \sum_{(+)} a_i\omega^{e_i}$.

□

Now suppose $\alpha, \beta \in \mathbf{EP}$ are as given before defn. 39. In the notation of defn. 21, let $(\mathbf{EP})^+_\beta =$ the $\prec$ –ordered system of polynomials of the form $e = \varepsilon + \sigma$, $\sigma \in (\mathbf{EP})_\beta$; and let $\mathcal{X} = (\mathbf{EP})_\alpha \cup (\mathbf{EP})^+_\beta$. By corol. 16, it is clear that $(\forall f, g \in \mathcal{X})\ f \prec g$ iff $f \prec g$ in one of the disjoint classes comprising $\mathcal{X}$, or if $f \in (\mathbf{EP})_\alpha$ & $g \in (\mathbf{EP})^+_\beta$.

THEOREM 2.41 $(\mathcal{X}, \prec) \simeq (\mathbf{EP})_\gamma$, where $\gamma = \alpha(+)\beta$.

Similarity Product

Repetitions of one normal form, determined by another, give rise to

DEFINITION 2.42 *1. For arbitrary* $e, f \in \mathbf{EP}$,

$$\omega^e(\cdot)\omega^f = \omega^{e(+)f}.$$

2. For α, β *as above,*

$$\alpha(\cdot)\beta = \begin{cases} \alpha & \text{if } \beta = 1 \\ a_r\omega^{\alpha_r}(\cdot)\beta^- + \alpha & \text{if } Suc(\beta) \\ a_r\omega^{\alpha_r}(\cdot)\beta = \sum_{i=s}^{1} a_r b_i \omega^{\alpha_r(+)\beta_i}, & \text{otherwise.} \end{cases}$$

□

Examples

1. $\omega^+(\cdot)\omega = \omega^2$; $\omega(\cdot)\omega^+ = \omega^2(+)\omega = \omega^2 + \omega$.
2. $\omega(\cdot)\{\psi + \omega^2 + \omega\} = \psi + \omega^3 + \omega^2$;
   $\{\psi + \omega^2 + \omega\}(\cdot)\omega = \omega^{\omega^+}$.

Now let $\alpha \in \mathbf{EP}$, and consider the progression

$$\omega \star (\mathbf{1}\omega),\ \omega \star (\mathbf{2}\omega), \ldots, \omega \star (\omega\omega), \ldots, \omega \star (\mu\omega), \ldots$$

for $\mu \prec \alpha$. Clearly these can be used in the same way as $\varepsilon = \omega \star (\mathbf{1}\omega)$ was used to construct $(\mathbf{EP})^+_\beta$ before thm. 41, to give the hierarchy

$$(\mathbf{EP})^+_\beta = (\mathbf{EP})^{+\mathbf{1}}_\beta,\ (\mathbf{EP})^{+\mathbf{2}}_\beta, \ldots, (\mathbf{EP})^{+\mu}_\beta, \ldots$$

of pairwise disjoint $\prec$ –ordered systems of polynomials. Taking $\mathcal{X} = \cup_{\mu \prec \alpha}(\mathbf{EP})^{+\mu}_\beta$

THEOREM 2.43

$$(\mathcal{X}, \prec) \simeq (\mathbf{EP})_\gamma, \text{ where } \gamma = \beta(\cdot)\alpha.$$

□

Similarity Exponentiation

The power $\alpha^{(n)}$, $n \in \mathbf{N}$ in **EP** may be defined by recursion: $\alpha^{(1)} = \alpha$; $\alpha^{(n+1)} = \alpha^{(n)}(\cdot)\alpha$. The general power $\alpha^{(\beta)}$ may then be defined constructively [27] by diagonalization. The CANTOR operations receive detailed treatment in Ch. XIV of SIERPIŃSKI's monograph [138]. [28]

Fixed Points

When 2 normal forms are 'pushed together' one may becomed 'covered', so that $\alpha(+)\beta = \beta$, as in exs 4 and 5, foll. defn. 39. In fact $\mathbf{1}(+)\alpha = \alpha$, for all $\alpha \succeq \omega$; $\omega(+)\alpha = \alpha$, for all $\alpha \succeq \omega^2$; $\omega^2(+)\alpha = \alpha$, for all $\alpha \succeq \omega^3, \ldots$.

As to the $(\cdot)$–fixed points, $\omega(\cdot)\omega^\alpha = \omega^{\mathbf{1}}(\cdot)\omega^\alpha = \omega^{\mathbf{1}(+)\alpha}$ by definition. Therefore the fixed points (with respect to ω) of $(\cdot)$ are:

$$\psi,\ \omega^{\omega+\mathbf{1}},\ \omega^{\omega+\mathbf{2}}, \ldots .$$

It is to be emphasized that the CANTOR operations $(+), (\cdot)$ will *not* be used in the sequel.

[27] CANTOR's original definition of similarity exponentiation, viz. in terms of 'ordinal' limits, is described by BIRKHOFF (*Selected Papers on Algebra and Topology* , Birkhauser, (1987) p290) as 'essentially non-constructive'.

[28] See also ABIAN, A., *The Theory of Sets and Transfinite Arithmetic* , publ. W. B. Saunders, Philadelphia and London, 1965.

2.3 A Fundamental Property of Eventual Domination in EP

The ordering $\prec$ of **EP**, linear by corol. 19, is also well-founded. These properties are the motivation for the construction of systems such as the proof theoretic rational numbers of GENTZEN. But the following theorem is the most remarkable arithmetic attribute of eventual domination in **EP**, without analogue in the CANTOR theory of similarity types. The result is easy to state: given $e \prec f$, then the value of x at which f passes e (defn. 13) depends only on the explicit coefficients (defn 23·3) of e : the value is never more than a single unit greater than its largest ec. The following is contained in McBETH [103], June, 1987:

DEFINITION 2.44 *The integer-valued function $N(e)$ on $\mathbf{EP_0}$ is given by:*

1. $N(\mathbf{0}) = 1$;
2. *For* $\mathbf{1} \preceq e$, $N(e) = 1+$*the maximum explicit coefficient of* e.

Note that $N(\alpha+\beta) \geq N(\alpha)$. The proof of the uniformity [29] result requires 2 lemmas. Let $m, h \in \mathbf{N}$, with $m \geq 2$.

LEMMA 2.45 *The number of $e \in \mathbf{EP_0}$ for which* $\mathrm{ht}(e) \leq h-1$, *and for which* $N(e) \leq m$, *is* $m \star h$.

Proof, by induction on $x-$height of e. If $\mathrm{ht}(e) = 0$, the $m = m \star 1$ functions

$$\mathbf{0, 1, 2, \ldots, m-1}$$

are exactly those that satisfy the conditions. Now suppose the result has been proved for functions of height $\leq h-1$, and $e = \Sigma a_i \omega^{e_i}$ is of height $\leq h$. Then there are $m \star h$ coefficients a_i, each of which may take any of the values $0, 1, 2, \ldots, m-1$, giving

$$m^{m \star h} = m \star (h+1)$$

possible e, and the lemma follows.

□

Now let $m, n \geq 2$ in $\mathbf{N}$,

$$m^n - 1 = mm^{n-1} - 1 = (m-1)m^{n-1} + m^{n-1} - 1 = \ldots .$$

Therefore

[29] The passing number bound is independent of the dominant f. The result is therefore uniform in the sense of e.g. uniform continuity. Thus on the interval $(0, \infty)$, $\sin x$ is, but $\sin \frac{1}{x}$ is not uniformly continuous, since in the ϵ, δ formulation of continuity, for $\sin x$, $\delta = \epsilon$ (i.e. as a function of ϵ alone) may be taken; while for $\sin \frac{1}{x}$, no matter how small the number $\delta > 0$, there are numbers a, b in the interval such that $| b - a | < \delta$, and for which $| \sin(\frac{1}{b}) - \sin(\frac{1}{a}) | > 1$.

LEMMA 2.46

$$m^n - 1 = \sum_{i=n-1}^{0} (m-1)m^i.$$

Passing now to the uniformity result, via thm. 22 consider the normal forms

$$e = \sum_{i=A}^{0} \omega^{e_i}; \qquad f = \sum_{i=B}^{0} \omega^{f_i};$$

where the $e_i, f_i \in \mathbf{EP_0}$ are nd in i wrt $\prec$; and $0 \le A, B$. Suppose $e \prec f@M,\ 1 \le M$.

THEOREM 2.47 [30]

$$M \le N = N(e).$$

Proof, by induction on x–height h of e. For $h = 0$, e is cst= $\mathbf{k}$, with $N = k+1$. If f is cst, $M = 1$. If f is nc, by thm. 4, $f(k+1) \ge k+1$, so $e \prec f@M \le k+1 = N(e)$. Now suppose the theorem has been proved for all functions of $\mathbf{EP}^{(h)}$, $0 \le h$, and let $e \in \mathbf{EP}$ be of height $\le h+1$, with e, f given as above. If f is of the form $e + \alpha$ in $\mathbf{EP}$, then $(\forall x)\ e(x) < f(x)$, and $M = 1 < N$. Thus suppose there is a smallest $I \ge 0$ such that

$$e_{A-I} \prec f_{B-I}.$$

Common terms can be disregarded, so assume $e_A \prec f_B$. Now setting $C^+ = N\star(h+1)$ and using lemma 45, let $\phi_C \succ \phi_{C-1} \succ \ldots \succ \phi_0$ be the functions ϕ of $\mathbf{EP}_0^{(h)}$, for which $N(\phi) \le N = N(e)$; the linear order being ensured by corol. 19. Then for some $K \le C, e_A = \phi_K$. Therefore by considering the possible terms which may occur in the normal form of e :

$$(\forall x)\ e(x) \le \sum_{i=K}^{0} (N-1)x^{\phi_i(x)}.$$

But by induction hypothesis,

$$(\forall i < C)_{i \ge 0}(\exists l \le N)\ \phi_i \prec \phi_{i+1}@l.$$

So for $x \ge N$,

$$\phi_K(x) > \phi_{K-1}(x) > \ldots > \phi_0(x) = 0.$$

Therefore for x in the above range,

$$\begin{aligned} e(x) &\le \sum_{i=K}^{0}(x-1)x^{\phi_i(x)} \le \sum_{i=\phi_K(x)}^{0} (x-1)x^i \ \text{ since } K \le \phi_K(x) \\ &< x^{1+\phi_K(x)} \ \text{ by lemma 46} \\ &= x^{1+e_A(x)} \le x^{f_B(x)} \ \text{ again by i.h.} \\ &\le f(x). \end{aligned}$$

[30] [103], thm. 3.7.

So, $\{\forall x \geq N(e)\}\ e(x) < f(x)$, proving the result for functions of height $h+1$. Therefore the theorem follows by induction.

□

Remarks

1. In the proof of thm. 14, a sum is replaced by a single term, much as in the application of lemma 46 in the above proof.

2. Thm. 47 gives a bound $N(e)$ independent of the dominating function f. On the other hand, for any nc f and natural number $M > 1$, put $n = f(M-1)$. Then by thm. 4, $\mathbf{n} \prec f@M$. So there is no bound $N'(f)$, dependent only on f, such that
$$(\forall e \in \mathbf{EP})\ e \prec f@M \Rightarrow M \leq N(f).$$

3. By lemma 46, $x^m > \sum_{i=m-1}^{0}(x-1)x^i$. Therefore
$$e = \lambda x \cdot \sum_{i=m-1}^{0} (m-1)x^i \prec f = \lambda x \cdot x^m @m.$$
But $e(m-1) \geq (m-1)(m-1)^{(m-1)} = f(m-1)$. Since $N(e) = m$, this example shows that the bound of thm. 47 is best possible.

4. Notwithstanding rem. 3, the upper bound N for the pass M is rather coarse. Even in **XP**, the 'lower ranking' ec's have markedly less influence on M. On the other hand, as with the polynomials of ex.2, §2·1, in a domination
$$a \cdot \omega^n + \ldots + A \cdot \omega^k + \ldots \prec \omega^{n+1},$$
as the degree n becomes large, the leading coefficient a may well be small in comparison with 'medial' coefficients. There appears to be no means, at once elegant and practical, to calculate/estimate M for the domination $e \prec \omega^{n+1}$ in **XP**; still less for the $e \prec f$ of thm. 47.

5. That $e \prec f@M$ for some large M does not imply that e, f 'lie close' wrt $\prec$. On the contrary, it merely indicated that e has a large ec: for given $k, r \geq 1$, the constant $\kappa = \lambda x \cdot k \star r$ satisfies $\kappa \prec \omega \star r@k+1$. Conversely, as seen after defn. 13, that $e \prec f@1$ does not imply e, f lie 'far apart'; although if they do, a value of M small compared with $N(e)$ may be expected.

6. By taking $K = C$ in the proof of thm. 47, it follows that $e(N) < N \star \{1 + x\text{-ht}(e)\}$. Therefore, if $e \prec f$ and $N \star \{...\} \leq f(1)$, it follows
$$e(1) \leq e(2) \leq \ldots \leq e(N) < f(1) \leq \ldots \leq f(N),$$
and thm. 47 gives: $e \ll f$.

THEOREM 2.48 $(\forall e \in \mathbf{EP})\ N(e) \leq 1 + e(2)$, *strict if* e *nc.*

Proof, by induction on $x-$height. If $e = \mathbf{k}$ is cst, then $N(e) = 1 + k = 1 + e(2)$. Now suppose the theorem has been proved for height ≥ 0, and let e be nc, of height $\leq h + 1$. Then assuming the normal forms of thm. 47,

$$\begin{aligned} -1 + N(e) &< \sum_{i=A}^{0} N(e_i) \leq \sum \{1 + e_i(2)\} \text{ by i.h.} \\ &\leq \sum 2^{e_i(2)} = e(2), \end{aligned}$$

and the theorem follows by induction.

□

Therefore $e \prec f$ in **EP** implies that for every $x > e(2)$, $e(x) < f(x)$.

2.4 A Construction of EP Using All Finite (Viz. Constant) Types

2.4.1 Higher Types

According to tradition, in the first millenium B. C., EPIMENIDES the Cretan claimed to be making a false utterance. If true, the claim would have been false. If false, it would have been true. RUSSELL [136], 1908 resolved this paradox or antinomy by introducing a hierarchy of predicate types: a primitive variable is of type [31] 0; a predicate which assumes a truth value for every predicate of type h, is of type $h+1$. The idea has very many variations, e.g. multiple arguments; arguments (and value) of mixed type. Then: '...if EPIMENIDES asserts "all first-order propositions asserted by me are false" he asserts a second-order proposition ...'. [32]

RUSSELL's distinction applies to functions as to predicates, and in this direction has been built on e.g. by HILBERT, 1925 ('Über das Unendliche', in English translation in van HEIJENOORT [154]) and later e.g. by TAIT. [33] Higher-type functions arise commonly, as in

$$\sum(f) = F \equiv (\forall x)\ F(x) = \sum_{i \leq x} f(i), \quad f \in \mathcal{N}.$$

Such objects are called *functionals.* [34]

A note of warning might be sounded, as to the use of higher types. The space $\mathcal{N}$ is vastly greater than **N**. Experience tends to show that the step up to that of functionals $F : \mathcal{N} \rightarrow \mathcal{N}$ is much greater. In ot er words, to hse all finite types to recover the (type-1 system) **EP** is to be utilizing objects which are found only in increasingly enormous dilution. This, therefore, from the viewpoint of hierarchy theory, is not the principal application of higher types.

It is to Dr. Peter ACZEL that I owe the following presentation of **EP**.

2.4.2 Constant Types

The arithmetic objects of pure type may be introduced as follows: let $\mathcal{N}_0 = \mathbf{N}$ and

$$\mathcal{N}_{h+1} = \text{ the space of functionals } F : \mathcal{N}_h \rightarrow \mathcal{N}_h,$$

$h = 0, 1, 2, \ldots$. The type h of $F \in \mathcal{N}_h$ may be indicated by a subscript. However the usual '$x \in \mathcal{N}_0$' will be used; and in the notation of I, §3·1, $\mathcal{N}_1 = \mathcal{N}$.

Well-Formed Strings

A finite string of letters, each denoting a type object, and associated by a fixed pattern of brackets is *well-formed* if the expression evaluates to an object of $\mathcal{N}$. Thus a string

$$\underline{F} = F_a F_b \ldots F_n F_1,$$

[31] The usages of 'type' in §§2·18 and 4 should not be confused.

[32] §IV: The Hierarchy of Types.

[33] TAIT, W. W., 'A characterization of ordinal recursive functions', abstract, *J. Symb. Logic,* **24** (1959) 325.

[34] The terminology was introduced by HADAMARD to replace terminology of V. VOLTERRA. Ref. introduction to *Theory of Functionals* , publ. Dover, 1959.

bracketed so that every juxtaposition ...)(... coincides with a unit decrement in type from left to right, is well-formed. In particular, for strictly descending subscripts such strings assume the form

$$((\dots((F_{h+1}F_h)F_{h-1})\dots F_3)F_2)F_1.$$

2.4.3 Iteration

An arbitrary object $F = F_{h+1}$, $h \geq 0$, transforms its argument to an object of type h, and may therefore be iterated: $F^{(1)} = F$; and

$$F^{(n+1)} = \lambda F_h \cdot F\{F^{(n)}(F_h)\}.$$

The higher iteration $F^{(\omega)}$ is defined by:

$$F^{(\omega)}_{h+1} = \lambda F_h \cdot \dots \cdot \lambda F_2 \cdot \lambda F_1 \cdot \lambda x \cdot F^{(x)}_{h+1} F_h \dots F_2 F_1(x).$$

Example. For $h = 1$, $F_2^{(\omega)} = \lambda F_1 \cdot \{\lambda x \cdot F_2^{(x)} F_1(x)\}$, a special case of which is

$$\Sigma^{(\omega)} = \lambda f \cdot \{\lambda x \cdot (\Sigma^{(x)} f)(x)\},$$

where the function of x is the x–fold sum

$$\sum_{l \leq x} \cdots \sum_{j \leq k} \sum_{i \leq j} f(i).$$

Notes

1. $\lambda F_1 \cdot \lambda x \cdot F_2 F_1(x) \neq \lambda F_1, x \cdot F_2 F_1(x)$, the object on the right acting on 2 arguments of mixed type.
2. The bracketing pattern for iteration of $F = F_{h+1}$ is $F^{(n)} F_h = F(F \dots (F(F_h))\dots)$, viz. opposite to that for the types.

2.4.4 Iterators of Constant Type

DEFINITIONS

Let the symbol $\{E_1^{(y)}\}^{(z)}(x) =_{df} yz$. Thus in particular, $E_1 = \mathbf{1}$. Furthermore, for $h \geq 1$, type symbols denoting the higher iteration are given by

$$E_{h+1} F_h =_{df} F_h^{(\omega)}.$$

E-STRINGS

Now let $\mathcal{X} \subset \mathcal{N}$ be the class of polynomials obtainable from E_1 by closure under

1. well-formed associations (E-strings) of the operations $E_2, E_3, E_4, \dots$; and
2. addition, defined by juxtaposition of well-formed strings.

That $\underline{E}(x)$ evaluates to $e(x)$ may be denoted by $\underline{E}(x) = \underline{E}(e)(x)$.

□

Remark.

According to §4·2, the highest (type) indices are eliminated first, in the calculation of $\underline{E}(x) = e(x)$. So the final string contains only the subscript 1. Therefore $\underline{E}(x) = E_1^{\{e(x)\}}$.

Examples

1. $E_1 = \mathbf{1}$; $E_1E_1 = \mathbf{2}$; $E_2E_1 = \lambda x \cdot E_1^{(x)}(x) = \omega$; $E_2E_1E_1E_1 = \omega + \mathbf{2}$; $E_2E_1^{(3)}(x) = \{E_1^{(3)}\}^{(x)}(x) = 3x = \{E_2E_1\}^{(3)}(x)$.
2. $(E_2E_1)^{(2)}E_1 = 2\omega + \mathbf{1}$; $E_2^{(2)}E_1 = \lambda x \cdot \{E_1^{(x)}\}^{(x)}(x) = \omega^2$.
3. $(E_2^{(2)}E_1)^{(5)}(E_2E_1)^{(2)}E_1^{(3)} = 5\omega^2 + 2\omega + \mathbf{3}$.
4. $E_3E_2E_1 = \lambda x \cdot E_2^{(x)}E_1(x) = \lambda x \cdot \{\ldots\{E_1^{(x)}\}^{(x)}\ldots\}^{(x)}(x) = \omega^\omega$; $E_3E_2E_2^{(2)}E_1 = \omega^{\omega+\mathbf{2}}$.
5. By the commutativity of '+,' different strings may give the same polynomial. E.g.
$$E_1(E_2E_1) = \omega^+ = (E_2E_1)(E_1).$$
6. Some polynomials may have no string whose subscripts are non-increasing. E.g. for $e = \omega^2 + \omega$, the only possibilities are
$$(E_2(E_2E_1))(E_2E_1) = e = (E_2E_1)(E_2(E_2E_1)).$$
In fact, only e involving addition of $\prec$ −at most a constant term have non-increasing form.

Note that for $e \in \mathcal{N}$,

LEMMA 2.49 $\lambda x \cdot E_2^{\{e(x)\}}E_1(x) = \omega^e$.

Proof. Let $e(x) = y$. Then

$$E_2^{(y)}E_1(x) = \{\ldots\{E_1^{(x)}\}^{(x)}\ldots\}^{(x)}(x) = x^y.$$

□

Thus, in particular $\lambda x \cdot E_4E_3E_2E_1(x) = \lambda x \cdot E_2^{(x^x)}E_1(x) = \omega^\psi$; and the above argument shows that, more generally,

LEMMA 2.50

$$E_{h+1}^{(y)}F_h\underline{F} = F_h^{(x^y)}\underline{F}.$$

Therefore by induction

COROLLARY 2.51

$$E_{h+1}E_hE_{h-1}\ldots E_2E_1 = \omega \star h.$$

Finite types, in this context thus give a natural occurrence of the operation $\star$ of recurrent exponentiation.

THEOREM 2.52

$$\mathcal{X} = \mathbf{EP}.$$

Proof (sketch). By lemma 50, the operations of $\mathcal{X}$ do not lead out of **EP**, so

$$\mathcal{X} \subseteq \mathbf{EP}.$$

Conversely, it has to be shown that $\mathcal{X}$ contains the closure operations of definition 2 (§2.1). Well, $\mathbf{1} \in \mathcal{X}$. And if $\underline{E}(e), \underline{E}(f)$ are strings in $\mathcal{X}$, then so is $\underline{E}(e)\underline{E}(f) = \underline{E}(e+f)$. As to exponentiation, suppose $e = E_a \ldots E_n = \underline{U}$, where $E_n = E_1$. $\underline{U}$ determines the iteration of

$$+\mathbf{1}$$

according to the exponential polynomial e. Now consider the well-formed string

$$\alpha = E_{a+1}E_{b+1}\ldots E_{n+1}E_1 = \underline{V}E_1 \in \mathcal{X}.$$

$\underline{V}$ determines the iteration of

$$\cdot x.$$

Since $\underline{V}$ applies to E_1 exactly as $\underline{U}$ iterates $+\mathbf{1}$, the pattern of application is identical, viz. iteration is according to the form e. Therefore

$$\alpha = \omega^e.$$

Therefore the converse is established, proving the theorem.

Note on Iterators

Suppose $F \in \mathcal{N}_h$, for some $h \in \mathbf{N}$. Then $F^{(2)}, F^{(3)}, \ldots, F^{(\omega)}$ etc. are well-defined. Viz. iteration acts on an object of arbitrary type. Thus the operations may be regarded as objects, of higher than any constant type.

Remarks

1. Aside from the individual applications of E_1, the E–string defines diagonalizations, and recurrences of diagonlizations in a way that clearly demonstrates the iteration needed to reach the higher rates of growth attained by the gradually ascending hierarchy **EP**.

2. Lemma 50 shows that $\omega \star h$ cannot be obtained, using only E–operators of type $\leq h$. HILBERT (loc. cit.) 1925 called type 'height'. Aside from a unit difference in value (owing to the initial function E_1 being cst), this coincides with the indexing of defn. 5.

3. There is evidently only one E–string which gives e in the normal (or reverse normal) form of thm. 22.

4. By 3. above, every expression $\underline{E}(e)(x)$ determines a unique *calculation*, the product of whose final step is an E–string, all of whose elements are E_1.

2.5 Enumeration of the System EP

2.5.1 Types of Enumeration

As in the case arising in Ch. I §3.1, an enumeration of an infinite structure $\mathcal{S}$ is an exhaustive list [35] $a_1, a_2, a_3, \ldots$ of the elements of $\mathcal{S}$. Given an enumeration, for every point of the vast space of permutations of **N**, there is a unique enumeration of $\mathcal{S}$. Interest centres, therefore, on the source of the original enumeration.

Finite Subclasses

Any $\mathcal{X}$ having a decomposition $\mathcal{X} = \cup_i \mathcal{X}_i$ into finite $\mathcal{X}_i$ has an easy type of enumeration; viz. as obtained by successively enumerating the finite classes $\mathcal{Y}_1, \mathcal{Y}_2, \mathcal{Y}_3, \ldots$, where $\mathcal{Y}_1 = \mathcal{X}_1$, and $\mathcal{Y}_{n+1} = \bigcup_{i=1}^{n+1} \mathcal{X}_i - \bigcup_{i=1}^{n} \mathcal{X}_i$. Such an enumeration for **EP** may be based on a bound for length (23·4); or via thm. 28, for height and ec's.

Combinatorial Enumeration

If the finite subclasses $\mathcal{Y}_1, \mathcal{Y}_2, \mathcal{Y}_3, \ldots$ are correctly chosen, it is possible that

$$F(x) = \sum_{i=0}^{\infty} \mid \mathcal{Y}_i \mid x^i$$

for some closed[36] function F, either as a formal power series, or for complex x within a disc of convergence.

2.5.2 Enumeration and Diagonalization

Diagonalization (as in I, §3.1, metatheorem 1.3.1, end of §2.18 and of the previous section) may be applied to sequences of functions, to machine languages, etc., in each case in many possible forms. If $f_1, f_2, \ldots$ is an enumeration of $\mathcal{X} \subset \mathcal{N}$, of particular importance is the diagonal function taking the basic form

$$f(x) = f_x(x), \quad 1 \leq x.$$

Constructively Defined Classes

It is possible to state in general terms why e.g. **EP**, $\mathcal{P}$ are constructively defined, while the class $\mathcal{R}$ is not. To be constructively defined, a class $\mathcal{X}$ has to

1. contain explicit initial functions;

2. be closed only under finitary rules of enlargement.

---

[35] Unless otherwise specified, the list contains no repetitions: $i \neq j \Rightarrow \alpha_i \neq \alpha_j$.

[36] 'Closed' here means either a finite algebraic expression, or an infinite sum or product of such expressions.

$\mathcal{R}$ satisfies 1., but not 2., since each formulation contains a rule of closure (e.g. 'μ') based on conditions which may never be fulfilled. Any $\mathcal{X}$ satisfying the above conditions may be constructively enumerated. Note however that e.g. the class $\mathcal{E}$ of I, §3.1 is spanned by **EP**, yet it is closed under e.g. substitution, is far more inclusive, and so is less easily enumerated. In the case of constructive hierarchies $\mathcal{X} = \bigcup \mathcal{X}_i$, $\mathcal{X}_{i+1}$ may contain the enumerator of $\mathcal{X}_i$, leading to results such as [23] cited at the end of I, §3.5. But the most important application of enumeration [37] is to non-constructive classes, as seen in I, §3·1. [38] Indeed, if the goal is to show that the diagonal discussed above satisfies $f \notin \mathcal{X}$, as with a pure growth hierarchy such as **EP**, it is usually possible to by-pass enumeration altogether.

Iteration and Diagonalization

Diagonalization, however is of cardinal importance in the study of arithmetic growth. The result may or may not produce a great gain in apparent growth; of key importance is that, under appropriate conditions, the diagonal eventually dominates every member of a given fundamental sequence. Thus e.g. ε (§2.4) is the diagonal of the **EP**- spanning sequence of corol. 17. Of course the challenge is to build up the hierarchy as far as possible, using iteration and diagonalization, without ever allowing any but finitarily specified constructions at each step.

[37] KLEENE [73], 'Extension of an effectively generated class of functions by enumeration', *Coll. Math.* **4** (1958), has applied non-constructive techniques of extension to constructively defined classes.

[38] C.f also I, fn 24.

2.6 Variants of EP

EP is a specialized, effectively defined infinite class of polynomial functions, radically susceptible to variation. It may be noted that while there is space here to consider only some classes of variations, none has the applicability of **EP** itself.

2.6.1 Extensions

Extensions (e.g. by extra rules of closure) will not be considered in Ch. 2, save to mention that $\mathbf{EP} \subset \mathcal{E}$, the CSILLAG-KALMAR class of elementary primitive recursive functions discussed in Ch. 1, §3. $\mathcal{E}$ contains such functions as $f(x) = x \dot{-} 1$, [39] so, as noted in §5.2, the inclusion is strict.

2.6.2 Inverses

The inverses of the arithmetic functions and their compositions ($x \dot{-} 1$, $[\sum_{i \leq x} \frac{1}{i}]$, $[\sqrt{x}]$, $[\log_2 x]$, etc.) play no part in the presentation and development of **EP**. However the inverse f^{-1} of a si f (Ch. 1, §3), no matter how fast-growing f, is unbounded. So

$$e \prec f \text{ in } \mathbf{EP} \Leftrightarrow f^{-1} \prec e^{-1},$$

with the relation [40] on the right in the class $\mathbf{EP}^-$ of inverses. Thus

$$(\mathbf{EP}, \prec) \simeq (\mathbf{EP}^-, \succ).$$

2.6.3 Non-Decreasing Functions

An extensive 'downward reflection' occurs between $\prec$ – consecutive polynomials of **EP**, viz. between $\alpha, \alpha^+ \in \mathbf{EP}$ and and, if Diag(α), between α and all polynomials preceding it.

---

[39] Where

$$x \dot{-} 1 = \begin{cases} 1 & \text{if } x = 1, \\ x - 1 & \text{otherwise.} \end{cases}$$

[40] The relation of defn. 8 has to be modified slightly: $\alpha \prec \beta$ in $\mathbf{EP}^-$ iff

$$(\exists N)(\forall x)\; x \geq N \Rightarrow \{\alpha(x) \leq \beta(x) \;\&\; (\exists u \geq x)\; \alpha(u) < \beta(u)\}.$$

That this modification is necessary, and that it is sufficient, may be seen by considering the value $e(k)$, where $2\omega \preceq e$ in **EP**. The new relation may be tried by taking the least polynomial in **EP** which dominates e, viz. e^+, and taking inverses:

1. for $x = e(k) + 1$, $(e^+)^{-1}(x) = k$, and

$$e^{-1}(x) = (\mu y)[x \leq e(y)] = k + 1,$$

while

2. for x in the range $e(k) + 1 < x \leq e(k+1)$,

$$e^{-1}(x) = (\mu y)[x \leq e(y)] = k + 1, \text{ and}$$

$$(e^+)^{-1}(x) = (\mu y)[x \leq e(y) + 1] = k + 1.$$

The reflection occurs without (explicitly) involving the operations of inverse. By thm. 4, $\alpha(x) \leq \alpha(x+1)$. Consider the class $\mathcal{X} = \mathcal{X}(\alpha)$ of nd $f \in \mathcal{N}$ for which

$$\alpha \prec f \leq \alpha^{+}.$$

$\mathcal{X}$ may be ordered lexicographically: $f < g$ in $\mathcal{X}$ iff

$$(\exists N)\ f(N) < g(N)\ \&\ (\forall x < N)\ f(x) = g(x).$$

As α increases with respect to $\prec$, there is 'room' for more non-decreasing functions, and $\mathcal{X}$ becomes larger. In fact, it may be shown

THEOREM 2.53 *The hierarchy* $(\mathcal{X}, >)$ *is similar to* $\{(\mathbf{EP})_e, \prec\}$, *where* $e = \omega^{\alpha}$.

2.6.4 Constant Base

A constant term may be introduced in clause $\mathbf{EP} \cdot 3$ of defn. 3. E.g. Let $\mathcal{X}$ be defined by: $\mathbf{1}, \omega \in \mathcal{X}$; the class is closed under sum; and $e \in \mathcal{X} \Rightarrow \lambda x \cdot 2^{e(x)} \in \mathcal{X}$.

Another variation is obtained as $\mathbf{K} : \omega$ belongs to $\mathbf{K}$. Closure under sum is deleted. For any natural number $n \geq 2$, if $e \in \mathbf{K}$, then $\lambda x \cdot n^{e(x)} \in \mathbf{K}$. Note that $\mathbf{K}$ is closed under composition. Evidently, ht $\alpha <$ht $\beta \Rightarrow \alpha \prec \beta$ in $\mathbf{K}$. While for equal (base-)height h, let

$$\alpha = m_h^{\cdot^{\cdot^{\cdot^{m_1^x}}}}, \ \beta = n_h^{\cdot^{\cdot^{\cdot^{n_1^x}}}}, \ e(x) = n_r^{\cdot^{\cdot^{\cdot^{n_1^x}}}},$$

where $m_i = n_i$; $i \leq r$ and $2 \leq m_{r+1} < n_{r+1}$. Now writing $f = m_{r+1}^{e(x)}$, $F = n_{r+1}^{e(x)}$; $\frac{F}{f} \to \infty$ with x; so for any $m \geq 2$ and k with $m \leq 2^k$,

$$\frac{2^F}{2^{kf}} \to \infty,$$

as at every step $r+1, r+2, \ldots, h$. Therefore the bases m_i, n_i may be compared lexicographically, i.e. the height$-h$ class $\mathbf{K}^{(h)}$ is similar to the $h-$words on $\mathbf{N}$, giving by rem. 1., §2·10

THEOREM 2.54

$$(\mathbf{K}, \prec) \simeq \mathbf{XP}.$$

Example. $\mathbf{K}^{(2)}$ is similar to the $\mathbf{XP}-$subclass $\mathbf{NL}$, of §2.1.

2.7 Notes on Ch. II

1. §2.1 COHEN [24] contains an interesting reference to the system **XP**.
2. §2·2 Note that a single level of exponentiation in **EP** results in eventual domination of all the forms of footnote 10: $\psi(x) = e^{x \ln(x)}$ gives $t^{\mu}e^{\lambda t} \prec \psi(t)$, for all μ, λ.
3. §2·4 NEDERPELT [117] uses a binary tree lemma of K. A. POST to find the number of evaluations of bracketings of a ladder of n 2's.
4. **The Classical Theory of Exponentiation**

   §2·5. The classical theory is founded on the *exponential theorem*, viz.: the function *defined* by $\exp(x) = \sum_{n=0}^{\infty} \frac{x^n}{n!}$ satisfies

   $$\exp(x) = e^x =_{df} \{\exp(1)\}^x.$$

   It is now a corollary that trigonometry and exponentiation possess an astonishing link, via $i = \sqrt{-1}$, as obtained in 1748 by EULER (in the form of $\cos\theta + i\sin\theta = e^{i\theta}$; and leading to formulae such as $\sin^{-1} x = -i\ln\{ix + \sqrt{1-x^2}\}$), to the FOURIER series $\Sigma_{k=-\infty}^{\infty} c_k e^{i2\pi kt}$ representing a function absolutely integrable on $[0, 2\pi]$, etc.). The fundamental result is proved by taking in the first instance x to be integral; and then by stages until x ranges over the complex field. In this way, the theorem rests on law 3 listed in §2·5 (as well as implicitly on the three others); so no analogue of the classical theory may be anticipated for $\star$-exponentiation.

Fractional Iteration

Besides e.g. the EULER-MAURER function '$x \star \infty$', work has been done on such forms as '$x \star \frac{1}{2}$'. Thus, taking the unique f for which $f\{f(x)\} = e^x$, $f^{2n}(x) = 2 \cdot 718 \ldots \star \{n; x\}$. In this context, it may be remarked that $y = e^x = h(y) = g(x)$ is but one of a homotopic family of such pairs (ξ, ζ), for which $\xi\{\zeta(x)\} = e^x$.

Chapter III

FUNDAMENTAL SEQUENCES FOR EXPONENTIAL FORMS

Section Headings

Section Headings (Cont'd)

3.1 Introduction

Fundamental sequences represent the skeleton or girder elements of construction, notably in the p-adic theory (see MAHLER [93], (1971), Ch. II), in functional analysis and, dating from the 'arithmetization of analysis' of the early to mid-19th century, in the CAUCHY-CANTOR theory of the real line. Fundamental sequences may be built for their own sake, and this is perhaps the main purpose of the exponential hierarchy **EP**.

The great problem of fundamental sequences, pointed out early in the Introduction, is not solved by **EP**. That the hierarchy may readily be extended establishes this. **EP** is merely a preliminary to the construction of a complete system of fundamental sequences, viz. one which enables measurement of the rate of growth of all arithmetic functions. However, by devoting effort to this preliminary, it is possible to excise all non-constructive ideas, and to bring to bear the available apparatus, to study fundamental sequences in detail. The result is a new landscape of exponential arithmetic.

A rather longer introductory section is included to provide a general setting; while §2 is confined to studies of the HARDY-BACHMANN (register-polynomial) assignment Ω of fundamental sequences (of functions for functions) in **EP**. This was effectively introduced in 1904 by G. H. HARDY: 'It is generally evident...' that a diagonal e is the diagonal of '...one particular' fundamental sequence ([57], foot of p92). It should be pointed out that **EP** had not been identified at this time. Moreover, HARDY's definition is left as above, besides a few examples. Nonetheless, it is quite possible that the main definition of §2 was essentially known to mathematicians working in the 19th century, including Du BOIS REYMOND, CANTOR and BOREL.

How is it possible for the introduction of the standard assignment Ω to antedate that of **EP**? **EP** is one of the most elementary of many structures for which an assignment of the kind Ω may be made, such as constructions based on the real line, or CANTOR's 'second number class', as used in HARDY's definition.

§3 includes proofs of the key density and BACHMANN initialization properties for Ω. Discussion of iterated canonical reduction in the subsequent § is via $\Omega-$generated finite subclasses of **EP**. The closing section considers non-standard assignments, none of which rivals the $\prec-$maximal assignment Ω in utility.

3.1.1 Du Bois REYMOND's Theorem

This key result uses the diagonalization of II, §5 · 2. Let

$$f_1 \prec f_2 \prec f_3 \prec \dots$$

be any strictly ascending sequence in the space $\check{\mathcal{N}}$ of si arithmetic functions.

THEOREM 3.1 *There is a* $\phi : \mathrm{N} \to \mathrm{N}$ *for which*

$$f_n \prec \phi,$$

for all $n = 1, 2, 3, \dots$.

Proof. ϕ may be taken as the 'upper envelope' of the f_n : let $f_n \prec f_{n+1} @ K_n$ (ref. II·13), and let

$$M_n = \max_{i \leq n} K_i.$$

Then $f_i(x) < f_{n+1}(x)$ for all $i \leq n$ and $x \geq M_n$. Thus define $\phi_1 = f_1$; and

$$\phi_{n+1}(x) = \begin{cases} \max\{f_{n+1}(x), \phi_n(x)\} & \text{for } x \leq M_n, \\ f_{n+1}(x) & \text{otherwise.} \end{cases}$$

By definition, $(\forall n, x)\ \phi_n(x) \leq \phi_{n+1}(x)$; and $x > M_n \Rightarrow \phi_n(x) < \phi_{n+1}(x)$. Let $\phi(x) =_{df} \phi_x(x)$. Then

$$\begin{aligned} (\forall n)\ x > \max\{n, M_n\} \Rightarrow \phi(x) &> \phi_n(x) \\ &\geq f_n(x). \end{aligned}$$

□

Remark. The construction may be extended to functions of a continuous variable.

3.1.2 A Theorem of HADAMARD

HADAMARD generalized the above theorem. Suppose

$$f_1 \prec f_2 \prec \dots \prec f_n \prec \dots \prec f$$

and

$$g \prec \dots \prec g_n \prec \dots \prec g_2 \prec g_1$$

are given ascending and descending scales in $\check{\mathcal{N}}$.

THEOREM 3.2 *1. There are functions* ϕ, γ *such that*

(a) $f_1 \prec f_2 \prec \dots \prec f_n \prec \dots \prec \phi$ *and* $\phi \prec f$.

(b) $\gamma \prec \dots \prec g_n \prec \dots \prec g_2 \prec g_1$ *and* $g \prec \gamma$.

2. *Suppose the given scales are such that* $f_m \prec g_n$*, for all* m, n*. Then there is a function* α *such that*

$$f_1 \prec f_2 \prec \ldots \prec f_n \prec \ldots \prec \alpha \prec \ldots \prec g_n \prec \ldots \prec g_2 \prec g_1.$$

Proof, by diagonalization.

□

These examples are intended to indicate the multiple diversities of rates of growth and diagonalization. An outstanding merit of **EP**, however, is its uncompromising specializedness. It is a tool explicitly designed for an attack on the problem of classifying rates of growth, and diagonal constructions built on it are characteristically free of obscurities and complications. The next section, therefore, outlines the canonical theory of **EP**.

3.2 Fundamental Sequences in EP

The treatment of the following 3 definitions is an expansion of that contained in McBETH [1] 1976, and in McBETH [96] §3, 1980.

Suppose $e \in \mathbf{EP}$ is non-constant.

DEFINITION 3.3 *A* fundamental sequence *for* e *is a sequence* $\mathcal{S} = (e_n)_{n=1,2,3,\ldots}$ *such that:*

1. $e_n \prec e_{n+1}$*, viz.* e_{n+1} *eventually dominates (II·8)* e_n*; and*
2. $e(x) = e_x(x)$, $x = 1, 2, 3, \ldots$.

This relation is expressed $e = \operatorname{diag}_n e_n$ *or* $e = \operatorname{diag} \mathcal{S}$.

Examples

1. Consider the unit squares that tile the 1st quadrant of the plane. If (x, y) is the square whose upper right corner has coordinates (x, y); and if the numbers $e_n(1)$, $e_n(2), \ldots$ are engraved respectively on squares $(n, 1), (n, 2), \ldots$; then the numbers $e(1)$, $e(2), \ldots$ are to be found in the squares which contain the *main diagonal.*
2. $\psi = \omega^\omega = \operatorname{diag} \omega^n$
3. $$\omega^\omega + \omega^2 + \mathbf{1} = \operatorname{diag} \begin{cases} \omega^n + \omega^2 + \mathbf{1} & \text{or} \\ \omega^\omega + n\omega + \mathbf{1} & \text{or} \\ \omega^n + n\omega + \mathbf{1} & \text{or} \\ \mathbf{n}^n + \mathbf{n}^2 + \mathbf{1} & \end{cases}$$

[1] 'Fundamental sequences for exponential polynomials', manuscript.

4. $\omega^{\omega^3+5\omega^2} = \text{diag}\ \omega^{\omega^3+4\omega^2+n\omega}$

5. $\omega^3+\omega^2+\omega = \text{diag}\{\omega^3+\mathbf{n}^2+\mathbf{n}\}_{1\leq n\leq N} \cup \{\omega^3+\omega^2+\mathbf{n}\}_{N<n}$, where $\cup$ denotes concatenation of sequences.

As in the last case of example (3), it is possible to represent any function on the natural numbers as the diagonal of a sequence of constant functions. The fundamental sequence of example (5) has an irregularity after the N^{-th} term.

DEFINITION 3.4 [2]

1. $\mathbf{1},\mathbf{2},\mathbf{3},\ldots$ *is a regular sequence* (rs)*.*
2. *Suppose* $(e_n),(f_n)$ *are* rs*'s, and that* $\alpha \in$ **EP**. *Then*
   (a) $(e_n+f_n) = (f_n+e_n)$; $(e_n+\alpha)$
   (b) $(n\omega^\alpha)$
   (c) (ω^{e_n})

   are each rs*'s.*

□

Note that every rs is $\prec$ −si.

THEOREM 3.5 *1. Every* rs *is a fundamental sequence for some e.*

2. Every nc e is the diagonal of a rs.

3. Every nc e is the diagonal of only finitely many rs*'s.*

Proof, by induction on order-of-definition in **EP** :

1. (a) ω is diag$(\mathbf{1},\mathbf{2},\mathbf{3},\ldots)$.
   (b) i. Suppose $e = \text{diag}\ e_n$, $f = \text{diag}\ f_n$. Then $\text{diag}(e_n+f_n) = (\text{diag}\ e_n) + (\text{diag}\ f_n) = e+f$.
   ii. $\text{diag}(e_n+\alpha) = e+\alpha$.
   iii. $\text{diag}\ n\cdot\omega^\alpha = \omega\omega^\alpha = \omega^{\alpha+\mathbf{1}}$.
   iv. $\text{diag}\ \omega^{e_n} = \omega^{\text{diag}\ e_n} = \omega^e$.

2. (a) ω is the diagonal of $(\mathbf{1},\mathbf{2},\mathbf{3},\ldots)$.
   (b) Suppose e, f are diagonals of sequences $(e_n),(f_n)$.
   i. Then $e+f$ is the diagonal of (e_n+f_n). If α is cst, $e+\alpha$ is diagonal, as in 1.(b) ii.

[2] This definition arose from remarks of Prof. C. A. ROGERS.

ii. As in 1.(b) iv.

3. (a) ω is the diagonal of no rs other than $(\mathbf{1}, \mathbf{2}, \mathbf{3}, \ldots)$.

   (b) Suppose e, f of given order-of-definition satisfy the condition. Then the three clauses of defn. 4·2 can give only finitely many sequences whose diagonals are the functions $e + f$, ω^e.

□

3.2.1 Constant Polynomials Do Not Have Fundamental Sequences

If $e \in \mathbf{EP}$ is a successor and eventually dominates the identity ω, then in fact e dominates ω from the first place: $(\forall x)\ x < e(x)$. Now although the successor polynomials which thus dominate the identity have fundamental sequences, the remaining successors i.e. the constants, do not. By Ch. II, thm. 4, any e is either constant or strictly increasing. Thus, if $(\forall x)\ e(x) = k$, and e is the diagonal of (e_n), then necessarily the functions e_{n+k}, $n = 1, 2, \ldots$ must themselves be the constant function $\mathbf{k}$, and so the condition 1. of defn. 3 is not satisfied. Thus the constant polynomials in this respect stand in a special place. On the other hand, if e is non-constant and a successor, then by II·31 $e = \delta + \mathbf{a}$, where δ is diagonal, so that $e = \text{diag}(\delta_n + \mathbf{a})$, where $\delta = \text{diag}\ \delta_n$. Therefore in $\mathbf{EP}$, the assignment of fundamental sequences reduces to the assignment of fundamental sequences to diagonal polynomials.

3.2.2 Canonical Fundamental Sequences

For a given non-constant e, from the choice of possible regular fundamental sequences for e, which is finite by thm. 5·3, by II·19 there is exactly one whose terms are largest with respect to $\prec$. In example 3. following defn. 3, this sequence is $(\omega^\omega + \omega n + \mathbf{1})$. It is possible therefore to make a canonical assignment of fundamental sequences to diagonal and to non-constant successor polynomials of $\mathbf{EP}$ according to this principle. That is, in the $x-$cluster representation of $\lambda x \cdot e(x)$, the single occurrence of 'x' associated with the register polynomial of e (II·33) is the site of introduction of 'n' in the cluster representing the nth fundamental sequence term. This is the assignment Ω informally introduced by G. H. HARDY in 1903. Inductive definitions appear in BACHMANN [9] and in many other sources, notably LÖB and WAINER [89], and WAINER [156], [157]. With 'diagonal' in place of 'limit', the definition is here aplied to $\mathbf{EP}$. For each non-constant e, the following notation is adopted: the nth function (in $\mathbf{EP}$) or *canonical term* in the fundamental sequence of functions assigned to e is denoted by

$$\Omega(e, n),\ n = 1, 2, 3, \ldots\ .$$

Since non-standard assignments $\Omega' \neq \Omega$ are of only scant interest, the 'Ω' is invariably dropped, leaving

$$(e, n)$$

as the nth canonical term of e. Thus for every natural number x,

$$(e,x)(x) = e(x).^3$$

The definition is by recursion on the order-of-definition of the diagonal polynomial e. It is convenient to include an assignment for the function $\varepsilon \notin \mathbf{EP}$ of Ch. II, notation 2.1:

DEFINITION 3.6 *For each* $n = 1, 2, 3, \ldots$,

1. *if* $e = \omega$, *the identity polynomial, then* $(e, n) = \mathbf{n}$.
2. *Now suppose* $\omega \prec e$, *and let* e *have the normal form of Ch. II, thm. 22. Then*

$$(e,n) = \sum_{i=r}^{2} a_i \cdot \omega^{e_i} + (a_1 - 1) \cdot \omega^{e_1} + (\omega^{e_1}, n),$$

   where

$$(\omega^{e_1}, n) = \begin{cases} n \cdot \omega^{e_1^-} & \textit{if } \mathrm{Suc}(e_1) \\ \omega^{(e_1, n)} & \textit{if } \mathrm{Diag}(e_1). \end{cases}$$

3. $(\varepsilon, n) = \omega \star n$.
4. *The polynomials* $(e,1), (e,2), \ldots$ *are the* canonical components *or* canonical terms *(ct's) of* $e \in \overline{\mathbf{EP}} = \mathbf{EP} \cup \{\varepsilon\}$.

□

Defn. 6 is by *recurrence.* Ch. II, thm. 26 therefore establishes that, given a diagonal e, and $n \geq 1$, after a finite number of steps the explicit expression (e, n), the nth canonical term of e, is obtained.

Remarks.

1. By II·4, $e \succeq 2\omega \Rightarrow (e, n)$ is si.
2. Broadly speaking, for a polynomial e of considerable height and involving many terms, the length of the linear system **EP** traversed by $(e,1), (e,2), \ldots$ is small in comparison to that defined by e.
3. The '$\prec$ –largest' choice of regular fundamental sequence of defn. 4 is achieved by restricting reduction to the minimal exponent at each inductive step. Of course, e may be the diagonal of a fundamental sequence not in **EP**, whose terms dominate the canonical terms. E.g. $\psi = \mathrm{diag}_n \lambda x \cdot n^x$, where

$$\mathbf{1} = \mathbf{1}^\omega \prec 2^\omega \prec 3^\omega \prec \ldots,$$

[3]C.f. the "$\overrightarrow{\equiv}$" notation of BACHMANN [9], **3**, p. 124.

and yet $(\forall n)\ (\psi, n) = \omega^n \prec 2^\omega$.

Following the construction to its conclusion, $e = \sum_{i=r}^{1} a_i\omega^{e_i}$ is the diagonal of the 'ex-**EP**' sequence $\xi_1, \xi_2, \xi_3, \ldots$, where

$$\xi_n = \sum_{i=r}^{2} a_i\omega^{e_i} + (a_1 - 1)\omega^{e_1} + n^{e_1}.$$

Examples.

1. (a) For $e = \sum + \omega^{\omega^+}$, $((e,2),1) = \sum + \psi + \omega$.
   (b) $(\omega \star (k+1), 1) = \omega \star k$.
2. For all e, by defn. 4 $(e,1)$, $(e,2), \ldots$ is regular.
3. Note that ω is the only diagonal e with $(e,1) = 1$.
4. In the normal form $e = \sum_{i=R}^{1} \omega^{e_i}$,

$$(e,n) = \sum_{i=R}^{k+1} \omega^{e_i} + (\sum_{i=k}^{1} \omega^{e_i},\ n),$$

   $1 \le k < R$; $n = 1, 2, 3, \ldots$.
5. Only in case 1. of defn. 6 [or of occurrences of this case carried inductively via Diag(e_1)] can there be a (unit) decrement in height, in the passage from e to a canonical term (e, n).

3.3 Properties of the Canonical Assignment

The basic properties of the canonical assignment are now summarized: for Diag(e),

THEOREM 3.7 *1.* $e = \text{diag}(e, n)$.

2. $(\forall n)\ (e,n) \ll (e, n+1)$; *in particular,* $(e,n) \prec (e, n+1)$.

3. $(\forall n)\ (e,n) \prec e$.

4. $\lambda n \cdot (e,n)(x)$ *is si, for* $x > 1$.

5. $(\forall n)$ n *is an explicit coefficient of* (e, n).

6. $(\forall n, x)_{x>1}\ (e,n)(x) \lesseqgtr e(x)$ *as* $n \lesseqgtr x$.

Proof. Let $n = 1, 2, 3, \ldots$. The proof is by induction on order-of-definition of (e, n). For $e = \omega$,

1. diag $(\mathbf{1}, \mathbf{2}, \mathbf{3}, \ldots) = \omega$.

2. $\mathbf{n} \ll \mathbf{n}^+$, $1 \leq n$.

3. $\mathbf{n} \prec \omega$, $1 \leq n$.

4. $\lambda n \cdot (\omega, n) = \omega$ is si.

5. A constant in **EP** is its own ec.

6. $\mathbf{n}(x) = n \lesseqgtr x$.

Now let

$$(e, n) = \Sigma + E_n, \qquad \Sigma = \sum_{r=2}^{2} a_i \omega^{e_i} + (a_1 - 1) \cdot \omega^{e_1},$$

as in defn. 4, and suppose for i.h. that 1.-6. hold for e_1, in case this is diagonal. Then

1. diag $(e, n) = \Sigma +$diag $E_n = \Sigma + \omega^{e_1} = e$.

2. (a) $n \cdot \omega^{e_1^-} \ll (n+1) \cdot \omega^{e_1^-}$ if Suc(e_1).
   (b) by induction hypothesis, $\omega^{(e_1, n)} \ll \omega^{(e_1, n+1)}$ if Diag(e_1).
   So in either case $(e, n) \ll (e, n+1)$.

3. If Suc(e_1), $n \cdot \omega^{e_1^-} \prec \omega^{e_1}$; if Diag$(e_1)$, by i.h. $\omega^{(e_1, n)} \prec \omega^{e_1}$. Therefore
$$(e, n) = \Sigma + E_n \prec \Sigma + \omega^{e_1} = e.$$

4. Suppose $x = k \geq 2$ is fixed. Then $(\exists M, N)$ for each $n = 1, 2, 3, \ldots$
$$(e, n)(k) = \Sigma(k) + E_n(k) = M + \begin{cases} n \cdot k^N & \text{if Suc}(e_1), \\ k^{(e_1, n)(k)} & \text{otherwise.} \end{cases}$$
In either case, a si function of n.

5. Note that e_1 is the minimal basic exponent (II, defn. 23·2(c)) of the diagonal $e = \Sigma + \omega^{e_1}$. If Suc$(e_1)$, then e_1^- is the minimal basic coefficient of $(e, n) = \Sigma + n\omega^{e_1^-}$, whose associated basic coefficient is n (*a fortiori* n is an ec of (e, n)). If Diag(e_1), then by the note of Ch. II, following thm. 24, n is an ec of $\omega^{(e_1, n)}$; and hence also of (e, n).

6. If Suc(e_1),
$$n x^{e_1^-(x)} \lesseqgtr x x^{e_1^-(x)} = x^{e_1(x)},$$
as $n \lesseqgtr x$. While if Diag(e_1), the relation follows directly from the i.h.

□

Remark. The basic property 7·1: $e =$diag (e, n) establishes that the assignment Ω satisfies the requirement of defn. 3·2. Ex. 1 following the definition, together with the conclusion of §2.1, reveal the provenance of the terminology 'diagonal' II, defn. 30·3.

3.3.1 Homology

The canonical assignment gives rise to the relation of homology between polynomials.

DEFINITION 3.8 *β is* homologous (immediately) above α *in* **EP**, $\alpha(\lhd)\beta$, *iff*

$$(\exists e)_{\text{Diag}(e)}(\exists n)\ \alpha = (e,n) \prec (e, n+1) = \beta \prec e.$$

The transitive closure of this relation is denoted by $f\ (\tilde{\lhd})\ e$.

□

Thus suppose β is imperfect, viz. $\text{rp}(\beta) = \beta_1^{(T)}$ is a successor. Then

THEOREM 3.9 *If α is obtained from β by replacing $\beta_1^{(T)}$ by the non-diagonal $\beta_1^{(T)-}$, then*

$$\alpha(\lhd)\beta.$$

□

Remarks.

1. Diag$\{\beta_1^{(T)-}\}$ rules out $\alpha(\lhd)\beta$.
2. Even allowing $\beta_1^{(T)} = \mathbf{1}$, the above replacement cannot be effected in $\mathcal{X}$, the class discussed in Ch. II after defn. 33.

3.3.2 Regularity

For diagonals $e, f \in$ **EP**,

$$(e+f,n) = \begin{cases} e+(f,n) & \text{or} \\ f+(e,n). \end{cases}$$

So by thm. 7·3

THEOREM 3.10 $(e,n)+(f,n) \prec (e+f,n)$.

□

It therefore follows that there are regular non-canonical sequences in **EP**. Thus

$$\mathcal{S} \qquad \psi^+,\ \omega^{\omega^2}+\mathbf{2},\ \omega\star(2,3)+\mathbf{3},\ \ldots$$

is the cfs of no $e \in$ **EP**; but diag $\mathcal{S} = \omega\star 3+\omega$. And such sequences may be obtained by the rule of addition, definition 4·2(a): let $\alpha \prec e$ in **EP**, where Diag(e). Then the rs

$$\{\alpha + (e,n)\}_{n=1,2,3,\ldots}$$

is non-canonical.

A Regular, Non-Canonical Assignment

The theorem suggests a non-canonical regular assignment for **EP** :

DEFINITION 3.11 *1.* $\Lambda(\omega, n) = \mathbf{n}$.

2. If $\mathrm{Diag}(e), \mathrm{Diag}(f)$, *then* $\Lambda(e+f, n) = \Lambda(e, n) + \Lambda(f, n)$.

3.

$$\Lambda(\omega^e, n) = \begin{cases} n\omega^{e^-} & \textit{if } \mathrm{Suc}(e), \\ \omega^{\Lambda(e,n)} & \textit{if } \mathrm{Diag}(e). \end{cases}$$

□

Thus, e.g., $\Lambda(2\omega, n) = \mathbf{2n}$; $\Lambda(\omega^2, n) = n\omega$; $\Lambda(5\omega^2, n) = 5n\omega$; $\Lambda(\psi, n) = \omega^{\Lambda(\omega,n)} = \omega^n$; $\phi = \sum_{1\leq i\leq k} \omega \star (i+1)$, $\Lambda(\phi, n) = \sum_{1\leq i\leq k} \omega \star (i;n)$.

3.3.3 Register Polynomial

The register polynomial $e_1^{(T)}$ of Ch. II, defn. 33 was identified (via the $\prec$ –lowest terms in repeated uses of the normal form) precisely because it is the standard canonical assignment Ω which operates on $e_1^{(T)}$, thereby ensuring a minimal $\prec$ – descent in the passage from e to any of its canonical terms. Thus, in the transformation from e to (e, n), $e_1^{(T)} = \Sigma_T + 1$ is replaced by its predecessor $\Sigma_T \in \mathbf{EP_0}$, and in the expression

$$e = \quad \cdot^{\cdot^{\cdot}} \Sigma_{T-1} + \omega^{e_1^{(T)}},$$

$\omega^{e_1^{(T)}}$ is replaced by $n \cdot \omega^{\Sigma_T}$.[4] Therefore it may be said that $\omega^{e_1^{(T)}}$ is the 'part' of e that changes; either in the passage from e to a particular (e, n), or along the cfs of e.

The above 'minimal descent' may be viewed as follows: A diagonal $e = \Sigma_{i\leq r} a_i\omega^{e_i}$ may be expected to have a larger value of r with increasing x-ht in **EP**, particularly in view of II, §2·17. And $(e, 1)$ shares the $r-1$ most dominant terms with e. Furthermore, this proportioning, although to a lesser extent, occurs again with e_1, with $(e_1)_1, \ldots$ until $e_1^{(T)}$ is reached.

Remarks

1. Note that, in the change from e to a particular (e, n),

   (a) by thm. 7·5, the canonical term (e, n) aquires n as an explicit coefficient.

[4] Note that in case $\Sigma_T = \mathbf{0}$, $\omega^{\mathbf{0}} = \lambda x \cdot x^0 = \mathbf{1}$, while if $\mathrm{Suc}\{\mathrm{rp}(e)\}$, then

$$\mathrm{rp}\{(e,n)\} = \begin{cases} \text{indeterminate} & \text{if rp}(e) \text{ is the successor of a diagonal} \\ \{\mathrm{rp}(e)\}^- & \text{otherwise.} \end{cases}$$

(b) The 'minimal ec' of the register polynomial of e is de-cremented by unity.
(c) Aside from cases (a),(b), no ec of e is altered.

2. If the diagonal e is perfect (II·33·2), then (e, n), $n \geq 2$, is imperfect. If each (e, n) is perfect (viz. of the form $\therefore + n \cdot \omega^{\Sigma}$, where Perf($\Sigma$) or $\Sigma = \mathbf{1}$), then e is imperfect.

Effect of Taking a ct

For $e = \varepsilon_N = \omega \star N$, the register polynomial is at maximal $x-$height. Therefore taking a ct (e, n) of e (no matter how large n) effects a great $\prec -$ descent. E.g.

$$\sum_{i \leq n}(e, i) \prec e \text{ and } \prod_{i \leq n}(e, i) \prec e, \quad n = 1, 2, 3, \ldots .$$

On the other hand, α^+ scarcely dwarfs its predecessor in **EP**, yet ω^{α^+} certainly dwarfs ω^{α}; so that given diagonal $e \prec f$ in **EP**, the comparative slightness of descent from $\beta = f + e$ to (β, n) is erased by taking instead that from ω^{β} to $\omega^{(\beta, n)}$.

The Compilation Class of $\alpha \in$ EP

Let the diagonal $\beta \succ \alpha = \Sigma_{i=r}^{1} a_i \cdot \omega^{\alpha_i} + a$ in **EP** be a *compilation* of α iff α is a ct of β. Then

THEOREM The class Cα of compilations of α is finite.

Proof. There are 2 main cases.

1. $a > 0: \ \alpha = (\Sigma + \omega, a)$, where $\beta = \Sigma + \omega$ is the unique (simple) diagonal for which α is a ct. Therefore $|\mathrm{C}\alpha| = 1$. Note that $\mathrm{C}\mathbf{1} = \{\omega\}$.
2. $a = 0, \ \alpha = \Sigma_r^2 + a_1 \omega^{\alpha_1}$. Then
   (a) for $a_1 \geq 1, \ \beta = \Sigma + \omega^{\alpha_1^+}, \ \alpha = (\beta, a_1)$; or
   (b) for $a_1 = 1, \ \beta_1 \in \mathrm{C}\alpha_1$ such that for some $n \geq 1, \ \alpha_1 = (\beta_1, n)$; where (if $r > 1$) $\beta_1 \preceq \alpha_2$, so for $\beta = \Sigma_r^2 a_i \omega^{\alpha_i} + \omega^{\beta_1}, \ \alpha = (\beta, n)$.

Now considering $\alpha = \Sigma_r^2 + a_1 \omega^{\alpha_1} + a$, α has a unique $\beta \in \mathrm{C}\alpha$ by 1. or by 2(a) as appropriate. Suppose $a = 0, a_1 = 1$. Then in addition $\beta \in \mathrm{C}\alpha$ may be found, one for each $\beta_1 \in \mathrm{C}\alpha_1$ which satisfies the conditions of 2(b). Since α has no other compilations,

$$|\,\mathrm{C}\alpha\,| \leq 1 + |\,\mathrm{C}\alpha_1\,| \qquad \text{(I)}.$$

The thm. now follows, by II·26.

□

Suppose the diagonal α is expanded in the explicit II·22-normal form as

$$\alpha = \Sigma_{(1)} + a_{(1)} \cdot \omega^{\cdots^{\Sigma_{(T-1)} + a_{(T-1)} \cdot \omega^{\mathrm{rp}\alpha}}} \qquad \text{(II)}$$

so that e.g. $\Sigma_{(1)} = \Sigma_r^2$ and $a_{(1)} = a_1$. Then taking $a_{(t)} = 1, \ t = 1, 2, \ldots T - 2$, consider the case $\ldots^{+a_{(T-1)} \cdot \omega^{\Sigma_{(T)} + a_{(T)}}}$.[5] If $a = a_{(T-1)} \geq 1$, then $(\ldots^{+\omega^{\Sigma_T + \mathbf{2}}}, a)$ determines a unique

[5]Note $\Sigma_{(T)} + a_{(T)} = \Sigma_T + \mathbf{1}$.

unit contribution to $|C\alpha\;|$. If $a = 1$, $(\ldots{}^{+1\cdot\omega^{\Sigma(T)+\omega}}, a_{(T)})$ may determine another unit contribution to $|C\alpha\;|$. Therefore, by taking all inequalities of the form (I) from the expression (II), an explicit bound is obtained: $|C\alpha\;| \leq T$.

DEFINITION

In the calculation of the above bound, a contribution via case 2(b) is precluded if $a_{(t)} > 1$ is encountered. Therefore, aside from one possible exception, for every $\beta \in C\alpha$, $\alpha = (\beta, 1)$. Thus the *explicit index* (ei) for α may be taken as the unique ec $a_{(t)} > 1$ for which t has the lowest value; or as 1 if every ec on the rightmost chain is unity.

Examples

1. $C\omega = \{\omega^2, \psi\}$. $C5\omega + 5 = \{6\omega\}$. $\alpha = \omega \star 3 + \psi + 1$; $\beta = \Sigma_{i\leq 3}\omega \star i$.
2. $\psi = (\omega^{\omega^+}, 1) = (\omega^{\omega^2}, 1) = (\omega \star 3, 1)$; $|C\psi\;| = 3$. $\omega^{\omega^2} = (\omega^{\omega^2+\mathbf{1}}, 1) = (\omega \star 3, 2)$.
3. Given any fixed value, a greater $|C\alpha\;|$ may be found. Thus, let $\beta = \omega \star (l; \omega^{\mathbf{1}+\omega\star m})$. Then $(\beta, 1) = \alpha = \omega \star N^+$, for all $l + m = N$, $l \geq 0$.

3.3.4 Properties of the Canonical Assignment

2 properties of Ω, important in applications, are considered here.

Density of Ω in $(\mathbf{EP}, \prec)$

THEOREM 3.12 (DENSITY OF $(\mathbf{EP}, \Omega)$) [6] *Suppose $\beta \prec e$ in* **EP**, *where e is diagonal. Then*

$$(\exists N)(\forall n)\; N \leq n \Rightarrow \beta \prec (e, n).$$

Proof, by induction on $x-$height h. For $h = 0$, $\mathbf{EP}^{(0)}$ contains no diagonals. Now suppose that the diagonals of $\mathbf{EP}^{(h)}$ have the property, and let $e \in \mathbf{EP}^{(h+1)}$ be diagonal, $0 \leq h$, with $\beta \prec e$. Put

$$e = \sum_{i=r}^{1} a_i\omega^{e_i}, \qquad \beta = \sum_{i=q}^{1} b_i\omega^{\beta_i} + \mathbf{b}$$

as in II·22. The canonical fundamental sequence, cfs, of e is determined by the cfs of ω^{e_1}. If, therefore, there is an i such that $r - 2 \geq i \geq 0$, for which $e_{r-i} \neq \beta_{q-i}$, or for which

$$e_r = \beta_q, \ldots, e_{r-i+1} = \beta_{q-i+1},\; e_{r-i} = \beta_{q-i}$$

and $a_{r-i} \neq b_{q-i}$, then $\beta \prec (e, n)$, $n = 1, 2, \ldots$.

Now suppose there is no such i. Write $s = q - r$, and compare $a_1\omega^{e_1}$ and $b_{s+1}\omega^{\beta_{s+1}}$. $\beta \prec e$ implies $\beta_{s+1} \preceq e_1$. In the case of equality, $\beta_{s+1} = e_1$, $\beta \prec e$ implies $b_{s+1} < a_1$.

[6] [96], thm. 3.4, p119.

Then $s = 0 \Rightarrow (\forall n)\ \beta \prec (e, n)$. Suppose therefore $r < q$, and suppose first $\mathrm{Suc}(e_1)$, say $e_1 = \gamma^+$. Then $\beta_s \prec \beta_{s+1}$ implies $\beta_s \preceq \gamma$, therefore $(\forall n > b_s)$

$$(\omega^{e_1}, n) = n\omega^\gamma \succ b_s\omega^{\beta_s},$$

viz., writing $\sum = \sum_{i=r}^{2} a_i\omega^{e_i}$,

$$\begin{aligned} \beta &= \Sigma + b_{s+1}\omega^{e_1} + b_s\omega^{\beta_s} + \text{ lower terms} \\ &\prec \Sigma + (a_1 - 1)\cdot\omega^{e_1} + (b_s + 1)\cdot\omega^{\beta_s} \\ &\prec \Sigma + (a_1 - 1)\cdot\omega^{e_1} + n\omega^\gamma, \quad \text{for } n > b_s + 1 \\ &= (e, n). \end{aligned}$$

Now suppose $\mathrm{Diag}(e_1)$. Then by the i.h., $\beta_s \prec e_1$ gives

$$(\exists N)\ n \geq N \Rightarrow \beta_s \prec (e_1, n),$$

giving, for each constant c,

$$c\omega^{\beta_s} \prec \omega^{(e_1, n)},$$

and therefore for $\beta_{s+1} = e_1$, provided $n > N$,

$$\begin{aligned} \beta &\prec \Sigma + (a_1 - 1)\cdot\omega^{e_1} + (b_s + 1)\cdot\omega^{\beta_s} \\ &\prec \Sigma + (a_1 - 1)\cdot\omega^{e_1} + \omega^{(e_1, n)} \\ &= (e, n). \end{aligned}$$

The easier case of inequality, $\beta_{s+1} \prec e_1$, is dealt with similarly. The theorem thus follows by induction.

□

Remarks

1. By Ch. II, thm. 14, the theorem remains true if '**EP**' is replaced by '$\overline{\mathbf{EP}}$'.
2. The thm. defines 2 possible cases: (i) $\beta \prec (e, 1)$, or $(\exists m)\ (e, m) \preceq \beta \prec (e, m + 1)$.
3. III·12 is an '∞' result, in the sense that β precedes a tail of the cfs for e.
4. At each inductive step of the proof of thm. 12, the bound N is built up from the basic coefficient b_s of β. It therefore follows, in the notation of the theorem, that
$$\beta \prec e \Rightarrow \beta \prec (e, N),$$
where $N = N(\beta)$, as defined in II·44.

Examples

1. Let $e = \omega\star(2;3)$ and $\beta = \omega^{3\omega^2} + p$, where $p \prec \omega^{3\omega^2}$ in **EP**. Then $\beta \prec (e,n) = \omega^{n\omega^2}$ for $n = 4, 5, 6, \ldots$.

2. The property of Ω, defined by thm. 12, is not shared by the regular assignment Λ of defn. 11. Thus $\omega^2 \prec 2\omega^2$ but $\Lambda(2\omega^2, n) = 2n\omega \prec \omega^2$, $n = 1, 2, 3, \ldots$.

3. Note that if e has the normal form $\sum + \gamma$ and $\beta \preceq \sum$ in **EP**, then $\beta \prec (e,n)$ for *every* $n = 1, 2, 3, \ldots$.

3.3.5 BACHMANN's Starting Property

In rapid growth applications, it is important that for consecutive canonical terms of a compound [7] diagonal $e \in$ **EP**, the sequence of the canonical terms of the upper term starts at or above the lower term. Ω has this property:

DEFINITION 3.13 *For e a compound diagonal of* **EP**; $m, n \in$ **N**,

$$(e, m, n) =_{df} ((e, m), n).$$

Note that the definition does not give, in general, $(e, m, n) = (e, n, m)$.

Now let e be as above. The following argument is given in McBETH [95] [8], 1980:

THEOREM 3.14 (BACHMANN) [9] *Suppose for some $n \geq 1$, there is a diagonal $\beta \in$* **EP** *such that*

$$(e, n) \prec \beta \preceq (e, n+1).$$

Then for all $m = 1, 2, 3, \ldots$,

$$(e, n) \preceq (\beta, m).$$

Proof. By thm. 7·2, it is sufficient to show

$$(e, n) \preceq (\beta, 1).$$

The proof proceeds by induction on x-height h. For $h = 0$, $\mathbf{EP}^{(0)}$ contains no diagonals, leaving nothing to prove. Now suppose $h \geq 0$, and that the result has been established for $\mathbf{EP}^{(h)}$. Let $e \in \mathbf{EP}^{(h+1)}$ be of the form $e = \Sigma + \omega^{e_1}$, where $e_1 \succ \mathbf{1}$ and

$$(e, n+1) = \Sigma + \begin{cases} (n+1)\cdot\omega^{e_1^-} & \text{if Suc}(e_1) \\ \omega^{(e_1, n+1)} & \text{if Diag}(e_1). \end{cases}$$

Let β be any diagonal of $\mathbf{EP}^{(h+1)}$ for which

$$(e, n) \prec \beta \preceq (e, n+1).$$

[7] Viz. for which $e = \Sigma a_i \omega^{e_i}$, and $e_1 \succ \mathbf{1}$. The ct's of e are thus diagonal.

[8] 'A generalization of ACKERMANN's function', *Zeitschr. math. Logik Grundl. Math.*, Bd. 26, 509-516, thm. 2.2.

[9] Proved for **EP**, [97], thm. 2.2, p510.

If Suc(e_1), then β is of the form $\Sigma + n\omega^{e_1^-} + p$, where $p \in \mathbf{EP}^{(h+1)}$ and satisfies $p \preceq \omega^{e_1^-}$. Now Diag(β) implies Diag(p), and

$$(e,n) = \Sigma + n\omega^{e_1^-} \prec \Sigma + n\omega^{e_1^-} + (p,1) = (\beta,1).$$

If Diag(e_1), then β is of the form $\beta = \Sigma + a\omega^q + p$, where $a \in \mathbf{N}$, $q \in \mathbf{EP}^{(h)}$, $p \in \mathbf{EP}_{\mathbf{0}}^{(h+1)}$ and $(e_1,n) \preceq q \preceq (e_1, n+1)$ (note that $a = 1$ and $p = \mathbf{0}$ if $q = (e_1, n+1)$) and where $p \prec \omega^q$. Since β is diagonal, p is itself diagonal, or equal to $\mathbf{0}$. If Diag(p), then as in the case Suc(e_1), $(e,n) \prec (\beta, m)$, $m = 1,2,3,\ldots$. $p = \mathbf{0}$ implies $a \geq 2$ or $q \succ (e_1, n)$, since

$$(e,n) = \Sigma + \omega^{(e_1,n)} \prec \Sigma + a \cdot \omega^q = \beta.$$

If $a \geq 2$, then the result is obtained as in all previous cases. So consider the case $a = 1, q \succ (e_1,n)$. Suppose first that q is a successor. Then $q^- \succeq (e_1, n)$ and

$$(e,n) \preceq \Sigma + m\omega^{q^-} = (\beta, m), \quad m = 1,2,\ldots .$$

Now suppose Diag(q). Then

$$\begin{aligned}(\beta,1) &= \Sigma + \omega^{(q,1)} \text{ since it is assumed } a = 1,\\ &\succeq \Sigma + \omega^{(e_1,n)} \text{ by i.h., since } q \text{ lies in the range}\\ &\quad\ (e_1,n) \prec q \preceq (e_1, n+1),\\ &= (e,n).\end{aligned}$$

Therefore all compound diagonals of $\mathbf{EP}^{(h+1)}$ have the BACHMANN starting property, and the result now follows by induction.

□

Remark. III·14 is a '$< \infty$' result, giving the (empty) segment of the cfs for β which precedes (e,n).

Examples.

1. $(\psi, n+1, 1) = (\psi, n)$; and in general, for $e = \omega \star (N+1)$,

$$(e, n+1) = \omega \star (N; n+1), \qquad (e, n+1, 1) = \omega \star (N; n).$$

2. $(\omega \star (3;\omega^2), 4, 1) = (\omega \star (3; 4\omega), 1) = \omega^{\omega^{\omega^{3\omega+1}}} \succ (\omega \star (3; \omega^2), 3)$.

3. The BACHMANN starting property, assured for Ω by theorem 14, is not shared by the assignment Λ of defn. 11. Thus, e.g., $\Lambda(\omega^2, n) = n\omega$. Taking $n = 2$, $\Lambda(3\omega, 1) = (3)1 \prec 2\omega = \Lambda(\omega^2, 2)$.

Note that in the final part of the proof of thm. 14, the equality in '$\preceq$' arises as a possibility in the case Suc(q), namely when $q^- = (e_1, n)$ (and $m = 1$); while in the case Diag(q), the possibility is merely carried by the inductive hypothesis. Now let $n \geq 1$. Then for e, β diagonals in **EP**,

COROLLARY 3.15

$$(e, n) \prec \beta \prec e \Rightarrow (e, n) \preceq (\beta, 1).$$

Proof. By thm. 12 $(\exists N)$ $(e, N) \prec \beta \preceq (e, N+1)$. So by thm. 14, $(e, N) \preceq (\beta, 1)$. But $n \leq N \Rightarrow (e, n) \preceq (e, N)$, and the result follows.

□

Remarks

1. Suppose $\alpha \prec e_1 \prec e_2$ in **EP** such that $\alpha = (e_1, m) = (e_2, n)$. Then $(e_2, n) \prec e_1$, so by the corol., $\alpha \preceq (e_1, 1)$, i.e. $m = 1$. So Cα may contain only *one* polynomial for which α is a ct of index > 1. This accords with the rem., after the result of §3·3.

2. The corol. combines the respective '∞,' '$< \infty$' results, III·12, 14.

Extension to $\overline{\textbf{EP}}$

Using defn. 6·3, let the diagonal β lie in the range $(\varepsilon, n) \prec \beta \preceq (\varepsilon, n+1)$. Then β has the form

$$\beta = \Sigma_1 + \omega^{\cdot^{\cdot^{\cdot}}} = \ldots \Sigma_{T-1} + \omega^{\Sigma_T + \mathbf{1}},$$

with ht $\beta = n$, $T \leq n+1$. If $T < n+1$, then $(\varepsilon, n) \prec (\beta, 1)$. If $T = n+1$, $\Sigma_T = \mathbf{m}$, $0 \leq m$. If $m = 0$, then in the notation of §3·3, $\beta \succ \omega \star n$ gives: $(\exists t < T)$ $a_{(t)} > 1$; so $(\varepsilon, n) \prec (\beta, 1)$. If $m \geq 1$, $(\varepsilon, n) \preceq_{\cdot^{\cdot^{\cdot}}} \Sigma_{T-1} + \omega^m = (\beta, 1)$. Thus, noting also rem. 1. after thm. 12, thms. 14, 15 remain true after replacement of '**EP**' by '$\overline{\textbf{EP}}$'.

3.3.6 Classification of Diagonals of EP Based on the Standard Assignment

By Ch. II, defn. 30·3, e is diagonal, if e is neither **1** nor a successor. Thus, taking e diagonal as above, the first refinement of the classification is

DEFINITION 3.16 *Suppose* $(\forall n)$ (e, n) *is a successor. Then* e *is called a* simple *diagonal, otherwise a* compound *diagonal (abbreviated* cd*).*

□

By defn. 6, the simple diagonals e are exactly the polynomials of the form

$$e = f + \omega, \ f \in \mathbf{EP}, \ \mathrm{Diag}(f) \text{ or } f = \mathbf{0}.$$

The cd's are therefore of the form

$$e = \Sigma + \omega^{\beta},$$

where $\Sigma \in \mathbf{EP_0}$, and $\beta = e_1 \succ \mathbf{1}$ in the normal form expression, II·22. This is the basis of the following 2nd refinement of the classification. Let $e = \Sigma + \omega^{\beta} \in \mathbf{EP}$ be compound.

DEFINITION 3.17 *1. If Suc(β), then e is of the* first *kind.*

2. If Diag(β), then e is of the second *kind.*

Examples

1. By defn. 6, if the diagonal e is compound, then $(\forall n)$ Diag(e_n).
2. Every diagonal of **XP** is either simple, or of the first kind.
3. Every $\omega \star n$, $2 \leq n$, is of the second kind.
4. If α is simple or of the first kind and β is of the form ω^{γ} in **EP**, then

$$(\alpha \cdot \beta, n) = (\alpha, n) \cdot \beta, \ n = 1, 2, 3, \ldots .$$

3.4 Canonical Reduction

By thm. 7·3, the taking of a canonical term of a diagonal e entails a $\prec$ −reduction. Similarly with the operation of taking the predecessor (of a successor). These reductions commonly occur together, and are therefore subsumed under a single definition.

Firstly, let $e, f \in \mathbf{EP}$, $\mathbf{1} \prec f$, and let $r \geq 1$.

DEFINITION 3.18 *1.*

$$e \lhd_r f \text{ iff } \begin{cases} e = f^- & \textit{if } \mathrm{Suc}(f) \\ e = (f, r) & \textit{if } \mathrm{Diag}(f) \end{cases}$$

2. The integer r is the canonical index.

3. (SCHMIDT) If $r = 1$, then e is the step-down *of f.*

Notes

1. For given canonical index, a unique e is always determined by f ($\succ \mathbf{1}$). Also $e \lhd_r f \Rightarrow e \prec f$.

2. If, in the notation used in §3·3, e is the step-down of the diagonal $f =_{\therefore} \Sigma_{T-1} + \omega^{\mathrm{rp} f}$; then $e =_{\therefore} \Sigma_{T-1} + \omega^{(\mathrm{rp} f)^-}$.

Examples

$$\omega \lhd_1 \omega \star 2 \lhd_1 \omega \star 3 \lhd_1 \dots ; \quad \mathbf{1} \lhd_{17} \dots \lhd_4 3\omega \lhd_3 \omega^2 \lhd_2 \psi.$$

The greater the canonical index r, the smaller the descent, from a diagonal f. The step-down is a somewhat drastic operation. E.g. after only 8 applications, $f = \omega \star 3 + \omega \star 2 + \omega$ is reduced to $\mathbf{1}$. The rate of descent is slowed, if f contains larger coefficients. Descent via $\lhd_r$, however, soon slows with increase in r. Note that, if $\omega \prec \beta$ in $\mathbf{EP}$, and $(\omega^\beta, 1) = \omega^{\beta_1}$, then β_1 is β^- if $\mathrm{Suc}(\beta)$, or $(\beta, 1)$ if $\mathrm{Diag}(\beta)$. Therefore $\omega^{\beta_1} = (\omega^\beta, 1) \Rightarrow \beta_1 \lhd_1 \beta$.

Now let e, f be as in III·18:

DEFINITION 3.19 *1. $e \lhd f$ iff $(\exists r)\, e \lhd_r f$.*

2. $\tilde{\lhd}$ is the transitive closure[10] *of $\lhd$.*

Examples

1. $\alpha \in \mathbf{XP} \Rightarrow \alpha \tilde{\lhd} \psi$.

2. $$e =_{\therefore} \omega^{\mathrm{rp}(e)} \;\&\; \alpha \;\tilde{\lhd}\; \mathrm{rp}(e) \Rightarrow_{\therefore} \omega^\alpha \;\tilde{\lhd}\; e.$$

---

[10]Viz.

$$e \tilde{\lhd} f \text{ iff } \begin{cases} e \lhd f, & \text{or} \\ (\exists \alpha \in \mathbf{EP})\, e \lhd \alpha \;\&\; \alpha \tilde{\lhd} f. \end{cases}$$

3.4.1 Finite Subclasses of EP

Recalling Ch. II, notation 2·1

DEFINITION 3.20 $(\forall e \in \overline{\mathbf{EP}})_{\mathrm{Diag}(e)}(\forall r \in \mathrm{N})$

1. *$\mathbf{EP}[e;r]$ is the smallest class $\mathcal{X} \subset \mathbf{EP}$ such that $(e,r) \in \mathcal{X}$; and $\mathcal{X}$ is closed under $\lhd_r$.*

2. *The class $\overline{\mathbf{EP}}[e;r]$ is obtained by replacing '$(e,r) \in \mathcal{X}$' above with: '$e \in \mathcal{X}$'.*

Thus (e,r) is the $\prec -$ greatest polynomial of $\mathbf{EP}[e;r]$; and $\mathbf{EP}[e;r]$ is closed under predecessor. $\mathbf{EP}[e;r]$ is thus the class of polynomials transitively below e with respect to $\lhd_r$.

By defn. 18, $(\exists! e_1)\ e_1 \lhd_r e$; $(\exists! e_2)\ e_2 \lhd_r e_1 \ldots$, so that $\mathbf{EP}[e;r]$ can be obtained as a descending sequence

$$\ldots \prec e_2 \prec e_1 (\prec e).$$

Let $e = \omega \star h$.

THEOREM 3.21 *The explicit coefficients of every $\alpha \in \mathbf{EP}[e;r]$, are each $\leq r$.*

Proof, by induction on order-of-definition: the largest ec of the polynomial (e,r) is r. Now suppose $\alpha \in \mathbf{EP}[e;r]$ has the required property. Then certainly, if Suc(α), the predecessor α^- has no ec exceeding r. While if Diag(α), by thm. 7·5 and §3·3, rem. 1(c), the greatest ec of (α, r) is the number r.

□

Plainly, the operations of closure of $\mathbf{EP}[e;r]$ do not increase the $x-$height of a polynomial (while in some cases decrease it). Therefore, taking $e = \omega \star h$ as above, by Ch. II, lemma 45,

COROLLARY 3.22 $\mathbf{EP}[e;r]$ *is finite, containing $\leq (r+1) \star h$ polynomials.*

□

Remark. The inequality of the corollary is in fact *strict*, since e.g. $e = \sum_{i=r}^{0} r\omega^i$ is of unit $x-$height, has all ec's $< r+1$, yet $e \notin \mathbf{EP}[\psi; r]$.

Examples

1. $\mathbf{EP}[\mathbf{1}; r] = \{\ \}$.

2. $\mathbf{EP}[\psi; 1] = \{\mathbf{1}, \omega\}$.

3. $\mathbf{EP}[\omega^{n+1}; 1] = (\mathbf{1}, \omega, \omega^2, \ldots, \omega^n)$.

4. The finite, sequentially ordered system $\mathbf{EP}[e;1]$ may be called the *step-down sequence* of e.

The **converse** to thm. 21 takes the form

THEOREM $\alpha \prec e = \omega \star h \ \&\ N(\alpha) \leq r \Rightarrow \alpha \in \mathbf{EP}[e;r]$.

Proof. By rem. 4 after thm. 12, $\alpha \prec (e,r) = e_1$, say. Then in case $\alpha \neq e_1$, replacing 'e' by 'e_1' leads to the descending sequence

$$\ldots \lhd_r e_2 \lhd_r e_1 \lhd_r e$$

given after defn. 20, where $\alpha \preceq e_i$, $i = 1, 2, 3, \ldots$. This is finite by corol. 22, so $(\exists I)\ \alpha = e_I$.

□

Therefore

COROLLARY

1. $\mathbf{EP}[\omega \star h; r^+] \supseteq \{\alpha \in \mathbf{EP} \mid \text{ht}\alpha < h \ \&\ N(\alpha) \leq r^+\} \supset \mathbf{EP}[\omega \star h; r]$.
2. $e \in \mathbf{EP} \Rightarrow \mathbf{EP}[e;r]$ is *finite*.

Proof. Part 1. follows by thm. 21 and its converse; for part 2, let $K = \text{ht}e + N(e) + r$, $\zeta = \omega \star K$ and consider $\mathbf{EP}[\zeta; K]$. Well, $e \prec \zeta$. Evidently by rem. 1 of §3·3

$$\alpha \in \mathbf{EP}[e;r] \Rightarrow N(\alpha) \leq \max\{N(e), r^+\} < K,$$

therefore $\alpha \in \mathbf{EP}[\zeta; K]$ by the above converse. Part 2. now follows by corol. 22.

□

Remarks

1. By Ch. II, thm. 47, $\alpha \in \mathbf{EP}[e;r]$ satisfies $\alpha \prec e@M \leq r+1$.

2. $\mathbf{EP}[e;r]$ contains $\mathbf{1}$. For, if

$$e_M \lhd_r e_{M-1} \lhd_r \ldots \lhd_r e_2 \lhd_r e$$

(different use of 'M') is the finite sequence of corol. 22, then $e_M = \mathbf{1}$, since otherwise there would be an $f \lhd_r e_M$, and e_M would not be the lowest term.

3. Let $\mathbf{EP}[e;r]$ be as in 2., with $r > 1$. Consider consecutive diagonals obtained in the $\lhd_r$–reduction of e. Thus suppose

$$\text{Diag}(e_i),\ i = 1, 2, 3, \ldots I,\ 1 \leq I;\ \ e_{i+1} \lhd_r e_i \text{ if } 1 < I.$$

Then by repeated application of thm. 14, $r > 1$ gives

$$(e, 1) \prec e_I.$$

But by corol. 22 the diagonal sequence is finite; in the above notation, $I < M$. Since $r > 1$, $(e_I, r) = \mathbf{1}$ is impossible, so there is a value of I for which (e_I, r) is a *successor* polynomial. Thus $e^{(-)} =_{df} e_I \preceq e$ is the least of the consecutive diagonals obtained in the reduction.

4. Under restriction to the finite subclasses, thm. 12 takes the (stronger) form

$$\beta \prec \alpha \text{ in } \mathbf{EP}[e;r] \Rightarrow \beta \preceq (\alpha, r)$$

for diagonal α.

5. Since $(\omega \star n^+, 1) = \omega \star n$ (§2·2, ex. 1 (b)), even if no ec of e exceeds unity, it is possible that $\mathbf{EP}[e;1]$ may contain many terms; e.g. $e = \omega^{\psi+\omega+\mathbf{1}}$, gives 6 polynomials. This number may be made to grow exponentially with ehe.

THEOREM 3.23 *If e is diagonal, then every (e,i), $i = 1,2,\ldots,r$ belongs to $\mathcal{X} = \mathbf{EP}[e;r]$.*

Proof. By definition $e_1 = (e,r) \in \mathcal{X}$. If $r = 1$, there is nothing further to prove; so take $r > 1$, and put

$$e_{i+1} = \begin{cases} e_i^- & \text{if } \mathrm{Suc}(e_i) \\ (e_i, r) & \text{if } \mathrm{Diag}(e_i), \end{cases}$$

$i = 1,2,3,\ldots$. The (e_i) is finite, with last term $\mathbf{1}$, by part 1 of the preceding corol. Thus

$$\mathbf{1} = e_M \prec \ldots \prec e_3 \prec e_2 \prec e_1 = (e,r),$$

is the sequence corresponding to that given after defn. 20. Now $(e,r-1) \prec (e,r)$, so $(\exists K)\ 1 \le K < M$

$$e_{K+1} \preceq (e,r-1) \prec e_K. \qquad (A)$$

Now suppose

$$e_{K+1} \prec (e,r-1) \prec e_K. \qquad (B)$$

Well, if $\mathrm{Suc}(e_K)$, $e_{K+1} = e_K^-$, and (B) is impossible. If $\mathrm{Diag}(e_K)$, then

$$\begin{aligned}(e,r-1) &\preceq (e_K, r) \text{ by BACHMANN's theorem 14} \\ &= e_{K+1},\end{aligned}$$

so again (B) is impossible and (A) gives $e_{K+1} = (e,r-1)$, viz. $(e,r-1) \in \mathcal{X}$. e_1 may now be replaced by $(e,r-1)$ and the argument repeated. In this way it is also proved that $(e,r-2), (e,r-3), \ldots, (e,1) \in \mathcal{X}$.

□

Notes

1. For $r > 1$, in the case $\mathrm{Diag}(e_K)$ above, $(e,r-1) \prec e_K \preceq (e,r)$ so thm. 14 gave $(e,r-1) \preceq (e_K,r)$. Well, thm. 14 also gives $(e,r-1) \preceq (e_K, r-1) \prec (e_K, r)$, and so in fact $(e,r-1) \prec (e_K,r)$. Therefore the condition (A) & $\neg$(B) can *only* be satisfied in the case $\mathrm{Suc}(e_K)$. When, therefore $e_{K+1} = e_K^-$.

2. For $r > 1$, if $e = \Sigma + a\omega$ is a simple diagonal, then $e_1 = (e, r) = \Sigma + (a-1)\omega + r$, $e_2 = e_1^- = \Sigma + (a-1)\omega + r - 1 = (e, r-1)$, so in this case $K = 1$; if e is a cd, then $(e, r) = e_1$ is diagonal, so by (a), $e_2 \preceq (e, r-1) \prec e_1$ is impossible, therefore $(e, r-1) \prec e_2$, and in this case $K > 1$.

3. BACHMANN's theorem, the key to the proof of thm. 23, also may be deduced, starting from thm. 23: if a ct violating the condition of theorem 14 could be found, then a counter-example to thm. 23 could be constructed. It is remarkable that, while thms. 14, 23 are of approximately equal strength, the insight afforded by thm. 14 makes it overwhelmingly more important.

From thm. 23 follows $\mathbf{EP}[e; r] \subset \mathbf{EP}[e; r+1]$; and given $s < r$,

$$(\forall e, f \in \mathbf{EP})\ e \lhd_s f \Rightarrow \mathbf{EP}[e; r] \subset \mathbf{EP}[f; r].$$

Therefore $e \mathrel{\widetilde{\lhd}_s} f \Rightarrow e \mathrel{\widetilde{\lhd}_r} f$.

General Remark

The remark about non-standard assignments made before defn. 6 may be given substance in the light of the present section. Thm. 21 for example is a property of the classes $\mathbf{EP}[e; r] = [e; r]_\Omega$: for the classes $[e; r]_\Lambda$, e.g. with Λ as in defn. 11, it fails.

3.4.2 Enumerations of EP

As in Ch II, §5, the decomposition

$$\mathbf{EP} = \bigcup_r \mathbf{EP}[\varepsilon; r],$$

for example, may be used to provide an enumeration of $\mathbf{EP}$: let $\mathcal{X}_r = \mathbf{EP}[\varepsilon; r]$, and consider the classes $\mathcal{Y}_r$ of II, §5. A canonical enumeration is obtained by taking e before f iff

1. $e \in \mathcal{Y}_r$, $f \in \mathcal{Y}_s$ and $r < s$; or
2. $e, f \in \mathcal{Y}_r$ and $e \prec f$.

Now write $\text{num}(e) = k$, the place of e in the enumeration. Then $e \in \mathcal{Y}_r \Rightarrow \text{num}(e) \geq r$ (strict for $1 < r$), since $|\mathcal{Y}_r| > 1$. Therefore $N(e) \leq \text{num}(e)$ by thm. 21 (in the notation of II·44). So by rem. 4 after thm. 12, if α is diagonal, $e \prec \alpha$ and $n \geq \text{num}(e)$, then WAINER's [11] relation

$$e \prec (\alpha, n)$$

holds. What is more, by the remarks of §3.3,

$$N\{(e, n)\} \leq N\{(e, n+1)\};$$

therefore using thm. 21 and its converse, since $(e, n) \neq (e, n+1)$,

$$\text{num}\{(e, n)\} < \text{num}\{(e, n+1)\}.$$

[11] Ref. [156], 1970, lemma 3.1.

3.4.3 Iterated Canonical Reduction

Given a sequence $\mathcal{S}$ of natural numbers $n(1)$, $n(2), \ldots$, and a function $e \in \overline{\mathbf{EP}}$, $e \neq \mathbf{1}$, it is possible to repeatedly apply $\lhd$, restricted by $\mathcal{S}$, to obtain the sequence

$$\ldots \lhd_{n(2)} \alpha \lhd_{n(1)} e$$

in **EP**. This gives rise to

DEFINITION 3.24 *For any* $e \in \overline{\mathbf{EP}}$,

1. $e_{(1)} = e$.

2.
$$e_{(i+1)} = \begin{cases} \mathbf{1} & \text{if } e_{(i)} = \mathbf{1} \\ \alpha \in \mathbf{EP} \text{ s.t. } \alpha \lhd_{n(i)} e_{(i)} & \text{otherwise.} \end{cases}$$

At the diagonal steps, $\ldots \alpha \lhd_{n(i)} \beta \ldots$ the larger is $n(i)$, the greater the coefficients that occur in $\alpha \prec \beta$. It therefore follows that a rapidly ascending $\mathcal{S}$ affords a slowly descending $\{e_{(i)}\}$.

The Descent Function

In the constant case $n(1) = n(2) = \cdots = n$ of the above iteration, an explicit function $D_e(n)$ of n is obtained:

DEFINITION 3.25 *The* descent function D_e *of the diagonal* $e \in (\overline{\mathbf{EP}}, \Omega)$, $e \succ \mathbf{1}$, *is the number of canonical reductions* (mod n) *required to reduce* e *to a successor polynomial.*

Thus, e.g., $e = 2 \cdot \omega^f$, $D_e(1) = 9$, in the ex. after defn. 18. For (diagonal) e far into the linear hierarchy, the descent function is evidently of fairly rapid growth, and there is the problem of estimating this. The approach is informally sketched. First note that for $\mathrm{ht}(e) = h$, $\mathrm{mult}(e) = m$, (e, n) is a subpolynomial of some $\lhd_N$–reduct of $\omega \star N$, where $N = 1 + \max\{n, m, h\}$. It suffices, therefore to consider the case $e = \omega \star N$, for $N = 1, 2, 3, \ldots$. But every such reduct is in $\mathbf{EP}[e; N]$, therefore by corol. 22, the number of steps is fewer than $(N+1) \star (N+1)$, so

THEOREM 3.26

$$D_e(n) \leq (N+1) \star (N+1).$$

□

Note that the above estimate is adequate for the variant $D'_e \gtrsim D_e$, for which reduction must be to the initial polynomial **1** rather than only to a successor.

3.5 Assignments Other Than Ω

3.5.1 Constant Assignment

Let $e \in \mathbf{EP}$ be nc, expressed in the usual x−cluster notation. Let the separate occurrences of 'x' in the e−expression be replaced by distinct integer variables $x_1, x_2, \ldots$. 'x' occurs only finitely many times, N say, so $\lambda x \cdot e(x)$ defines another function $\widetilde{e}(x_1, x_2, \ldots, x_N)$ of N variables: in general, $\widetilde{e} \notin \mathbf{EP}$. E.g. $e = \omega \star 3$ gives x^{y^z}. Since '+' and exponentiation are si, $\widetilde{e}$ is si in all variables. Returning to the e−expression, the functions e_n obtained by replacing one or more ocurrences of 'x' by 'n' in the expression, $n = 1, 2, 3, \ldots$, determine a fundamental sequence for e, since $\widetilde{e}$ is si, and $e(x) = e_x(x)$. In general, the $e_n \notin \mathbf{EP}$. E.g. as noted in rem. 3 after defn. 6, $\mathbf{1}^\omega, \mathbf{2}^\omega, \mathbf{3}^\omega, \ldots$ is a fundamental sequence for $\omega \star 2$, while only the term $\mathbf{1}^\omega$ actually belongs to $\mathbf{EP}$. In the extreme case, in which *every* x−occurrence is replaced, the constant function equal to $e(n)$ is obtained. Thus a remarkable assignment defined by $\overline{\Omega}(e, n) = \lambda y \cdot e(n)$, $e \succeq \omega$, arises naturally. e is si, with $\overline{\Omega}(e, n) \prec \overline{\Omega}(e, n+1)$ and of course $e(x) = \{\lambda y \cdot e(x)\}(x)$: the conditions of defn. 3 are satisfied. [12] $\overline{\Omega}$ is not a regular assignment, since e.g. $\{\overline{\Omega}(\omega^2, n)\}_{1 \leq n}$ cannot be obtained from the fundamental sequence $\mathbf{1}, \mathbf{2}, \mathbf{3}, \ldots$ by the addition of a constant; while the other rules lead only to non-constant polynomials.

3.5.2 Fundamental Sequences for Real Numbers

Of first priority for this text are fundamental sequences for (non-constant) generalized polynomials. However the assignment of fundamental sequences to real numbers makes an interesting comparison.

Denoting the semi-open interval $0 < x \leq 1$ by $(0, 1]$, let the sequence $\mathcal{X}_0 \subset (0, 1]$ consist of the points $\frac{n}{n+1}$, $n = 1, 2, 3, \ldots$. Mark these *red*. By a *transcription* of a sequence $\mathcal{Y} \subset (0, 1]$ to an arbitrary interval $(a, b]$ is meant an embedding $\Phi : (0, 1] \to (a, b]$ such that for $x \in \mathcal{Y}$, $\Phi(x) = a + (b - a)x$.

Form $\mathcal{X}_1$ by transcribing the points of $\mathcal{X}_0$ to each interval $(\frac{n}{n+1}, \frac{n+1}{n+2}]$, determined by a point $\frac{n}{n+1}$ of $\mathcal{X}_0$. Mark the image points of the transcription red, and (with the exception of the least point, $\frac{1}{2}$) mark all points of $\mathcal{X}_0$ *black*.

Generally, given $\mathcal{X}_i$, $i = 1, 2, 3, \ldots$, form $\mathcal{X}_{i+1}$ by taking all consecutive points $a < b$, $a, b \in \mathcal{X}_i$, $\frac{i+1}{1+2} \leq a$, then transcribing the points of $\mathcal{X}_0$ into the interval $(a, b]$. Mark the image points red; and the points $x \geq \frac{i+1}{i+2}$ of $\mathcal{X}_i$ black. On the basis of similarity with hierarchies of polynomials, $\mathcal{X}_0, \mathcal{X}_1, \mathcal{X}_2, \ldots$ may be regarded as constant, linear, quadratic, ... sequences respectively. Consider the system $\mathcal{X} \subset (0, 1]$ whose points are those of $\mathcal{X}_i$ which

[12] Note that $m < n \Rightarrow \lambda y \cdot e(m) \ll \lambda y \cdot e(n)$.

lie in the interval $[\frac{i}{i+1}, \frac{i+1}{i+2})$, $i = 1, 2, 3, \ldots$: quite evidently, none of the $\mathcal{X}_i$ contains an

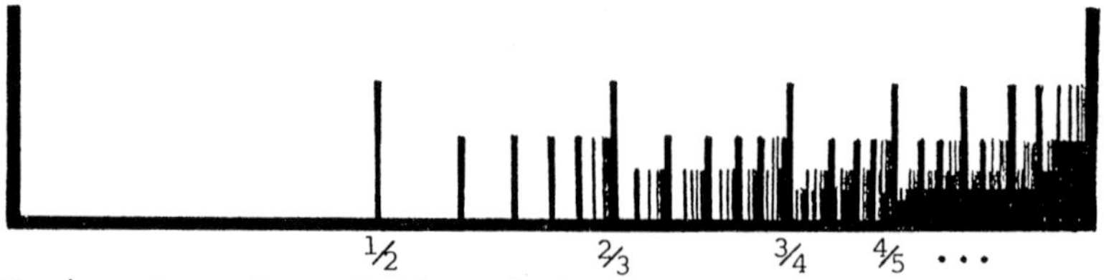

Regions beyond resolution of sketch shown shaded.

endlessly descending sequence

$$x_1 > x_2 > x_3 > \ldots$$

(of real numbers). Since, in the linear order, there is no such descending sequence of $\mathcal{X}_i$, it follows that the polynomial system of points $\mathcal{X}$ also has this property. Furthermore the isolated [13] points of $\mathcal{X}$ are red; the limit points which $\mathcal{X}$ contains are black.

An infinite sequence $\mathcal{S} = x_1 < x_2 < x_3 < \ldots$ in $\mathcal{X}$ is a *fundamental sequence* for $x \in \mathcal{X}$ iff

1. $(\mathcal{S}, <) \simeq (\mathbf{N}, <)$.
2. $x = \lim \mathcal{S}$.

Note that $x \notin \mathcal{S}$. Any black point is the limit of endlessly many different fundamental sequences. But, as noted by HARDY, it is plain that one particular assignment is outstanding. Indeed, if $\mathcal{X}_0$ is transcribed into $(a, b]$, then b may be assigned the fixed sequence (x_n), where $x_n = a + \frac{n(b-a)}{(n+1)}$, $n = 1, 2, 3, \ldots$. [14]

Remarks

1. $\mathcal{X} \subset \mathbf{Q}$, the rational field.
2. A similarity $\Psi : (\mathcal{X}, <) \to (\mathbf{XP}, \prec)$ may be set up by:
   (a) $\Psi\{\frac{n}{n+1}\} = \lambda x \cdot x^{n-1}$, $n = 1, 2, 3, \ldots$.
   (b) Suppose $x \in \mathcal{X}$ is a limit point, for which $\Psi(x)$ has been defined. Then $\Psi(x_n) = (\Psi(x), n)$, via defn. 6.

   Thus $\alpha = \mathrm{diag}_n(\alpha, n)$ in $\mathbf{XP}$ via $\Psi^{-1}(\alpha) = x$ corresponds to $x = \lim_n x_n$ in $\mathcal{X}$.
3. The above process of transcription may plainly be carried further, to obtain a more complex well-founded ordered system $\mathcal{Y}$, for which $(\mathcal{X}, <)$ is an initial segment.

---

[13] x is *isolated* if $\{\exists a, b \in (0,1]\}\ a < x < b$ is the only point of $\mathcal{X}$ in the neighbourhood (a, b); is a *limit point* if $\{\forall a, b \in (0,1])\}\ a < x < b \Rightarrow (\exists w \in \mathcal{X})\ w \neq x, w \in (a, b)$.

[14] The endpoint 1 of $(0,1]$ may be assigned the fixed sequence $(\frac{n}{n+1})$.

4. $(\mathcal{X}, <) \simeq (\mathcal{Y}, >)$, where $\mathcal{Y}$ is obtained by a device of MITTAG-LEFFLER, viz. by taking all points

$$\frac{1}{2^k} + \frac{1}{2^{k+n_1}} + \frac{1}{2^{k+n_1+n_2}} + \ldots + \frac{1}{2^{k+n_1+n_2+\ldots+n_k}},$$

where $k, n_1, n_2, \ldots, n_k \in \mathbf{N}$ (ref. E. W. HOBSON [65], Vol. I, §68, p99).

Properties of the Real Assignment

Contrary to the case of diagonalization in **EP**, the result of taking the limit of a convergent sequence $x_1, x_2, x_2, \ldots$ of real numbers is not affected by changes in any finite initial segment of (x_i). One consequence of this is the extraction of an assignment which satisfies the BACHMANN property, from a 'straggling' assignment: viz. for which some assigned terms violate the condition of thm. 14. Thus, in particular the condition demands that if $x_1 < x_2 < x_3 < \ldots$ is the fundamental sequence assigned to the real number x; and if, for $n > 1$, $y = x_{n+1}$ has been assigned the fundamental sequence $y_1 < y_2 < y_3 \ldots$, then

$$x_n \leq y_1.$$

Suppose this is not the case. E.g.

$$x_{n-1} < y_1 < y_2 < x_n < y_3 < \ldots.$$

This particular violation of BACHMANN's condition can easily be redressed: the points y_1, y_2 may be eliminated from the fundamental sequence assigned to y : the tail $y_3 < y_4 < y_5 < \ldots$ still has the limit y. Furthermore, the elimination of violating points may be exhaustively applied to any assignment of $\mathcal{X}$. [15] In fact the eliminations made at a given step make the removal of violating points at later steps, in general easier: with x, y as in the example, suppose $z < y_2$ is assigned $z_1 < z_2 < z_3 < \ldots$, where $z_1 < y_1 < z_2 < z_3 < \ldots$. Then the previous removal of y_1 obviates the need to shorten the fundamental sequence for z at the next step.

3.5.3 The General Recursive Theory of Well-Order

For the purposes of this exposition, a 'denumerable ordinal number' will be a point on the real unit interval $(0, 1]$ defined by a procedure to be given. KLEENE [**72**]1938 instigated a general recursive theory of ordinal numbers $x_1, x_2, x_3, \ldots$, calling i the 'ordinal notation' of x_i.

CANTOR's Theory of Fundamental Sequences

The definition of a limit point x of $\mathcal{X}$ of the previous section is usually given by:

$$(\forall \epsilon > 0)(\exists a \in \mathcal{X})\ y \neq x \ \&\ \mid x - a \mid < \epsilon.$$

This may be formulated differently. Thus, given a fundamental sequence (x_n) in $\mathcal{X}$ for x, if

$$(\forall \epsilon > 0)(\exists N)(\forall n)\ N \leq n \Rightarrow \mid x - x_n \mid < \epsilon,$$

then x is a limit point of $\mathcal{X}$; and this form of definition may be used to found theories of convergence and continuity for the reals.

[15] Note that the *choice* of $\mathcal{X}$ (within, say $(0, 1]$) and the *assignment* of $\mathcal{X}$ are 2 distinct constructions.

Recursively Constructed Hierarchies

As noted in rem. 2 of §5·2, the canonically assigned, well-founded hierarchy $(\mathcal{X}, <)$ may be extended to say $(\mathcal{Y}, <)$. KLEENE's $\mathcal{R}$-classification of such hierarchies rests on the theory of fundamental sequences which according to the preliminary, may be assumed to be real:

1. $x \in [0,1]$ *corresponds* to the nd function $f \in \mathcal{N}$ iff $x = \Sigma_{n=1}^{K} \frac{a_n}{2^n}$, where K is a fixed value $1, 2, 3, \ldots$, or ∞; and where $a_n = \begin{cases} 0 & \text{if } f(n) = f(n+1), \\ 1 & \text{otherwise.} \end{cases}$ In each case the series for x converges.

   Examples

   (a) $f = \mathbf{1}$, $x = 0$; $f = \omega; x = 1$.

   (b) Any given $x \in [0,1]$ corresponds to a certain f.

2. Let the denumerable sequence of points $\mathcal{Y} \subset (0,1]$ be enumerated without repetition by $x_1, x_2, x_3, \ldots$; where

   (a) x_1 is the point nearest the origin; and

   (b) $\mathcal{Y}$ is well-founded, viz. every descending sequence $\ldots < x_{n(3)} < x_{n(2)} < x_{n(1)}$ is *finite*.

   (c) $\mathcal{Y}$ is *internally closed*: if x ($\neq \sup \mathcal{Y}$) is a limit point of $\mathcal{Y}$, then $x \in \mathcal{Y}$.

3. $(\mathcal{Y}, <)$ is *general recursive* if

   (a) $(\exists f \in \mathcal{R})$ x_i corresponds to $\phi_{f(i)}$, in the enumeration of Ch. I, §3·1, $i = 1, 2, 3, \ldots$.

   (b) $(\exists g \in \mathcal{R})$ if x_i is an isolated point of $\mathcal{Y}$, then $\phi_{g(i)}$ is the constant function, equal to i; and if x_i is a limit point of $\mathcal{Y}$, then $\{x_{(k)}\}_{k=1,2,3,\ldots}$ is a fundamental sequence for x_i, where $(k) = \phi_{g(i)}(k)$. Then in either case, $x_i = \lim_{k<\infty} x_{(k)}$.

Remarks

1. 3.(a) above alone is not enough to ensure that $\mathcal{Y}$ is a recursive hierarchy. Thus, let the points $x_1, x_2, x_3, \ldots$ be distributed in $(0,1)$ in any manner whatsoever, but so as to satisfy 2. above. Then a (piecewise) order-conserving transformation Φ_1 of $(0,1)$ may be found, such that $\Phi_1(x_1)$ corresponds to some $f_1 \in \mathcal{R}$. Now suppose $(0,1)$ has been repeatedly transformed, so that x_m has been moved so as to correspond to a known $f_m \in \mathcal{R}$, $m = 1, 2, \ldots i$. Well, $x = x_{i+1}$ now either corresponds to a function $f_{i+1} = \phi_j \in \mathcal{R}$, for which the index j may be obtained from that of f_i by a recursive algorithm; or can be made to do so, by moving x up or down $(0,1)$ via the appropriate Φ_{i+1}, which may be assumed to leave $x_1, x_2, \ldots, x_i$ fixed [16] (the same may not be said of $x_{i+2}, x_{i+3}, \ldots$.) Then since $f_i \in \mathcal{R}$, $i = 1, 2, \ldots$, by the choice of the Φ_i, it follows that $(\exists f \in \mathcal{R})(\forall i)$ $f_i = \phi_{f(i)}$. Thus the arbitrary denumerable hierarchy $\mathcal{Y}$ is similar to one which satisfies condition 3(a).

2. Any si Φ whatever, i.e. not necessarily recursive, on $(0,1)$ will transform a given recursive hierachy $\mathcal{Y}$ to a recurive hierarchy.

---

[16] Whether or not $x = x_{i+1}$ is a limit point of $\mathcal{Y}$, x has a neighbourhood, upon which Φ_{i+1} may be taken to act, in which no $x_1, x_2, \ldots, x_i$ occurs.

Incompleteness of Recursive Hierarchies

Suppose $(\mathcal{Y}, <)$ has been constructed recursively. Suppose $\mathcal{Y}$ has a greatest element $x = x_i < 1$. Then to $\mathcal{Y}$ may be added the point $y = x + \frac{1}{2}(1-x)$, to give $\mathcal{Y}^+$ say, enumerated by $y_1 = x_1, y_2 = x_2, \ldots, y_i = x_i, y_{i+1} = y, y_{i+2} = x_{i+1}, \ldots$, where $\mathcal{Y}$ is a proper segment of $\mathcal{Y}^+$. Now suppose $\mathcal{Y}$ has no greatest x_i, so that $x_i < 1$, $i = 1, 2, 3, \ldots$, where it may be supposed $\lim \mathcal{Y} = 1$. Now let $x_{n(1)} = x_k$, where $k = (\mu i)[\frac{1}{2} < x_i]$; and given $x_{n(i)}$, let

$$x_{n(i+1)} = (\mu i)[\max\{\frac{i+1}{i+2}, x_{n(i)}\} < x_i].$$

Then evidently $\{x_{n(i)}\}$ is a fundamental sequence for $x = 1$, recursively defined via the 'μ-'operator. If $\mathcal{Y} \cup \{x\}$ is now (recursively) transcribed into, say $(0.5, 0.75)$, then as above

$$y_1 = \Phi x_1 = 0.5, y_2 = 0.75, y_3 = \Phi x_2, \ldots$$

gives a hierarchy $\overline{\mathcal{Y}}$ for which $\mathcal{Y}$ is a proper sub-hierarchy. Thus, with respect to the natural order

COROLLARY

There is no maximal general recursive hierarchy $\mathcal{Y}$.

3.5.4 A Theorem of de BRUIJN

de BRUIJN [27], 1961 gave a remarkable proof of a converse to CAUCHY's condition. The infinite array (δ) is *strong* if every sequence (x_n) for which

$$(\forall m, n \geq N) \mid x_m - x_n \mid < \delta_{mn}, \tag{I}$$

converges. The CAUCHY condition is

$$\lim_{m,n\to\infty} \delta_{mn} = 0. \tag{II}$$

The condition (II) is more than is necessary for convergence as demonstrated in [28]; but it is shown that if the strong array (δ) satisfies

1. $\delta_{mn} \geq 0$,
2. $\delta_{mn} = \delta_{nm}$, (III)
3. $\delta_{ln} \leq \delta_{lm} + \delta_{mn}$,

then condition (II) becomes necessary for convergence. The proof uses an infinite variant of F. P. RAMSEY's theorem on partitions: if (III) holds, but (II) does not, then a divergent sequence which satisfies (I) may be constructed. The infinite RAMSEY theorem depends on the axiom of choice, even in the special case used in de BRUIJN's theorem. That is, as it is proved from the 'finite' version, the monochrome sequence $X = \{a_1, a_2, a_3, \ldots\}$ is obtained by making an endless sequence of choices from such sequences $X_1 \subset X_2 \subset \ldots$, where $a_1 \in X_1, a_2 \in X_2, \ldots$. (Ref. BOLLOBÁS [14], 1988, *Graph Theory*, pp105-6.)

3.5.5 The GIRARD-VAUZEILLES Construction

From the pioneering work of HARDY [57], 1904 onwards, CANTOR's 'ordinals' have been used to build hierarchies within $\mathcal{N}$. The distinguishing feature of **EP** is that the construction is undertaken *without* reliance on an infinitistic concept.

GIRARD [43], 1981, has used the language of category theory to re-formulate the association of the so-called 'direct limit', viz. a point in $\mathcal{N}$ with an 'ordinal'. This *requires* an assignment Λ of fundamental sequences; if the morphisms used in the construction satisfy laws of composition and directness, then thm. 12, and the corollary before the general remark at the end of §4.1, follow. These laws are set out in the exposition of DENNIS-JONES and WAINER [30], 1983. Theorem 14, however, does not follow and as a property must be included in the definition, in order for the construction to work.

3.6 Notes on Ch. III

1. §2·2. CANTOR applies the notion of 'limit' to the abstract idea of 'ordinal' by analogy with limit of a linear point set. This has led e.g. to VEBLEN's theory of 'normal' functions, viz. si 'continuous' functions on these abstact objects. A 'continuous' f satisfies

$$f(\beta) = \lim_{\alpha<\beta} f(\alpha)$$

for all 'limit' β. But such an f is scarcely continuous: it has in common with a function continuous in the sense of real analysis only the 'ϵ-δ' form of definition, not the real metric on which the classical notion of continuity actually rests. This semantical problem cannot arise in $(\mathbf{EP}, \prec)$: a diagonal is unequivocally the diagonal of its canonical fundamental sequence.

2. §5·4. The 'finite' RAMSEY theorem states that for every $m, n \in \mathbf{N}$, there is an $r = r(m,n)$ such that if G is a finite graph, $\mid G \mid=$ vertex order $\geq r$, then either

   (a) K_m, the complete m-graph is a subgraph: $K_m < G$; or

   (b) $K_n < \overline{G}$, the edge-complement of G.

   By the theorem of ERDÖS-SZEKERES, the RAMSEY numbers are dominated by the binomial coefficients: $r(m,n) \leq \binom{m+n-2}{n-1}$. Despite this, the only 'non-trivial' $r(m,n)$ known are $r(3,i)$, $i \leq 9$; and $r(4,4) = 18$ (see Appx. C, §2·2), some of which were only found after hard computing. It should be noted that there is no mystery in this; $r(m,n)$ depends on checking a number of cases which is a rapidly-growing function of m, n.

Chapter IV

NORMAL TREES AND THE DCC

Section Headings

Section Headings (Cont'd)

4.1 Introduction

The most striking feature of an exponential polynomial is its x−height h (II, defn. 5). It may be noted, however, that as h increases, the number [1] of sites at which coefficients may occur rapidly increases. Therefore in contrast to a 'typical' integer polynomial of degree 10 say, which might be found in $\mathbf{EP}[\omega^{10}, 10^{10}]$ (III·17), a 'typical' $e \in \mathbf{EP}$ of edge-height 10 might be found in $\mathbf{EP}[\omega \star 10,\ 10 \star 10]$.

The present Chapter tacitly assumes the great lengths of the higher exponential polynomials: attention is directed to the ramified, or *tree* structure of the x−cluster symbol of $e \in \mathbf{EP}$. The trees have combinatorial properties essentially dating back to work of EULER [35], 1777 through CAYLEY [18], 1857 and to HALL [55] in the 20th century. These are entirely independent of arithmetic properties of the polynomial function e.

The following is an exposition of the tree T of $e \in \mathbf{EP}$. T is finite, contains a main vertex (called by CAYLEY the *root*) and posseses plane structure. [2] T belongs to a proper infinite subsystem $\mathfrak{T}$ of the hierarchy $\mathfrak{S}$ of all such trees. The distinguishing property of T is that the trees defined by the successors of every vertex, end vertices excepted, possess the (recurrent) non-strict dictionary order $\preceq$ with respect to the representation in the plane. As a result, $(\mathfrak{T}, \prec)$ itself is an ordered system: the bijection $e \to T$ establishes $(\mathbf{EP}, \prec) \simeq (\mathfrak{T}, \prec)$, where for $\mathbf{EP}$, $\prec$ denotes the relation of eventual domination.

4.2 Plane Structure Oriented Trees

4.2.1 Finite Trees

Infinite trees have a place in finitist interpretations, no less than infinite structures, such as $\mathrm{N} = (1, 2, 3, \ldots)$. For example consider the finite multiple-edge rooted di-

[1] Ref. the indexing of basic coefficients, II, §2.17; also III, §3·3.

[2] This is the plane structure of the element e of the linear structure $\mathbf{EP}$.

graph

$\Gamma =$ 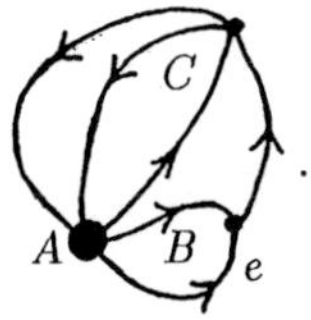

Γ defines an infinite (recurrent) tree:

$T =$ 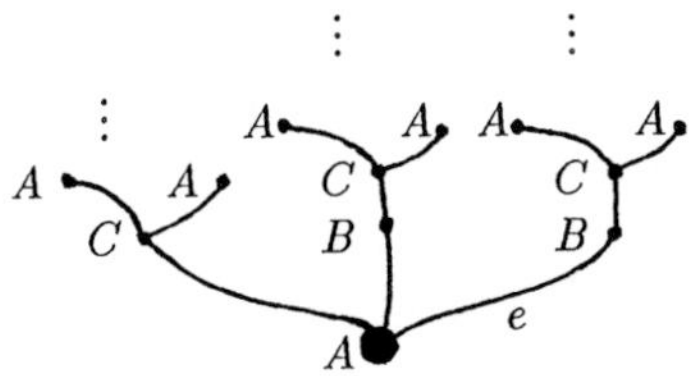

where the root A of Γ is that of T; and the edge marked e is taken as the rightmost edge incident on A in T. The plane-structure of T implies e.g. that any path $ACA\ldots$ lies to the left of any path $ABCA\ldots$. However infinite trees play no part in what follows. Therefore, by 'tree T' is to be understood 'finite tree', viz. constituted of finitely many (fm) indistinguishable *vertices*, or *points*, or *nodes*, or *knots* (CAYLEY). '$v \in T$' is used to mean '$v \in V(T)$, the finite set of vertices of T'. Every pair u, v of distinct vertices of T either does or does not posess (one and only one) *edge*, or line. If u, v do possess an edge e, then u, v are said to be *adjacent* and e to be *incident on* u, v. If no edge is incident on v, v is *isolated.* Every edge posessed by a pair of vertices of T is regarded as an edge of T. The predicate notation $\mathrm{Adj}(u, v)$ is used iff u, v as above are adjacent. Note $\mathrm{Adj}(u, v) \Rightarrow \mathrm{Adj}(v, u)$. Also for every v of T, $\neg\mathrm{Adj}(v, v)$; in particular, if T has a single vertex, then T has no edges. Given a vertex v of T, the *valency* (or *degree*) $\mathrm{va}(v)$ of v is the number of edges incident on v.

If the tree S has just a single vertex, then this is the single *endvertex* or *endv.* If S has n vertices; and v is an isolated vertex, and u is any vertex of S, then the object T, obtained by adding v to the vertices of S, and assigning u, v an edge, is also a tree. v is an endv of T; u is an endv of T iff S has the single vertex u; otherwise every endv $w \in S$, $w \neq u$, is an endv of T. Those and only those objects obtainable by the above inductive definition are trees.

This definition is more fundamental than the common definition [3] of a tree as a graph which contains no cycles, this property following from the definition by induction.

Since each new vertex, in building up T from an isolated vertex, requires one edge, for any tree: number of vertices = 1+ number of edges. The *(vertex-)order* $\mid T \mid$ of T is the number of vertices in $V(T)$.

---

[3] Ref. Note 1., end of this Ch.

Trees posessing no structure [4] other than that provided by the definition are sometines called *free trees.* The manner in which a free tree T may be asigned additional structure is highly various. T may be asssigned a **principal vertex** $v^\sharp$, or **pv** or *root*; T may be assigned a *principal edge* $e^\sharp$. In either case, T becomes *oriented.* [5] The tree form is very common both in science and elsewhere, as in decay patterns of fundamental particles, discharges of lightning, river confluences, arboreal trees, the arterial trees of higher plants and animals, the taxonomic hierarchy, disk filing systems, etc., etc.

Plane Structure

Any free tree T may be embedded in the plane: in the inductive definition of T, $v^\sharp$ alone, is embeddable as a point. And if S is embeddable, then given any point vertex u of S, the point vertex v may be joined by a planar arc defining the edge uv to u, by placing v so that uv cuts no arc of S, this always being possible, since S contains no cycles. Indeed, it is evident that the vertices may be placed so as to allow straight edges. Even allowing for rotation in the plane, for variation of lengths of edges, and of angles between edges, the representation of T as a figure of lines and points embedded in the plane is not unique, as may be seen from the example:

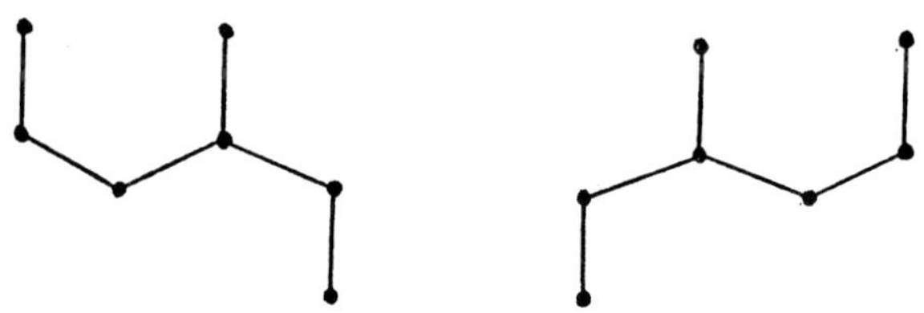

The 2 figures represent the same free tree, but neither may be obtained from the other without passing one edge across another. This remains true whether or not any vertex is taken as distinguished.

[4] The *edge-distance* $d(w,v)$ between vertices w,v in the (free) tree T may be defined using the above inductive definition: $d(w,w)=0$; if $d(w,u)=k$ in S, then $d(w,v)=k+1$ in T. The *radius* of T is

$$r(T)=\min_w \max_v d(w,v).$$

Thus it may be deduced that the *centre* of T, viz. $\{w\in T : \max_v d(w,v)=r\}$, consists either of a single vertex, or of a pair of vertices, joined by an edge. (Ref. e.g. BEHZAD, CHARTRAND and LEZNIAK-FOSTER, *Graphs and Digraphs* (1981)).

[5] A tree is *directed* if each edge e is assigned an order for the vertices of e. This may be taken e.g. to increase with edge-distance from $v^\sharp$. Thus the oriented trees determine a proper subclass of the directed trees.

Horizon

In accordance with the natural growth of trees 'upwards', in the above plane-embedded metric realization of an oriented tree T, it is convenient to take a line Λ, the *horizon* through the pv, such that all the $v \in T, v \neq v^{\sharp}$, lie on the same side ('above') Λ; and that the greater the edge-distance of v from $v^{\sharp}$, the greater the perpendicular distance of the point v from Λ. The example of the next figure contains a possible elaboration of the notion of 'root':

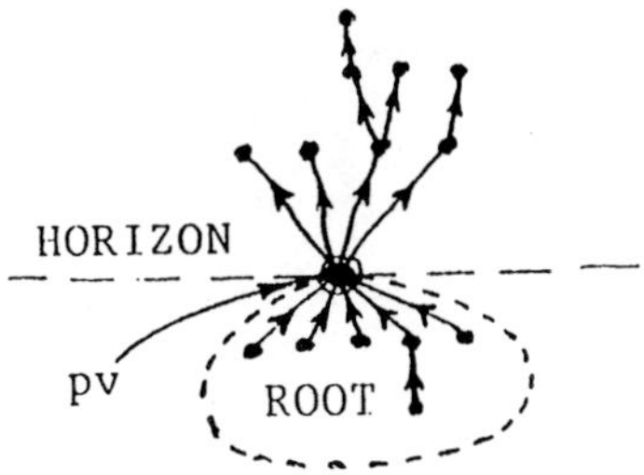

Orientation, plane structure, etc. are just a few of the remarkable and quite different ways in which a free tree may be assigned structure.

CANTOR-Type Normal Form

In the following, in order to keep to the convention established by CANTOR, enumerations are from right to left.

Square Bracket Notation

If the oriented plane structure tree T has pv-valency r, then the vertices $v_r, v_{r-1}, \ldots, v_1$ linked by edges to $v^{\sharp}$, may themselves be taken as pv's of trees $T_r, T_{r-1}, \ldots, T_1$ defined by T. **Therefore the notation**

$$T = [T_r, T_{r-1}, \ldots, T_2, T_1]$$

may be adopted for T. [ ] denotes the isolated vertex $v^{\sharp}$.

Indices

1. Suppose the condition that the vertices of a tree T, $\mid T \mid = n$, be indistinguishable is dropped, so that each is assigned a unique *index* (or *label*). This may be a number, letter, polynomial etc. In particular, the v's of T may be *enumerated*: if T is pv-oriented with plane structure, each enumeration gives one of the $n!$ indexed trees associated with T.

   On the other hand, consider the indexed *free* trees which may be formed on n vertices. It is well-known (CAYLEY [19], 1889; for more recently-discovered proofs, see

CLARKE [23]) that for $n \geq 1$, the number of such trees is n^{n-2}. Now suppose $1 < n$, and let an endv v be removed. Then v may have n possible values. And the remainder may be any of $(n-1)^{(n-3)}$ trees; in each case v may have been attached to any of $n-1$ possible vertices. Therefore the total number of endv's of the indexed trees on n vertices is $n(n-1)^{n-2}$. So the proportion p of endv's is obtained by dividing by the total n^{n-1} :

$$p = (\frac{n-1}{n})^{n-2}, \qquad \text{and} \qquad \lim_{n\to\infty} p = \frac{1}{e} = 0.367879\ldots .$$

This proof of RÉNYI's result is based on that by ROBINSON and SCHWENK [130] *Disc. Math.* **12**, (1975,) 359-372.

2. If the vertices of T are indexed, then the tree obtained from T by ignoring the indices is the *outline* of T.

3. With one or two exceptions, the trees of Ch. IV are non-indexed, and may be regarded as providing a general setting for the more specialized theory [6] of trees whose v's are assigned e.g. an integer value, and whose edges e.g. are interpreted metrically.

4.2.2 The Class $\mathfrak{S}$

The class $\mathfrak{S}$ contains the (finite) pv-trees with plane structure. The following (McBETH [99], 1979; see also [106], 1990) exploits the well-known correspondence between trees and balanced patterns of brackets, discussed in the previous section:

DEFINITION 4.1 $\mathfrak{S}$ *is the smallest class closed under*

1. *(a)* $[\,] \in \mathfrak{S}$.

   (b) The isolated vertex $[\,]$ *is denoted by* $v^{\sharp}$.

2. *(a) Let* $T_r, \ldots, T_2, T_1 \in \mathfrak{S}$, $1 \leq r$. *Then*

$$T = [T_r, \ldots, T_2, T_1] \in \mathfrak{S}.$$

   (b) The trees T_i, $i = r, \ldots, 2, 1$ *are the* principal branches *of* T.

□

Thus, with every vertex of T is associated a unique pair of brackets in the representation of T : Adj(u, v) in T iff either (i): $[\,]$ are the brackets of u such that $T = \ldots[X_s, \ldots, X_1]\ldots$ where for some $i \leq s$, the outermost brackets of X_i are those of v, or (ii): the same as (i), but with u, v exchanged.

[6] See particularly BARTHÉLMY, J-P and GUÉNOCHE, *Trees and Proximity Relations* , transl. G. LAWDEN, Wiley (1991).

Examples

1. S = [[ ],[ ]] =  ; T = [[[ ]],[[ ],[ ]]] =  ; U = [S,T,[[ ]]] = 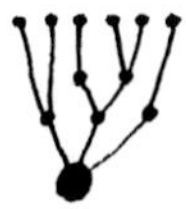

   are trees of $\mathfrak{S}$.

2. A *chain*, or tree without bifurcations, in $\mathfrak{S}$ has only a single term enclosed [...] at each step of the inductive definition.

3. The derivations or *proof figures* of GENTZEN's sequent calculus ([40], section 5.22, in SZABO [143], p151) determine trees of $\mathfrak{S}$. Each sequent corresponds to a vertex of S , in such a way that basic sequents determine endv's, such that the conclusion of each inference is linked by edges [7] of S to each premiss, and such that the endformula determines the pv of S. Thus, e.g. if $T_r, T_{r-1}, \ldots, T_1$ are the trees determined by derivations of the upper sequents of the final inference of a derivation, then this derivation determines the tree

$$T = [T_r, T_{r-1}, ..., T_2, T_1] \in \mathfrak{S}.$$

4. The substitution of a function of m variables for a variable of a function of n variables (one of the operations of Ch. I, §3.1) corresponds to the insertion of an $\mathfrak{S}$-tree between a pair [ ] arising from defn. 1.

5. Let $\cdot$ denote concatenation of trees. Thus, e.g. $[Q]\cdot[S,T] = [Q,S,T]$. Then $\mathfrak{S}$ may alternatively be characterized as the smallest class closed under

   (a) $[\,] \in \mathfrak{S}$
   (b) $S, T \in$ S $\Rightarrow S \cdot T \in \mathfrak{S}$
   (c) $T \in \mathfrak{S} \Rightarrow [T] \in \mathfrak{S}$.

   These bracket patterns (and their rules of production) are in correspondence with the words on $\{1, a, b\}$ given by SKOLEM [141], p58.

6. Allowing a small change in notation: suppose $v \in T = [T_r, \ldots, T_1, T_0] \in \mathfrak{S}$. Then either $v = v^\sharp$, or $(\exists i = i_1)_{\geq 0}\ v \in T_i$. If v is not the pv of T_i, the analysis may be continued, to give a sequence $i_1, i_2, \ldots, i_k$ of (non-negative) integers, finite since eh$T < \infty$. On the other hand, since every $\xi \in \mathbf{XP}$ has a unique expression in the form $\xi(x) = (\ldots(a^+x + b)x + \ldots + c)x + d$, $v \in T$ may be indexed by $\mathbf{0}$ if $v = v^\sharp$; otherwise with the natural polynomial ξ for which the coefficients $a = i_1, b = i_2, \ldots, c = i_{k-1}, d = i_k$. Thus in particular, $T = n$ has indices $\xi_v = \mathbf{i}$, $i = n, n-1, \ldots, 1$, for the $v \neq v^\sharp$.

There follow a number of definitions applying to oriented plane-structure trees, it being convenient to group them together, although they are intended for use in different areas such as reductions, partitions and fundamental sequences.

[7] These edges being determined by a *line of inference*.

Notation

When the meaning is clear, numerals $0, 1, 2, \ldots, n, \ldots$ are used to denote, respectively, the trees [ ]; [[ ]]; ...; [[ ], ..., [ ]]; ... of $\mathfrak{S}$. The notation may be combined with the bracket notation. E.g.

$$[2, [0, 1, 0, 2], 0, 0, [0, 1], 0] =$$ 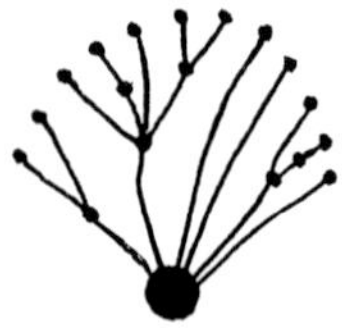 .

The *edge-height* function eh$T$ of $T \in \mathfrak{S}$ is given by

DEFINITION 4.2 *1.* eh[ ] = 0

2.

$$\text{For } T = [T_r, \ldots, T_2, T_1], \ \ \text{eh}T = 1 + \max_{i \leq r} \text{eh}T_i.$$

□

Examples

1. The trees of ex. 1 following defn. 1 have eh$S = 1$, eh$T = 2$, eh$U = 3$.

2. Let T be of eh $\leq h = 1, 2, \ldots$ in S , where va$(v) \leq r$ in T. Then the number of such T is finite, $\leq 2^{r^h}$.

DEFINITION 4.3 *The* principal vertex *or* pv *of $T \in \mathfrak{S}$ is denoted by $v^\sharp$.*

□

In the special case $| T |= 1$, the defn. reduces to defn. 1·1 (b).

Notes

1. It is the outermost pair of brackets, in the representation of T, that is uniquely associated with $v^\sharp$.

2. Valency of $v^\sharp$ = number of principal branches of T.

3. Suppose $T = [T_r, \ldots, T_2, T_1] \in$ S , $T \neq v^\sharp$. Let the pv's of T_i, $i = r, r-1, \ldots, 1$ be denoted by $v_i^\sharp$. Then Adj$(v^\sharp, v_i^\sharp)$, and $\neg$Adj$(v_i^\sharp, v_j^\sharp)$, $i, j \leq r$.

DEFINITION 4.4 *1. A vertex of T of* maximal *edge-distance from $v^\sharp$ is called an* endvertex *or* endv *of T.*

2. Any vertex of T not an endv is interior.

□

Thus, in the inductive definition of a tree, §2·1, when T is formed from S, an endv is added. The addition may or may not eclipse an $S-$endv. Note that only if $\mathrm{va}(v^\sharp) = 0$ is $v^\sharp$ reckoned an endv. IV 4·1 thus differs slightly from the inductive defn. of endv, viz. in the case $\mathrm{va}(v^\sharp) = 1$. Therefore by IV defn. 1, every endv of an $\mathfrak{S}$-tree occurs as a pair [ ].

Now let $T \in \mathfrak{S}$. The successors of $v \in T$ are defined inductively.

DEFINITION 4.5 *1. (a) For $v \in T$, if $\mathrm{Adj}(v^\sharp, v)$, then v is a successor of $v^\sharp$, $v \;\mathrm{Sc}\; v^\sharp$.*

(b) For $u, v \in T$, suppose $v \;\mathrm{Sc}\; u$ has been defined. Then for all $w \in T$, if $\mathrm{Adj}(v, w)$ and $w \neq u$, then w is a successor of v, $w \;\mathrm{Sc}\; v$.

2. The transitive closure of Sc *is denoted by* $\widetilde{\mathrm{Sc}}$.

□

Thus Adj is the 'symmetrization' of Sc: $\mathrm{Adj}(u, v)$ iff $u \;\mathrm{Sc}\; v$ or $v \;\mathrm{Sc}\; u$. If follows immediately from the definition of Adj that the *predecessor* of $v \neq v^\sharp$, viz. the vertex u for which $v \;\mathrm{Sc}\; u$, is *unique*. Note that for $v \in T$, $v = v^\sharp$, or $v \;\widetilde{\mathrm{Sc}}\; v^\sharp$; and for all $u, v \in T$, $v \;\mathrm{Sc}\; u \Rightarrow \neg u \;\mathrm{Sc}\; v$.

Now let $T = [T_r, \ldots, T_2, T_1]$, $\in \mathfrak{S}$, $T \neq v^\sharp$.

LEMMA 4.6 *For $i, j \leq r$, $i \neq j$,*

$$u \in T_i,\ v \in T_j, \Rightarrow \neg u \;\widetilde{\mathrm{Sc}}\; v.$$

Proof. By the inductive definition of T, a unique vertex-edge path $\ldots x\ e\ y \ldots$ links u to v. In order for $u \;\widetilde{\mathrm{Sc}}\; v$, $x \;\mathrm{Sc}\; y$ is necessary at every step. Since $u \in T_i$, $v \in T_j$, in the notation of note 3 after defn. 3, the path passes through $v_i^\sharp$ and $v_j^\sharp$. Well $v_i^\sharp \;\mathrm{Sc}\; v^\sharp$, but $\neg v^\sharp \mathrm{Sc} v_j^\sharp$. Therefore $\neg u \;\widetilde{\mathrm{Sc}}\; v$.

□

Now let T be as above, and let $u, v, w \in T$, $u \neq v$. Then 'no birfurcations appear, looking down towards the pv':

THEOREM 4.7 $(w \;\widetilde{\mathrm{Sc}}\; u \;\&\; w \;\widetilde{\mathrm{Sc}}\; v) \Rightarrow (u \;\widetilde{\mathrm{Sc}}\; v \vee v \;\widetilde{\mathrm{Sc}}\; u)$.

Proof. Suppose the number $i = i(1)$ is such that

$$w \in T_i = T^{(1)} = [T^{(1)}_{r(1)}, \ldots, T^{(1)}_2, T^{(1)}_1].$$

Similarly, suppose the number $i(2)$ is such that

$$w \in T^{(1)}_{i(2)} = T^{(2)} = [T^{(2)}_{r(2)}, \ldots, T^{(2)}_2, T^{(2)}_1].$$

In this way a sequence $T = T^{(0)}, T^{(1)}, T^{(2)}, \ldots$ of trees in $\mathfrak{S}$ is obtained, for which the respective pv's $v^{\sharp} = v^{(0)}, v^{(1)}, v^{(2)}, \ldots$ satisfy

$$\ldots \text{ Sc } v^{(2)} \text{ Sc } v^{(1)} \text{ Sc } v^{(0)}.$$

Since T contains no cycles, the sequence is finite, so $(\exists N)\ w = v^{(N)}$. And by lemma 6,

$$(\exists m, n < N)\ u = v^{(m)} \ \&\ v = v^{(n)}.$$

Then $v \ \widetilde{\text{Sc}}\ v \vee u \ \widetilde{\text{Sc}}\ v$, as $m < n \vee n < m$.

□

Remark. The notion of edge-height of defn. 2 may be extended (as assumed in IV·4·1) to vertices of $T \in \mathfrak{S}$. Thus, for w as above, eh $w = N$; for ξ the index of $v \in T$ (defn. 1, ex. 6), then eh$v$ = 1+degξ.

Now let $T \in \mathfrak{S}$.

DEFINITION 4.8 *1. $S \in \mathfrak{S}$ is a* subtree *of T, $S \subseteq T$, if*

(a) for each vertex $v \in S$, $v \in T$

(b) for all vertices u and v in S, $v \ \widetilde{\text{Sc}}\ u$ in $T \Rightarrow v \ \widetilde{\text{Sc}}\ u$ in S.

2. $S \subseteq T$ is a pv-subtree, *$S \leq T$, if* pv *of S* =pv *of T.*

□

Note. $S \subset T \Rightarrow |S| < |T|$.

LEMMA 4.9 $S \subseteq T \Rightarrow$ eh$S \leq$ eh$T$ *in* $\mathfrak{S}$.

Proof. $u^{\sharp}$ is the pv of S.

□

Notes

1. If
$$[S_q, \ldots, S_2, S_1] = S \leq T = [T_r, \ldots, T_2, T_1],$$
then S contains $q \leq r$ of the pv's of the T_i. There are therefore principal branches $T_{i(q)}, T_{i(q-1)}, \ldots, T_{i(2)}, T_{i(1)}$ of T such that $S_j \leq T_{i(j)}$, $j = q, q-1, \ldots, 1$.

2. Associated with $\mathfrak{S}$ is a certain arithmetic function $\gamma(T)$ = the number of distinct subtrees of T. Thus, $\gamma < 2^{r^h}$ (ref. 2, ex. 2).

Suppose now $v \ \widetilde{\text{Sc}}\ u$ in T.

DEFINITION 4.10 *1. A subtree is a* chain $C = C(u,v)$, *if it is the unique vertex-edge path in* T *from* u *to* v.

2. The chain C *is an* r-chain *if* C *contains* r *edges; is* maximal *or* complete *if it is a proper subchain of no chain of* T.

□

By lemma 9, a chain satisfies eh$C$ $\leq$eh$T$.[8]

Examples

1. The maximal chains of T are of the form $C(v^\sharp, v)$, where v is an endv.

2. $\max_v$eh$C(v^\sharp, v)$ =eh$T$; eh v =eh$C(v^\sharp, v)$.

LEMMA 4.11 *The endv's of* $T \in \mathfrak{S}$ *are linearly ordered.*

Proof. Let u, v be endv's of T. Consider $C(v^\sharp, u)$ and $C(v^\sharp, v)$. If $u \neq v$, these cannot be the same. That is, $(\exists w)$ w is the last common vertex. Thus u is counted before v iff the path to v turns to the right of the path to u, at w. The (left-to-right) order is linear; in particular the transitivity follows from thm. 7.

□

Thus, taking $C = C(v^\sharp, v)$ as above, C dissects T into 2 subtrees, for which C is a common chain: those whose endv's respectively lie not after, and not before v in the linear order. In the extreme cases:

DEFINITION 4.12 C_L, C_R *are respectively the leftmost and rightmost complete chains of* T.

□

Now let u be any vertex of T.

DEFINITION 4.13 *1. The* branch of T at u, Branch(u, T), *is the subtree* B *of* T *such that*

$$v \in B \Leftrightarrow v = u \vee v \ \widetilde{\mathrm{Sc}}\ u \quad \text{in} \quad T.$$

2. The edge-height *of* B *in* T *is* eh u.

□

[8] Given $h \geq 0$, an *anti-chain* of $T \in \mathfrak{S}$ may be taken as $\{v \in V : \text{eh}v = h\}$.

Thus, the pv of B =Branch(u,T) is u, and $u = v^\sharp \Rightarrow B = T$. On the other hand, (ref. note 2 after defn. 3) for $u \neq v^\sharp$, the valency of u = 1+number of principal branches of Branch(u,T). Note that the branches and vertices of a tree are in 1-1 correspondence.

$V(T)$ may be partitioned as follows. Suppose B =Branch$(u,T) = [B_s,\ldots,B_1]$, and that for $s \geq j > i \geq 1$, $B_j = B_i$. Then the tree S obtained from T by interposing these 2 principal branches of B has: $S = T$. In this sense, vertices such as those of B_i, B_j are *indeterminate*; while those remaining are the *determinate* vertices of T, viz. the pv $v^\sharp$, or the v which determine a unique principal branch of Branch(u,T), where u is the predecessor of v; and such that every vertex of $C(v^\sharp, u)$ is determinate. Thus, e.g. an endv may or may not be determinate.

Suppose now that $T \in \mathfrak{S}$ is represented in the plane, by means of line edges and point vertices in the usual manner, viz. with

1. edges of equal height of equal length

2. $(\forall v \in T)$ the successors of v drawn *above* v.

Now suppose u, w Sc v in T (note that ehu, w = 1+eh$v$).

DEFINITION 4.14 [9] *w L u iff w stands to the* left *of u in the above representation.*

□

Examples

1. Returning to the tree U of ex. 1, following defn. 1, and writing w = pv of S, u = pv of [[ ]], w L u is true.

2. Suppose $v^\sharp \neq u \in T \in \mathfrak{S}$, and that the successors v_i of u satisfy $v = v_r$ L ... L v_1 L v_0. Then the index ξ of defn. 1, ex. 6 satisfies $\xi_v = \omega \cdot \xi_u + \mathbf{r}$.

The relation is transitive: wLv & vL$u \Rightarrow w$Lu in T. The weakened relation $u\tilde{\mathrm{L}}v$ may be defined by

$$u \ \tilde{\mathrm{L}} \ v \ \text{ iff } \ u \ \mathrm{L} \ v, \ \text{ or eh}u = \text{eh}v \ \& \ (\exists x, y)$$

$$u \ \widetilde{\mathrm{Sc}} \ x \ \& \ v \ \widetilde{\mathrm{Sc}} \ y \ \& \ x \ \mathrm{L} \ y;$$

and may be obtained as: u, v belong to the same anti-chain, and u follows v in the ordering determined by that of lemma 11. Thus, if ξ_u, ξ_v are the respective indices of $u, v \in T$ of 1, ex. 6, then $\xi_v \prec \xi_u$ iff eh$u$ > eh$v$, or $u \ \tilde{\mathrm{L}} \ v$.

[9] Mc BETH [106] (1990), 'Partitions and normal trees', §2, defn. 2, *Discrete Mathematics* **79**, 95-101.

4.2.3 KÖNIG's Lemma

In 1936, Dénes KÖNIG published [10] a result equivalent to: if a pv-plane-structure tree T is finitely branching (i.e. the valency of each vertex is finite), and if every chain of T is finite, then $T \in \mathfrak{S}$ (viz. T is finite). No constructive proof of this result has been found.

Notation

If C is a chain and v a vertex of T, then $v(C)$ denotes that v is a vertex of C ('C passes through v').

THEOREM 4.15 *Suppose T is as in the above premiss. Then $T \in \mathfrak{S}$.*

Proof. If T has only fm C's, then T is finite; but in any case since each C is finite, starting from C_L, the ordering of lemma 11 may be applied to obtain a sequence $\mathcal{S}$:

$$C_L = C_1, C_2, C_3, \ldots .$$

of C's of T. At this point of the proof, $\mathcal{S}$ may be finite or infinite; and if infinite, it may or may not contain each C of T.

Now consider the fm successors $(r), (r-1), \ldots, (1)$, say, of the pv $v^\sharp$ of T. Each chain passes through a unique successor of $v^\sharp$, and such that for all i, j,

$$i < j \;\&\; (x)(C_i) \;\&\; (y)(C_j) \Rightarrow y \leq x.$$

There is therefore a fixed x_1, $r \geq x_1 \geq 1$, such that for some $N \geq 1$,

$$(\forall n \geq N)\,(x_1)(C_n).$$

If (x_1) is an endv, then $x_1 = 1$, and T is finite. Otherwise, by replacing $v^\sharp$ by the vertex (x_1) and repeating the above argument, the corresponding successor (x_1, x_2) of (x_1) may be found. The vertex sequence

$$(x_1), (x_1, x_2), \ldots$$

so obtained defines a chain C, which is *finite*, by the defining properties of T. Therefore only fm chains lie to the left of the rightmost chain C of T. Therefore (again $x_1 = 1$ and) T is finite.

□

I am indebted to Professor Horst LUCKHARDT for pointing out to me that the principle on which the above rests, that a $\leq$-sequence of natural numbers is convergent, is non-Intuitionistic.

[10] *Theorie der endlichen und unendlichen Graphen* [79], Akad. Verl. Leipzig. Reprinted, 1950, Chelsea, New York.

4.2.4 Specialized Subclasses of $\mathfrak{S}$

The so-called rooted binary trees, on assignment of plane structure, yield the *dyadic trees.* [11]

DEFINITION 4.16 *A tree T is* dyadic *if*

1. $T \in \mathfrak{S}$
2. $v^{\sharp}$ *is an isolated vertex, or* $v^{\sharp}$ *has valency 2.*
3. *The interior vertices (ref. defn. 4·2)* $\neq v^{\sharp}$ *of T have valency 3.*

The class of dyadic trees is denoted by $\mathfrak{D}$.

□

A dyadic T is built up from $v^{\sharp}$ by the combinatorial unit

$$\{[v^{\sharp}, v^{\sharp}] - v^{\sharp}\}.$$

Therefore

THEOREM 4.17 *For $T \in \mathfrak{D}$, T has $2n$ edges, and $2n+1$ vertices, $n+1$ of which are endv's, for some $n \geq 0$.*

Remarks

1. The valencies in a dyadic T are either 0, 1 (endv's), 2 or 3.
2. The notations for $T \in \mathfrak{D}$ of defn. 1 give balanced patterns of brackets built up from $v^{\sharp}$ and the unit [,].
3. An association of n factors by brackets corresponds to a dyadic tree of n endv's. Therefore by thm. 17, the number of such associations is the number of T in $\mathfrak{D}$ of $2n-1$ vertices.
4. de BRUIJN, KNUTH and RICE [116] proved in 1972 [in *Graph Theory and Computing*] that the mean eh of $T \in \mathfrak{S}$, where T has n vertices, is
$$\sqrt{\pi n} - \frac{3}{2} + O(\frac{\ln n}{\sqrt{n}}).$$
5. As pointed out before thm. 15, a most remarkable unsolved problem of logic arises in connection with the proof of KÖNIG's lemma. The problem appears in its sharpest form with the class $\mathfrak{D}$. Consider the pv-oriented trees of $\mathfrak{D}$-valency, of *finite* chain length. Is this class $\mathfrak{D}$? No constructive proof has been found. The assumption that such a T is infinite leads immediately to an endless chain, an abrupt proof by contradiction.

[11] See Intn. to SMULLYAN, *First-Order Logic* (1960).

6. It is well-known that the number of T in $\mathfrak{D}$ with $2n$ edges is

$$\mathfrak{c}_{n+1} = N/(n+1),\ N = \binom{2n}{n}$$

viz. the $(n+1)^{th}$ CATALAN number. [12] Therefore by rem. 3 above, n factors may be associated in $\mathfrak{c}_n$ different ways.

4.2.5 Post Order

The vertices of a tree of $\mathfrak{S}$ may be counted in post (i.e pv-*last*) order. E.g. $T = [u, v]$ gives and $v, u, v^\sharp$ in right-to-left post order. There follows a formal inductive definition of the right-to-left post order relation $<_{\text{post}}$ on V^2, where V is the vertex set of any $T \in \mathfrak{S}$.

DEFINITION 4.18 *1. If T is a vertex $v^\sharp$, $v^\sharp$ is both the first and the last in the post order.*

2. *Suppose $<_{\text{post}}$ has been defined on the vertices of all $T \in \mathfrak{S}$ of edge-height $h \geq 0$. Suppose $T = [T_r, \ldots, T_2, T_1] \in \mathfrak{S}$ is of edge-height $h+1$. Then for u, v in T, $u <_{\text{post}} v$ in T iff*

   (a) $(\exists i, j \leq r)\ i < j\ \&\ u \in T_i\ \&\ v \in T_j$, or

   (b) $(\exists i \leq r)\ u, v \in T_i\ \&\ u <_{\text{post}} v$ in T_i, or

   (c) $u \neq v\ \&\ v = v^\sharp$.

□

Remarks

1. $<_{\text{post}}$ is *linear:* each v is counted exactly once. Via defn. 12, the first v to be counted is the endv of C_R.

2. The R-to-L post order is taken, so that in the case $T \in \mathfrak{T}$, the convention of the CANTOR normal form is adopted. In general, very many 'intermediate' post orders for T, based on permutations wrt the relation of defn. 14 may be defined, before reaching the L-to-R form.

3. Pre- and post-orders are commonly encountered in operations with trees. See KNUTH [77], paras 2.3.1,2. See also McBETH [97], where the calculations considered are executed in pre-order.

---

[12] The subscript of this number is usually taken to be less by 1; for further background see e.g. F. HARARY, 'Combinatorial problems in graphical enumeration' in *Applied Combinatorial Mathematics* , Wiley, New York (1964); and TAMARI [150] (1970) and SINGMASTER [139] (1978). For values, see appendix C.

Conservation of Post Order

By defn. 18·2(b), if $T \in \mathfrak{S}$ and $u \in T$, then for all $w, v \in$Branch$(u, T) = B$, $v <_{\text{post}} w$ in B iff $v <_{\text{post}} w$ in T. More generally, suppose

$$S = [S_q, S_{q-1}, \ldots, S_1] \subseteq T = [T_r, T_{r-1}, \ldots, T_1]$$

in S .

THEOREM 4.18a $(\forall u, v \in S)\ u <_{\text{post}} v$ in $S \Leftrightarrow u <_{\text{post}} v$ in T.

Proof, by induction on h =eh$S$, considering the various cases of definition 18. For $h = 0$, $S = v^\sharp$, and there is nothing to prove. Now let $h \geq 0$, suppose the result has been obtained for all S with eh$S = h$, and now consider S for which eh$S = h+1$, and let $u, v \in S$, with $u <_{\text{post}} v$ in S.

1. Given $(\exists i < j \leq q)\ u \in S_i$ and $v \in S_j$, by note (2) after lemma 9, there are integers $k(j) > k(i)$ such that $S_i \subset T_{k(i)}$, $S_j \subset T_{k(j)}$. So $u \in T_{k(i)}, v \in T_{k(j)}$ and $u <_{\text{post}} v$ in T by 18·2(a).
2. Given $(\exists i \leq q)\ u, v \in S_i$ & $u <_{\text{post}} v$ in S_i, then by the note cited in (a) $(\exists k \leq r)\ S_i \subseteq T_k$, so by induction hypothesis, $u <_{\text{post}} v$ in T_k. Thus by defn. 18·2(b) $u <_{\text{post}} v$ in T.
3. Given $u \neq v, v = v^\sharp$. Consider B =Branch(v, T). Now $u \in B$, so by defn. 18·2(c), $u <_{\text{post}} v$ in B. So by part (b) of the definition, $u <_{\text{post}} v$ in T.

Therefore in all cases, $u <_{\text{post}} v$ in T, and the theorem is proved by induction.

□

Pre Order

Suppose a single change is made to defn. 18·2(c): $u \neq v$ & $u = v^\sharp$. Then the order defined is pre (i.e pv first) order: $u <_{\text{post}} v$ in T becomes $u <_{\text{pre}} v$ in T. Pre-order is obtained (with repetitions) by passing round the 'outsides' of the v's and e's of T. This may be taken from left, or right, as with $<_{\text{post}}$. The vertices of T therefore possess a family of 4 duality related orders. In particular, an instance of thm. 18a holds in each case.

Remark. The reverse of L-to-R post- is R-to-L pre-order, and vv.

(Left-to-right) pre-order determines the endv ordering of lemma 11; and decides (somewhat indirectly) the bracket pattern of $T \in \mathfrak{S}$ of §2.2: a left bracket '[' is added the first time an internal vertex is passed; a ']' the second [13] time; a pair '[ ]' is added at each pass of an endv. This is described most clearly in de BRUIJN and MORSELT [29] (1964), §2.

[13] Note that for u L v; u, v Sc w in T; a pass from u to v is not to be counted as a pass of w.

ŁUKASZIEWICZ Notation

An important application of pre-order occurs in the ŁUKASZIEWICZ 'operator-first' (or 'forward Polish' notation), in which the bracketing of note 4 after thm. 17 associates operands with endv's, and (binary) operations with interior vertices of $T \in \mathfrak{D}$. (The dual 'reverse Polish' notation determines post order).

Analysis of $<_{\text{post}}$ into L and Sc

Suppose $u <_{\text{post}} v$ in $S \in \mathfrak{S}$. Then evidently u Sc v, u L v, or u, v are related by some finite combination of these relations as in the figure; and vv.

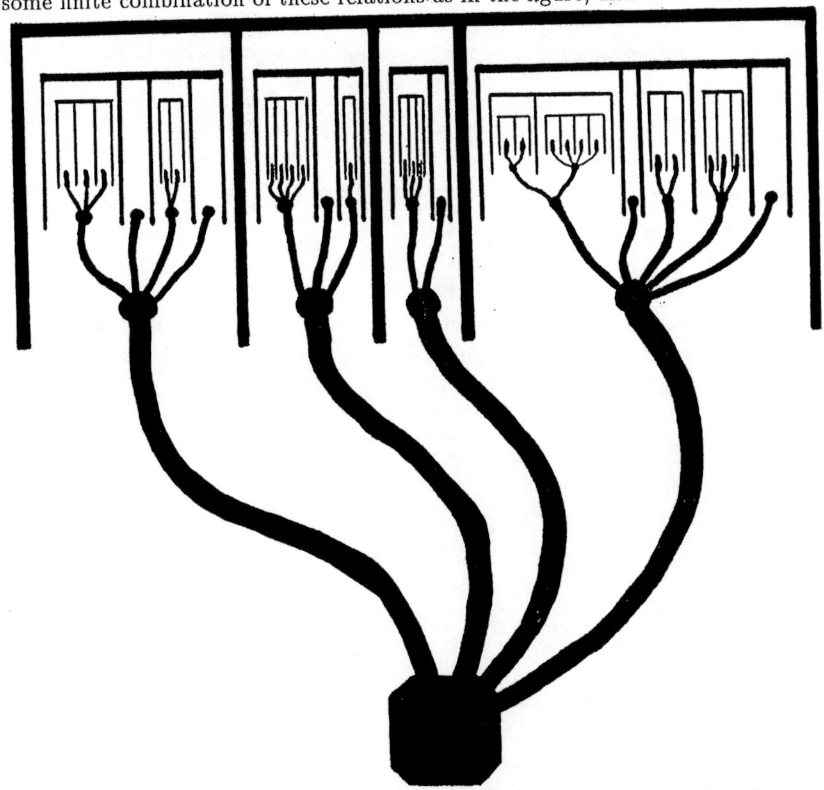

Thus, (assuming the convention of §2·1) provided S is drawn so that the successors of v in the tree are placed sufficiently closer, as eh$v$ increases,

$u <_{\text{post}} v$ iff v is 'on a common chain with, and below; otherwise, to the left of' u

in S. Substituting 'above' in this formula gives $u <_{\text{pre}} v$.

4.3 The Class $\mathfrak{T}$ of Normal Trees

Definitions

The ordered proper subclass $\mathfrak{T}$ of the oriented plane structure trees $\mathfrak{S}$ (McBETH, 1979 [14]) is given by

DEFINITION 4.19 *1. (a)* $v^{\sharp} = [\,] \in \mathfrak{T}$.

(b) $(\forall T \in \mathfrak{T})\, T \neq v^{\sharp} \Rightarrow v^{\sharp} \prec T$.

2. *Let $h \geq 0$, and suppose the trees of $\mathfrak{T}$ of edge-height $\leq h$ have been defined, together with their connected ordering $\prec$:*

   (a) Let $r \in \mathrm{N}$, and let $T_r, \ldots, T_2, T_1 \in \mathfrak{T}$ be of edge-height $\leq h$. Then

   $$T = [T_r, \ldots, T_2, T_1] \in \mathfrak{T}$$

   iff $T_r \succeq \ldots T_2 \succeq T_1$.

   (b) Suppose $T = [T_r, \ldots, T_2, T_1]$ and $S = [S_q, \ldots, S_2, S_1]$ are trees of $\mathfrak{T}$ of edge-height $h+1$. Then $S \prec T$ iff

      i. $S_q \prec T_r$, or
      $S_q = T_r$ & $S_{q-1} \prec T_{q-1}$, or
      ⋮ ; or

      ii. $S_{q-i} = T_{r-i}$, for $i = 0, 1, 2, \ldots, q-1$ & $q < r$.
      Since $S \prec T, S = T$ or $T \prec S$, the ordering is connected on trees of $\mathfrak{T}$ of edge-height $h+1$.

This completes the inductive definition of the system $(\mathfrak{T}, \prec)$. In part 2. above, T_r is the leading branch *of T.*

□

Remark

By this definition, the relation $\prec$ is the recursive principal-branch lexicographic order on $\mathfrak{T}$.

Examples

1. Let $S = [[1,0],2,1,1,0]$, $T = [[1,0],2,2,0]$ (using the notation introduced before defn. 2). Then $S, T \in \mathfrak{T}$, with $S \prec T$.

2. $[[2,[\,]]] \succ [[2],[1]]$; suppose $U =$Branch(u,T); $V =$Branch(v,T). Then u L v in $T \in \mathfrak{T} \Rightarrow V \preceq U$. Note: $\mathfrak{T}$ is closed under *principal branch excision.*

[14] [99] 'Normal trees', manuscript, originally entitled 'Number-words, finite trees and exponential polynomials'.

3. The $\mathfrak{T}$-sequence $T_1, T_2 = v^\sharp$; $T_{n+2} = [T_{n+1}, T_n]$ determines the FIBONNACI sequence $1, 1, 2, 3, 5, 8, 13, 21, 34, \ldots$ via # of endv's.

4.

5. Suppose $T \in \mathfrak{S}$ has equal-height chains. Then if the valencies in T are nd wrt R-to-L pre-order (endv's excepted), then $T \in \mathfrak{T}$.

6. Following the remark after defn. 1 of the class $\mathfrak{S}$, the sequence $\mathfrak{n}_1, \mathfrak{n}_2, \mathfrak{n}_3, \ldots$ of numbers of $T \in \mathfrak{T}$ of respectively $1, 2, 3, \ldots$ vertices may be found to be

$$1, 1, 2, 4, 9, 20, \ldots.$$

Edge-Height in $\mathfrak{T}$

THEOREM 4.20 *Suppose* eh$S <$ eh$T$ *in* $\mathfrak{T}$. *Then* $S \prec T$.

Proof. Set

$$X_1 = [[\,]]; \text{ and for } h > 1,\ X_h = [X_{h-1}].$$

First note that 19·1,2(b) $X_1 \prec X_2 \prec X_3 \prec \ldots$. Now $v^{\natural} \prec X_1$ by defn. 19·1(b). Let $h \geq 1$, and suppose that eh$T < h \Rightarrow T \prec X_h$ in $\mathfrak{T}$. Then if $T = [T_r, T_{r-1}, \ldots, T_1] \in \mathfrak{T}$ is of edge-height h, by hypothesis $T_r \prec X_h$, since eh $T_r + 1 \leq$eh$T = h$. Therefore $T \prec X_{h+1} = [X_h]$, by defn. 19·2(b). Thus by induction, every T of edge-height h in $\mathfrak{T}$ satisfies $T \prec X_{h+1}$. On the other hand, again by induction on 19·2(b), given eh$T = h$, $X_h \preceq T$. Therefore

$$\text{eh}S < h \Rightarrow S \prec X_h \preceq T.$$

□

Therefore, for $T = [T_r, \ldots, T_1]$ in $\mathfrak{T}$, by thm. 20 and by defn. 19, it follows

$$\text{eh}T_i \leq \text{eh}T_{i+1},\ i = 1, 2, \ldots, r-1.$$

Viz. the leading branch T_r is of maximal height, and therefore for any $T \in \mathfrak{T}$, and any chain C of T, the leftmost chain C_L satisfies

$$\text{eh}C \leq \text{eh}C_L.$$

Since every $S \in \mathfrak{S}$ has at least one chain C for which eh$C$ =eh$S$,

$$\text{eh}C_L = \text{eh}T.$$

From these observations follow

$$\text{eh}T = 1 + \text{eh}T_r,$$

and generally

COROLLARY 4.21 $T_i \prec T$, $i = r, r-1, \ldots, 1$.

□

Note that, conversely to the theorem $S \prec T \Rightarrow$ eh$S \leq$eh$T$ in $\mathfrak{T}$. Note also that $S \subseteq T \Rightarrow S \preceq T$ in $\mathfrak{T}$.

Remark

Strictly, defn. 19 defines $T \in \mathfrak{T}$ together with one of the 2 halves into which the embedding plane Π of T divides space. This *chirality* is not part of the structure of T, only a convention as to which side of Π the observer lies. The class $\mathfrak{T}'$ of trees of opposite chirality may be obtained by taking all reflections of $T \in \mathfrak{T}$ in a vertical line (an operation under which $\mathfrak{S}$ is closed), say to the left of T. Thus $\mathfrak{T}'$ is obtained if instead the non-decreasing left-to-right convention is adopted in defn. 19. $\mathfrak{T} \cap \mathfrak{T}'$ is not empty; some of the trees $T \in \mathfrak{S}$ for which every complete chain of T is of pre-assigned length, are common to the 2 classes.

4.3.1 Dominating Branch

Suppose $S = v^\sharp$, or

$$S = [S_q, \ldots, S_1] \prec [T_r, \ldots, T_1] = T$$

in $\mathfrak{T}$. Then as in 2(b) of defn. 19, either (i) T has an 'excess (or right extension) of branches' over S :

$$S_q = T_r,\ S_{q-1} = T_{r-1}, \ldots, S_1 = T_{r-q+1} \text{ and } q < r,$$

or (ii) T posesses a dominant branch:

$$(\exists i < q)_{\geq 0}\ S_q = T_r,\ S_{q-1} = T_{r-1}, \ldots, S_{q-i+1} = T_{r-i+1},\ S_{q-i} \prec T_{r-i}.$$

In case (ii) the analysis may be continued (cf. proof of thm. 7). Writing

$$i = i_1, \text{ and } S_{q-i_1} = S^{(i_1)},\ T_{r-i_1} = T^{(i_1)} = [T^{(i_1)}_{r_1}, \ldots, T^{(i_1)}_1],$$

$$(\exists h)\ S^{(i_1)} \prec T^{(i_1)},\ S^{(i_1,i_2)} \prec T^{(i_1,i_2)}, \ldots, S^{(\mathrm{i})} \prec T^{(\mathrm{i})},\ (\mathrm{i}) = (i_1, i_2, \ldots, i_h)$$

and $S^{(\mathrm{i})} = v^\sharp$, the isolated vertex, or

$$S^{(\mathrm{i})}_{q_h} = T^{(\mathrm{i})}_{r_h}, \ldots, S^{(\mathrm{i})}_1 = T^{(\mathrm{i})}_{r_h-q_h+1},$$

and $q_h < r_h$, for some natural number h.

DEFINITION 4.22 *The tree*

$$B = \begin{cases} T & \text{if } S = v^\sharp, \text{ or (i) above; or} \\ T^{(\mathrm{i})} = [T^{(\mathrm{i})}_{r_h}, \ldots, T^{(\mathrm{i})}_1] & \text{otherwise} \end{cases}$$

is the dominating branch of T for S.

□

Note that if $B \neq T$, then (a) h is the height of B in T, and (b) if C_L passes through B, then B is of the form $B = T^{(0,0,\ldots,0)}$.

Examples

1. $S = [[[3]],[[2,0],[1,0,0]],0,0] \prec [[[3]],[[2,0],[1,0,0,0]],0] = T$, where $h = 2$, and the dominating branch $B = [1,0,0,0]$ of T for S is $B = T^{(1,1)} = [T_4^{(1,1)},\ldots,T_1^{(1,1)}]$.
2. $S \subset T$; or $S^{(\mathrm{i})} \subset T^{(\mathrm{i})}$. If S or $S^{(\mathrm{i})} \neq v^\sharp$, then $S^{(\mathrm{i})}, T^{(\mathrm{i})}$ have the same leading branch.

Separating Chain and Departure Vertex

Let $S \prec T \in \mathfrak{T}$, and let B be the dominating branch of T for S.

DEFINITION 4.23 *1. The* departure vertex *of T for S is the pv $v^\star$ of B.*

2. Let B_S =Branch($v^\star, S$). Then the separating chain $C(S,T)$ *of T for S is the unique chain obtained by taking first the path from $v^\sharp$ to $v^\star$, then (via defn. 12) taking the chain*

$$\begin{cases} \text{(i) } C_L \text{ of } B & \textit{if } S \textit{ or } S^{(\mathrm{i})} = v^\sharp \textit{ before defn. 22;} \\ \text{(ii) the IV} \cdot 11 \text{ successor in } T \text{ of the } C_R \text{ of } B_S & \textit{otherwise.} \end{cases}$$

□

Note: C lies in T, since $B_S \subset B$.

There are thus 2 kinds of explicit $\prec$-domination in $\mathfrak{T}$: case (i) viz. domination of a vertex by a proper tree; case (ii), domination of a proper tree via an excess of branches.

Example.

$S = [[[1]],[3,2],4] \prec T = [[[1]],[3,3],2]$, gives a separating chain of edge-height 3.

Union of Chains

For any $T \in \mathfrak{S}$, given complete chains $C_1, C_2, \ldots, C_h$, the union

$$\bigcup_{i=1}^{h} C_i,$$

whose vertices and incidences are those of T, is itself a tree of $\mathfrak{S}$. Now let $C_1 = C_L, C_2, \ldots, C_R$ be the (complete) chains of T in pre-order. Let $C = C_r$, for some r in the range $2, 3, \ldots, R$.

DEFINITION 4.24 *The tree $T_C = \bigcup_{i=1}^{r-1} C_i$ is the* upper part; $T^C = \bigcup_{i=r}^{R} C_i$ the lower part of T cut by C.

Thus, for S not of height $<$ eh$T$ (in which case $C(S,T) = C_L$, and the upper part is void), $S \prec T$ in $\mathfrak{T} \Rightarrow T_{C(S,T)} \subset S, T$; the largest upper part with this property.

THEOREM 4.25 $T \in \mathfrak{T} \Rightarrow T_C \in \mathfrak{T}$.

□

4.3.2 Normal Form Theorem

The following establishes the fundamental relation between the classes $\mathfrak{S}$ and $\mathfrak{T}$:

THEOREM 4.26 [15] $(\forall S \in \mathfrak{S})(\exists! T \in \mathfrak{T})$ *T is the $\prec$ – normal form of S.*

Proof, by induction on edge-height. For zero edge-height, the isolated vertex $v^\sharp$ may be taken as its own normal form. Now suppose the result has been proved for all $S \in \mathfrak{S}$ for which eh$S \leq h$, $0 \leq h$. Let $S = [S_q, \ldots, S_2, S_1] \in \mathfrak{S}$ be of edge-height $h+1$. By hypothesis, $(\forall i \leq q)$ S_i has a unique $T_i \in \mathfrak{T}$. Now consider $S' = [T_q, \ldots, T_2, T_1]$. Since '$\prec$' is connected on $\mathfrak{T} \times \mathfrak{T}$, there exists a permutation σ of degree q such that

$$T_{\sigma(q)} \succeq \ldots \succeq T_{\sigma(2)} \succeq T_{\sigma(1)}. \qquad \text{(I)}$$

Therefore $T_\sigma = [T_{\sigma(q)}, \ldots, T_{\sigma(2)}, T_{\sigma(1)}]$ is a normal tree. Now suppose there exists a permutation ρ of degree $q, \rho \neq \sigma$, for which

$$T_{\rho(q)} \succeq \ldots \succeq T_{\rho(2)} \succeq T_{\rho(1)}. \qquad \text{(II)}$$

Then there exists a unique permutation θ (of degree q) for which $\rho = \theta\sigma$. But

$$(\forall i, j \leq q)\ T_{\sigma(j)} \succ T_{\sigma(i)} \Leftrightarrow T_{\theta\sigma(j)} \succ T_{\theta\sigma(i)},$$

since otherwise the application of θ to σ would not in every case produce the non-increasing sequence (II) from the non-increasing sequence (I). It therefore follows that θ is the product of permutations $\theta, \theta', \theta'', \ldots$ each of which is fixed $(\theta^{(k)}(i) = i)$ except on a set of indices of *equal* trees. So

$$T = T_\sigma = T_{\theta'\sigma} = T_{\theta''\theta'\sigma} = \ldots = T_\rho,$$

and T is the unique normal tree of S. Thus the theorem follows by induction.

□

Examples

1. Taking again ex. 1 after defn. 1, the normal tree of T is $W = [2, 1]$; and that of U is $[W, S, 1]$.

2. Let

$$S = [9, S_{11}, [\,], S_9, S_8, 2, S_6, S_5, S_4, [\,], [\,], S_1] \in \mathrm{S}\,,$$

where

(a) $S_{11} = [7]$,

(b) $S_9 = [3, [[\,], 1, [\,]], [\,], 3, 3]$,

[15] [99], thm. 2.5.

(c) $S_8 = [2,1,2,[1,1]]$,
(d) $S_6 = [[\not{2},[1,\not{2}],2],[[[\,],[\not{2},1]],\not{5}],\not{3},[\not{2},1],[[\,],[1,[\,]]],2,[\not{2},2,2,[[\,],1]],4,[1,2]]$.
(e) $S_5 = [4,[\,],4]$,
(f) $S_4 = [[\,],2,[\not{2},2,1,[\,]],2,\ [\,],\ [\,][\not{5},2],[\,],1,1]$,
(g) $S_1 = [1,3,7]$.

Then the normal form of S is

$$T = [T_{12}, T_{11}, \ldots, T_6, 9, 2, \not{3}] \in \mathfrak{T},$$

where

(a) $T_{12} = [[[[1,\not{2}],[\,]],\not{5}],\ [[1,\not{2}],2,\not{2}],[[1,[\,]],2,2,\not{2}],\ [[1,[\,]],[\,]],\ [2,1],[1,\not{2}],4,2,\not{3}]$,
(b) $T_{11} = [2,1,\not{3}],[2,\not{5}],2,2,1,1,\not{4}]$,
(c) $T_{10} = [[1,1],2,2,1]$; $T_9 = [[1,\not{2}],3,3,3,[\,]]$;
(d) $T_8 = [7,3,1]$; $T_7 = [7]$ $T_6 = [4,4,[\,]]$,

viz. the tree of ex. 3 following defn. 19, where the stroke through a numeral indicates a run of blank pairs, e.g. $[2,\not{2}] = [[[\,][\,]],\ [\,],\ [\,]]$.

3. The commutative bracketings on $3,4,5,6,\ldots$ indistinguishable factors are enumerated by the WEDDERBURN-ETHERINGTON numbers

$$1,2,3,6,11,23,46,98,\ldots\ .$$

(See e.g SLOANE [142], p53)[16]. Consider an arbitrary $S \in \mathfrak{D}$(ref. §2.7). Within every balancing pair $[\cdots]$ of the representation of S, the order of the expressions (if any) may be deranged, so as to give a unique *normal* dyadic tree, T, say. It therefore follows that the W-E numbers enumerate by vertices the class $\mathcal{X} = \mathfrak{D} \cap \mathfrak{T}$. Since the order of factors is of no significance, the commutative bracketings are in 1-1 correspondence with the dyadic trees without plane structure, viz. in 1-1 correspondence with the binary trees. Now consider a tree $T \in \mathcal{X}$. Suppose the endv's of T are indexed, to give $v_1, v_2, \ldots, v_R$ in the ordering of lemma 11. Clearly, any commutation defined by a pair of brackets of T deranges the order of the v_i. Although a great many derangements are possible, for $1 < h$ their number is $\leq 2^{-1+2^{\mathrm{eh}T}}$ and $< R!$ Since they may be inverted by commutations, they form a closed system. As permutations of bracketed factors do not in general equal, under composition, the composition of the corresponding permutations of the factors, they do not necessarily define a subgroup of the symmetric group of degree R. In the special case where T is complete (viz. if the chain C is complete, then eh$C$ =eh$T$), however, induction on eh establishes that the commutations of T do define such a subgroup. [17]

[16] COMTET,L, 1970, *Analyse Combinatoire*, Vols I, II, Presses Universitaires de France, Paris.(ref. Vol I, p58)

[17] See HOFFMANN, 1979, *Group-Theoretic Algorithms and Graph Isomorphisms*, Springer Lecture Notes in Computer Science, # 136.

Note that, given the number h, there is the separate problem of finding the number of $T \in \mathcal{X}$ with eh$T = h$.

Remarks

1. As a special case of the observation of §2.1, any oriented tree X may be embedded in the plane, to give some $S \in \mathfrak{S}$. By thm. 26, S has a unique $T \in \mathfrak{T}$. Now the oriented trees X, X'; $X \neq X'$ give, by the above means, $T, T' \in \mathfrak{T}$. $T \neq T'$, since otherwise e.g. X could be superimposed on X'. Also, another embedding for X may give $S' \neq S$ in $\mathfrak{S}$. But the normal form of S' is T, since otherwise X could not be superimposed upon itself. It therefore follows that **the pv-oriented trees without plane structure are in 1-1 correspondence with the normal trees.** As a consequence of this, and in the light of rem. 1 following def. 18, the canonical post order of a pv-oriented tree X may be taken to be that of its unique normal form tree, viz. the *normal post order* of X. Under the CANTOR convention, the R to L order is then 'ascending'.

   Suppose X above, $| X |= n$, is vertex-indexed, via the enumeration $v_1, v_2, \ldots, v_n$. Then a plane-embedding of X to give S, and the normalization of S to give T, results in an *indexed normal tree*. Now, associated with each $v \in T$, $v \neq$ endv, is a set of permutations θ, as in the proof of thm. 26. Let $B =$Branch$(v, T) = [B_s, \ldots, B_1]$. Then there is exactly one permutation θ of the principal branches $B_s, \ldots, B_1$ such that the indices of their pv's are R-to-L si with respect to the given vertex enumeration. The canonical post-order of the indexed pv-oriented tree X may now be taken to be that of the indexed $\mathfrak{T}$-tree T, all of whose branches have been permuted according to the above prescription.

2. Suppose, in thm. 26, that the principal branches of S' fall into $p \leq q$ classes, each containing n_i, $i = 1, 2, \ldots, p$ identical trees. Then the number of different trees, of $\mathfrak{S}$, obtainable by deranging the order of the S'−branches, is $N = q!/(n_1!n_2! \ldots n_p!)$. Therefore the number of permutations σ, ρ (of degree q) which may be used to obtain T is $q!/N$.

Exercise. Find the normal tree of the solar system, given the interpretation u Sc v iff u orbits v.

4.3.3 Normal Reduction

This section, based on Mc BETH [99] (1987), considers the reduction of a normal tree by the removal of endv's. Thus, suppose S is obtained from $T = [T_r, \ldots, T_2, T_1] \in \mathfrak{T}$, $T \neq v^\sharp$, by the removal of an endv v (and its corresponding edge).

LEMMA 4.27 $S \in \mathfrak{T} \Rightarrow S \prec T$.

Proof. Let $u \in T$ be the unique vertex for which v Sc u. Let B =Branch(u,T). The proof is by induction on the edge-length l of $C(v^\sharp, u)$. In the initial case, $l = 0, u = v^\sharp$, i.e. $T = B$ and [18] $T_1 = v$, so

$$S = [T_r, \ldots, T_2] \prec [T_r, \ldots, T_2, T_1] = T.$$

Now suppose the result has been proved for all trees for which $l \leq k,\ 0 \leq k$, and consider the case for which $l = k+1$. Suppose v is removed from T_i, for some $i \leq r$, to give $S_i \in \mathfrak{T}$. Then by i.h., $S_i \prec T_i$. Now given $S \in \mathfrak{T}$ (note that this implies $T_{i-1} \preceq S_i$ when $1 < i$),

$$S = [T_r, \ldots, T_{i+1}, S_i, T_{i-1}, \ldots, T_1] \prec [T_r, \ldots, T_i, \ldots, T_1] = T$$

and the lemma is proved by induction.

□

Remark. More generally, if $U \subset T \in \mathfrak{T}$, and U is normal, then $U \prec T$.

Given $T \in \mathfrak{T}$, there is a natural stepwise reduction of T to the isolated vertex $v^\sharp$. That is, the removal, at each step, of a single vertex-edge unit, such that the tree obtained is $\prec$ – maximal. Let C_R be the rightmost (complete) chain of $T \in \mathfrak{T}$, as in definition 12. Let e be the edge of C_R farthest from $v^\sharp$. Let v be the endv of C_R. The combinatorial unit consisting of e and v may be called the *minimal ev* of T.

Suppose $T \neq v^\sharp$ in $\mathfrak{T}$.

DEFINITION 4.28 *T' is the tree obtained from T by removing the minimal ev.*

□

In particular, $[T_r, \ldots, T_2, [\ ]]' = [T_r, \ldots, T_2]$. Now let $T = [T_r, \ldots, T_2, T_1] \in$ T, $T_1 \neq [\ ]$. Then

LEMMA 4.29 $T' = [T_r, \ldots, T_2, T_1']$.

Proof. For the tree T, C_R passes through T_1. Since $T_1 \neq v^\sharp$, the endv of C_R is not the pv of T_1. Therefore the minimal ev of T is the minimal ev of T_1. Therefore $T' = [T_r, \ldots, T_2, T_1']$.

□

Now let T be as in defn. 28.

THEOREM 4.30 $T' \in \mathfrak{T}$.

[18] Strictly speaking, v may be taken to be T_1, in the case where T_i defines an isolated vertex, for $i > 1$.

Proof. Let $T = [T_r, \ldots, T_2, T_1]$ be a normal tree. Suppose v is the endv of C_R. Let u be the predecessor of v : v Sc u. The proof is by induction on edge-length, l, of $C(v^\sharp, u)$. In the initial case, $l = 0$, i.e. $u = v^\sharp$. Then $T' = [T_r, \ldots, T_3, T_2]$. But since $T \in \mathfrak{T}$, $T_r \succeq \ldots \succeq T_3 \succeq T_2$, and so $T' \in \mathfrak{T}$. Now let $h \in \mathrm{N}$, and suppose it has been proved that for all $T \in \mathfrak{T}$ for which $l \leq h$, $T' \in \mathfrak{T}$. Let T be any normal tree for which $l = h + 1$. Well, $T_2 \succeq T_1$, by defn. 19. Let w be the pv of T_1. Then the length of $C(w, u) = h$, so by induction hypothesis, $T_1' \in \mathfrak{T}$. Therefore $T_1' \prec T_1$ by lemma 27. So $T_2 \succ T_1'$, and $S = [T_r, \ldots, T_2, T_1'] \in \mathfrak{T}$ by definition 19. But $S = T'$ by lemma 29, therefore T' is normal. Thus it has been proved by induction that for any $T \neq v^\sharp$ in $\mathfrak{T}$, T' is normal. Thus, by lemma 27,

COROLLARY 4.31 $(\forall T \in \mathfrak{T})\, T \neq v^\sharp \Rightarrow T' \prec T$.

□

Now let $T \in \mathfrak{T}$, and let S be obtained from T by removal of some ev unit other than the minimal ev. S may or may not be a normal tree; however

THEOREM 4.32 (Maximality of T') $S \in \mathfrak{T} \Rightarrow S \preceq T'$.

Proof. Suppose $S \in \mathfrak{T}$, and let v be the minimal vertex of T. Let w be the vertex removed (together with it's edge) from T to give S. Let u be the vertex of greatest edge-distance from $v^\sharp$ such that

$$w\ \widetilde{\mathrm{Sc}}\ u\ \&\ v\ \widetilde{\mathrm{Sc}}\ u.$$ [19]

Let B =Branch$(u, T) = [X_s, \ldots, X_2, X_1]$, $X_i \in \mathfrak{T}$, $i = 1, 2, \ldots, s$. If X_1 is a mere vertex, $w \neq v \Rightarrow w \notin X_1$. And generally, $v \in X_1$, while $w \in X_i$ for some $i = 1, 2, \ldots, s$. Well, $i \neq 1$. For, suppose $v, w \in X_1$. Then

$$v, w\ \widetilde{\mathrm{Sc}}\ (\text{pv of } X_1)\ \mathrm{Sc}\ u,$$

contradicting assumption that u is at greatest edge-distance from $v^\sharp$. So, in all cases, $w \notin X_1$. Let X_i be reduced to X_i^*, in the reduction of T to S. Since S is normal, $X_i^* \in$ T, with [20] $X_i^* \succeq X_{i-1}$. To compare S and T', in the first case compare lexicographically A =Branch(u, S) and B' =Branch(u, T'). If X_i is an isolated vertex, then $A = B'$. Otherwise taking $v^\sharp \prec X_i$,

$$A:\ X_s \succeq \ldots X_{i+1} \succ X_i^* \succeq \ldots X_2 \succeq X_1,$$

$$B':\ X_s \succeq \ldots X_{i+1} \succeq X_i \succeq \ldots X_2 \succ X_1';$$

[19]Note that $u = v^\sharp$ is possible.

[20]Cf. footnote 14: it is possible that X_i is an isolated vertex (viz. which vanishes in the formation of X_i^*), so that either X_{i-1} is the leading branch, or $X_{i+1} \succeq X_{i-1}$ in the normal tree A.

(if $X_1 = v^\sharp$, then X_2 becomes the minimal branch of B'). By lemma 27, $X_i^* \prec X_i$. But $i > 1$, so by defn. 19,

$$A \prec B'.$$

The proof is completed by induction on the edge-length l of $C(v^\sharp, u)$. The case $l = 0$ has just been established. Let $1 \leq k$ and suppose that for all S, T' for which $l \leq k$, the result has been proved. Suppose S, T' are such that $l = k + 1$, where $T = [T_r, \ldots, T_2, T_1]$. Now $v, w \in T_1$, since in any case $v \in T_1$, and $w \in T_i$, $i > 1$, would imply $l = 0$. Well, let S_1 be the tree obtained from T_1, in the reduction of T to S. Then by induction hypothesis $S_1 \preceq T_1'$, so

$$\begin{aligned} T' &= [T_r, \ldots, T_2, T_1'] \text{ by lemma 29,} \\ &\succeq [T_r, \ldots, T_2, S_1] \text{ by definition 19} \\ &= S. \end{aligned}$$

Thus the case l=k+1 is established, and the maximality theorem follows by induction.

□

Normal Sequences

Repeated removal of the minimal ev, starting with $T \in \mathfrak{T}$, leads in a finite number of steps to the initial tree of $\mathfrak{T}$. Thus

DEFINITION 4.33 [21]

1. *$T^{(0)} = T$. Now suppose $0 \leq k$. Then $T^{(k+1)} = S'$, where $S = T^{(k)} \neq v^\sharp$.*
2. *$T^{(0)}, T^{(1)}, T^{(2)}, \ldots$ is the* normal sequence *of T.*

□

THEOREM 4.34 *For each T of $\mathfrak{T}$, the normal sequence is finite.*

Proof. T is finite.

□

Remarks

1. If T has n vertices, $1 \leq n$, then the normal sequence of T has n terms. Defn. 33 includes the case $T = v^\sharp$ (whose normal sequence has just 1 term).
2. By corol. 31, $T^{(0)} \succ T^{(1)} \succ T^{(2)} \succ \ldots$, a descending sequence which has just been found finite.

[21] Mc BETH [99] (1987), §3.

Let $T = [T_r, \ldots, T_2, T_1]$ in $\mathfrak{T}$.

THEOREM 4.35 *The normal sequence of* T *contains each tree*

$$[T_r, \ldots, T_{i+1}, T_i], \qquad \text{for } i = 1, 2, \ldots, r.$$

Proof. For $i = 1, 2, \ldots, r-1$, all the chains of T which pass through T_i, lie to the right of those which pass through T_{i+1}.

□

Edge-Heights in Normal Sequences

The last chain to be reduced, in the normal sequence of T is C_L, of maximal length. Let $S \prec T$ in $\mathfrak{T}$. The respective normal sequences have at least $v^\sharp$ in common; and if the tree U is common, then $U^{(0)}, U^{(1)}, U^{(2)}, \ldots$ is a common remainder of the 2 sequences. Therefore taking U as $\prec$ −maximal,

THEOREM 4.36 $\mathrm{eh}U = \mathrm{eh}S$.

□

Note that $T_{C(S,T)} \subseteq U$ is in some cases strict (ref. defn. 14).

Normal Order is Post Order

Suppose $T \in \mathfrak{T}$ has n vertices. In the normal sequence of T, let v_{i+1} be the vertex removed from $T^{(i)}$ in the formation of $T^{(i+1)}$. By rem. 1 following thm. 34, the sequence has n terms, therefore there are $n-1$ reductions $T^{(i)}$ to $T^{(i+1)}$. Let v_n be the isolated vertex $v^\sharp$.

THEOREM 4.37 [22] *The vertex sequence* $v_1, v_2, \ldots, v_n$ *is identical with that obtained by taking the vertices of* T *in post order.*

Proof, by induction on edge-height. For eh$= 0$, $T = v^\sharp$, and all orders coincide. Let $h \geq 0$, and suppose for induction hypothesis that the result has been proved for the normal sequences of trees of edge-height $\leq h$. Let $T \in \mathfrak{T}$ be of edge-height $h+1$, with $T = [T_r, \ldots, T_2, T_1]$. Let u, v be vertices of T. There are 3 cases to consider, corresponding to those of defn. 18·2.

1. $(\exists i, j \leq r)\ [i < j \ \&\ u \in T_i \ \&\ v \in T_j]$. Then by thm. 35, in the normal sequence, the vertices of T_i are removed before those of T_j. Therefore for case 18·2(a), u is counted before v iff $u <_{\text{post}} v$.

[22] [99], thm. 3·9.

2. $(\exists i \leq r)\ u, v \in T_i$. Let $T_i^{(0)}, T_i^{(1)}, \ldots$ be the normal sequence of T_i. Well, since the C_R of

$$T_{ij} = [T_r, \ldots, T_{i+1}, T_i^{(j)}]$$

passes through $T_i^{(j)}, j = 0, 1, 2, \ldots$, the trees of $T_{i0}, T_{i1}, \ldots$ are consecutive terms of the normal sequence of T. But by induction hypothesis, in that of T_i, the 2 orders coincide. Therefore by defn. 18·2(b), u is counted before v via the normal sequence of T iff $u <_{\text{post}} v$ in T.

3. $u \neq v$ & $v = v^\sharp$. $v^\sharp$ is the last vertex in the normal sequence, i.e. u is counted before v; and by defn. 18·2(c), $v^\sharp$ is the last vertex in the post ordering of T.

So in all cases the orderings are identical, for trees of edge-height $h + 1$, and the theorem follows by induction.

□

Remarks

1. Let $T \in \mathfrak{S}$. Then T posesses a rightmost chain C_R, so T' may be defined as in defn. 28. Since, in the proofs of thms. 35, 37 the normality of T is not required, these results apply to all trees of $\mathfrak{S}$. That is, *normal order for* $T \in \mathfrak{S}$ = *post order for* T.

2. Let v, v' be endv's of $T \in \mathfrak{S}$. Then in the ordering of lemma 11,

$$v < v' \Leftrightarrow v <_{\text{post}} v'.$$

3. The maximal chains of $T \in \mathfrak{T}$ taken in R-to-L post order are not necessarily in nd edge-length order. E.g [[2,0],[1]] has chain-length vector $(3, 3, 2, 3)$.

4. The vectors of 3. above under the dictionary order do not correspond to the ordering '$\prec$' of $\mathfrak{T}$. E.g. $S = [[2], [1]], T = [[2, 0]]$ give respectively $(3, 3, 3)$ and $(3, 3, 2)$, while $S \prec T$.

THE DCC PROPERTY FOR $(\mathfrak{T},\prec)$

4.3.4 Lexicographic Descent in $\mathfrak{T}$ is Finite

This § considers what is probably the most fundamental property of the system $(\mathfrak{T},\prec)$.

DEFINITION 4.38 *1. An ordered system, for which only finite descent is possible, is said to satisfy the* descending chain condition *(DCC)* [23].

2. A linear system, which satisfies the DCC, is said to be sequentially ordered.

□

Use of the stronger assumption of well-order, viz. that every subsequence of the linear system is either void, or has a first term, is prohibited in this volume. This is because, as discussed in §2.3 of Ch. 1, while proofs resting on sequential order escape dependence on the axiom of choice, the same claim cannot be made for those resting on well-order. Thus in the case of $\mathfrak{T}$, in order to show that an arbitrary sequence $\mathfrak{T}_1 \subseteq \mathfrak{T}$ is void, or has a first term, by showing that the contrary implies $\mathfrak{T}_1$ contains an endlessly descending subsequence $\mathfrak{T}_2 = \ldots \prec T_2 \prec T_1$, means are required to pick out the terms $T_1, T_2, \ldots$ of $\mathfrak{T}_2$. However, all the great applications of well-order, e.g. GENTZEN's consistency proof for N, KRUSKAL's and NASH-WILLIAMS's proofs of VAZSONYI's conjecture, use only the properties of well-order contained in sequential order.

The following argument is due to myself. I subsequently read the version SKOLEM published in 1956 ([141], thm. 2, p57). This differs only in that it proves well-order, and therefore assumes the axiom of choice.

THEOREM 4.39 [24] $(\mathfrak{T},\prec)$ *satisfies the DCC.*

Proof. Given a descending sequence

$$\mathcal{S} = \ldots \prec T_3 \prec T_2 \prec T_1$$

in $\mathfrak{T}$, it is required to show that $\mathcal{S}$ is finite. The proof is by induction on $\mathrm{eh}T_1$. For $\mathrm{eh}T_1 = 0$, $v^{\sharp}$ defines an ordered, finite system.

Suppose now it has been proved that for all $\mathcal{S}$ for which $\mathrm{eh}T_1 \leq h$, $0 \leq h$, $\mathcal{S}$ is finite, and consider an $\mathcal{S}$ for which $\mathrm{eh}T_1 = h+1$. For $i = 1, 2, 3, \ldots$, let

$$T_i = [T_{i,1}, T_{i,2}, \ldots, T_{i,r}],$$

[23] ref. HIGMAN [62] and KRUSKAL [81]); the terminology was introduced in 1940 by BIRKHOFF [14] who made no distinction between the present notion, and that of well-order.

[24] Mc BETH, 1979, in [99], thm. 3.2.

where $r = r(i)$ (to facilitate the proof, reverse CANTOR convention is adopted). Since $T_{i+1} \prec T_i$, for each i, there is a least j for which $T_{i+1,j} \prec T_{i,j}$, or for which T_i has a jth principal branch, and T_{i+1} has not. Let this be expressed $i \downarrow j$. Either (i) there are numbers $i(1) < i(2) < \ldots$ for which $i(1) \downarrow 1, i(2) \downarrow 1, \ldots$, or (ii) there are no such numbers. In case (i), by the induction hypothesis, the sequence

$$\ldots \prec T_{i(2),1} \prec T_{i(1),1}$$

ends in a smallest term, say $T_{i(K_1),1}$. In case (ii), put $K_1 = i(K_1) = 0$. Similarly, either there are numbers

$$i(K_1+1) < i(K_1+2) < \ldots < i(K_1+K_2)$$

for which $i(K_1+1) \downarrow 2,\ i(K_1+2) \downarrow 2, \ldots, i(K_1+K_2) \downarrow 2$; or there are no such numbers, with $K_2 = 0$, and $i(K_1+K_2) = 0$ if also $i(K_1) = 0$. Furthermore,

$$T_{i(K_1+1),2} \preceq T_{i(K_1)+1,2} \preceq T_{i(K_1)+1,1} \prec T_{i(K_1),1}.$$

Thus, for $j = 1, 2, 3, \ldots$, there is a sequence $\mathcal{R}$:

$$\ldots \prec T_{i(K_1+1),2} \prec T_{i(K_1),1} \prec \ldots \prec T_{i(1),1}.$$

By induction hypothesis, $\mathcal{R}$ ends with a lowest term T, viz.

$$(\exists J, N \in \mathrm{N})\ N = K_1 + K_2 + \ldots + K_J \ \&\ T = T_{i(N),J}$$

is the last term of $\mathcal{R}$. It therefore follows that $\mathcal{S}$ is finite, containing $i(N)+1$ terms. The theorem now follows by induction.

□

Remarks

1. In the course of the above proof, the induction hypothesis is used *twice:* once with respect to $\prec$ - descents obtained by $\prec$ - descents of letters; and also with respect to $\prec$ - descents obtained by increasing word-length.

2. Thm. 39 is of key importance in definitions which must be shown to lead, in a finite number of steps, to a definite value.

3. The system $\mathfrak{S}$ could be inductively assigned the lexicographic order, as in the defn. 19 of $(\mathfrak{T}, \prec)$, by removing the restriction 19·2(a); and by applying part (b). The system $(\mathfrak{S}, \prec)$ obtained, however, is not well-founded. E.g. in the notation introduced before defn. 2:

$$[1] \succ [0,1] \succ [0,0,1] \succ \ldots.$$

On the other hand, this system, freed of the $\mathfrak{T}$- $\succeq$ - restriction, more resembles the standard dictionary order.

4. DCC vs. Well-Order in Mathematics

As noted in footnote 23, of II, §2·18, it cannot be proved without the AC that $(\mathrm{N}, <) = (1, 2, 3, \ldots)$ is well-ordered. The nomenclature 'well-ordered' is correct, but the concept with which it is associated in the established usage, is not. That there can be any doubt that the natural numbers are well-ordered is truly absurd. Furthermore, there are few if any instances of a theorem of mathematics (i.e. excluding the remaining results of metamathematics), whose proof depends on well-order, such that the proof cannot still be effected if no more than the assumption of the theorem of the DCC is made.

Intuitionistic Proof of the DCC for $\mathfrak{T}$

The proof as given above rests on the convergence principle, used in §2.6 to obtain KÖNIG's lemma. For example, in the case where T_1 is of the form

$$T_1 = [n_r, n_{r-1}, \ldots, n_1, n_0] = [n_{11}, n_{12}, \ldots, n_{1r+}],$$

$n_r, \ldots, n_1 \geq 1$ if $r > 0$; $n_0 \geq 0$; i.e. $\mathrm{eh}T_1 \leq 2$. In this case $\mathcal{S}$ gives

$$n_{i(1)} > n_{i(1)+1} = \ldots = n_{i(2)} > n_{i(3)} > \ldots .$$

Certainly the N$-$sequence $n_{i(1)} > n_{i(2)} > n_{i(3)} > \ldots$ is finite, having no more than $n_{i(1)}$ terms. But unless $(\exists K)\ n_{i(K)} = 1$, there is no means (given in the proof), of knowing when the lowest term has been reached.

This problem can be overcome, by *assuming at each step* that $n_{i(K)}$ has been reached. If a further $\downarrow$ is encountered, then the whole pattern of descents below $T_{i(K)}$ is re-calculated. In this way at every step, either there is a $\downarrow$ in some column, or the final polynomial is reached.

SKOLEM's Application of KÖNIG's Lemma

The following is [141], proof 2, p57; and is by induction on edge-height. Thus, let $\mathcal{S}$ in $\mathfrak{T}$ be as in IV·39, where $\mathrm{eh}T_1 = h + 1$. Let $\mathcal{S}$ be maximally partitioned into disjoint segments $\ldots, \mathcal{S}_n, \ldots, \mathcal{S}_2, \mathcal{S}_1$: the $\prec$-smallest tree T_i say, of $\mathcal{S}_n$ and the $\prec$-greatest tree T_{i+1} of $\mathcal{S}_{n+1}$ are such that either the leading branch (lb) of T_{i+1} is $\prec$ than that of T_i; or it occurs to lower multiplicity in T_{i+1} than in T_i. Since each multiplicity is $< \infty$, by the i.h. there are only fm such segments. Now let $\mathcal{S}_n$ be partitioned into disjoint $\ldots, \mathcal{S}_{n2}, \mathcal{S}_{n1}$ as was $\mathcal{S}$ itself, except via comparison of the leftmost principal branches (pb's), corresponding to cases (i), (ii) before defn. 22, in place of lb's. So long as the above case (ii) provides a new branch, this partitioning may be iterated, to give the general partition $\ldots, \mathcal{S}_{n\cdots s2}, \mathcal{S}_{n\cdots s1}$. This iterated partitioning defines a tree τ, for which $\mathcal{S}$ is the pv; and whose vertices v have as successors the partitions of v. (The indexing of the parts may be taken as the ordering of defn. 14). Each v has, thus, only fm successors. Each edge of τ specifies a certain pb and its multiplicity, for a candidate member T of $\mathcal{S}$. But T has only fm pb's. Therefore every chain C of τ is finite; the endv of C is a segment $\mathcal{S}_{n\cdots s\cdots t}$ of $\mathcal{S}$ containing exactly one normal tree. But τ satisfies the 2 finiteness conditions of §2·3. So by thm. 15 τ is finite. Hence $\mathcal{S}$ is finite, and as before the thm. follows by induction.

4.3.5 Strictly Normal Trees

There are many ways in which strict subclasses may be recovered from $\mathfrak{T}$, as $\mathfrak{T}$ itself was from $\mathfrak{S}$. E.g. a restriction may be placed on the height or valency. Or, for trees of eh$\geq$ 1, the relation '$\succeq$' of defn. 19·2 may be replaced by '$\succ$'. The strictly normal trees are obtained by imposing a chain-length condition. As already seen, for $T_1, T_2 \in \mathfrak{T}$,

THEOREM 4.40 *1.* ehT_1 <eh$T_2 \Rightarrow T_1 \prec T_2$ *(theorem 20).*

2. $T_1 \prec T_2 \Rightarrow$eh$T_1 \leq$eh$T_2$ *(converse following corollary 21).*

□

Now let $T \in \mathfrak{S}$, $V = V(T)$, and define the function $f : V \to \mathrm{N}_0$ by $f(v)$ =eh$C$, where C is the longest chain of T, with v a vertex of C. Let $u, v, w \in V$ be such that u, v Sc w and v L u (ref. defn. 14).

THEOREM 4.41 $T \in \mathfrak{T} \Rightarrow f(u) \leq f(v)$.

Proof, by induction on edge-height, using thm. 40.

□

The converse to the theorem does not hold. E.g. let T = [[[ ]], [[ ], [ ]]]. Then $T \in \mathfrak{S}$ and T satisfies the $f(u) \leq f(v)$ condition, but $T \notin \mathfrak{T}$. Normality implies that a valency, as well as a height condition may have to be fulfilled. On the other hand, since height takes priority over valency, given T and u, v, w as above

THEOREM 4.42 $f(u) < f(v)$ *in all cases implies* $T \in \mathfrak{T}$.

Proof, by induction on edge-height, using thm. 40·1.

□

Let the trees of S which satisfy the $f(u) < f(v)$ condition be called *strictly normal* trees.

DEFINITION 4.43 *The class of strictly normal trees is denoted by* $\mathfrak{N}$.

Since $\mathfrak{T}$ contains trees, all of whose chains are of equal length, viz. which do not satisfy the $f(u) < f(v)$ condition, by thm. 42,

COROLLARY 4.44 $\mathfrak{N} \subset \mathfrak{T}$.

□

It may be noted that

THEOREM 4.45 *$(\mathfrak{N}, \prec)$ has the order-type of the natural numbers.*

Proof, by induction on edge-height, noting that the subclasses of $\mathfrak{N}$ containing trees T with eh$T \leq k$, $k = 0, 1, 2, \ldots$, are finite.

□

Remark. The DCC property of $(\mathfrak{N}, \prec)$ is a corollary of thm. 39. But thm. 45 affords a somewhat simpler proof of this result.

Finally, let $T \in \mathfrak{S}$. Suppose for all vertices $u, v, w \in T$, if u, v Sc w and v L u, then

1. $f(u) \leq f(v)$ and
2. valency(u) $\leq$valency(v). Then

LEMMA 4.46 $T \in \mathfrak{T}$.

□

4.4 The Normal Tree of a Function in EP

4.4.1 Preliminary

Here the relationship of similarity which exists between the normal trees $(\mathfrak{T}, \prec)$ and the system $(\mathbf{EP}, \prec)$, from which the notion of normal tree originated, is established. In the first instance, it may be noted that the inductive defn. II·3 of **EP** makes no reference to the (linear) relation $e \prec f$ of eventual domination for $e, f \in \mathbf{EP}$. This is given separately, II·8. The inductive defn. 19 of $(\mathfrak{T}, \prec)$ could not work, without including the recurrent dictionary order '$\prec$'. The **EP** definition is possible because addition is commutative: $e + f = f + e$ in **EP**, so that the functions are 1-1 with the normal forms of II·22. If the order of summands in a polynomial were to be taken into consideration, then Ch. II defn. 3 and rem. 5 following defn. 1 would become closely analogous. Finally, a similar analogy links x−height of $e \in \mathbf{EP}$ with edge-height of $T \in \mathfrak{T}$.

4.4.2 The Tree of a Recurrent Exponential

THEOREM 4.47 [25] *Between the classes* $\mathfrak{T}$ *and* $\mathbf{EP_0}$ *may be established a unique bijection given by*

$$(\forall e \in \mathbf{EP_0})(\exists! T = T(e) \in \mathfrak{T}).$$

The mapping has the property

$$(\forall e, f \in \mathbf{EP_0})\; e \prec f$$

(eventual domination by value) if and only if

$$T(e) \prec T(f)$$

(recurrent lexicographic order).

Proof, by induction on height. For zero height, $T(\mathbf{0}) = [\,]$. Now suppose the bijection has been established for all polynomials of $\mathbf{EP_0}$ of x−height $\leq h,\ 0 \leq h$. Let

$$e = \omega^{e_q} + \ldots + \omega^{e_2} + \omega^{e_1}$$

$$f = \omega^{f_r} + \ldots + \omega^{f_2} + \omega^{f_1}$$

be the expressions, in the II·22-normal form, of e, f of height $h+1$ in **EP**. Now applying the definition of '$\prec$,' $(\forall \alpha, \beta \in \mathbf{EP})$

1. $\alpha \prec \beta \Rightarrow (\exists N)(\forall x \geq N)\; \alpha(x) + 1 \leq \beta(x)$, and
2. $(\forall a \in \mathrm{N})\; a\omega^{\alpha} \prec \omega^{\alpha^+}$.

[25] Mc BETH, [99] (1979), thm. 4.4.

Therefore f eventually dominates e, $e \prec f$, iff

1. (a) $e_q \prec f_r$ or
   (b) $e_q = f_r \,\&\, e_{q-1} \prec f_{r-1}$ or

   $\vdots$; or

2. $e_{q-i} = f_{r-i}$, $i = 0, 1, 2, \ldots, q-1$ and $q < r$.

But this is the condition of defn. 19·2 (b), therefore applying the inductive hypothesis, and putting

$$T(e) = [T(e_q), \ldots, T(e_2), T(e_1)]; \;\; T(f) = [T(f_r), \ldots, T(f_2), T(f_1)],$$

it follows that $e \prec f \Leftrightarrow T(e) \prec T(f)$, and the theorem is proved by induction.

□

Examples.

1. For $e = (\omega^2)^+$, $f = \omega \star n$, $T(e) = [2, [\,]]$; $T(f) = [\ldots[\,]\ldots]$, $n+2$ pairs of brackets occurring, with $T(e) \prec T(f)$, for $2 \leq n$; $T(f+e) = [T(\omega \star (n-1)), 2, [\,]\,]$.

2. $e = \Sigma_{i=r}^{1} a_i \omega^{e_i} + \mathbf{a}$ in **EP** gives

$$T(e) = [T(e_r), \ldots T(e_i), \ldots, T(e_1), [\,], \cdots [\,]],$$

   where e.g. there are a occurrences of '[ ]'. In particular, the leading branch of T is $T(e_r)$. The normal tree of ex. 4 following defn. 19 is $T(e)$, where

$$e = \sum_{i=9}^{1} a_i \omega^{e_i} + \mathbf{a},$$

   and

   (a) $a = 3$, $a_i = 1$, and
   (b) $e_9 = \omega^\alpha + \omega^\beta + \omega^\gamma + \omega^\phi + f$, where
      i. $\alpha = \omega^{\omega^{\omega+2}+1} + \mathbf{5}$,
      ii. $\beta = \omega^{\omega+2} + \omega^2 + 2$,
      iii. $\gamma = \omega^{\omega+1} + 2\omega^2 + \mathbf{2}$,
      iv. $\phi = \omega^{\omega^+} + \mathbf{1}$
      v. $f = \omega^{\omega^2+\omega} + \omega^{\omega+2} + \omega^4 + \omega^2 + \mathbf{3}$;
   (c) $e_8 = \omega^{\omega^2+\omega+3} + \omega^{\omega^2+5} + 2\omega^2 + 2\omega + \mathbf{4}$,
   (d) $e_7 = \omega^{2\omega} + 2\omega^2 + \omega$,

(e) $e_6 = \omega^{\omega+2} + 3\omega^3 + \mathbf{1}$,

(f) $e_5 = \omega^7 + \omega^3 + \omega$,

(g) $e_4 = \omega^7$,

(h) $e_3 = 2\omega^4 + \mathbf{1}$,

(i) $e_2 = \mathbf{9}$,

(j) $e_1 = \mathbf{2}$.

3. x−height of $e \in \mathbf{EP} = -1+\mathrm{eh}T(e)$.

4. Let $\omega, \omega \star 2, \omega \star 3, \ldots$ in **EP** be the cfs assigned to ε (III, 6·3). Then the corresponding sequence in T is $X_1, X_2, X_3, \ldots$ of thm. 20.

5. The e whose explicit exponents (II, 27·2) take the normal form

$$\omega^{\alpha} + \omega^{\beta}$$

in $\mathbf{EP}_0$ are exactly those for which $T(e) \in \mathfrak{D}$.

6. Let $e \in \mathbf{EP}$ be a compound diagonal. Let C_R be the rightmost chain of $T = T(e)$, in the ordering of lemma 11 (viz. in the post ordering of chains, ref. rem. 2 after theorem 36). Now let v be the endv of C_R, with v Sc u in C_R. Form $S \subseteq T$ from $C(v^{\sharp}, u)$ by affixing all w $\widetilde{\mathrm{Sc}}$ u in T. Then if e is perfect (II, 33·2), S corresponds to a function

$$\varepsilon_h = \omega \star h,$$

for some $h \geq 1$ ($h > 1$ if $e \neq \omega$); otherwise S corresponds to a function $\omega\star(h;\alpha)$, $1 \leq h$, and α a successor in **EP**. In particular, if e is of the 1st kind, then $h = 1$.

7. e may be taken as a 'subpolynomial' of f iff $T(e) \subseteq T(f)$ in $\mathfrak{T}$; in which case $e \ll f$ (cf. rem. after lemma 27).

Thus the lexicographic, or dictionary order which arises so often in combinatorics and in logic, as indeed in everyday life, arises in **EP**, in the most explicitly arithmetic context: if

$$\ldots \prec e_2 \prec e_1$$

is a descending sequence in **EP**, then by thm. 47

$$\ldots \prec T(e_2) \prec T(e_1)$$

is a descending sequence in $\mathfrak{T}$. Therefore by thm. 39

THE HIERARCHY EP IS WELL-FOUNDED:

COROLLARY 4.48 *The system* $(\mathbf{EP}, \prec)$ *satisfies the DCC.*

□

Thus it follows, in the terminology Ch. I, §2.3, that the similar, linear systems $\mathbf{EP}, \mathfrak{T}$ are sequentially ordered. And as a further corollary: by II, §2·2: $(\mathbf{XP}, \prec)$ is well-founded.

The corollary has many consequences for later constructions; and many applications to notions of Chs. II, III. E.g. the relation $\lhd$ of Ch. III, defn. 19 of arbitrary canonical reduction implies eventual domination, so given a sequence

$$\mathcal{S}: \quad \ldots \lhd e_2 \lhd e_1$$

in **EP**, corol. 48 gives

COROLLARY 4.49

$\mathcal{S}$ is finite.

□

Remarks.

1. The proof given here of corol. 48 in no way depends on the assignment of fundamental sequences in **EP**.

2. As to the correspondence, of thm. 47, note that given $S, T \in \mathfrak{T}$, if $S \prec T$ arises as a result of defn. 19·2(b)(2), then the unique $e, f \in \mathbf{EP}$ such that $T(e) = S, T(f) = T$ satisfy a stronger relationship than eventual domination, viz.

$$(\forall x \in \mathbf{N})\ e(x) < f(x).$$

So in particular, $e \ll f$ in the notation of II·35. Note that $e \ll f$ in **EP** does not necessarily imply $\mid T(e) \mid < \mid T(f) \mid$. E.g. $\omega^{10} \ll \omega \star 5$ but the orders of the respective trees are 12 and 7.

4.4.3 Dominating Exponent

Suppose $S, T \in \mathfrak{T}; S \prec T$; and let B be the dominating branch of T for S. Now by thm. 47, $(\exists! e, f \in \mathbf{EP})$ S, T are the respective normal trees of $e \prec f$. This correspondence extends to the exponents (II·23(2)) and principal branches (definition 1·2). Therefore either $B = T$, or the sequence

$$T, T^{(i_1)}, T^{(i_1,i_2)}, \ldots, T^{(i_1,i_2,\ldots,i_h)} = B$$

of defn. 22 determines unique $f, f_{(1)}, f_{(2)}, \ldots, f_{(h)}$ (viz. [26] an initial segment of such a sequence as obtained in II·26), for which B is the normal tree of $f_{(h)} \in \mathbf{EP}$. Thus

[26] Allowing a unit downward shift of indices.

DEFINITION 4.50 *1. If $B \prec T$, then*

(a) $f_{(1)}$ *is the* dominating basic exponent of f for e.

(b) $f_{(h)} \prec f$ *is the* dominating explicit exponent (dee) of f for e.

2. If $B = T$, then dee of f for e is f itself.

Examples

1. For the exponentials whose normal trees are the S, T of the ex. after defn. 22, the dee is $\omega + \mathbf{3}$.

2. The relation $(e, n) \prec e$ (III·7·2) gives: the dee of e for (e, n) is rp(e) (ref. II·33). Note that for the eventual domination of a ct, the case of IV·50 corresponding to $S^{(\mathrm{i})} = v^{\sharp}$ (or $S = v^{\sharp}$) in IV·22 arises iff Perf(e).

3. If $\alpha \prec e$ s.t. the diagonal e has $\alpha \prec (e, 1)$, there is no fixed relation between the dee and rp(e). E.g. $e = \omega \star 3 + \omega^{\omega^2 + \mathbf{2}}$, rp$= \omega^2 + \mathbf{2}$; $\alpha = \omega \star 3 + \psi$, dee$= \mathbf{2}$; $\alpha = \omega \star 3$, dee$= e$. However, if *height* is taken into consideration, then in this sense the dee is never dominated by the rp.

4. The cases (i), (ii) of defn. 23 give 2 possible kinds of 'explicit domination' of the ee $e_{(h)}$ of e subordinate to $f_{(h)}$. Thus case (i) arises when $\mathbf{1} = \omega^{\mathbf{0}}$ is used to give $\mathbf{1} \prec \omega^e$, for all $e \in \mathbf{EP}$. E.g. $\mathbf{1} \prec \omega^2$; $\omega \star n \prec \omega \star n^+$. And case (ii) when $f_{(h)}$ has an 'excess of terms' over $e_{(h)} \succeq \mathbf{1}$: $e_{(h)} \prec e_{(h)} + \alpha = f_{(h)}$, $\alpha \in \mathbf{EP}$. E.g. $\mathbf{1} \prec \mathbf{2}$. If $\alpha + \beta$ is a normal form in $\mathbf{XP}$, $\omega \preceq \beta$, then case (i) gives $\alpha + \mathbf{a} \preceq \alpha + \beta$, $1 \leq a$, in $\mathbf{XP}$; all others being case (ii).

5. Let $\alpha \prec e$ be an arbitrary eventual domination in $\mathbf{EP}$, with $S = T(\alpha), T = T(e)$. Let $C = C_{(S,T)}, T_C$ be as in §3·1. Then (the polynomial corresponding to) $U = T_C$ is the 'common part' of the exponential forms α, e. As described in III, §3·3, it will account for 'most of' e in the case of ex. 2 above; and C will be nearer the C_L of e in that of ex. 3.

6. Let $e \prec f@M$. Although the exponent sequence, culminating in $f_{(h)}$, most explicitly traces the part of the tree structure of f upon which the relation $e \prec f$ depends (in this sense, the portion T^C is redundant): the analysis gives no means of calculating M, the value of which is likely to be pushed up by ec's in the part of e corresponding to S^C.

4.4.4 Term-Order Correspondence

By the preliminary to §4, if the relation $\prec$ is ignored, and the order of distinct summands in $e = \Sigma a_i \omega^{e_i} \in \mathbf{EP}$ is taken into consideration, then another 1-1 correspondence $e \to S(e) \in \mathfrak{S}$ may be set up, via II·3, and rem. 5 foll. def. 1. Thus e.g. $S(\omega \star 3 + \omega + \psi + 1) = [[[1]], 1, [1], [\ ]] \notin \mathfrak{T}$. In case e is in normal form, then $S(e) = T(e)$.

4.5 Trees and Integer Partitions

The purpose of this section is to establish that the partitions of $n \in \mathrm{N}$, taken in their natural or dictionary order, happen to define a normal tree. This dictionary order is entirely different from the '$\preceq$' which occurs in definition 19, being nearer that of lemma 11. The treatment of partitions is based on that in McBETH [106] [27], 1990.

4.5.1 Partitions

A *normal partition* of a natural number n is an ordered partition

$$n = n_r + \ldots + n_2 + n_1,$$

where $n_{i+1} \geq n_i, i = 1, 2, \ldots, r-1$. The normal partitions of n are equinumerous with the unordered partitions of n. [28] Thus in the common notation, n has $p(n)$ normal partitions.

4.5.2 The Strictly Normal Trees $T[n]$

A simple process of identification leads from the list of partitions of n, to a strictly normal tree (ref. §3.5). $v^1, v^2, \ldots$, etc. denote different vertices, in the present section.

For each natural number n, a finite vertex-indexed tree $T[n]$, whose outline (§2.4, note 1) belongs to $\mathfrak{S}$, is obtained. (This new usage of square brackets is only occasional, and no confusion should arise, with the notation introduced in §2.4). It is shown that the chains of $T[n]$, taken in right-to-left (post) order, correspond to the normal partitions of n, taken in lexicographic order. Furthermore, it is shown that the outline of $T[n]$ belongs to the class $\mathfrak{N}$ of strictly normal trees. $T[n]$ is introduced via the following specialized definitions: let $u, v \in T$ in $\mathfrak{S}$.

DEFINITION 4.51

$$v\ (\mathrm{Sc}, 0)\ u \text{ iff } v = u,$$

and for $k = 0, 1, 2, \ldots,$

$$v\ (\mathrm{Sc}, k+1)\ u \ \text{ iff } \ (\exists w \in T)\ v\ (\mathrm{Sc}, k)\ w\ \&\ w \text{ Sc } u.$$

□

Now let $n \geq 1$.

[27] 'Partitions and normal trees', *Disc. Math.*

[28] $n \in \mathrm{N}$ has 2^{n-1} *ordered* partitions, as may be seen by taking n as a finite set, viz. as a string of n 1's. Any subset S defines a unique alternating pattern of strings of O's and 1's, and therefore an ordered partition of n. Since the pattern determined by the compliment of S determines the same ordered partition, their number is half the number of subsets.

DEFINITION 4.52 *1. $v^\sharp$ is the pv of $T[n]$. $v^\sharp$ is assigned the index $pt(v^\sharp) = 0$. $v^\sharp$ has the n successors*

$$v_n \mathrm{~L~} v_{n-1} \ldots \mathrm{~L~} v_1,$$

where v_i is assigned the index $pt(v_i) = n - i + 1, i = n, n-1, \ldots, 1$. The pv, and its successors, form the initial case of the inductive definition.

2. If $n = 1$, then $v^\sharp$ has the unique successor v_1. Now suppose $n > k > 0$, and that $(\forall v \in T[n])(\forall i)_{0 \leq i \leq k}\ v\ (\mathrm{Sc}, i)\ v^\sharp \Rightarrow v$ has been defined, and, for $v \neq v^\sharp, v$ has been assigned an index $pt(v) \geq 1$. Now let

$$v = v^k \mathrm{~Sc~} \ldots v^2 \mathrm{~Sc~} v^1 \mathrm{~Sc~} v^0 = v^\sharp$$

be the (in general incomplete) chain linking v to $v^\sharp$. Let $K =_{df} n - \sum_{i=1}^{k} pt(v^i)$: and $R = \min\{K, pt(v)\}$. If $R = 0$, then v is an endv of $T[n]$. If $R > 0$, then for $i = R, R-1, \ldots, 1$; the successors of v are the vertices v_i, where $v_{i+1} \mathrm{~L~} v_i, R > i \geq 1$, and where $pt(v_i) = R - i + 1,\ i \leq R$.

3. The vertices of $T[n]$ (together with their adjacency relations) as given in the above inductive definition, are the only vertices of $T[n]$.

4. The tree $T[n]$ is the partition tree *of n.*

□

Example. The outline of $T[4]$ is $[[[1]], [1,0], 1, 0]$. Note that $T[n]$ has $n+1$ vertices. Now suppose $n > 2$, and let

$$C = v^s \mathrm{~Sc~} \ldots v^2 \mathrm{~Sc~} v^1 \mathrm{~Sc~} v^0 = v^\sharp$$

be a chain of $T[n]$.

LEMMA 4.53 $(\forall i \leq s)\ i \geq 1 \Rightarrow pt(v^{i+1}) \leq pt(v^i)$.

Proof. In the inductive definition of $pt(v_i)$ of the successor v_i of v in defn. 52·2,

$$pt(v_i) = R - i + 1 \leq R = \min\{K, pt(v)\},$$

hence $pt(v_i) \leq pt(v)$, so that in the given chain $pt(v^{i+1}) \leq pt(v^i)$, $i = 1, 2, \ldots, s-1$.

□

Let n be as in defn. 52. Let '$\prec$' denote the left-to-right lexicographic order of the normal partitions of n. Thus, e.g., the normal partitions of 4 give:

$$1111 \prec 211 \prec 22 \prec 31 \prec 4.$$

Let $<_{\text{post}}$ denote the left-to-right post ordering of the (maximal) chains of $T[n]$. For example, putting $<_{\text{post}}=<$, the vertex indexings of $T[4]$ give

$$1111 < 211 < 22 < 31 < 4.$$

The following theorem is a statement that such an equivalence holds for all trees $T[n]$.

THEOREM 4.54 [29]

1. *Each chain C of $T[n]$ determines a unique normal partition P_C of n.*
2. *$T[n]$ contains $p(n)$ chains.*
3. *If C, C' are chains of $T[n]$, then*

$$C <_{\text{post}} C' \Leftrightarrow P_C \prec P_{C'}.$$

Proof.

1. Let C be a chain of $T[n]$. Suppose the vertices of C are related by

$$v^s \text{ Sc } \ldots v^2 \text{ Sc } v^1 \text{ Sc } v^0 = v^\sharp.$$

   Then v^s is an endv, therefore by the definition of v^s in $T[n]$, the parameter $R = 0$. Since $pt(v^s) \geq 1, \min\{K, pt(v^s)\} = 0$ gives $K = 0 = n - \sum_{i=1}^{s} pt(v^i)$, and by lemma 53, $P_C = \sum_{i=1}^{s} pt(v^i)$ is the unique normal partition of n determined by C.

2. Let $P = n_s + n_{s-1} + \ldots + n_1$ be a normal partition, so that $n_s \geq n_{s-1} \geq \ldots n_1$. A chain C of $T[n]$ such that $P = P_C$ may be obtained by induction on the left-to-right term index of P. Thus, in the initial case, the pv of $T[n]$ has a successor v^1 of index $pt(v^1) = n_s \leq n$. Now for r in the range $1 \leq r \leq s$, for inductive hypothesis suppose vertices $v^1, v^2, \ldots, v^r$, of indices $n_s, n_{s-1}, \ldots, n_{s-r+1}$ respectively, for which

$$v^r \text{ Sc } \ldots v^2 \text{ Sc } v^1$$

   have been found. If $r = s$, let C be this chain. Then $P = P_C$. If $r < s, K = n - \sum_{i=s}^{s-r+1} n_i > 0$; and $pt(v^r) \geq 1$, so $R > 0$ and v^r is not an endv. Now $n_{s-r} \leq n_{s-r+1} = pt(v^r)$, since P is normal. Therefore a unique successor v^{r+1} of v^r is defined, such that

$$pt(v^{r+1}) = n_{s-r} \leq R = \min(n_{s-r} + n_{s-r-1} + \ldots + n_1, n_{s-r+1}).$$

   And so by induction it follows that a unique chain $C = v^s \text{ Sc } \ldots v^2 \text{ Sc } v^1$ in $T[n]$ may be found, whose vertices are of respective indices $n_s, \ldots, n_2, n_1$.

[29] [106], thm. 3.4, p100.

Hence $P = P_C$. Therefore, taking into consideration part (1), it follows that the partitions and chains are equinumerous. Therefore by a remark at the beginning of the section, n has $p(n)$ normal partitions, and $T[n]$ has $p(n)$ (maximal) chains.

3. Now suppose C, C' are chains of $T[n]$, with $C <_{\text{post}} C'$. Let v be the vertex common to C, C', at maximum edge-distance from $v^{\sharp}$. The successors u, u' of v respectively in C, C' satisfy u L u', since $C <_{\text{post}} C'$. But by defn. 52·2, $pt(u) < pt(u')$. Therefore $P_C \prec P_{C'}$. But $T[n]$ contains $p(n)$ chains, by part (2), therefore part (3) follows.

□

The above result may be summarized by: the partitions of $n \geq 1$, in their natural order determine a (vertex-indexed) tree of $\mathfrak{S}$. This is in fact a normal tree:

THEOREM 4.55 $T[n] \in \mathfrak{N}$.

Proof. Let v be an interior vertex of $T = T[n]$ (viz. $v \neq$ endv). Let u, u' be successors of v such that u L u'. Let u' be a vertex of a chain C' of T. In the notation of normal partitions of thm. 54, let

$$P_{C'} = n_s + n_{s-1} + \ldots + n_k + \ldots + n_1 = n,$$

where $k < s$ is such that $pt(u') = n_k$. Now $pt(u) < pt(u')$, since u L u'. Therefore u is a vertex of a chain C with

$$P_C = n_s + n_{s-1} + \ldots + n_{k+1} + m_t + m_{t-1} + \ldots + m_1,$$

where t is a natural number $< n$, and where

$$m_t = pt(u) < pt(u') = n_k, \qquad \text{and} \qquad m_t \geq m_{t-1} \geq \ldots m_1.$$

Indeed, the vertices of C may be chosen so that

$$m_{t-1} \leq n_{k-1},\ m_{t-2} \leq n_{k-2}, \ldots m_{t-k+1} \leq n_1.$$

Thus $\sum_{i=k+1}^{s} n_i + \sum_{i=t}^{t-k+1} m_i < n$. But P_C is a partition on n, so $k < t$, and P_C has the additional terms $\sum_{t=k}^{1} m_i$. I.e. ehC' <eh$C$, and the chains of $T[n]$ satisfy the strict condition of thm. 42.

□

Therefore by corol. 44, the partition tree $T[n]$ of a natural number n is a normal tree. $T[n]$ becomes fairly complex as n increases, as may be expected from e.g. $p(100) = 190\ 569\ 292$.

4.5.3 Application to Partitions into r Parts

Let $n, r \geq 1$. Then writing

$$p_r(n) = \begin{cases} 0 & \text{if } r > n \\ \text{the number of partitions of n into r parts} & \text{if } r \leq n. \end{cases}$$

Thus $p(n) = \sum_{r=1}^{n} p_r(n)$. $P_r(n)$ is the number of r-chains in $T[n]$. For small r, n, a glance at $T[n]$ may be sufficient to obtain the value of $p_r(n)$. For example, $p_3(9) = 7$. The following standard identity is easily proved in terms of the partition tree $T[n]$:

THEOREM 4.56 *For n, r; $p_r(n)$ as above,*

$$p_{r+1}(n+1) = p_r(n) + p_{r+1}(n-r).$$

Proof. Consider an $(r+1)$-partition of $n+1$. This is P_C, for C a unique (maximal) chain in $T[n+1]$, of edge-length $r+1$. If the endv v of C has $pt(v) = 1$, then all possible such C are obtained by preceding v by a chain C' (of edge-length r), where $P_{C'}$ is any partition of n into r parts. This accounts for the term $p_r(n)$. If $pt(v) > 1$, then every u of C has $pt(u) > 1$. The $(r+1)$-partitions of $(n+1)$ in this case are therefore $1-1$ with the partitions $P_{C''}$, where C'' is obtained from C by replacing every $pt(u)$ by $pt(u) - 1$. This accounts for the term $p_{r+1}(n-r)$.

□

The above proof is wholly based on a sketch, consisting of 2 FERRERS diagrams, shown to me in May, 1981, by Mr. Rob STACEY.

Similarly, it may be shown that the # of partitions of n into parts greater than 1 is

$$p(n) - p(n-1); \text{ and also } 0 \leq p(n+2) - 2 \cdot p(n+1) + p(n).$$

4.6 Recurrent Factorization

A. A. MULLIN [114], 1964 considered the recurrent prime factorization of a natural number x, viz. if $x = \prod p_i^{x_i}$, then the factorizations of the exponents x_i, of the exponents of the $x_i, \ldots$, are performed, as in the characterization of explicit exponents of Ch. II, §2. The final expression, which is reached after fm factorizations by the DCC for $(\mathrm{N}, <)$, is called by MULLIN the *mosaic* $\nu(x)$ of x. E.g. $\nu\{2 \star (3 \star 4; 5)\}$ contains exactly the primes $2, 5$. The correspondence betwen sums and products of exponents at once gives, by the analysis of §4, a tree $S \in \mathfrak{S}$ corresponding to $\nu(x)$. Thus the *tree* $S(x)$ of x is given recurrently by

1. $S(1) = v^{\sharp}$.

2. If $S(x_i)$ have been defined, for $i = 1, 2, \ldots, R$, then $S(x) = [S(x_1), \ldots, S(x_R)]$.

Following [115] [30] $\nu(a)$ is a *submosaic* of $\nu(b)$ if $S(a) \subseteq S(b)$ (ref. defn. 8); and if the primes at vertices corresponding under the embedding are equal.

Suppose the integers d, x, y have $\mathrm{GCD}(x, y) = 1$, and $d \mid xy$. Then there are unique i, j for which $d = ij$, such that $i \mid x$, $j \mid y$ and $(i, j) = 1$. Now let $f \in \mathcal{N}$ be multiplicative, viz. for x, y as above, $f(xy) = f(x)f(y)$; and let

$$F = \lambda u \cdot \sum_{d \mid u} f(d).$$

Then, as is well-known, F also is multiplicative:

$$F(xy) = \sum_{d \mid xy} f(d) = \sum_{i \mid x, j \mid y} f(i)f(j) = \sum_{i \mid x} f(i) \sum_{j \mid y} f(j) = F(x)F(y).$$

Now suppose the relation '$(a, b) = 1$' is replaced by the stronger condition '$\nu(a)$ and $\nu(b)$ have no prime in common'. Then 'multiplicative' is weakened to 'generalized multiplicative' ([113], p118); and if $a \mid b$ is replaced by the different condition '$\nu(a)$ is a submosaic of $\nu(b)$', then the above gives

LEMMA 4.57 (MULLIN) *f gen. mult. $\Rightarrow$ F gen. mult.*

□

Notes.

1. Unless $S(d) < S(xy)$ (ref. $2 \cdot 8$), i.e. $\nu(d)$ is a 'pv-submosaic' of $\nu(xy)$, either $i = 1$ or $j = 1$. In this sense, the lemma is only non-trivial for basic, rather than for recurrent prime factorizations.

[30] A. A. MULLIN, 'A contribution towards computable number theory', *Zeitschr. math. Logik Grundl. Math.*, 11, 117 - 119 (1965).

2. In order for $S(x) \in \mathfrak{T}$, it is necessary for (in the reverse-CANTOR convention) $S(x_1) \succeq S(x_2) \succeq \ldots \succeq S(x_R)$. But, in plain language x is more likely to be divided (without remainder) by a smaller prime, than by a larger, so that a smaller prime will produce a larger exponent. Since larger numbers generally have more complex mosaics, it is therefore expected that the above condition will usually be satisfied. Viz. it is conjectured that the density [31] of natural numbers whose MULLIN tree is normal is unity.

[31] If P is a predicate on **N**, then the density of the numbers satisfying P is taken as

$$\liminf \frac{1}{x} \cdot \sum_{\mathrm{P}(i)}^{i=1 \text{ to } x} 1.$$

(Ref. HALBERSTAM, W. and ROTH, K. F., *Sequences*, C.U.P., 1966).

4.7 Fundamental Sequences for Normal Trees

Chapter III was devoted to fundamental sequences for exponential polynomials. The diagonal of a fundamental sequence of arithmetic functions has a clear interpretation. E.g. the canonical assignment in **EP** gives

$$e = \omega \star (4;3) + \omega \star (3;3), \quad e_n = \omega \star (4;3) + \omega \star (2; n \cdot \omega^2), \quad e = \mathrm{diag}_n \, e_n.$$

On account of the similarity of the sequentially ordered systems $\mathbf{EP}, \mathfrak{T}$, it follows that the assignment $(\mathbf{EP}, \Omega)$ induces an assignment $(\mathfrak{T}, \Omega)$. Therefore, what form do the canonical terms $\Omega(T,n) = (T,n)$, $n \in \mathrm{N}$, take? What is the interpretation of the relation between T and the (T,n)? It turns out to be easy enough to define the terms, and then prove they correspond. The answer to the second question seems rather harder to find; there may be no satisfactory interpretation. However the reduction of canonical terms from exponential polynomials to trees renders them into simpler objects, furthering the analysis of fundamental sequences. The following ideas are contained in McBETH, 1986. [32]

As a preliminary, the trees of $\mathfrak{T}$ are classified as follows:

DEFINITION 4.58 *1. The* **initial tree** *is* $[\,] = v^\sharp$.

2. *A* successor [33] tree *is of the form* $T = [T_r, \ldots, T_2, v]$, $r \geq 1$. *This is indicated by* $\mathrm{Suc}(T)$.

3. *The* predecessor, T^-, *of a successor tree* T *is* $T^- = [T_r, \ldots, T_2]$.

4. *A* diagonal tree *is of the form* $T = [T_r, \ldots, T_2, T_1]$, $\quad T_1 \neq$ a vertex. *This is indicated by* $\mathrm{Diag}(T)$.

□

Thus the predecessor is formed by the removal of the minimal ev (§3·3) of the successor T : $T^- = T'$. If $T = [v]$, then $T^- = [\,]$. In all other cases, $\mathrm{Suc}(T) \Rightarrow \mathrm{eh}T = \mathrm{eh}T^-$. If $T \prec U$ in T, then by defn. 19, T is immediately followed by its successor $T^+ \preceq U$.

4.7.1 Assignment

Let $T = [T_r, \ldots, T_2, T_1]$ be a diagonal (tree) of $\mathfrak{T}$. Then T may be assigned a unique fundamental sequence, using induction on edge-height, as follows:

DEFINITION 4.59 *For* $n \geq 1$,

[32] [101] 'Fundamental sequences for normal trees', manuscript.

[33] Note that the successor of a vertex (defn. 2·5) is a purely combinatorial notion, while the successor of a tree is an arithmetic notion.

1. *Suppose* Suc(T_1). *Then* $(T,n) = [T_r,\ldots,T_2,T_1^-,\ldots,T_1^-]$ *(viz. n occurrences of* T_1^- *).*

2. *Suppose* Diag(T_1). *Then* $(T,n) = [T_r,\ldots,T_2,(T_1,n)]$.

Remark. By the minimal property of the successor, for Suc(T_1); and by induction for Diag(T_1), defn. 19 gives: T *is the* $\prec$*-lowest normal tree* for which

$$(T,1),(T,2),(T,3),\ldots \prec T.$$

4.7.2 Correspondence of Assignments

The terms of the assignment are chosen to conform to the 'standard' assignment introduced by G. H. HARDY in 1903, viz. that of Ch. III, defn. 6. For the tree $T = T(e) \in \mathfrak{T}$ of $e \in \mathbf{EP}$, the definition contained in the proof of thm. 47 gives

LEMMA 4.60 *For* $e = \sum_{i=r}^{1} \omega^{e_i}$, $T(e) = [T_s,\ldots,T_2,T_1]$,

$$s = r, \text{ and } T_i = T(e_i),\ i = r, r-1,\ldots,1.$$

□

LEMMA 4.61 *Suppose* $T \in \mathfrak{T}$ & Suc(T). *Then* $T(e^-) = T(e)^-$.

Proof. $T(e) = [T_r,\ldots,T_2,v]$.

□

THEOREM 4.62 Diag(e) $\Rightarrow (\forall n \in \mathbf{N})\, T\{(e,n)\} = (T,n)$.

Proof, by induction on x−height. In the initial case, $e = \omega$ and $T(\omega) = [1]$, so $(T,1) = [0] \neq v$, and $(T,n) = [0,\ldots,0] = T(\mathbf{n})$. Now suppose the result has been obtained for x−height $\leq h$, and suppose $e = \sum_{i=r}^{1}$ is of x−height $h+1$. Then for all n, if Suc(T_1),

$$\begin{aligned}(T,n) &= [T_r,\ldots,T_2,T_1^-,\ldots,T_1^-] = [T_r,\ldots,T_2,T(e_1^-),\ldots,T(e_1^-)] \\ &= T(\sum_{i=r}^{2}\omega^{e_i} + n\cdot\omega^{e_1^-}) = T\{(e,n)\}.\end{aligned}$$

If Diag(T_1),

$$\begin{aligned}(T,n) &= [T_r,\ldots,T_2,(T_1,n)] = [T_r,\ldots,T_2,T\{(e,n)\}] \text{ by i.h.} \\ &= T\{(e,n)\}.\end{aligned}$$

□

Examples.

1. For $e = \omega\star(2;3)$, $T(e) \in \mathfrak{T}$ has the fundamental sequence $[[2]], [[2,2]], [[2,2,2]], \ldots$.
2. For $T \in \mathfrak{T}$, Diag$(T) \Rightarrow$eh(T, n) =eh$(T, n+1)$, for all $n \geq 1$.
3. Let $n(1), n(2), \ldots$ be a si natural number sequence. Then given $T_{(1)} = T(e)$ and $T_{(i+1)} = (T_{(i)}, n_{i+1})$, the iterated cr sequences of III, §4·3 and that derived via §7·1 above correspond:

$$T_{(i)} = T(e_{(i)}), \quad i = 1, 2, \ldots .$$

By thm. 39,

COROLLARY *the sequence* $T_{(1)}, T_{(2)}, \ldots$ *is* **finite**.

Now let Diag(T) in $\mathfrak{T}$.

COROLLARY 4.63 *1.* $(T, n) \prec (T, n+1)$; *2.* $(T, n) \prec T$.

Proof, follows from thm. 62, from the similarity transform of thm. 47, and from the standard properties of fundamental sequences, III·7.

□

Diagonal Trees

The proper subclass of diagonal trees T of $\mathfrak{T}$ may further be classified as follows. Let $T = [T_r, \ldots, T_2, T_1]$.

DEFINITION 4.64 *1. Suppose* $T_1 = [v]$. *Then* T *is a* simple diagonal.

2. Suppose Suc$(T_1), T_1 \neq [v]$. *Then* T *is a* compound diagonal cd of the 1st kind.

3. Suppose Diag(T_1). *Then* T *is a* cd of the 2nd kind.

□

Therefore, in this definition, for (1), (2), eh$C_R = 2$; while for (3), eh$C_R \geq 3$. The purpose of defn. 64 is to classify diagonals, as in Ch. III, defns 16, 17 according to properties of their fundamental sequences. Let $T = T(e)$, as in thm. 62.

THEOREM 4.65 *The diagonal* $e \in$ **EP** *is simple, or compound of the first or second kind as is* $T(e) \in \mathfrak{T}$.

Proof, by induction on $x-$height.

□

Thus, the canonical terms (T, n) of a simple diagonal T are successor trees. Those of a cd are diagonal. The diagonal e has an infinite domain $1, 2, 3, \ldots$; although the form e itself is built up from fm unary and binary operations. $T = T(e)$ is wholly finite; yet $(T, 1), (T, 2), \ldots$ is infinite: by 59·1, there is an n for which $\mid T \mid < \mid (T, n) \mid$.

4.7.3 The Register Branch

The correspondence of §4 defines a bijection $\Psi : \mathfrak{T} \to$ **EP** (viz. $\Psi\{T(e)\} = e$). Now let $T \in \mathfrak{T}$ be diagonal. Let u, v be the vertices of C_R of T for which

1. v Sc u
2. the edge-distance of v to $v^\sharp$ is maximum.

(Note that $u \neq v^\sharp$.)

DEFINITION 4.66 *The register branch of* T =Br=Branch(u, T).

THEOREM 4.67 *If* $T(e)$ *is the tree of thm. 47, and* B *is the register branch of* $T(e)$, *then the subpolynomial* $\Psi(B)$ *is the* rp *of* e.

□

Thus, the interest of the register branch is that, just as discussed in III, §3.3 the inductive defn. 59 of (T, n) reduces to the explicit replacement of $B = [\ldots,\text{Br}]$ in T, by $(B, n) = [\ldots,\text{Br}^-,\text{Br}^-,\ldots,\text{Br}^-]$ $(\ldots + n$ terms). It may be noted that for no $T \in \mathfrak{T}$ is the register branch diagonal. Thus, let $T \in \mathfrak{T}$, and let T be reduced to T', as in definition 28. Then the respective register branches B, B' of T, T' satisfy $B' = B^-$, if B is neither **1**, nor the successor of a diagonal.

The ei (end of the above §) of an exponential e may be visualized within $T = T(e) = [T_r, \ldots, T_1]$, say. If $r > 1$, either $T_1 \prec T_2$; or $(\exists i \leq r)_{\geq 2}\ T_1 = T_2 = \ldots = T_i \prec T_{i+1}$ (with the inequality excluded if $i = r$), in which case $t = 1$ and the ei= $a_{(t)} = i$. If $r = 1$ or $T_1 = T^{(1)} \prec T_2$, then the rightmost principal branches (pb's) of T_1 may be checked, as were those of T. Therefore, continuing the checks through the pb's $T^{(1)}, T^{(1,1)}, \ldots$ (in the notation of §3·1), either some pb= $v^\sharp$, with ei= 1; or for some least $t > 1$, the rightmost pb's occur to multiplicity ei= $a_{(t)} > 1$.

4.7.4 Normal Sequence

Let $\lhd$ be the step-down relation of Ch. III, defn. 18 (viz. $\lhd_1$). Thus the correspondence $T = T(e)$ and lemma 61 give $(\forall S, T \in \mathfrak{T})\ S \prec T \Rightarrow$

$$S \lhd T \Leftrightarrow \begin{cases} S = T^- & \text{in case Suc}(T) \\ S = (T, 1) & \text{in case Diag}(T). \end{cases}$$

Let $\tilde{\lhd}$ be the transitive closure of $\lhd$. Then by corol. 63 and the DCC property of T, thm. 39,

LEMMA 4.68 $(\forall T \in \mathfrak{T})_{\neq[\,]}\ v^\sharp \mathrel{\tilde{\lhd}} T$.

□

Therefore $(\exists! k \geq 1)(\exists! S_1, S_2, \ldots, S_k \in \mathfrak{T})$

$$v^{\sharp} = S_k \lhd \ldots \lhd S_2 \lhd S_1 = T.$$

This sequence may be called the *step-down* sequence of the normal tree T, as in Ch. III, ex. 4, following corol. 22. Using defn. 33:

THEOREM 4.69 [34] *The step-down sequence of $T \in \mathfrak{T}$ is the normal sequence of T.*

Proof, by induction on edge-height. It is sufficient to prove that, in the notation of defn. 28,

$$(\forall S, T \in \mathfrak{T})\ S \lhd T \Leftrightarrow S = T',$$

so that, in particular, $(\forall i < k)\ S_{i+1} = S_i'$. For eh$=0$, the 1-term sequences are identical. Now let $h \geq 1$, and suppose for induction hypothesis that for trees T for which eh$T \leq h$, the equivalence holds. Let $T = [T_r, \ldots, T_2, T_1]$ be of edge-height $h+1$ in $\mathfrak{T}$. Then

1. if Suc(T), $T_1 = v$ and $[T_r, \ldots, T_3, T_2] = T' \lhd T$;
2. If Diag(T) & Suc(T_1), $T' = [T_r, \ldots, T_2, T_1']$, by lemma 29. But $T_1' \lhd T_1$ by hypothesis, i.e. $T_1' = T_1^-$. Therefore $T' = [T_r, \ldots, T_2, T_1^-] = (T, 1) \lhd T$;
3. If Diag(T) & Diag(T_1), $T' = [T_r, \ldots, T_2, T_1'] = [T_r, \ldots, T_2, (T_1, 1)]$ by i.h. $= (T, 1) \lhd T$.

Thus the result is obtained for all trees of edge-height up to $h+1$, and so follows in the general case by induction.

□

In the notation of III, §4·3, $T = T(e)$ and rem. 1 after thm. 34 give:

COROLLARY $D_e(1) = -2 + \mid T \mid$.

Remarks

1. The $B' = B^-$ result noted after thm. 67 shows that a 'register sequence' obtained by iterated reduction of this form coincides with a segment (in general interior) of the normal sequence of T.
2. Using thm. 34, thm 69 affords a proof of lemma 68, without using the well-foundedness of $\prec$ on $\mathfrak{T}$ (alternatively, the DCC may be avoided via III, §4·1).

By thm. 36,

COROLLARY 4.70 *the step-down sequence numerates the vertices of T in post order.*

---

[34] [101], thm. 4.7.

4.7.5 Density and BACHMANN Properties in $(\mathfrak{T}, \Omega)$

The main results of III, §3·4 may be proved via the similarity Ψ.

(i) Density. Suppose S precedes the diagonal T in the linear system $\mathfrak{T}$. Then by the remark after defn. 59, S cannot exceed every canonical term of T. Viz. either $(\forall n)\ S \preceq (T,n)$; or $(\exists n)\ (T,n) \prec S \preceq (T,n+1)$.

(ii) Starting Property. Suppose $(T,n) \prec S \preceq (T,n+1)$ for the diagonals S,T in $\mathfrak{T}$. If $S = (T,n+1)$, removal of the minimal ev of the 'extra' Br^- results in a tree $(S,1) \succeq (T,n)$. Thus, suppose $S \prec (T,n+1)$.

Let w be the predecessor of pv(Br) in T. From the above inequality, it follows that the separating chain $C = C\{(T,n),S\}$ cannot lie to the left of the chain $C(v^\sharp, w)$ in T. Therefore w occurs as a vertex of S. Thus, suppose first $w \in C$. Branch(w,S) is of the form $[\ldots,\mathrm{Br}^-,\ldots,\mathrm{Br}^-,S_q^*,\ldots,S_1^*]$ (n occurrences of Br^-), where $\mathrm{Br}^- \succ S_q^* \succeq \ldots \succeq S_1^*$. There are various cases, but in each it is clear that the branches S^* ensure that Branch$\{w,(S,1)\} \succeq [\ldots,\mathrm{Br}^-,\ldots,\mathrm{Br}^-] =$ Branch$\{w,(T,n)\}$, i.e. $(T,n) \preceq (S,1)$. The same argument may now be applied to the appropriate vertex, lower down $C(v^\sharp, w)$ than w, in the case where C lies to the right of this chain.

4.7.6 Further Applications of the Canonical Assignment

The Hydras of KIRBY and PARIS

In defn. 59·1, a branch of a tree $T \in \mathfrak{T}$ is '$\prec$ −reduced'; and then a number of copies of the reduced branch are introduced. The 'hydras' [71] are the finite pv-trees. Thus, by the 1-1 correspondence of rem. 1 following thm. 26, each hydra determines a unique $T \in \mathfrak{T}$. Let $u \in T, B =$Branch(u,T), say $B = [S_t,\ldots,S_i,\ldots,S_1]$, for $T \neq v^\sharp$. An 'excision' is an excision from T of S_i, for some $i \leq t$, transforming B to $B' \in \mathfrak{T}$, and T to a tree S' in $\mathfrak{S}$. Again, by thm. 26 (the basis of the above correspondence), S' determines a unique $T' \in \mathfrak{T}$. Either $u =$pv$T$, or $(\exists v \in T)\ u$ Sc v, and

$$\mathrm{Branch}(v,T) = [\ldots,B,\ldots]$$

$$\mathrm{Branch}(v,T') = [\ldots,B',\ldots].$$

If the 2 positions correspond, since $B' \prec B$, in either case the argument of lemma 27 gives $T' \prec T$. Otherwise, a finite number of $S_j \succ B'$ have to move to the left to make way for B', so again $T' \prec T$. A 'growth' is the introduction of arbitrarily many copies of B' into Branch(v,T'), as principal branches, transforming T' to a tree S'' of $\mathfrak{S}$. By the above thm, S'' determines a unique T'' of $\mathfrak{T}$. But since $\prec$ is lexicographic, the presence of the copies has no effect; that is, $B' \prec B$ still gives a $\prec$ −lower normal tree:

$$T'' \prec T.$$

Since the DCC result, thm. 39, holds for $\mathfrak{T}$, the above finitary argument establishes a (generalization of) thm. 2(i) of [71]: given any hydra, an arbitrary sequence of excision-growth pairs (with null growth when $u = v^\sharp$) leads in a finite number of steps to the isolated vertex $v^\sharp$.

4.8 Homeomorphic Embedding

KRUSKAL [81], 1960 established a result for finite trees, at the same time bringing together a number of lines of enquiry by VAZSONY, HIGMAN, etc. The survey is continued in [82] (1972)

DEFINITION 4.71 *A partially ordered* [35] *system* $(\mathcal{X}, \subseteq)$ *is* well-quasi-ordered *(wqo) iff all sequences of pairwise distinct terms have the ERDÖS-RADO property, viz.*

$$x_1, x_2, \ldots \text{in } \mathcal{X} \Rightarrow (\exists i, j \in \mathbf{N})_{i<j}\ x_i \subseteq x_j.$$

The KRUSKAL, NASH-WILLIAMS tree theorem is: the oriented trees are wqo under homeomorphic embedding. The result is obtained here via a new characterization of homeomorphic embedding. Let $S \in \mathfrak{S}$, $S \neq v^\sharp$. Let $u \in S$ be not an endv. Let B =Branch$(u, S) = [B_q, \ldots, B_2, B_1]$. Now let $E \in \mathfrak{S}$ be such that $(\exists l, m \leq q)\ l \leq m$ and

$$E = [B_q, \ldots, B_{m+1}, [B_m, \ldots, B_l], B_{l-1}, \ldots, B_1]$$

(with suitable conventions in the extreme cases. E.g. if $m = q$, $l > 1$, then E takes the form

$$E = [[B_q, \ldots, B_l], B_{l-1}, \ldots, B_1])$$

DEFINITION 4.72 1. *If E satisfies the above conditions, then E is said to be obtainable from B by* single principal edge introduction, *written* $B \sqsubset_{\text{Branch}} E$.

2. *Suppose T is obtained from S, by replacing B in S by E. Then T is said to be obtainable from S by* single edge introduction, *written* [36]

$$S \sqsubset T.$$

3. *The transitive closure of $\sqsubset$ is $\tilde{\sqsubset}$. Thus, for all $U, V \in \mathfrak{S}$, $U \tilde{\sqsubset} V$ iff V is obtainable from U by fm single edge introductions.*

---

[35] Viz.

1. $A \subseteq A$,
2. $A \subseteq B \ \&\ B \subseteq A \Rightarrow A = B$,
3. $A \subseteq B \ \&\ B \subseteq C \Rightarrow A \subseteq C$,

for every A, B and C of $\mathcal{X}$.

$(\mathcal{X}, \subseteq)$ is a *lattice* if for every $A, B \in \mathcal{X}$, there is a lub $A \cup B \in \mathcal{X}$, viz.

$$(\forall C)\ A \subseteq C \ \&\ B \subseteq C \Rightarrow A \cup B \subseteq C,$$

and dually for a glb $A \cap B$.

[36] S is an *edge-contraction* or *minor* of T.

4. *Let* $U, V, W \in \mathfrak{S}$. *Then if* $U = V$, *or* $U \tilde{\sqsubseteq} V$ *and* $V \subseteq W$, *then* U *is said to be* homeomorphically embeddable in W. *This is written*

$$U \subseteq_{\text{homeo}} W.$$

□

Notes.

1. $U \sqsubset V \Rightarrow \text{pv}V = \text{pv}U$.

2. Regarding defn. 71·2, edge introduction removes no vertex of S. This operation does, however, in general alter the valencies of (some or all of) the vertices of S.

3. Each edge-introduction introduces a single new vertex, corresponding to the inner pair of brackets in the notation for E. Therefore by (2), $U \subseteq_{\text{homeo}} W \Rightarrow | U | \leq | W |$.

Note also that, in the definition, $B, E \in \mathfrak{T}$ & $B \sqsubset_{\text{Branch}} E \Rightarrow B \prec E$. Similarly $S \sqsubset T \Rightarrow S \prec T$, and $U \tilde{\sqsubseteq} V \Rightarrow U \preceq V$ in $\mathfrak{T}$. By the remark following lemma 27, $V \subseteq W \Rightarrow V \preceq W$, and by the transitivity of eventual domination,

LEMMA 4.73 $S \subseteq_{\text{homeo}} T \Rightarrow S \preceq T$.

The following fundamental property of homeomorphic embedding has been taken as the defining property:

THEOREM 4.74 *pre and post orders are conserved under homeomorphic embedding.*

Proof. Suppose $U, W \in \text{S}, U \subseteq_{\text{homeo}} W$. The case for post order is considered. Applying the notations of definition 71, it is sufficient to consider the single principal edge introduction inclusion $\sqsubset_{\text{Branch}}$. Thus suppose $u, v \in B$, where $u <_{\text{post}} v$. Then by note (2) following defn. 71, $u, v \in E$. Considering the cases of defn. 18:

1. If $u = v^\sharp$, there is nothing to prove.

2. (a) Suppose $u \in B_i, v \in B_j, i < j$. let B, E be as given in defn. 71. Let $\tilde{B} = [B_m, B_{m-1}, \ldots, B_l]$. Suppose e.g. $m \geq j \geq l$; and $i < l$. Then $u \in B_i, v \in \tilde{B}$, and since $\tilde{B}$ lies to the left of B_i in the E-expression, by defn. 18·2(a), $u <_{\text{post}} v$ in E. The various remaining possibilities, e.g. $j > m \geq j > i$, all follow by the same clause of defn. 18.

   (b) Taking $\tilde{B}$ as in case (2)(a), suppose e.g. $m \geq i \geq l$. Then since $u \leq_{post} v$ in B_i, by 18·2(b), $u <_{\text{post}} v$ in $\tilde{B}$, therefore by the same clause, $u <_{\text{post}} v$ in E.

(c) If $u \neq v, v =$ pv of B, then $(\exists i \leq q)\ u \in B_i$, therefore u cannot be the pv of E. But by note (1) following defn. 71, pv$E$ =pv$B = v$. So $u <_{\text{post}} v$ in E.

Thus in all cases $u <_{\text{post}} v$ in B iff in E. I.e. $B \sqsubset_{\text{Branch}} E \Rightarrow \{u <_{\text{post}} v$ in $B \Rightarrow u <_{\text{post}} v$ in $E\} \Leftrightarrow$Post(B, E), say. It therefore follows by induction that $S \sqsubset T \Rightarrow$Post$(S,T)$; by a further induction that $S \tilde{\sqsubset} T \Rightarrow$Post$(S,T)$; and finally by thm. 37 that $U \subseteq_{\text{homeo}} W \Rightarrow$Post$(U, W)$. Similarly, $U \subseteq_{\text{homeo}} W \Rightarrow$Pre$(U, W)$.

□

Let $T \in \mathfrak{S}$, $T = [T_r, \ldots, T_2, T_1]$, $T \neq v^\sharp$. T is dissected into 2 parts as follows

DEFINITION 4.75 (STACEY)

$$\mathrm{L}(T) =_{df} T_r; \qquad \mathrm{R}(T) =_{df} [T_{r-1}, \ldots, T_{r-2}, \ldots, T_2, T_1].$$

Note that pv of $\mathrm{R}(T) = v^\sharp =$ pv of T. Also that if $r = 1$, $\mathrm{R}(T) = v^\sharp$.

Remark. $T \neq v^\sharp \Rightarrow |\mathrm{L}(T)|, |\mathrm{R}(T)| < |T|$.

Example. [STACEY.] $T = [[2], 2, 0, [3]]; \mathrm{L}(T) = [2]; \mathrm{R}(T) = [2, 0, [3]]$.

Now let $S, T \in \mathfrak{S}$.

LEMMA 4.76 $S \subseteq_{\text{homeo}} T$ *if either*

1. $S = v^\sharp$, *or*

2. $S, T \neq v^\sharp$ & $\mathrm{L}(S) \subseteq_{\text{homeo}} \mathrm{L}(T)$ & $\mathrm{R}(S) \subseteq_{\text{homeo}} \mathrm{R}(T)$.

Proof. If the pv's of S_q, T_r (and similarly for $\mathrm{R}(S)$,$\mathrm{R}(T)$) do not coincide, then the trees

$$S'_q \subseteq_{\text{homeo}} T_r \text{ and } \mathrm{R}(S)' \subseteq_{\text{homeo}} \mathrm{R}(T)$$

obtained from S_q,$\mathrm{R}(S)$ by successive edge-introductions, so that the pv's *do* coincide, may be constructed. Now construct S' by adjoining S'_q as the left principal branch of $\mathrm{R}(S)'$. Then

$$S \subseteq_{\text{homeo}} S' \subseteq_{\text{homeo}} T.$$

□

Let $T_1, T_2, \ldots$ be any endless sequence in $\mathfrak{S}$, where $T_i \neq T_j$, $i \neq j$; $i, j \in \mathbf{N}$.

THEOREM 4.77 (KRUSKAL, NASH-WILLIAMS, STACEY)

$$(\exists i, j)\ i < j \ \&\ T_i \subseteq_{\text{homeo}} T_j.$$

Suppose there are sequences for which the thm. is false (i.e. one or more). Let T_1 be chosen so that $| T_1 |$ is the least such that T_1 starts a sequence which contains no embedding $T_i \subseteq_{\text{homeo}} T_j$. Now given $T_1, T_2, \ldots, T_i$ chosen minimally, let T_{i+1} be such that $| T_{i+1} |$ is the least such that $T_1, T_2, \ldots, T_i, T_{i+1}$, are the 1st $i+1$ terms of an embedding-free sequence. Thus a 'minimal embedding-free sequence' is defined inductively.

1. No $T_i = [\,]$, since $[\,]$ is embeddable in any $T \in \mathfrak{S}$.

2. There cannot be a subsequence $T_{n_1}, T_{n_2}, \ldots$, for which $\mathrm{L}(T_{n_1}), \mathrm{L}(T_{n_2}), \ldots$ is an embedding-free sequence, since otherwise, by the remark following defn. 74, and by 1.,

   $$T_1, T_2, \ldots, T_{n_1-1}, \mathrm{L}(T_{n_1}),\ \mathrm{L}(T_{n_2}), \ldots$$[37]

   would also be an embedding-free sequence, contradicting the minimality of T_{n_1}.

3. There cannot be endlessly many i for which

   $$\mathrm{L}(T_i) \not\subseteq_{\text{homeo}} \mathrm{L}(T_{i+1}), \mathrm{L}(T_{i+2}), \ldots \qquad \text{(I)}$$

   since otherwise, if such values were $i = n_1, n_2, \ldots$, a subsequence $T_{n_1}, T_{n_2}, \ldots$ contradicting (2) would be obtained. There are, therefore, only fm i for which (I) is satisfied. Thus, let $i = n$ be the greatest value. Now pick $N_1 > n$. Then $(\exists N_2 > N_1)\ \mathrm{L}(T_{N_1}) \subseteq_{\text{homeo}} \mathrm{L}(T_{N_2})$. And generally an endless sequence of numbers $N_1, N_2, N_3, \ldots$ can be found such that

   $$\mathrm{L}(T_{N_1}) \subseteq_{\text{homeo}} \mathrm{L}(T_{N_2}) \subseteq_{\text{homeo}} \mathrm{L}(T_{n_3}) \subseteq_{\text{homeo}} \cdots.$$

4. By lemma 75,

   $$\mathrm{R}(T_{N_i}) \not\subseteq_{\text{homeo}} \mathrm{R}(T_{N_{i+1}}), \mathrm{R}(T_{Ni+2}), \ldots$$

   for $i = 1, 2, \ldots$ otherwise $T_{N_i} \subseteq_{\text{homeo}} T_{N_j}$, contrary to $T_1, T_2, \ldots$ being an embedding-free sequence. Therefore

   $$T_1, \ldots, T_{N_1-1}, \mathrm{R}(T_{N_1}), \mathrm{R}(T_{N_2}), \ldots$$

   is embedding-free. But no $T_i = v^\sharp$, so $|\mathrm{R}(T_{N_1})| < | T_{N_1} |$, contradicting minimality of T_{N_1}.

□

Remarks

[37]Note that $T_{n_1-1} \subseteq_{\text{homeo}} \mathrm{L}(T_{n_1})$ is ruled out by $T_{n_1-1} \not\subseteq_{\text{homeo}} T_{n_1}$.

1. A subtree-embedding version of thm. 76 is impossible, since lemma 75 fails in this case; a pv-subtree version (for which the lemma holds) is prevented by the possibility that

$$T_{n_1-1} \leq \mathrm{L}(T_{n_1})$$

(ref. defn. 8·2) in part 2. of the proof.

2. The proof is non-constructive, as is the theorem.

4.9 The Lattice $(\mathfrak{T}, \cup, \cap)$

In this short section, Gabor SZASZ' monograph [144] 'Introduction to Lattice Theory' (1963) will be used as a source of terminology and standard results. Binary operators on $\mathfrak{T}$ may be obtained from defn. 19 as follows: if $S, T \in \mathfrak{T}$, then by the connexity of $\prec$, $\min_{\prec}(S,T)$ and $\max_{\prec}(S,T)$ are always defined. Let $S = [S_q, S_{q-1}, \ldots, S_1]$, $T = [T_r, T_{r-1}, \ldots, T_1]$.

DEFINITION 4.78 *1.* $v^{\sharp} \cup T = T$; $v^{\sharp} \cap T = v^{\sharp}$.

2. $S \cup T =_{df} [U_M, U_{M-1}, \ldots, U_1]$ *and* $S \cap T =_{df} [V_m, V_{m-1}, \ldots, V_1]$, *where* $M = \max(q,r)$, $m = \min(q,r)$; *and*

$$U_{M-i} = \begin{cases} T_{r-i} & \textit{if } q < r \;\&\; q \le i < r; \\ S_{q-i} & \textit{if } r < q \;\&\; r \le i < q; \\ W & \textit{if } i < m; \end{cases}$$

for $W \equiv \max_{\prec}\{S_{q-i}, T_{r-i}\}$; *and* $V_{m-i} = \min_{\prec}\{S_{q-i}, T_{r-i}\}$, $i < m$.

Thus

THEOREM 4.79 $\mathfrak{T}$ *is closed under the operations* $\cup$, $\cap$.

□

Note that $S \le_{pv} T \Rightarrow S \cup T = T$. By checking the standard laws,

THEOREM 4.80 *The system* $(\mathfrak{T}, \cup, \cap)$ *is a distributive lattice.*

□

4.9.1 Lattice Order

Define the partial order $\trianglelefteq$ by:

$$(\forall S, T \in \mathfrak{T})\; S \trianglelefteq T \Leftrightarrow S = S \cap T.$$

(With $S \angle T$ iff $S \trianglelefteq T \& S \neq T$). In the notation of defn. 78,

$$\begin{aligned} S \angle T \;&\Rightarrow\; S_{q-i} \preceq T_{r-i},\; i = 0, 1, 2, \ldots, q-1 \\ &\qquad \text{(strict for at least one value of } i\text{)} \\ &\Rightarrow\; S \prec T. \end{aligned}$$

So by thm. 39,

COROLLARY 4.81 *The ordered system* $(\mathfrak{T}, \angle\,)$ *satisfies the DCC.*

□

Note. $(\mathfrak{T}, \trianglelefteq\,)$ is non-linear.

4.9.2 Filters in $(\mathfrak{T}, \trianglelefteq)$.

A filter $\mathcal{F}$ [38] in $\mathfrak{T}$ is *principal* if $\mathcal{F} = \{S \in \mathfrak{T} \mid T \trianglelefteq S\}$, for some $T \in \mathfrak{T}$. By the corollary $(\mathfrak{T}, \angle)$ satisfies the DCC. Therefore by a theorem of DILWORTH (ref. SZASZ [144], Ch. VIII, p160)

THEOREM 4.82 *Every filter of $(\mathfrak{T}, \trianglelefteq)$ is principal.*

On the other hand, as already observed, $(\mathfrak{T}, \trianglelefteq)$ does not satisfy the ACC; so by the dual DILWORTH theorem, the lattice contains non-principal ideals.

[38] Viz. a dual ideal. Thus $\mathcal{F}$ is a non-void subsystem of $\mathfrak{T}$ which satisfies

1. $S, T \in \mathcal{F} \Rightarrow S \cap T \in \mathcal{F}$.
2. $S \in \mathcal{F}, T \in \mathfrak{T} \Rightarrow S \cup T \in \mathcal{F}$.

4.10 Notes on Ch. IV

1. §2. The tree notion is amenable to many definitions. KIRK, 'A characterization of the finite tree frames which are adequate for the Intuitionistic logic', *Zeitschr. math. Logik u. Grundl. d. Math.*, **26** (1980), 498, defines a tree as 'a finite lattice in which no element has incomparable predecessors'.

2. §2.3 T. E. CONLON (1980, preprint), has applied KÖNIG's lemma to obtain the following finiteness property of trees. Let τ be the tree obtained by indexing the vertices v of $T \in \mathfrak{S}$ with values $in(v) = i,\ 1 \leq i \leq n \in \mathbf{N}$. Now let $u \in T,\ in(u) = k$, and $S \in \mathfrak{S}$. Then *local substitution* at u is the transformation of τ to $ls(\tau)$, by

   (a) removing from T all $w\ \widetilde{\mathrm{Sc}}\ u$ (note that u remains);
   (b) introducing an edge, incident on u and on pv$S$;
   (c) identifying each endv of S with pv of a copy of Branch(u, T),

   the indices of $ls(\tau)$ are defined by:

   (d) (i) $in(u) = k + 1$; (ii) $w \in S \Rightarrow in(w) = k + 1$ (this includes the endv's of S).

   Now let $\tau_1 = \tau;\ \tau_{h+1} = ls(\tau_h),\ h = 1, 2, 3, \ldots$ denote any sequence of local substitutions.

   THEOREM 4.83 *For any fixed index bound n, and any given choice of S (viz. these may be different at each step) $(\exists N)$ for the tree τ_N all indices equal n (viz. another ls cannot be accomodated).*

   Proof. In the first instance, suppose T, S are chains. Then only finitely many substitutions are posible, as these may be represented by a finitely-valency (fv) tree of eh$\leq n$. Now construct the tree Γ pv-upwards: the successors of the pv are vertices for each chain of τ. Now suppose every chain C of τ_h is represented by a v of eh$= h$ in Γ, $1 \leq h$. Then the successors of v if any are vertices, one for each chain of τ_{h+1}, which is a ls of C. Evidently every chain of τ_{h+1} takes this form. Now, Γ is of fv, and by the preliminary, every chain of Γ is finite. Therefore from thm. 15 follow the finiteness of Γ and of the substitution sequence.

   □

   Note. In general, $<_{\text{post}}$ is not conserved.

3. §2.5. Let $t_n(h)$ denote the number of vertex-labelled oriented trees on n vertices of eh$\leq h$. It is well-known (RIORDAN [128] 1960, 'The enumeration of trees by height and diameter', *IBM J. Research and Development*, **4**, 473-478; and RÉNYI and SZEKERES[127]) that the generator

$$G_h(x) = \sum_{n=1}^{\infty} \frac{t_n(h)}{(n-1)!} x^n$$

   satisfies the remarkable recurrence

$$G_{h+1}(x) = xe^{G_h(x)},\ h = 0, 1, 2, \ldots,$$

   where $G_0(x) = x$. Thus e.g. $\lambda x \cdot G_3(x) = xe^{xe^{xe^{x}}} \approx \omega \star 4$, and so is of the order of a higher polynomial of **EP**.

4. §3·2, rem. 3. The theory of (the class $\mathcal{X}$ of) pv-oriented trees *without* plane structure coincides exactly with that of the *finite* sets. Thus

   (a) $[\,] \equiv \{\}$, the empty set.

   (b) Let the sets $x, y, \ldots, z \equiv$ the trees $S, T, \ldots, U$. Then $\{x, y, \ldots, z\} \equiv \check{[S, T, \ldots, U]}$, where '$\check{}$ ' indicates that the order of terms enclosed by '[,]' is to be ingnored. Evidently the objects of $\mathcal{X}$ satisfy the usual axioms of ZF set theory, less the 'axiom of infinity':

$$(\exists y)\ (\{\} \in y \ \&\ (\forall x)\ (x \in y \Rightarrow \{x \cup \{x\}\} \in y)).$$

   Thus (ref. II, §2·18) if ZFF denotes the depleted theory, $(\mathcal{X}, \mathrm{Sc}) \models \mathrm{ZFF}$.

5. §3.3. $(\forall T \in \mathfrak{T})$ there is a fixed number, N, say, of $S \in \mathfrak{S}$ for which the normal form of S is T.

6. §2.1.The evolution of the terminology 'plane structure': a planar graph can be flattened without edge-intersection; every tree has this property, therefore 'every tree is planar'. A tree may or may not, however,possess plane structure.

7. §5. Let $S, U, \ldots$ be the 'odd (parts) subtree', the 'unequal subtree', ..., of $T[n]$. Are $S, U, \ldots$ normal trees?

8. §9. The operation of edge-elimination

$$T = [T_r, \ldots, T_1],\ T_i = [T_{(q)}, \ldots, T_{(1)}],\ S = [T_r, \ldots, T_{r-q}, T_{(q)}, \ldots, T_{(1)}, T_{q-1}, \ldots, T_1],$$

   gives $S \sqsubset T$.

Chapter V

A SECOND NORMAL FORM IN EP

Section Headings

5.1 Introduction

The exponentials which arise inductively, as in Ch. II, defn. 3, spontaneously align in the normal form of II·22. This form is the basis of the assignment Ω of fundamental sequences for diagonal exponentials e. By Ch. III, thm. 12, any $\beta \prec e$

in **EP** is either smaller than all the canonical terms (ct's) of e, or can be trapped between consecutive ct's:

$$(e,n) \preceq \beta \prec (e,n+1);$$

and similarly if $(e,n+1)$ is diagonal. On the other hand, the canonical reduction of III, §4 produces an inexorable $\prec$ – descent. These 2 procedures, together with the finiteness result, Ch. IV, corol. 48, suggest not only trapping but pinpointing any $e \in \mathbf{EP}$, by iterated canonical reduction, starting at the diagonal ε (ref. II, §2; and III, defn. 6·3). This approach is contained in McBETH [98] (1984) [1], and is taken in the present Chapter. It turns out that the indices obtained are (all but) the ec's of the II·22-normal form; that the coefficients naturally occur as a tree, the post-order of which is the above order of $\prec$ –descent. To begin, specialized notation is introduced. Let e be diagonal. Then for indices ranging over **N**,

DEFINITION 5.1 *1.* $(e,n) = \Omega(e,n)$, *as in Ch. III.*

2. Suppose that $(e,n_1,n_2,\ldots,n_k)$ *has been defined as a function* $\phi \in \mathbf{EP}$. *If* ϕ *is diagonal then* (ϕ, n_{k+1}) *is denoted by the symbol* $(e,n_1,n_2,\ldots,n_k,n_{k+1})$.

3. Where the $n_i = 1$, *it is often convenient to introduce suffices. E.g.* $(e,2,1,1) = (e,2,1_1,1_2)$.

4. For $e \in \mathbf{EP} \cup \{\varepsilon\}$, *by defn.* $(e,0) = \mathbf{1}$.

□

Note that $(e,\underline{n})$ is not defined for every sequence $\underline{n}$. Also, if $(e,n_1,\ldots,n_k)$ is defined, then

$$\ldots \prec (e,n_1,n_2) \prec (e,n_1) \prec e,$$

as by III·7·3, a diagonal eventually dominates each of its canonical terms.

Examples

1. (Ref. II, §2) $(\psi,4) = \omega^4$; $(\psi,4,5) = 5\omega^3$; $(\psi,4,5,3) = 4\omega^3 + 3\omega^2$; $(\psi,4,5,3,1) = 4\omega^3 + 2\omega^2 + \omega$; $(\psi,4,5,3,1,2) = 4\omega^3 + 2\omega^2 + 2$.
2. Let $e = \omega \star 4$ (ref. II, defn. 6). Then $(e,4,5,3) = \omega \star (2; 4\omega^3 + 3\omega^2)$.
3. $(\varepsilon,3) = \omega \star 3$; $(\varepsilon,3,1_1,1_2,1_3) = \mathbf{1}$.
4. $(\psi,4,1)(3) = 27$.

[1] *Zeitschr. math. Logik Grundl. Math.*, Bd. 30, 393-400.

5.2 Compound Diagonals

By Ch. III, defn. 16, a diagonal e is *simple* if $(\forall n)$ (e,n) is a successor, viz. $e = \Sigma + \omega$, $\Sigma \in \mathbf{EP_0}$, according to II·22. Otherwise e is *compound*, viz. $e = \Sigma + \omega^\beta$, $\beta \in \mathbf{EP}$, $\beta \succ \mathbf{1}$, in normal form. The subsequent defn. in Ch. III is a further classification of the compound e :

DEFINITION 5.2 *If* Suc(β), *then* e *is of the* first kind; *if* Diag(β), e *is of the* second kind.

Note the correspondence between the 2 cases above, and the respective successor and diagonal cases of defn. III 6·2 of the ct (ω^{e_1}, n).

Further Examples

1. By III·14, e compound $\Rightarrow (e,n) \preceq (e, n+1, 1)$, $n = 1, 2, 3, \ldots$.
2. The respective $\prec$-lowest cd's of the 1st and 2nd kinds are ω^2 and ψ.
3. A diagonal of the first kind may be 'further' than a diagonal of the second kind from a non-diagonal. e.g. $f = \omega \star (2; \psi)$, $e = \omega^{\psi+2}$ give

$$e \prec f \;\&\; (e, 1_1, 1_2, 1_3, 1_4, 1_5) = \mathbf{1} = (f, 1_1, 1_2, 1_3, 1_4).$$

Broadly speaking, **EP** falls naturally into a progression of non-diagonals, simple diagonals, compound firsts, compound seconds. An alternative classification of diagonals might be based on the class $\mathcal{X}$ discussed in Ch. II, after defn. 33. For the aims set out in §1, the classification of defn. 2, however, proves adequate.

5.2.1 An Elementary Structural Property of the Diagonal

Consider $\alpha = 2\omega = \lambda x \cdot 2x$ and $\beta = \omega^2 = \lambda x \cdot x^2$. In the notation of III, defn. 3, these functions are related to their canonical terms by

$$\alpha = \text{diag}(\omega + \mathbf{n});\; \beta = \text{diag}\; n\omega.$$

In the sense of III·4, these [2] are the only *regular* sequences for which the respective functions are the diagonals. Now $(\beta, 1) = \omega$; $(\beta, 2) = 2\omega = \alpha$. But $(\alpha, 1) = \omega + \mathbf{1}$, i.e. $(\beta, 1) \prec (\beta, 2, 1)$. Thus the standard assignment Ω cannot give

$$(e,n) = (e, n+1, 1)$$

for all cd's e, $n = 1, 2, 3 \ldots$. In the ($\prec$ −minimal) example above, $e = \beta$ is of the first kind (viz. $\beta = \mathbf{1}$). It will be shown that for *every* such cd e,

$$(e,n) \prec (e, n+1, 1).$$

[2] Aside from the constant assignment $2\omega = \text{diag}_n\,(\mathbf{n} + \mathbf{n})$; see III, §5·1.

At first glance, it may be thought that equality could be obtained always by allowing n to range over $\mathbf{N}_0$, [3] and canonical terms to range over $\mathbf{EP_0}$, in Ch. III, defn. 6. This would give e.g.

$$(\beta, 1) = \omega = (\beta, 2, 0).$$

But

$$(\psi, 1) = \omega \succ \mathbf{0} = (\psi, 2, 0).$$

Not only does the $(e, n+1, 0)$–equality therefore fail, but worse, the BACHMANN starting property is lost. Similarly in the minimal case: $(\psi, 0) \succ (\psi, 1, 0)$.

A Property of Compound Diagonals

Let $\beta = {\cdot^{\cdot^{\cdot}}}\omega^{\Sigma+\omega}$ be perfect (II·33), (viz. $\Sigma \in \mathbf{EP_0}$). Then

$$(\beta, n+1, 1) = ({\cdot^{\cdot^{\cdot}}}\omega^{\Sigma+\mathbf{n}+\mathbf{1}}, 1) = {\cdot^{\cdot^{\cdot}}}1 \cdot \omega^{\Sigma+\mathbf{n}} = (\beta, n).$$

Now let $\beta = {\cdot^{\cdot^{\cdot}}}\omega^{\Sigma^+}$, $\Sigma \in \mathbf{EP}$, be imperfect. Then

$$(\beta, n+1, 1) = ({\cdot^{\cdot^{\cdot}}}(n+1) \cdot \omega^{\Sigma}, 1) = {\cdot^{\cdot^{\cdot}}}n \cdot \omega^{\Sigma} + (\omega^{\Sigma}, 1) \succ (\beta, n).$$

Therefore for any compound diagonal,

$$(\beta, n+1, 1) = (\beta, n) \text{ iff } \beta \text{ is perfect.}$$

5.2.2 Reduction Sequences

By III·7·3 of the result summarizing defining properties of Ω, and as noted in §1, the taking of a ct entails a reduction with respect to $\prec$. Indeed, given $e \prec f$ in **EP** it is possible to reach e, starting from f after only finitely many (fm) canonical descents, or operations of taking the predecessor. Thus, taking the relation $\lhd$ of canonical component and its transitive closure $\tilde{\lhd}$ of III·19,

THEOREM 5.3 [4] *The relations $\tilde{\lhd}$, and $\prec$ on* $\mathbf{EP} \times \mathbf{EP}$ *are one and the same.*

Proof. V·3 holds for the constants, so take $\omega \preceq f$. Certainly $e \mathrel{\tilde{\lhd}} f \Rightarrow e \prec f$, since by the above-mentioned thm. of Ch. III (and in the successor case, by Ch. II, ex. foll. defn. 8), $e \lhd f \Rightarrow e \prec f$. Now let $e \prec f$. Suppose $f = \delta + \mathbf{a}$, $a \geq 1$ is a successor, in the form of II·31. Then either $e = \delta + \mathbf{k}$, $0 \leq k < a$, so that $e \mathrel{\tilde{\lhd}} f$; or $e \prec \delta$. Thus suppose now Diag(f). Then by III, thm. 12, there is a least n such that $e \preceq (f, n)$. If $e = (f, n)$, then $e \lhd f$ and $e \mathrel{\tilde{\lhd}} f$. Let

$$f_1 = \begin{cases} f^- & \text{if Suc}(f) \\ (f, n) & \text{if Diag}(f). \end{cases}$$

[3] And ignoring the convention, in §2.1 only, of defn. 1·4.
[4] [97], §2, thm. 2.5.

Thus, suppose $e \prec f_1$. Now $f_1 \lhd f$. Furthermore, either Suc(f_1) or Diag(f_1); and f_1 gives f_2 as f gave f_1, so that $e \preceq f_2 \lhd f_1$. Thus the sequence

$$(e = f_1) \vee (e \prec f_1 \ \& \ e = f_2) \vee (e \prec f_1 \ \& \ e \prec f_2 \ \& \ e = f_3) \vee \ldots$$

is obtained. But $\ldots \lhd f_3 \lhd f_2 \lhd f_1$. Therefore by the DCC for $(\mathbf{EP}, \lhd)$, IV, corol. 49, the sequence has only fm terms. So $(\exists m \geq 1)\ e = f_m$, and $e \tilde{\lhd} f$.

□

Note that the DCC need not be invoked. Thus, $f_1, f_2, \ldots \in \mathcal{X} = \mathbf{EP}[f, N(e)]$ by rem. 3 after III·12. But by the corol. after III·22, $\mathcal{X}$ is finite.

Examples

1. $e = \omega^4, f = \psi^+$, and $e = (f^-, 4)$; $\psi + 7\omega = (\omega \star (3;3), 1, 1, 2, 2, 2, 1, 2, 2, 7)$.
2. $(\omega \star 1\ 000, 1_1, 1_2, \ldots, 1_{1000}) = \mathbf{1}$.

Remarks

1. By III·12, ex. 2, thm. 3 does not hold for reductions wrt non-canonical assignments, e.g. Λ.
2. After the possible occurrence of fm predecessions at the beginning of the determinate sequence $f_1, f_2, f_2, \ldots$, either e is reached, or there is a first occurrence, f_i say of a diagonal. Now either every term $f_i, f_{i+1}, f_{i+2}, \ldots$ is diagonal (as is, therefore e) or there is a first subsequent occurrence, f_I say, $I > i$ of a successor. In this case $\phi = f_{I-1}$ is a simple diagonal. e is not necessarily equal to f_I, as it may be smaller than every ct of ϕ. However, in this case since

$$f_I = (\phi, 1) = \delta + \mathbf{1}$$

where $f_{I+1} = \delta$ is diagonal (rem. after III, defn. 16), with a single reduction by predecessor, the sequence returns to reduction via ct's. It therefore may be concluded that, in thm. 3, aside from the beginning of the sequence all reductions by predecessor occur *singly*.

3. In the notation of III, defn. 24, given an arithmetic sequence $\mathcal{S} = (n_r)$, and a diagonal $e \in \mathbf{EP}$, $e_{(r+1)} = (e, n_1, n_2, \ldots, n_r, n_{r+1})$ for all r for which Diag$\{e_{(r)}\}$. As noted e.g. before thm. 3, $\ldots \prec e_{(2)} \prec e_{(1)}$, so by the DCC, the functional sequence

$$e_{(1)}, e_{(2)}, e_{(3)}, \ldots$$

is *finite*. Therefore

(a) in particular, given any diagonal e and si function $F \in \mathcal{N}$, no matter how fast-growing, there is an $r \geq 1$ for which the exponential $(e, 1+F(1), F(2), F(3), \ldots, F(r)) \succ (e, 1 + F(1), -1 + F(2))$ is a successor;

(b) it is possible to generalize the 'descent function' of III, defn. 25 to $D_e = D_e(\mathcal{S}) =$ the (least) number $k \geq 1$ such that $(e, n_1, n_2, \ldots, n_k)$ is a successor.

5.3 Recursive Starting Property

5.3.1 Introduction

As in the example $e = \sum + \omega^{\omega^6+1}$, it is evident that, via iteration of the step-down of Ch. III, it is possible to pass downwards from one ct to its homologue:

$$(e, n+1, 1_1, 1_2, \ldots, 1_7) = (e, n)^+, \ 1 \leq n.$$

The reduction naturally falls into 2 cases, as in the following treatment.

5.3.2 Iterated Step-Down

Let $\lhd$ be the relation of Ch. III, defn. 18·3, viz. for all $e, f \in \mathbf{EP}, e \lhd f$ iff $e = f^-$ or Diag(f) & $e = (f, 1)$. Let $\tilde{\lhd}$ be the transitive closure of $\lhd$.

LEMMA 5.4 $(\forall e \in \mathbf{EP})\ \mathbf{1} \prec e \Rightarrow \mathbf{1}\ \tilde{\lhd}\ e.$

Proof, via corollary after III·22.

□

There follow 2 specialized thms. on the 'starting' properties of the standard assignment Ω. They are special cases of BACHMANN'S thm. 14 of Ch. III, and contain additional information about Ω as well.

NOTATION

Suppose for $n, k \in \mathbf{N}$, for the compound diagonal (cd) $e \in \mathbf{EP}$, $(e, n+1, 1_1, 1_2, \ldots, 1_k)$ is defined. Then $\mu_k = \mu_{e,n+1,k}$ will denote this function. E.g. taking $e = \omega\star(3; \omega^2 + \omega + \mathbf{1})$, $n = 4$; $\mu_3 = \omega \star (2; 4\omega^{\omega^2+\omega} + \psi)$.

THEOREM 5.5 [5]

Let e be a cd of the 1st kind, with $e = \Sigma + \omega^{\beta+1}$. Then

$$(\exists K)(\forall n)\ (e, n+1, 1_1, 1_2, \ldots, 1_{K+1}) = (e, n) + \mathbf{1}.$$

Proof. For $\beta = \mathbf{1}, (\Sigma + \omega^2, n+1, 1) = \Sigma + n\omega + \mathbf{1} = (e, n) + \mathbf{1}$; and for $\mathbf{1} \prec \beta$, by lemma 4 $(\exists K)\ \mathbf{1} = \beta_K \lhd \beta_{K-1} \lhd \ldots \lhd \beta_1 \lhd \beta$. So

$$\mu_1 = (\Sigma + (n+1)\cdot\omega^\beta, 1) = \Sigma + n\cdot\omega^\beta + (\omega^\beta, 1) = (e, n) + \omega^{\beta_1}.$$

And generally, for $1 \leq k \leq K$, $\mu_k = (e, n) + \omega^{\beta_k}$ implies

$$\begin{aligned} \mu_{k+1} = (e, n) + (\omega^{\beta_k}, 1) &= (e, n) + \begin{cases} \omega^{\beta_k^-} & \text{if } \beta_k = \mathbf{1} \text{ or if Suc}(\beta_k) \\ \omega^{(\beta_k, 1)} & \text{if Diag}(\beta_k) \end{cases} \\ &= (e, n) + \omega^{\beta_{k+1}} \text{ if } k < K \ \text{ (i.e. if } \beta_k \succ \mathbf{1}). \end{aligned}$$

[5] [98], thms. 2.2,3, p394.

Therefore by induction,

$$\mu_{K+1} = (e, n+1, 1_1, 1_2, \ldots, 1_{K+1}) = (e, n) + (\omega^{\beta_K}, 1) = (e, n) + \mathbf{1}.$$

□

Example. $e = \omega^{\omega^2+3}, \quad n = 4; \quad \mu_5 = (e, 4) + \mathbf{1}$.

In contrast to the case for $\mathrm{Suc}(e_1)$:

THEOREM 5.6 *Suppose e is a cd of the 2nd kind. Then*

$$(\forall n)(\exists K)\,(e, n+1, 1_1, 1_2, \ldots, 1_{K+1}) = (e, n).$$

Proof. Let $e = \Sigma + \omega^{e_1}$, $\Sigma \in \mathbf{EP_0}$, where $\mathrm{Diag}(e_1)$. Let the functions $e_1^{(t)}$, $t = 1, 2, 3, \ldots$ be defined recursively by: $e_1^{(1)} = e_1$; if $e_1^{(t)}$ has been defined, with normal form

$$e_1^{(t)} = \Sigma_t + \begin{cases} \omega^\xi & \text{if } \mathrm{Diag}(e_1^{(t)}) \\ \mathbf{1} & \text{if } \mathrm{Suc}(e_1^{(t)}), \end{cases}$$

$\xi \in \mathbf{EP}$, $\Sigma_t \in \mathbf{EP_0}$. Then if $\mathrm{Diag}(e_1^{(t)})$, $e_1^{(t+1)}$ is defined to be ξ. Well, by II·26, $e_1^{(1)} \prec e_1^{(2)} \prec e_1^{(3)} \prec \ldots \prec e_1^{(t)} \prec \ldots$, is finite, therefore

$$(\exists T) \begin{cases} 1.\ e_1^{(T)} = \mathbf{1} & \text{or} \\ 2.\ \mathrm{Suc}(e_1^{(T)}). \end{cases}$$ [6]

Since, by Ch. III, rem. 3 following defn. 6, $e^{(t+1)}$ is the minimal exponent of $e^{(t)}$, by II, defns 23·2(c) and 33, $e_1^{(T)}$ is the register polynomial of e. Now for $n \geq 1$, according to its inductive definition, (e, n) may be obtained by replacing, in the explicit expression of e

$$\omega^{e_1^{(T)}} \text{ by } \begin{cases} n & \text{if 1.} \\ n \cdot \omega^{\Sigma_T} & \text{if 2.} \end{cases}$$

Consider now $(e, n+1)$ obtained by the appropriate replacement in the e-expression.

1. rp$= \mathbf{1}$. $(e, n+1) = \therefore \Sigma_{T-2} + \omega \cdot \omega^{\Sigma_{T-1}+n}$ and $(e, n+1, 1) = \therefore \Sigma_{T-2} + 1 \cdot \omega^{\Sigma_{T-1}+n} = (e, n)$,

   and $K = 0$.

2. $\Sigma_T \succeq \mathbf{1}$. Let $\zeta = \Sigma_{T-1} + n \cdot \omega^{\Sigma_T}$. Then $(e, n) = \therefore \zeta$, $\quad (e, n+1) = \therefore \zeta + \omega^{\Sigma_T}$, $(e, n+1, 1) = \therefore \zeta + (\omega^{\Sigma_T}, 1)$.

[6] $T \geq 2$ since e is of the 2nd kind. But in case 2. T may take the value 2.

Thus $(e, n+1, 1)$ is of the form

$$\zeta + \beta, \text{ where } \beta \in \mathbf{EP}. \tag{I}$$

$\therefore$

The result for case (2) is now obtained by induction on k. Suppose, for some k it has been shown that the polynomial normal form of μ_k takes the form (I), so that $\mu_k \succ (e, n)$. Then

$$\mu_{k+1} = \Sigma_{T-2} + \begin{cases} 1 \cdot \omega^{\zeta} = (e, n) & \text{if } \beta = \mathbf{1} \\ 1 \cdot \omega^{\zeta+\beta^-} \succ (e, n) & \text{if Suc}(\beta) \\ \omega^{\zeta+(\beta,1)} \succ (e, n) & \text{if Diag}(\beta). \end{cases} \tag{II}$$

Thus it is proved by induction that: $(\forall k \geq 1)\ \mu_k \succ (e, n) \Rightarrow \mu_k$ is of the form (I).

Now write $\beta = \beta_k$ in the expression (I) for μ_k, for k such that $\mu_k \succ (e, n)$. Then since (II) holds for all such k, $\mu_{k+1} \succ (e, n) \Rightarrow \beta_{k+1} \lhd \beta_k$. Therefore by lemma 4 $(\exists K)\ \beta_K = \mathbf{1}$. Therefore $\mu_{K+1} = (e, n)$. Thus, for case (2) the required expression is found, and the thm. is proved.

□

Example. The $\prec$ −minimal case is: $(\psi, n) = \omega^n = (\psi, n+1, 1)$. Generally, if the cd $e = \Sigma + \omega^{e_1}$ is perfect, by §2·1,

$$(e, n+1, 1_1, \ldots, 1_n) = (e, 1).$$

In the special case $\Sigma_T = \Sigma_{T-1} = \ldots = \Sigma_1 = \mathbf{0}$, μ_{n+T-1} is $\mathbf{1}$ or a successor, as $\Sigma =$ or $\succ \mathbf{0}$.

COROLLARY 5.7 *If e is a diagonal of* **EP**, *and $n \geq 1$, then $(e, n) \tilde{\lhd} (e, n+1)$,*

Proof, by thms. 5 and 6.

□

If $\beta = (e, n+1, m)$ is defined, then (taking $m \geq 2$) the corol. gives:

$$(e, n) \tilde{\unlhd} (e, n+1, 1) \tilde{\lhd} \beta$$

And generally, if $\beta = (e, n+1, m_1, \ldots, m_R) \succ (e, n)$ is defined, $m_i \geq 1$, then

$$(e, n) \tilde{\lhd} \beta.$$

Note that corol. 7 immediately implies a special case of BACHMANN's thm., viz. if e is compound,

$$(e, n) \preceq (e, n+1, 1).$$

Remarks

1. In the above proof (of thm. 6) attention is directed to the *right-side* or $\prec$ *–lowest* term of the normal form of e (viz. the term which contains the register polynomial). The opposite is the case in the lexicographic proof of the DCC for $(\mathfrak{T}, \prec)$ of Ch. IV, thm. 39, in which attention is directed to the leading term.

2. In the (I)-expression for μ_1 in the above proof, $\beta_1 = (\omega^{\Sigma_T}, 1)$ is independent of n. Therefore using $\beta_{k+1} \lhd \beta_k$, $k = 1, 2, \ldots$, the number K is independent of n. Therefore the quantifiers may be exchanged, to give the stronger (uniform) version

COROLLARY 5.8

$$e \text{ of the 2nd kind } \Rightarrow (\exists K)(\forall n)\, \mu_K = (e, n).$$

3. As to the register polynomial $e_1^{(T)} = \Sigma_T + \mathbf{1}$ of e, $\Sigma_T \in \mathbf{EP_0}$:

$$e = \Sigma_1 + \omega^{\Sigma_2 + \omega^{\cdot^{\cdot^{\cdot}}}} = {}^{\cdot^{\cdot^{\cdot}}} \Sigma_{T-1} + \omega^{\Sigma_T + \mathbf{1}}.$$

$\omega^{e_1^{(T)}}$ is the 'part' of a diagonal e which changes in the passage from e to (e, n), for any $n = 1, 2, 3, \ldots$. The dee of f for e in **EP** of Ch. IV, defn. 50·2 depends on both e and f; while the register polynomial of f is determined by f alone. Thus, the 2 may be distinct, as in $e = \omega \star 4 + \psi + \omega$, $f = \omega \star 4 + \omega \star (\omega; \omega^2 + \mathbf{1}) + \omega$. $f_1^{(2)} = \mathbf{1}$, dee of f for $e = \omega^2 + \mathbf{1}$. Or they may coincide, as in $e = \psi$, $f = \omega^{\omega + \mathbf{1}}$, dee$= \omega + \mathbf{1} = f_1^1$. Another possibility is that the dee is a proper subpolynomial of rp$f$, as in

$$e = \Sigma + \omega \star (n+1) \prec \Sigma + \omega^{\omega \star (n;2)^+} = f,$$

where $f_1^{(1)} = \omega \star (n; 2)^+$, dee$= \mathbf{2}$.

From the forgoing, it is clear that if $e \prec f$ are both diagonals, then their cfs's give no information on the dee, α, of f for e. Conversely, α gives no information on the distance between intervals of **EP** swept out by the fundamental sequences.

4. Let $T \in \mathfrak{T}$ be diagonal, and consider any (T, n), $n > 1$. Then IV·69 and corol. 7 give as a further corollary: the normal sequence of (T, n) contains every (T, i), $i = 1, 2, \ldots, n-1$; viz. applying III·8 to IV·62: contains every $S\ (\tilde{\lhd})\ (T, n)$.

5. The iterated $\lhd_1$ occurs in the verification of rem. 1 after III corol. 15. Thus, suppose β is compound of the 2nd kind, with $\alpha = (\beta, n)$. Then $\alpha \prec (\beta, n+1)$, so $(\exists r)\ \beta' = (\beta, n+1, 1_1, \ldots, 1_r) \prec \beta$ and $\alpha = (\beta', 1)$. This has a clear representation in

the linear diagram

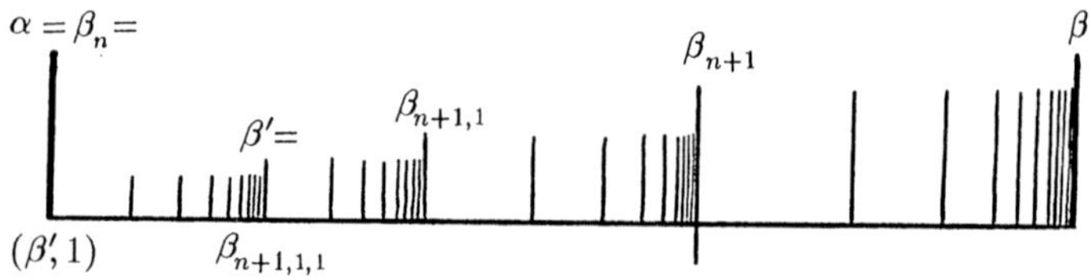

In the case $n = 1$, it may be asked: can $\gamma \succ \beta$ be found, s.t. $\alpha = (\gamma, 1)$, $\beta = (\gamma, 2, 1_1, \ldots, 1_R)$ for some $R \geq 1$? This would essentially replace β by γ and β' by β in the figure. By the result of III, §3·3, this could happen only fm times. Such a case is given by ex. 3 of the section. Let the exponentials $\beta_{l,m} = \omega \star (l; \omega^{\mathbf{1}+\omega\star m})$ be as in the ex. Then for $l = 1, 2, \ldots, N-1$ the forms $\beta = \beta_{l^-,m^+} \prec \beta_{l,m} = \gamma$ satisfy $(\gamma, 2) = \omega \star (l; 2 \cdot \omega \star m^+)$. But $\mathbf{1} \lhd_1 \omega \star m^+$, therefore

$$(\exists R)\ (\gamma, 2, 1_1, \ldots, 1_R) = \omega \star (l; \mathbf{1} + \omega \star m^+) = \beta,$$

while (for each l), $\alpha = (\beta_{l,m}, 1)$, so that

$$\alpha = (\gamma, 2, 1_1, \ldots, 1_R, 1_{R+1}).$$

5.4 ε–Normal Form for Functions in EP

Let $\star$ be the operation of iterated exponentiation of Ch. II, defn. 6, and recall the subsequent notation II, 2.1:

DEFINITION 5.9 *The arithmetic function ε is defined by $\varepsilon = \lambda x \cdot x \star x = \omega \star \omega$.*

□

By II, corol. 16, ε dominates every member of **EP**. Other nd arithmetic functions $f \prec \varepsilon$ have this property. E.g. $f = \omega \star l$ (vide ex. 3, after II·2) whose growth is considerably slower than that of ε : thus, $f = 1, 1, 3, 4, 5, 6, 7, 2^{24}, \ldots, 17^{17}, 18^{18^{18}}, \ldots$ as $x = 1, 2, 3, 4, 5, 6, 7, 8, \ldots, 17, 18, \ldots$. The initially sedate growth of f eventually exceeds that of every constant-height exponential. Nonetheless, ε appears as the smallest dominating function for **EP**, not constructed from inverse operations and as such may be considered the supremum. Recall Ch. II, notation 2·1·2:

DEFINITION 5.10 $\overline{\mathbf{EP}} = \mathbf{EP} \cup \{\varepsilon\}$.

□

ε has the evident assignment of canonical terms

$$\varepsilon = \text{diag } \omega \star n,$$

as in III·6·3. The notation ε is chosen, following CANTOR. Return now to III, defn. 19, viz. $e \lhd f$ in **EP** iff

1. $e = f^-$ if Suc(f);

2. $(\exists n)\ e = (f, n)$ if Diag(f).

Well, by thm. 3, $e \prec f \Leftrightarrow e \ \tilde{\lhd}\ f$. But in general the route $e = e_1 \lhd e_2 \lhd \ldots \lhd e_k = f$ is not unique. For example, let $e = \omega \star 2, f = \omega \star 3$. Then $e \ \tilde{\lhd}\ f$, with $e = (f, 1) = (f, 3, 1, 1) = (f, 2, 3, 1_1, 1_2, 1_3, 1_4)$ etc. Nonetheless, by choosing the canonical term of minimal index at each downward step, a unique route may be picked out. By taking $f = \varepsilon$, this route determines a second normal form for each e. The finiteness of the above reduction rests on the DCC thm. for $(\mathbf{EP}, \prec)$.

THEOREM 5.11 [7] *($\varepsilon-$normal form theorem for exponential polynomials) for every* $e \in \mathbf{EP}$, *there is an* R, *and a unique sequence* $n_1, n_2, n_3, \ldots, n_R$ *such that*

$$e = (\varepsilon, n_1, n_2, \ldots, n_R).$$

Proof. If $e = \mathbf{1}$, then $e = (\varepsilon, 1, 1)$. I.e. in this case, $R = 2, n_1 = 1, n_2 = 1$. Now let $\mathbf{1} \prec e$. By the boundedness of height of e, $(\exists! n_1)\ (\varepsilon, n_1 - 1) \prec e \preceq (\varepsilon, n_1)$. If $e = (\varepsilon, n_1)$, a unique expression is found. Otherwise, by III, thm. 12, $(\exists! n_2 \geq 1)$

$$(\varepsilon, n_1, n_2 - 1) \prec e \preceq (\varepsilon, n_1, n_2),$$

and the inequality '$\preceq$' may be treated as before. Thus a $\prec -$ descending sequence

$$\ldots \prec (\varepsilon, n_1, n_2) \prec (\varepsilon, n_1) \qquad \text{(I)}$$

may be obtained, for which

$$(\varepsilon, n_1 - 1), \prec (\varepsilon, n_1, n_2 - 1), \prec (\varepsilon, n_1, n_2, n_3 - 1) \ldots \prec e. \qquad \text{(II)}$$

Now by the corol. after III·22, the sequence (I) lies in $\mathbf{EP}[\varepsilon, N(e)]$, which is finite, by III·20. Viz. (I) has a lowest term, μ say. Thus

$$(\exists R)\ e \preceq (\varepsilon, n_1, n_2, \ldots, n_R) = \mu.$$

It is required to prove that $e = \mu$. The facile cases are dealt with first. Suppose Diag(μ). Then if $e \prec \mu$, by the density result III·12,

$$(\exists! n)\ (\mu, n - 1) \prec e \preceq (\mu, n),$$

[7] [98], §3, thm. 1.

so that μ is not the lowest term of the sequence (I), contrary to supposition. Therefore $e = \mu$. Now suppose $\mathrm{Suc}(\mu)$ [8]; and to begin, suppose $n_R > 1$. Then since $\mathrm{Suc}(\mu)$,

$$\bar{\mu} = (\varepsilon, n_1, n_2, \ldots, n_{R-1})$$

is a simple diagonal. But by (II),

$$(\bar{\mu}, n_R - 1) \prec e \preceq (\bar{\mu}, n_R) = \mu.$$

Therefore, as $\bar{\mu}$ is simple, $e = \mu$. Now, still under the assumption $\mathrm{Suc}(\mu)$: suppose

$$n_R = n_{R-1} = \ldots = n_{Q+1} = 1 \text{ and } n_Q > 1$$

(there must be such a $Q < R$, as $\mathbf{1} \prec e \preceq \mu$, $(\varepsilon, 1, 1) = \mathbf{1}$ and $(\varepsilon, 1_1, \ldots, 1_R)$ is not defined for $R > 2$). Thus

$$\mu = (\varepsilon, n_1, \ldots, n_Q, 1_{Q+1}, \ldots, 1_R).$$

Write $(\bar{\mu}, n_Q) = (\varepsilon, n_1, \ldots, n_Q)$, where $\bar{\mu}$ is a cd. Now $\bar{\mu}$ cannot be of the second kind, since if it were, by thm. 6 there is an $N \geq 1$ such that

$$(\varepsilon, n_1, \ldots, n_Q, 1_{Q+1}, \ldots, 1_{Q+N}) = (\varepsilon, n_1, \ldots, n_Q - 1) \prec e$$

by (II), and where $R < Q + N$ by the inequality on the right. And for $1 \leq M < N$, by the argument of thm. 6

$$(\varepsilon, n_1, \ldots, n_Q, 1_{Q+1}, \ldots, 1_{Q+M})$$

is diagonal, contradicting $e \preceq \mu$ and $\mathrm{Suc}(\mu)$. It therefore follows that $\bar{\mu}$ is of the 1st kind. So by thm. 5 $(\exists N)$

$$(\varepsilon, n_1, \ldots, n_Q, 1_{Q+1}, \ldots, 1_{Q+N}) = (\varepsilon, n_1, \ldots, n_Q - 1) + \mathbf{1},$$

while, by the argument of thm. 5, for $1 \leq M < N$,

$$(\varepsilon, n_1, \ldots, n_Q, 1_{Q+1}, \ldots, 1_{Q+M})$$

is diagonal. But by (II), $(\varepsilon, n_1, \ldots, n_Q - 1) \prec e$. Therefore, under the stated assumptions, $R = Q + N$, and $\mu = (\varepsilon, n_1, \ldots, n_{Q-1}, n_Q - 1) + \mathbf{1}$, with $e = \mu$, and the thm. is proved.

□

DEFINITION 5.12 (SECOND NORMAL FORM) *For* $e \in \mathbf{EP}$, *if the integers* $n_1, n_2, \ldots, n_R$ *are those determined by thm. 11, then*

$$e =_{df} \varepsilon_{n_1, n_2, \ldots, n_R}.$$

[8] Note that, since (ε, n) is diagonal, $n = 1, 2, 3, \ldots$, this assumption implies $R > 1$.

□

Defn. 3·6 ·3 assigns ε a cfs via $\star$-exponentiation, viz. differently from the assignment of diagonals below ε. Thus, by a partial analogue with decimal notation, n_1 may be distinguished as the *characteristic*, and the string $n_2, n_3, \ldots, n_R$ as the *mantissa* of the expression. In numerical examples, this may be indicated in the usual way.

Examples

1. The constants take the normal form $\mathbf{n} = \varepsilon_{1,n}$; $\omega \star \mathbf{n} = \varepsilon_n$. Let $e = \psi + \omega + 1$. Then $(\varepsilon, 3) = \omega \star 3; \ldots; (\varepsilon, 3, 2, 2, 1, 2, 2) = \psi + 2\omega$ and $e = \varepsilon_{3\cdot 221221}$.

2. Suppose e is of the first kind. Then by the final case of the proof,

$$(\exists K)\,(e, n)^+ = \varepsilon_{\underline{n}, 1_1, \ldots, 1_{K+1}}, \quad \text{where } (e, n+1) = \varepsilon_{\underline{n}}.$$

3. The example $(\mathcal{X}, <)$ given under the heading 'Fundamental sequences for real numbers' in Ch. III, §5·2 affords a clear view of the working of the present thm. 11. Thus, suppose $x \in \mathcal{X}$ is chosen, a black point, say. Then certainly $x = \frac{n}{n+1}$ for some n, or

$$a_1 = \frac{n}{n+1} < x < \frac{n+1}{n+2} = b_1,$$

in which case the search for x may be transcribed to the interval $(a_1, b_1]$ (in either case $n > 1$). Since $(\mathcal{X}, <)$ satisfies the DCC, after fm such transcriptions, x is reached. The indices $n_1, n_2, \ldots$ of the canonical terms $b_1, b_2, \ldots$ define x, as in the thm.

Remarks

1. In the canonical descent from ε to e,

   (a) If Diag(e), every term is diagonal.

   (b) If Suc(e), each one but the final term, e, is diagonal.

2. Suppose $e = (\varepsilon, n_1, n_2, \ldots, n_R)$, where the n_i are all > 1. Then for $1 \le i < R$, $e \prec (\varepsilon, n_1, \ldots, n_i)$ and $(\varepsilon, n_1, \ldots, n_i - 1) \preceq (\varepsilon, n_1, \ldots n_i, n_{i+1} - 1) \prec e$, the left inequality following from corol. 7. Considering now the definition of e, (I) and (II) of thm. 11 are satisfied, so

$$e = \varepsilon_{n_1, n_2, \ldots, n_R}.$$

3. $e = \varepsilon_{n_1, \ldots, n_R} \Rightarrow \omega^e = \varepsilon_{n_1^+, n_2, \ldots, n_R}$; while $e^+ = \varepsilon_{n_1, \ldots, n_R^+}$ iff $\neg$Diag(e).

4. Let $e = \varepsilon_{n_1,n_2,\ldots,n_R}$, $1 \le R$. For $i \le R$, let $\alpha = (\varepsilon, n_1, n_2, \ldots, n_i)$. Then, in reaching α by taking canonical terms of successive indices $n_1, n_2, \ldots, n_i$, by the expression for e, the indices have been chosen minimally. Therefore

$$\alpha = \varepsilon_{n_1,n_2,\ldots,n_i},$$

in the 2nd normal form.

5. Let $T \in \mathfrak{T}$(IV·19) be any normal tree. Then by the 1-1 correspondence IV·47, e, α of the previous rem. may be chosen such that

$$T = T(e), \text{ with } T_i = T(\alpha) \text{ in T, as } i = 1, 2, \ldots, R.$$

Thus

$$T = T_R \prec T_{R-1} \prec \ldots \prec T_1 = \text{ the chain } C \text{ on } 2 + n_1 \text{ vertices.}$$

Therefore the arbitrary normal tree T may be obtained from C by a unique succession of increasingly gradual $\prec$ −descents in $\mathfrak{T}$, each being a canonical reduction

$$T_{i+1} = (T_i, n_{i+1}),\ i < R,$$

as determined by IV, thm. 62. In particular, by rem. 1 after IV·26, each 'rooted tree' is assigned a unique finite integer sequence.

Taking for example $f = \omega^2 = \varepsilon_{2,2}$, gives $(f, 1) = \omega = \varepsilon_1$, so $(\varepsilon, 2, 2, 1)$ is not the ε−normal form of $(f, 1)$. However, for diagonal $e = \varepsilon_{\underline{n}}$,

LEMMA 5.13 $(\forall N > 1)\,(e, N) = \varepsilon_{\underline{n},N}$.

Proof. It is on account of thm. 6 that a canonical reduction expression may not be a normal form, viz. $(\alpha, k+1, 1_1, \ldots, 1_R) = (\alpha, k)$. Therefore, e.g. $e = (\alpha, k+1, 1_1, \ldots, 1_{R-1})$ gives

$$(\alpha, k) \prec (e, N) \prec (\alpha, k+1, 1_1 \ldots, 1_i),\ \ i = 1, 2, \ldots, R-1;$$

and the result follows.

□

5.5 ε−Indices and Explicit Coefficients

5.5.1 Numerical Values of the ε−Indices

Examples show that there is a connection between the explicit coefficients (ec's, II·23·3) and the ε−coefficients (or indices) determined by thm. 11. This connection is wholly determinate. Let $e = \varepsilon_{n_1,n_2,\ldots,n_R}$ in **EP**. Let each ec of e defined by a non-leading constant exponent be a *trailing* ec.

THEOREM 5.14 [9] *Disregarding occurrences of* 0, *each of the non-trailing* ec*'s of* e *occur among the positive integers* $n_2^-, n_3^-, \ldots, n_{R-1}^-, n_R$.

Proof, by induction on R. Any $\varepsilon_n = \omega \star n$ has only the coefficient 1. Thus let $R > 1$. By rem. 4 above, $\varepsilon_{n_1, n_2, \ldots, n_{R-1}}$ denotes the normal form of f for which $e = (f, n_R)$. Assume for induction hypothesis that the ec's of f are as in the thm., and that f is is either of the form

$$\begin{array}{l} \Sigma + \omega^{\Sigma' + n_{R-1}} \\ \iddots \end{array} \qquad \text{(I)}$$

viz. the n_{R-1}^{th} ct of a perfect diagonal, or

$$\begin{array}{l} \Sigma + n_{R-1}\omega^{\beta} \\ \iddots \end{array} \qquad \text{(II)}$$

where $\Sigma, \Sigma' \in \mathbf{EP_0}$ and $\beta \in \mathbf{EP}$. In case (I),

$$e = (f, n_R) = \begin{array}{l} \Sigma + n_R \cdot \omega^{\Sigma' + n_{R-1}^-} . \\ \iddots \end{array}$$

Therefore if $\Sigma' = \mathbf{0}$ and $n_{R-1} = 1$, then e is of the form (I); otherwise of the form (II). Taking into consideration the new coefficient n_R, and using the induction hypothesis, the non-trailing ec's of e occur among $n_2^-, n_3^-, \ldots, n_{R-1}^-, n_R$. Note that if $n_{R-1} = 1$, this accounts for an index not an ec. In case (II), consider the register polynomial α of f. $e = (f, n_R)$ may be obtained by replacing, in the explicit expression of f

$$\omega^{\alpha} \text{ by } \begin{cases} n_R & \text{if } \alpha = \mathbf{1} \\ n_R \cdot \omega^{\alpha^-} & \text{if } \mathrm{Suc}(\alpha). \end{cases}$$

But the register polynomial of β is α, and

$$e = (f, n_R) = \begin{array}{l} \Sigma + n_{R-1}^- \omega^{\beta} + (\omega^{\beta}, n_R). \\ \iddots \end{array}$$

Therefore e is of the form (I), (II) as $\alpha = \mathbf{1}$ or $\mathrm{Suc}(\alpha)$. And, as before, e takes the required ec's. The thm. thus follows by induction.

□

Notes.

[9] See [104] (1988), thm. 2.1.

1. By the thm., unit ec's are determined by ($\varepsilon-$)indices= 2 (n_R excepted). And in addition to the indices which determine ec's, indices= 1 correspond to the 'step-down' and occur in patterns determined by the length of the descent, as in:

$$\varepsilon_4^+ = \varepsilon_{5\cdot 221211211121111}.$$

2. The trailing ec's are excepted in V·14 on account of repeated $\triangleleft_1-$descents. Thus, if e.g.

$$\mu = \{\Sigma + \omega^{\Sigma'+n} + \omega^{\Sigma'+m}\} \text{ where } m < n,$$

then $\underline{n}$ in $\varepsilon_{\underline{n}} = \Sigma + 2\omega^{\Sigma'+n}$ is a segment of the $\varepsilon-$string for μ. Viz. by the previous note, the trailing ec m does not appear in the $\mu-$string:

$$\mu = (\varepsilon_{\underline{n}}, 1_1, \ldots, 1_{n-m}).$$

3. For e as in thm. 14, with the exception of $e = \varepsilon_n$, $n = 1, 2, 3, \ldots,$ $\mathrm{eh}T(e) = n_1$.

4. Thm. 14 expresses a difference between the 2 normal forms: the polynomial form is recursive; the $\varepsilon-$form is explicit.

Examples

1. The **XP**$-$polynomial $e = 4\omega^2 + 7\omega + 5$ has explicit coefficients $2, 4, 7, 5$; and $e = \varepsilon_{2\cdot 3585}$ in accordance with the thm.

2. Let $e = \omega^\psi + 8\omega \star (2; 3) + 5\omega^4 + 6\omega^2 + 7$. Then $e = \varepsilon_{4\cdot 22121124211191156171 7}$.

3. With ref. to note 2, suppose $1 < n_1$; $R - 2 \leq n_2$ in the ε-form. Then

$$e = \omega \star (n_1^-; \sum_{i=3}^{R} n_i^- \cdot \omega^{n_2 - i + 2})$$

(for such an e, rp= const. iff $n_2 > R - 2$).

Now let e take the normal form of defn. 12. Thus (ref. II·44),

$$N(e) = 1 + \max\{n_2^-, n_3^-, \ldots, n_{R-1}^-, n_R\}.$$

As was seen in III, §4·1, the classes $\mathbf{EP}[\varepsilon, r]$ bear a close relation to the $\mathcal{X}_r \subset$ **EP**, $r \geq 1$ for which

$$e \in \mathcal{X}_r \Leftrightarrow \mathrm{ht}(e) \leq r \;\&\; a = ec \text{ of } e \Rightarrow a \leq r.$$

COROLLARY 5.15 *1.* $\mathcal{X}_r \subset \mathbf{EP}[\varepsilon, r+1]$;

2. $\varepsilon(r) \leq \mid \mathbf{EP}[\varepsilon, r] \mid$.

Proof.

1. By thm. 14, and III, thm. 23;
2. by part (1), and II, lemma 45.

□

Thus by III, §3·3, defn. 12 and III·7 ·5 give

COROLLARY The explicit index (ei) for $e = \varepsilon_{n_1,\dots,n_R}$ is the ec n_R.

E.g. for $e = \varepsilon_{4\cdot 453(20)25}$, rp $e = 4\omega^3 + 2\omega^2 + 19\omega + \mathbf{1}$, ei$= 5$.

5.5.2 ε–Indices Occur in Post Order

According to thm. 14, the ε–indices of e are 'almost' the explicit coefficients of e. There remains the question of the order in which the ec's occur in the ε–string. The $\prec$ –normal form $e = \Sigma a_i \omega^{e_i}$ determines the normal tree $T(e)$ (IV·47). The plane-structure trees $\mathfrak{S}$ contain a proper subclass obtained as follows. e determines the *tree (of coefficients)* $\tau(e)$, each non-principal vertex of which corresponds to a coefficient of e. Thus, let e' be obtained from e by setting each of the non-unit ec's of e, constant exponents excepted, equal to unity. $e' \preceq e$; distinct e may have the same e', so $(e')' \neq e'$ is possible. The setting may give e' not in II·22-normal form. However, using the correspondence of IV, §4·4:

DEFINITION 5.16 $\tau(e) = S(e')$.

Thus $\tau(e) \subseteq T(e)$. In terms of IV, §2, τ may be defined as follows. $\tau = T$ if $T = n \geq 0$ is of eh$\leq$ 1. Now suppose $T = [n_r, \dots, n_{s+1}, v^\sharp_s, \dots, v^\sharp_1]$ is of eh$=$ 2 in $\mathfrak{T}$, $n_{s+1} \geq 2$. Then $\tau = [m_q, \dots, m_{s+1}, v^\sharp]$, where $m_q > m_{q-1} > \dots > m_{s+1}$ are the distinct n_i, $i = r, \dots, s+1$. Now suppose for eh $h \geq 2$, τ has been defined, and suppose T is of eh$=$ $h+1$ in $\mathfrak{T}$. Suppose $T(e) = [T_s, \dots, T_1]$, $T_i \preceq T_{i+1}$. Then writing $\sigma_i = \tau(T_i)$, $i \leq s$, consider the tree $\sigma = [\sigma_s, \dots, \sigma_1]$. From σ, $\tau(e)$ is obtained as

$$\tau = [\tau_r, \dots \tau_1], \quad \text{where } \tau_i \preceq \tau_{i+1},\ i < r,$$

and the τ_i are the σ_i corresponding to distinct T_i.

Example.

$$e = 4\omega^{\omega^4+\mathbf{3}} + 25\omega^{10\omega^3+\mathbf{5}} + 9\omega^{7\omega^3+\mathbf{2}} + 5\psi + 6\omega^3 + \mathbf{19}; \quad e' = \omega^{\omega^4+\mathbf{1}} + 2\omega^{\omega^3+\mathbf{1}} + \psi + \omega^3 + \mathbf{1};$$

$$\tau(e) = [[4,[\,]],[3,[\,]],[3,[\,]],[1],3,[\,]].$$

Note that eh$\{\tau(e)\}$ =eh$\{T(e)\}$.

5.5.3 Indexing of τ

In the above construction of τ from σ, suppose τ_i corresponds to a occurrences of σ_i. Then the vertex in τ which is the pv of τ_i is *indexed with the coefficient a of e*. Now in the passage from f to $e = (f, n_R)$ in the proof of thm. 14, the new coefficient n_R occurs by a replacement in the expression containing the register polynomial α of f. But by definition this is the rightmost expression of greatest level (taking 'level' to be the edge-height of the vertex in $T(e)$ corresponding to the register polynomial). Therefore the n_R–introduction is the alteration or replacement of the register branch (IV·66) in $\tau(f)$, to give $\tau(e)$. Therefore

THEOREM 5.17 [10] *The order in which the ec's a of e occur (as $a+1$ or as a) in the ε–form of e is the left-to-right post order in $\tau(e)$.*

□

Notes

1. If $e = (f, n)$, then still $\tau(f) \prec \tau(e)$ is possible: taking e as ex. 1 after thm. 14, $e = (f, 5)$, where $f = \varepsilon_{2358}$ and
$$\tau(f) = [[[\,],[\,]],[[\,]]] \prec [[[\,],[\,]],[[\,]],[\,]] = \tau(e).$$
2. A remarkable inversion occurs in the calculation of the ε–string of e. Thus if, e.g. $\tau(e) = [\tau_2, \tau_1]$, then the ε–expression of e contains a run of 1's and 2,'s corresponding to repeated reductions leading from τ_2 to τ_1. By Ch. IV, thms 36 and 69 the steps determine a vertex sequence in τ_2 in **right-to-left** post order.
3. By e.g. the remark before defn. 16, τ does not always satisfy the $f(u) < f(v)$ condition of IV, defn. 43, so it is possible that $\tau \notin \mathfrak{R}$, the class of strictly normal trees.

5.5.4 Iteration Coefficients

The E–string of Ch. II, §4 which determines e in the II·22- normal form contains the ec's of e, as may be seen in an example: if $e = a\omega^{b\omega^n + c\omega^m} + 2\omega^3 + 4$, then
$$e = \{(E_4^{(n)}E_3)^{(b)}(E_4^{(m)}E_3)^{(c)}E_2\}^{(a)}(E_3^{(3)}E_2)^{(2)}E_2^{(4)}E_1$$
(see II, §4.4, rem. 3).

[10] [104], thm. 2·4.

5.6 Applications of the 2nd Normal Form.

5.6.1 Comparison of Functions in EP.

The well-founded linear ordering $\prec$ of II·8 may be analysed as follows. Let

$$e = \varepsilon_{m_1,m_2,\ldots,m_R} \quad \text{and} \quad f = \varepsilon_{n_1,n_2,\ldots,n_S}.$$

Suppose

1. $S < R$ and $m_1 = n_1, m_2 = n_2, \ldots, m_S = n_S$, or
2. $\{\exists i \leq \min(R,S)\}\ m_1 = n_1 \ \&\ m_2 = n_2 \ \&\ldots\&\ m_{i-1} = n_{i-1} \ \&\ m_i < n_i$.

Then, in either case $e \prec f$ by the basic properties of fundamental sequences, III·7, and by the BACHMANN property, III·14. On the other hand, given e, f as above, and $e \prec f$, if neither 1. nor 2. holds, then either $R \leq S$, $m_1 = n_1, \ldots, m_S = n_S$ and $e \succeq f$; or $m_1 = n_1, \ldots, m_{k-1} = n_{k-1}$ & $m_k > n_k$, and $f \prec e$; in either case contrary to assumption. Therefore

THEOREM 5.18 [11] *$e \prec f$ in* **EP** *if and only if one of the above conditions 1., 2. holds.*

□

In case 1 above, $S + 1 < R$ and $m_{S+1} = 1$ give

$$e = \varepsilon_{n_1,\ldots,n_S,1_{S+1},m_{S+2},\ldots,m_R},$$

viz. $e \prec (f, 1)$. But if $m_{S+1} > 1$, for $S + 1 \leq R$,

$$(f, 1) \preceq e.$$

In this further case, e is 'just below' $f = \Sigma' + \omega^{\Sigma+1}$ (vide III, §3·3). III·14 now gives

$$e = \Sigma' + m_{S+1}^{-} \cdot \omega^{\Sigma} + (\omega^{\Sigma}, m_{S+2}, \ldots, m_R).$$

Therefore dee=rp f, generalizing ex. 2 after IV·50.

In case 2, consider the 'common root' $\alpha = \varepsilon_{n_1,n_2,\ldots,n_{i-1}} = \Sigma' + \omega^{\Sigma+1}$ of e, f. Note that

$$e \preceq (\alpha, m_i) \preceq (\alpha, n_i^{-}) \preceq (f, 1),$$

so $e = (f, 1)$ is possible: dee=rp f. Now suppose $i < R$ and $m_i > 1$. Then as above,

$$e = \Sigma' + m_i^{-} \cdot \omega^{\Sigma} + (\omega^{\Sigma}, m_{i+1}, \ldots, m_R) \prec \Sigma' + n_i^{-} \cdot \omega^{\Sigma} + (\omega^{\Sigma}, n_{i+1}, \ldots, n_S) = f,$$

so that dee (of f for e) $\preceq \text{rp}(\omega^{\Sigma}) = \text{rp}\{(\text{rp}\alpha)^{-}\}$, with equality if $m_{i+1} > 1$. If $R = i = S$, dee$= \Sigma' + n_S \cdot \omega^{\Sigma}$. If $R = i < S$, whether $m_i \leq n_i^{-}$, dee$= \Sigma' + n_i^{-} \cdot \omega^{\Sigma} + (\cdots)$. If $m_i = 1$, as in ex. 3 after IV·50, a dominance relation for the dee cannot be fixed.

Theorem 18 provides a sharpened version of thm. 3. Using the $\tilde{\lhd}_n$ notation of III, defn. 18,

[11] [104], thm. 2 ·2.

THEOREM 5.19

$$e \prec f \ \& \ N(e) \leq m^+ \Rightarrow e \ \tilde{\lhd}_m \ f.$$

Proof. Suppose $e \prec f$, and also that case 1 of thm. 18 holds. Then a $\lhd-$ passage downwards, from f to e, may be obtained via successive values $m_{S+1}, m_{S+2}, \ldots, m_R$ of the modulus. Therefore the result follows, by III, §4 and by thm. 14.

On the other hand, if case 2 holds, $\varepsilon_{m_1,m_2,\ldots,m_i} \lhd_1 f$ by the result after corol. 7, and if necessary by further repeats of the corol. V·19 then follows as above by thm. 14.

□

Thus $e \prec f$ are related by

$$e \lhd_{m(1)} \alpha \lhd_{m(2)} \cdots \lhd_{m(k-1)} \beta \lhd_{m(k)} f,$$

where $m(1), m(2), \ldots, m(k) \leq N(e)^-$. But by ex. 1 foll. II·8; and by III·7·6, $\{\forall x > m(k)\}\ \beta(x) < f(x)$. Therefore

$$x > m \Rightarrow f(x) > \beta(x) > \ldots > \alpha(x) > e(x).$$

Therefore thm. 19 provides a second proof of II, thm. 47:

COROLLARY 5.20

$$e \prec f \text{ in } \mathbf{EP} \Rightarrow \{\forall x \geq N(e)\}\ e(x) < f(x).$$

□

Remark. A special case of thm. 19 arises modulo a si arithmetic sequence $\mathcal{S} \equiv n(1), n(2), \ldots$ (vide III, §4·3; and rem. 3(b) after thm. 3). Thus, if $n(1) \geq m$, $(\exists f_1)\ e \preceq f_1 \ \lhd_{n(1)} \ f$. Therefore if $e \neq f_1$, $(\exists f_2)\ e \preceq f_2 \ \lhd_{n(2)} f_1, \ldots$. So as in the proof of thm. 3, but in this case of necessity using the DCC, $e \ \tilde{\lhd}_{\mathcal{S}} \ f$ viz. $(\exists k)$

$$e \ \lhd_{n(k)} \ f_{n(k-1)} \ \lhd_{n(k-1)} \ \cdots \ \lhd_{n(2)} \ f_1 \ \lhd_{n(1)} \ f.$$

Dependence on Ω

The above requires the ideas of Ch. III, and in this way differs from the more primitive proof of Ch. II, thm. 47.

5.6.2 Proof of the BACHMANN Property

The second normal form may be used to prove III·14 (thus strengthening corol. 7):

LEMMA 5.21 *Suppose: e is a diagonal, f a cd of* **EP** *and $n \in \mathbf{N}$ are such that $(f,n) \prec e \preceq (f,n+1)$. Suppose $e = \varepsilon_{m_1,\ldots,m_R}$ and $f = \varepsilon_{n_1,\ldots,n_S}$. Then* (i) $S < R$ *and* (ii) $n_i = m_i$ *for $i = 1,2,3,\ldots,S$ (viz. 18·1 holds).*

Thus the 'address' of e in the GOODSTEIN-SKOLEM hierarchy lies within that of f.

Proof. For (ii), by induction on $s \le S$, where $1 \le S$. In the initial case ($s = 1$) let k denote the characteristic n_1. Then $\varepsilon_{k-1} \prec f \preceq \varepsilon_k$. Since f is a cd, $k > 1$. Therefore if $f = \varepsilon_k, (f,n) \succeq (f,1) = \varepsilon_{k-1}$, and if $f \prec \varepsilon_k$,

$$f = a \cdot \omega \star (k-1) + \Sigma$$

in the polynomial normal form, where $a > 1$, or $\Sigma \succ \mathbf{0}$. Therefore $(f,n) \succ \omega \star (k - 1) = \varepsilon_{k-1}$. But in either case $e \succ (f,n)$, giving $e \succ \varepsilon_{k-1}$. Also $e \preceq (f, n+1) \prec f$, so $e \prec \varepsilon_k$ and $m_1 = n_1$. Now for the mantissa, suppose $S > 1$, and that the property demanded by the lemma has been established for all $s < S$. Suppose

$$f = \varepsilon_{n_1,n_2,\ldots,n_Q,1_{Q+1},1_{Q+2},\ldots,1_S},$$

where $n_Q > 1$ (there must be such an n_Q, as $f \succ \omega$). If $Q = S$, writing $f = (\alpha, n_S)$,

$$(\alpha, n_S - 1) \preceq (\alpha, n_S, 1) \preceq (\alpha, n_S, n) = (f,n),$$

using corol. 7. So

$$\begin{aligned} \varepsilon_{n_1,n_2,\ldots,n_{S-1},n_S-1} &= \varepsilon_{m_1,m_2,\ldots,m_{S-1},n_S-1} \text{ by i.h.,} \\ &\preceq (f,n) \prec e \preceq (f,n+1) \\ &\prec f = \varepsilon_{m_1,m_2,\ldots,m_{S-1},n_S}, \end{aligned}$$

therefore $n_S = m_S$.

If $Q < S$,

$$\varepsilon_{n_1,n_2,\ldots,n_{Q-1},n_Q-1} = \varepsilon_{m_1,m_2,\ldots,m_{Q-1},m_Q-1} \prec e,$$

applying i.h. But

$$e \prec f \preceq \varepsilon_{m_1,m_2,\ldots,m_{Q-1},m_Q,1_1,1_2,\ldots,1_q},$$

for $q = 1,2,\ldots,S-Q$, therefore also $n_{Q+q} = 1 = m_{Q+q}$, for q in this range. Thus (ii) follows by induction.

(i) now follows from $e \prec f$.

□

Remark. In the above lemma, taking $e = (f, n+1)$ gives $(f,n+1) = \varepsilon_{n_1,n_2,\ldots,n_S,n+1}$, and generally $m_{S+1} = n+1$; in accordance with lemma 13.

Passing now to the BACHMANN property, which is of such serious consequences for constructions resting on fundamental sequences, it may be remarked that the present proofs of this thm., and of the lemmata on which it depends, are technically rather intricate. This may be attributed to the structural property of the diagonal of §2.1; and to the more explicit form than that of e.g. III·14.

THEOREM 5.22 *Let e, f and the index n be as in lemma 21. Then*

$$(f, n) \preceq (e, 1).$$

Proof, by induction on the difference $r = R - S$ where, by the lemma

$$e = \varepsilon_{m_1, m_2, \ldots, m_S, n+1, m_{S+2}, \ldots, m_R}, \quad (f, n+1) = \varepsilon_{m_1, m_2, \ldots, m_S, n+1}.$$

For $r = 1, R = S + 1, e = (f, n + 1)$, the result is corol. 7. Now suppose $r > 1$ (so that $e \prec (f, n+1)$), and the property demanded has been established for all e, f for which the above difference is less than r.

1. Suppose $m_{S+2} = \ldots = m_R = 1$. Then $(e, 1) = (f, n + 1, 1_{S+2}, \ldots, 1_R, 1)$, and the result follows from thm. 5 or 6, whichever is appropriate.

2. Suppose $S + 2 \leq K \leq R$ and

   $$e = \varepsilon_{m_1, \ldots, m_S, n+1, m_{S+2}, \ldots, m_K, 1_{K+1}, \ldots, 1_R} = (\alpha, m_K, 1_{K+1}, \ldots, 1_R) \text{ say},$$

   where $m_K > 1$. Since e is in ε−normal form,

   $$(\alpha, m_K - 1) \prec \varepsilon_{m_1, \ldots, m_{K-1}, m_K, 1_{K+1}, \ldots, 1_R} = e. \qquad \text{(I)}$$

   Now, using the expression for $(f, n + 1)$

   $$(f, n) \prec e \prec \alpha = \varepsilon_{m_1, \ldots, m_S, n+1, \ldots, m_{K-1}} \preceq (f, n + 1),$$

   (noting $m_{R-1} = n + 1$ if $r = R - S = 2$). Therefore, since $S + 2 \leq K$,

   $$\begin{aligned} (f, n) &\preceq \varepsilon_{m_1, \ldots, m_{K-1}, 1_K} \text{ by induction hypothesis} \\ &\preceq \varepsilon_{m_1, \ldots, m_{K-1}, m_K - 1} \text{ since } m_K > 1 \\ &\preceq (\alpha, m_K, 1_{K+1}, \ldots, 1_R, 1_{R+1}), \\ &\quad \text{by thm. 5 or 6 applied to inequality (I)} \\ &= (e, 1). \end{aligned}$$

Thus, all cases are dealt with, and the thm. follows by induction.

□

Example. Let $f = \omega \star 4$; $e = \varepsilon_{4221211} = \omega^{\psi^+}$. Then $(f, 1) \prec e \prec (f, 2)$ with $R = 7,\ K = 5$ and $m_S = 4$; and $(f, 1) = \varepsilon_3 = (e, 1)$.

5.6.3 An Enumeration Based on ec's

The ε−normal form gives a quick enumeration (ref. II, §5.1) viz. an exhaustive list

$$\varphi_1, \varphi_2, \ldots, \varphi_N, \ldots$$

of the exponential polynomials. Suppose $e = \varepsilon_{m_1,\ldots,m_R}$, $f = \varepsilon_{n_1,\ldots,n_S}$. Then e is listed in the enumeration before f iff $\Sigma m_i < \Sigma n_i$ or

$$\Sigma m_i = \Sigma n_i \;\&\; (m_1, \ldots, m_R) \prec (n_1, \ldots, n_S),$$

where $\prec$ denotes the usual left-to-right lexicographic order.

5.6.4 Very Rapid Descent

The descending sequence, containing (the diagonal) α alone has already reached its lowest term. On the other hand, if e.g.

$$\alpha = (20\omega) \star (20; 20),$$

then $| T\alpha |> 20^{20}$, and the sequence $\alpha \succ (\alpha, 1) \succ (\alpha, 1, 1) \succ \ldots \succ 1$ contains (ref. IV, thm. 69) more terms than there are molecules in a tonne of water. Is it possible to obtain a sequence by other than canonical descent, which 'spans' $\mathbf{EP}_\alpha$?

Convex Term Reduction

In the example above,

$$\alpha = \varepsilon_{21\cdot 21,21,1_1,1_2,\ldots,1_{20},21,1_1,1_2,\ldots,1_{400},21,1_1,\ldots\;\ldots,1_{1+20^{19}},20}$$

where the index string contains 21 occurrences of '21'. This suggests the following: let $\alpha = \varepsilon_{m,n,\ldots,N} \succ \omega$. Then

DEFINITION $\alpha_1 = \alpha$; and for $t = 1, 2, 3 \ldots,$ $\alpha_{t+1} = \varepsilon_{\underline{n},n}$, where

$$\alpha_t = \begin{cases} \text{either } \varepsilon_{\underline{n},n+1} & \text{or} \\ \varepsilon_{\underline{n},n+1,1_1,1_2,\ldots,1_r}, & \text{for some } r \geq 1. \end{cases}$$

□

The descent $\alpha_1 \succ \alpha_2 \succ \ldots \succ \omega$ is convex in sense that it gets 'steeper' as the terms nearer the left of the ε-expression are reduced. The polynomials $\prec \alpha$ are spanned by (α_t) in the following sense: suppose in IV·19, at the inductive stage the restriction $T_r = T_{r-1} = \ldots = T_1$ is imposed. In this case let $T \in \mathfrak{T}$ be called *symmetric*. Now suppose $S \subset T\alpha$ is the maximal symmetric upper part (IV, defn. 24). Then if $v \in S$ is interior, va $v = 1 + k$ and Branch$(v, S) = [B_k, \ldots, B_1]$ (where of course, the B's are equal), and finally $T = T_{(v),l}$ is obtained by joining $C(\mathrm{pv}S, v)$ to $[B_l, \ldots, B_1]$, $l \leq k$, then for some i, $T = T\alpha_i$.

Remark. The sequence $\omega\star(m-1; n) \succ \omega\star(m-1; n-1) \succ \ldots \succ \omega\star(m-1) \succ \omega\star(m-2) \succ \ldots \succ \omega$ always is a remainder of (α_i).

5.7 Notes on Ch. V

1. §3. The proof of thm. 22 itself from corol. 7 may be sketched as follows. Let e be a cd, and let $\beta \in \mathbf{EP}$ lie in the range $(e, n) \prec \beta \preceq (e, n+1)$. By the argument of thm. 11, $\beta = (e, n+1, m_1, \ldots, m_R)$ may be expressed uniquely in the form of the comment after corol. 7, viz.

$$(e, n) \mathrel{\tilde{\lhd}} \beta,$$

giving $(e, n) \preceq (\beta, 1)$. Note that this proof by finite descent, like that of III·14, evades the DCC.

The above form of β shows that the condition satisfied by β implies that no predecession is necessary in the $\lhd_{\text{III}\cdot 19}$–descent from e to β, in contrast to the more general case of thm. 3. In ROSE ***Subrecursion*** [131], corol. 7 is given in place of the BACHMANN result.

KADOTA, AOYAMA (Hiroshima) [12] have classified non-canonical assignments for which corol. 7 is weakened to:

$$(e, n) \mathrel{\tilde{\lhd}_r} (e, n+1),$$

viz. in any given case, up to $r-1$, $r \geq 1$ violations of III·14 are allowed.

[12] Ref. seminar notes, Oxford, 1990.

Chapter VI

GENERALIZED INDUCTION AND RECURSION

Section Headings

6.1 Introduction

6.1.1 The Nature and Importance of Mathematical Induction

Proof by mathematical induction: the establishment of the initial case; and then from the assumption of a given case, reasoning the establishment of the following case: is an essentially arithmetic principle. Nonetheless, in its simplest form; or as one of its many possible variants, it is of the greatest importance not only in arithmetic (e.g. of the fundamental theorem of arithmetic) but also in algebra (e.g. of the fundamental theorem of algebra, of the JORDAN-HOLDER theorem, of SYLOW's theorems on prime-power subgroups, etc.) in graph theory (CAYLEY's theorem for indexed free trees, Ch. IV, §1; the ERDÖS-SZEKERES theorem on RAMSEY's $r(m,n)$ [1] etc.), and in many other areas. However mathematical induction is of outstanding importance in formal logic. In the first instance, objects of logic are usually *defined* by induction (or by recursion [2]). This is the case with any so-called formal system, notably PEANO's system $\underline{\mathrm{N}}$.

The main aim of this chapter is to try to introduce a lucid generalization of induction, of the kind required for a finitary diagonalization of ACKERMANN's function, based on the hierarchy **EP**. §1 contains summary examples of kinds of mathematical induction. The general form is defined in §2, while §3 deals with other generalizations. This survey is included as a basis for discussion of the entry of the axiom of choice (AC) into proofs and constructions, and is not intended to be comprehensive. §§4,5 treat generalizations of primitive and k–fold recursion.

6.1.2 Induction Scheme

Let P be a unary predicate on **N**. Then from P(1) and $(\forall n \in \mathrm{N})$ $\mathrm{P}(n) \Rightarrow \mathrm{P}(n+1)$ may be deduced $(\forall n)$ $\mathrm{P}(n)$. This is the most common variant of many equivalent schemata. It is usually included as the 5th axiom of the system $\underline{\mathrm{N}}$ (ref. Appendix B), and was adopted by PEANO in his original 1897 formulation. The Notes (§6) contain remarks on combinatorial notions of inclusion, foundation, etc. which have enabled the development of the first order theory of sets.

6.1.3 Arithmetic and Tree Induction

Consider the class **K** of Ch. II, §6·4. The IV, §1 -terminology may be applied to the cumulative *definition tree* τ of $\mathcal{X}$. Thus, let the initial functions $\mathbf{1}, \omega$, define the pv $v^{\sharp}$ of τ. Let the exponential rules of closure define right-to-left edges:

[1] See note 2, end of Ch. III.

[2] The validity of definition by recursion follows at once from the axiom of mathematical induction.

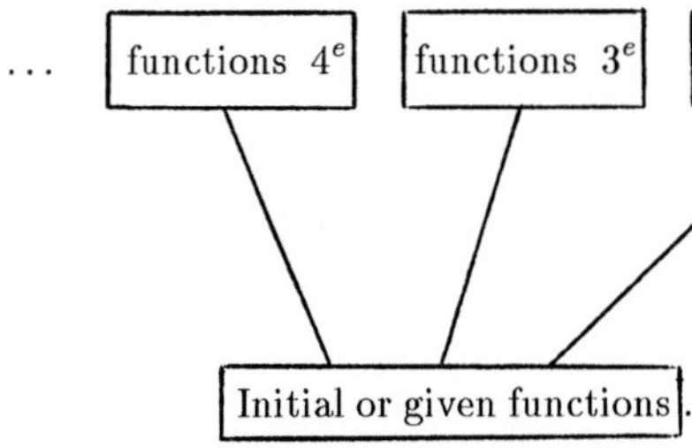

Every vertex in the infinite ramified system typified by the figure represents a unique (proper) subclass of **K**. Thus an induction on order-of-definition in **K** is essentially a *tree-induction*, or an arithmetic induction on *levels* of a tree whose vertices are of unlimited valency. Every given formal system is reducible to the above form of tree recursion.

6.1.4 Induction and the DCC

Consider the DCC for **N**. Every

$$\mathcal{S} \equiv n > m > \ldots$$

is finite. This theorem has a short proof: $\mathcal{S}$ has at most n terms. The 'axiom' of induction is also a theorem: let P be a predicate, and suppose

1. P(1); and
2. $(\forall x)\ \mathrm{P}(x) \Rightarrow \mathrm{P}(x+1)$.

Given any natural number n, if $n = 1$, then P(n). Otherwise

$$\frac{\mathrm{P}(1)\ \&\ \{\mathrm{P}(1) \Rightarrow \mathrm{P}(2)\}}{\mathrm{P}(2),}$$

by the ancient rule of *modus ponens*. This reasoning, repeated $n-1$ times, therefore gives P(1),P(2),...P(n). Since n was arbitrary, from (1) and (2) is proved

$$(\forall n)\ \mathrm{P}(n).$$

6.1.5 Downwards Induction

Suppose

1. $(\forall n)\ \mathrm{P}(n+1) \Rightarrow \mathrm{P}(n)$, and
2. $(\exists n)\ \mathrm{P}(n)$.

Then P(1).

Examples.

EUCLID's algorithm. The contrapositive of the schemata is the 'method of infinite descent' [3] pioneered by Pierre FERMAT: putting $A(n)$ iff $\neg P(n)$,

$$\frac{(\forall n)\ A(n+1) \Rightarrow A(n),\quad \neg A(1)}{(\forall n)\ \neg A(n).}$$

[3] Used, e.g. by EULER to show that (in particular)

$$x^4 + y^4 = a^4$$

has no solution in integers (ref. S. HOLLINGDALE, *Makers of Mathematics*).

6.2 Generalized Induction

Besides the above most basic intervention by induction in logic, the principle arises in generalized form, a form which assumes many variants. These may be classified most readily in terms of the DCC. Thus, ordinary induction is equivalent to the DCC for **N**. The generalization to be treated here is equivalent to an instance of the DCC for the system of next higher type, $\mathcal{N} = \mathbf{N}^{\mathbf{N}}$, viz. for $(\mathbf{EP}, \prec)$. VANAANEN (1984) has considered e.g. a generalized induction equivalent to assumption of well-foundedness of $\mathcal{N}^{\mathcal{N}}$, where this is *not* assumed for $\mathcal{N}$.

6.2.1 Generalized Induction in EP, GEN IND

Introduction

The purpose of the following is

1. to introduce the theorem of generalized induction, for later use; and
2. to expose the foundation of the theorem on the DCC for $(\mathbf{EP}, \prec)$, where $\prec$ is the relation of eventual domination (Chs. II·22 and IV·48).

The canonical assignment Ω may be given either directly by induction, as in III·6, or as below, using the assignment for $(\mathfrak{T}, \prec)$ of Ch. IV·59, and the 1-1 correspondence $T : \mathbf{EP} \to \mathfrak{T}$ of §4 of the same Ch.

DEFINITION 6.1 $(\forall n \geq 1)\ (e, n) = T^{-1}\{(T(e), n)\}$.

□

Every $n \in \mathbf{N}$ is either $n = 1$; or $n = m + 1$ for some $m \in \mathbf{N}$. This provides the basis for mathematical induction: if R is a unary relation (on **N**), then from R(1) and $(\forall n \in \mathbf{N})\ [\mathrm{R}(n) \Rightarrow \mathrm{R}(n+1)]$ may be inferred

$$(\forall n)\ \mathrm{R}(n).$$

According to the classification of II, §2, and by the assignment Ω, every e is either $e = \mathbf{1}$; or $e = f + \mathbf{1}$ for some f, or $e = \mathrm{diag}_n\ (e, n)$; in **EP**. This provides the basis [McBETH, 1980 [4]] of

AXIOM 6.2.1 *If R is a unary relation defined on* **EP**, *then from*

$$\mathrm{R}(1)\ \&\ (\forall e \in \mathbf{EP})\ \mathrm{R}(e) \Rightarrow \mathrm{R}(e^{+})\ \&$$

$$\&\ (\forall e \in \mathbf{EP})_{\mathrm{Diag}(e)}\ [(\forall n)\ \mathrm{R}\{(e, n)\}] \Rightarrow \mathrm{R}(e)$$

may be inferred

$$(\forall e \in \mathbf{EP})\ \mathrm{R}(e).$$

---

[4] *Zeitschr. math. Logik Grundl. Math.,* Bd. 26, S. 115-122, Axiom 3.5.

□

The above generalized induction for $\mathbf{EP} = (\mathbf{EP}, \Omega)$ can be accepted as a self-evident truth, and used as such as a tool of proof. But since it is equivalent to the DCC for $\mathbf{EP} = (\mathbf{EP}, \prec)$, this is not necessary. The DCC is a more elemental principle; moreover the DCC has the most compelling derivation, obtained as in Ch. IV by considering descending sequences of normal trees. The following proof of GEN IND is that contained in McBETH [102] [5], 1987.

6.2.2 Segments and Remainders

By Ch. II, corol. 19, the system $(\mathbf{EP}, \prec)$ is linear. It may be cut, therefore, in various ways, into a segment [6] $\mathcal{S}$ and a remainder $\mathcal{R}$. [7]

DEFINITION 6.2 *1. The non-void (proper) subsystem $\mathcal{S}$ (of* $\mathbf{EP}$*) is a (proper)* segment *iff*

$$(\forall f \in \mathbf{EP})(\exists e \in \mathcal{S})\ f \prec e \Rightarrow f \in \mathcal{S}.$$

2. The subsystem $\mathcal{R}$ (of $\mathbf{EP}$*) is a (proper)* remainder *if $(\exists \mathcal{S})$ $\mathcal{S}$ is a (proper) segment, and $\mathcal{R} = \mathbf{EP} - \mathcal{S}$.*

Remark. If $\mathcal{R} = \mathbf{EP} - \mathcal{S}$ is the remainder corresponding to $\mathcal{S}$, then for any $e \in \mathbf{EP}$ smaller than every polynomial of $\mathcal{R}$, $e \in \mathcal{S}$.

Examples.

1. For any $\mathbf{XP}$–polynomial (II·4) e, if $f \in \mathbf{EP}$, and $f \prec e$, then $f \in \mathbf{XP}$; $\mathbf{XP}$ and generally each of $\mathbf{EP}^{(h)}$, $h = 0, 1, 2, 3, \ldots$ is a proper segment of $\mathbf{EP}$.
2. Given any proper $\mathcal{S}$, $\mathcal{R} = \mathbf{EP} - \mathcal{S}$ is 'longer' than $\mathcal{S}$ in that $\mathcal{S}$ may be subtracted from $\mathcal{R}$ endlessly.

The Length of a Fundamental Sequence

Consider e.g. $e = \omega \star 3 + \psi$. $(\forall n)\ \omega \star 3 + \omega \preceq (e, n) \prec \omega \star 3 + \psi$; and generally the interval of the linear system $\mathbf{EP}$ swept out by the cfs for $e = \Sigma + \omega^{e_1}$ is short in comparison with $(\mathbf{EP})_e$. [8] Thus

DEFINITION 6.3 *The length $\mathcal{L}(e)$ [9] of the cfs $(e,1), (e,2), \ldots$ is the least $\lambda \in \mathbf{EP}$ such that $(\mathcal{X}, \prec) \simeq ((\mathbf{EP})_\lambda, \prec)$, where $\mathcal{X}$ is the interval of $\alpha \in \mathbf{EP}$ for which*

$$(\exists n)\ (e, n) \preceq \alpha \prec (e, n+1).$$

[5] *SIGACT News* **18**(3), Spring 1987.

[6] The term 'segment' is used in Ch. VI, §2 in place of 'initial segment'.

[7] Not to be confused with the '$\mathcal{R}$' of Ch. 1, §3.1.

[8] C.f. rem. 3, after V·8.

[9] Not to be confused with the length $\mathrm{lh}(e) \in \mathbf{N}$ of the polynomial itself (II, def. 23·4.

Then Chs. II, III defns. 38, 6 respectively give $\mathcal{L}(e) = \omega^{e_1}$, for e in II·22-form. The relation $e = \text{diag}_n(e,n)$ gives the appearance that $(e,1),(e,2),\ldots$ 'converges' to e. In the sense of III·12 this is affirmed: every $\alpha \in \mathbf{EP}$, $\alpha \prec e$ has $\alpha \prec (e,n)$ for some n. On the other hand, for all cd's (V·4,5) of the 2nd kind:

THEOREM 6.4

$$\mathcal{L}\{(e,n)\} \prec \mathcal{L}\{(e,n+1)\}.$$

□

6.2.3 Least Upper Bound of a Segment

THEOREM 6.5 *Every proper remainder $\mathcal{R}$ of* **EP** *has a $\prec$ −lowest polynomial.*

Proof. Using the ε−form of the previous Ch., thm. 11, there are smallest [10] numbers $n_1, n_2, \ldots$ for which

$$\varepsilon_{n_1,n_2,\ldots,n_r} \in \mathcal{R},\ r = 1,2,3,\ldots\ .$$

But for $r = 1,2,\ldots$, the ε−expressions are strictly $\prec$ −descending, ending therefore, by the DCC for **EP** at some $r = R$. Since every e has a unique expression in the 2nd normal form,

$$\varepsilon_{n_1,n_2,\ldots,n_R}$$

is this expression for the $\prec$ −lowest polynomial of $\mathcal{R}$.

□

Thus (ref. II·21):

COROLLARY 6.6 *Every segment is of the form* $(\mathbf{EP},\prec)_e$.

□

Now as in V·10, let $\varepsilon = \omega \star \omega$, $\overline{\mathbf{EP}} =_{df} \mathbf{EP} \cup \{\varepsilon\}$.

COROLLARY 6.7 [11] *Every $\prec$ − increasing sequence $\alpha(1) \prec \alpha(2) \prec \ldots$ in* **EP** *has a least upper bound* [12] *α in* $\overline{\mathbf{EP}}$.

[10] Note that the 'least number principle' (II, §2.18) is not used here, on account of the distinction between arbitrary arithmetic properties and arbitrary *remainders* of **N** (viz. $\varepsilon_{n_1}, \varepsilon_{n_1+1}, \varepsilon_{n_1+2}, \ldots \in \mathcal{R}$; and similarly for more highly recurrent canonical terms of ε).

[11] [101], thm. 3, p211.

[12] α is the *least upper bound* (lub) of $\{\alpha(i)\}_{i=1,2,3,\ldots}$, $\sup_i \alpha(i)$, if

1. $(\forall i)\ \alpha(i) \prec \alpha$, and
2. $\{(\forall i)\ \alpha(i) \prec \beta\} \Rightarrow \alpha \preceq \beta$

in **EP**.

Proof. Suppose the subsystem $\mathcal{S}$ is such that

$$(\forall\beta)\ \beta \in \mathcal{S} \Leftrightarrow (\exists i)\ \beta \prec \alpha(i) \text{ in } \mathbf{EP}.$$

Thus $\gamma \prec \beta \in \mathcal{S} \Rightarrow \gamma \prec \alpha(i)$, so $\gamma \in \mathcal{S}$, and by defn. 2·1, $\mathcal{S}$ is a segment. Now let the remainder $\mathcal{R} = \mathbf{EP} - \mathcal{S}$. If $\mathcal{R}$ is void, $\sup_{i \in \mathbf{N}} \alpha(i) = \varepsilon$. If $\mathcal{R}$ is proper, by thm. 5 $\mathcal{R}$ has a $\prec -$ least polynomial α. Thus, by the remark after defn. 2,

$$\alpha = \sup_i \alpha(i).$$

□

Note that if $\alpha = e^+$ were the sup., then $e \prec \alpha$ could not be, so

$$(\exists n)\ e \preceq \alpha(n) \prec \alpha(n+1) \succeq \alpha,$$

contrary to assumption. Thus α is diagonal.

Example.

By III·12, the sequence $\alpha(1), \alpha(2), \ldots$ of the corollary is cofinal with $(\alpha, 1), (\alpha, 2), \ldots$; viz. *$\alpha$ is the* lub *of its canonical sequence.*

6.2.4 The Canonical Form

The canonical assignment of Ch. III may be used to aquire new exponential polynomials from given ones, either by succession or by diagonalization:

DEFINITION 6.8 *$\mathcal{X}$ is the smallest class for which*

1. $\mathbf{1} \in \mathcal{X}$.
2. $e \in \mathcal{X} \Rightarrow e^+ \in \mathcal{X}$.
3. $e \in \text{diag}\mathbf{EP}\ \&\ \{(\forall n \geq 1)\ (e, n) \in \mathcal{X}\} \Rightarrow e \in \mathcal{X}$.

□

Since $\mathbf{EP}$ contains $\mathbf{1}$, and is closed under the operations of $\mathcal{X}$, it follows that $\mathcal{X} \subseteq \mathbf{EP}$. Since $\mathcal{X}$ is the *smallest* class satisfying 8·1.-3.,

COROLLARY 6.9 *$e^+ \in \mathcal{X} \Rightarrow e \in \mathcal{X}$; and if $e \in \mathcal{X}$ is diagonal, then $(\forall n)\ (e, n) \in \mathcal{X}$.*

□

Now let $\lhd$ be the relation of canonical component, III·19; and let $\tilde{\lhd}$ be the transitive closure.

THEOREM 6.10 *$\mathcal{X}$ is a segment (proper or improper) of* **EP**.

Proof. The lowest polynomial, **1**, belongs to $\mathcal{X}$. Suppose $e \succ \mathbf{1}$ is a polynomial of $\mathcal{X}$, and $f \prec e$ in **EP**. By V, thm. 3, this is logically equivalent to $f \mathrel{\tilde{\lhd}} e$. Therefore there are $e_1, e_2, \ldots, e_m \in \mathbf{EP}$ for which

$$f \lhd e_1 \lhd e_2 \lhd \ldots \lhd e_m = e.$$

By corol. 9 and the defn. of $\lhd$:

$$e \in \mathcal{X} \Rightarrow e_{m-1} \in \mathcal{X} \Rightarrow \ldots \Rightarrow e_1 \in \mathcal{X} \Rightarrow f \in \mathcal{X}.$$

Thus the theorem is proved.

□

6.2.5 Generalized Induction

THEOREM 6.11 $\mathcal{X} = \mathbf{EP}$.

Proof. Suppose that $\mathcal{X}$ is a proper segment. Using thm. 5, let e be the lowest polynomial of $\mathbf{EP} - \mathcal{X}$. By defn. 8·1, $e \neq \mathbf{1}$. If Suc(e), then $e^- \in \mathcal{X}$, giving $e \in \mathcal{X}$, by 8·2, contrary to assumption; so as with the note after corol. 7, e cannot be a successor. Similarly, if Diag(e), $(e, n) \prec e$ by III·7·4, so $(e, n) \in \mathcal{X}$, $n = 1, 2, 3, \ldots$, giving $e \in \mathcal{X}$, by 8·3. The supposition therefore is false, and $\mathcal{X} = \mathbf{EP}$.

□

The principle of generalized induction is now obtained as:

THEOREM 6.12 (GEN IND, Axiom 6.2.1) *Let* R *be any unary relation defined on* **EP**. *Suppose*

1. R(**1**); *and* $(\forall e \in \mathbf{EP})$:
2. $\mathrm{R}(e) \Rightarrow \mathrm{R}(e^+)$
3. $[(\forall n \in \mathbf{N})\, \mathrm{R}\{(e, n)\}] \Rightarrow \mathrm{R}(e)$, *when* Diag($e$).

Then $(\forall e \in \mathbf{EP})\, \mathrm{R}(e)$.

Proof. By the previous thm. 11, $\mathcal{X} = \mathbf{EP}$. But the clauses 1.-3. of defn. 8 coincide with the respective hypotheses of theorem 12. Therefore the functions satisfying the relation R exhaust the class **EP**, and the conclusion follows.

□

Examples

1. Ω-r-Reduction Tree

The following notions from earlier Chs. lead to a proof using GEN IND. Let $r = 1, 2, 3, \ldots$.

DEFINITION 6.13 *The (r–canonical reduction) tree $T = T(r)$ (in* **EP***) is defined inductively by*

1. *The pv $v^\sharp$ of T is indexed by $\lambda x \cdot x \star x = \varepsilon$.*
2. *Suppose $v \in T$ has been defined, and indexed by $e \in \overline{\mathbf{EP}}$.*
   (a) *If $e = \mathbf{1}$, or* Suc(e), *then v is an endv.*
   (b) *If* Diag(e), *then v has successors*

$$v_r \text{ L } v_{r-1} \text{ L } \ldots \text{ L } v_1;$$

where v_i is indexed by (e, i); $i = r, r-1, \ldots, 1$.

□

Thus, eh$T(1) = 2$; eh$T(2) = 4$. Generally, eh$T(r) = D_\varepsilon(r)$ (ref. III, §4·3). The endv of C_R of T has index **1**; the remaining endv indices are successors (in the sense of II·30 ·2). In the following, it may be assumed that $r > 1$.

No chain in T is longer than

$$\varepsilon \succ (\varepsilon, r) = \varepsilon_r \succ \varepsilon_{r,r} \succ \ldots,$$

which is finite, by V·15. So it follows that the finite-valency, finite height tree T is *finite*, without the need to appeal to KÖNIG's lemma (IV, §2·6). Therefore

$$(\forall r)\; T \in \mathfrak{S}.$$

(Note that T is r–fold). Now suppose $v \in T$ is indexed by $e \in \mathbf{EP}$. Then the (generally incomplete) chain linking $v^\sharp$ to v defines a sequence $(m_1, m_2, \ldots, m_s)$, $m_i \leq r$, $i = 1, 2, \ldots, r$ such that

$$e = (\varepsilon, m_1, m_2, \ldots, m_s).$$

THEOREM 6.14 *Disregarding vertex indices, $(\forall r)$ T is normal.*

Proof, by generalized induction on diagonals. Given an interior vertex $v \in T$, of index α, it is sufficient to show

$$B_r \succeq B_{r-1} \succeq \ldots \succeq B_1, \qquad \text{(I)}$$

where B_i =Branch$\{v_i, T\}$, $i = r, r-1, \ldots, 1$, and v_r L v_{r-1} L ... L v_1 are the successors of v. Note that the indices of the $v \in T$ play no part in determining the lexicographic order. For $e = \omega$, $(\omega, i) = \mathbf{i}$, $i = r, r-1, \ldots, 1$; and $B_i = [\,]$, so (I) is satisfied. Now let $e \in \overline{\mathbf{EP}}$ be diagonal, and suppose that for all $i \leq r$, it has been shown that B_i is normal. If e is simple, then $B_i = [\,]$, and (I) is satisfied, as in the first case. If e is compound, of the 1st kind, with $e = \Sigma + \omega^{\beta^+}$, then $(e, i) = \Sigma + i \cdot \omega^\beta$. But, disregarding vertex indices, by defn. 13·2,

$$B_{i+1} = B_i$$

for $i = r, r-1, \ldots, 1$; by g.i.h. this is a normal tree, and again (I) is satisfied. If $e = \varepsilon$, or is compound of the second kind, then by V·6

$$(\exists K \in N)\ (e, i+1, 1_1, 1_2, \ldots, 1_{K+1}) = (e, i) \qquad \text{(II)}$$

for $i = r-1, r-2, \ldots, 1$. By (II), $B_i \subset B_{i+1}$ (viz. B_i is a *proper* branch of B_{i+1}); by g.i.h., B_i, B_{i+1} are normal. Therefore

$$B_i \prec B_{i+1},$$

and again condition (I) is satisfied. Thus it is proved that in all cases, the defining condition of IV·19 is satisfied, and the theorem follows by generalized induction.

□

Generally, e has > 1 expressions as given before VI·14, viz. one for each occurrence of e as a vertex index in T. Exactly one chain defines the normal form

$$e = \varepsilon_{n_1, n_2, \ldots, n_R}$$

of V, defn. 12. Thus consider the tree $\tau = \tau(r)$ obtained from T as follows. If $v \in T$ is indexed by e, and the chain $v^\sharp, \ldots, v$ does not define the indices of the ε–normal form of e, then Branch$\{v, T\}$ is excised from T.

THEOREM 6.15 $\tau \in \mathfrak{T}$.

Proof. As remarked in the proof of V, lemma 13, it is only on account of V·6 that a non-normal form may occur, viz. when e is compound, of the 2nd kind. But $\mathrm{eh}B_{i+1} - \mathrm{eh}B_i \geq r$; and by V·13, the excision applied to B_{i+1} is removal of (part of) the rightmost of the r principal branches of B_{i+1}, to give B^-_{i+1}, say. Therefore, since the height is unaltered,

$$B_i \prec B^-_{i+1},$$

and τ is normal.

□

Suppose u L v in τ, and $w \in$ Branch(τ, u). Then by III·7 ·2 and III·14, ind$(v) \preceq$ ind(w). Since by definition the τ–indices are unique, the $\lhd_r$-reduction of ε thus enumerates the indices in L-to-R pre-order; or applying III·7 ·3:

$$(\{\varepsilon\} \cup \mathbf{EP}[\varepsilon; r], \prec) \simeq (\tau, <^{\text{R-to-L}}_{\text{post}}).$$

2. GEN IND in Real Analysis

(Positive) real numbers x_α may readily be associated with $\alpha \in \mathbf{EP}$. E.g. if $M = 1 +$ max ec in α, no pass $\alpha \prec \beta$ can occur after the value $y = \alpha(M)$. Thus the given integer y may be used as in: $x_\alpha = \frac{1}{y}$. Furthermore, real sums $\Sigma_\beta = \Sigma_{\alpha \prec \beta} x_\alpha$, $x_\alpha \in \mathbf{R}$, extending over (segments of) $\mathbf{EP}$ may be defined by:

$$\text{(i) } \Sigma_{\mathbf{2}} = x_{\mathbf{1}}; \quad \text{(ii) } \Sigma_{\beta+} = x_{\beta+} + \Sigma_\beta;$$

$$\text{(iii) } \Sigma_\beta = \lim_{n=1,2,3,\ldots} \Sigma_{(\beta,n)} \text{ if Diag}(\beta), \text{ if the rhs } < \infty.$$

Note that $\beta = \omega$ gives the classical sum; also that there may be no ready means of evaluating the limit (iii), even in cases where $\lim < \infty$ is easily proved.

Now let $(\varphi_m)_{m=1,2,3,\ldots}$ be a (standard) enumeration of **EP**, as in II, §5·1, or V, §6·3. Then each $\alpha \prec \beta$ has $\alpha = \varphi_m$ for some m; so for a (si) sequence $m(1), m(2), \ldots$ the (classical) sum $\Sigma_{i=1}^{\infty} x_{\varphi_{m(i)}}$ contains exactly the terms of Σ_β. There is no difficulty in establishing (by GEN IND):

THEOREM [HOBSON] $\Sigma_\beta = \Sigma_i x_\varphi$ iff the rhs converges absolutely.

□

Thus while the finite construction of II §2·17 for integer polynomials builds, via polynomial exponents, into a complex hierarchy; for real numbers, the limiting construction above (via polynomial suffices) collapses to identity with the classical theory.

Remark. A given property P of α (e.g. Diag(α), Perf(α)) may readily define a real sum

$$S = \sum_{\mathrm{P}(\alpha)} \frac{1}{\alpha(m)},$$

where e.g. $\alpha = \varphi_m$; but in few cases will it be possible to express S in terms of familiar quantities.

6.2.6 Remarks on the Class $\mathcal{X}$

Returning to the proof of thm. 11, viz. that the 'generalized induction class' $\mathcal{X}$ is in fact **EP**, the essential contradiction is obtained only by recourse to thm. 5: every remainder has a $\prec$ –lowest polynomial. And this theorem in turn rests on the DCC, IV, corol. 48; which is founded on IV, thm. 39. This last cited result, which is the foundation of the train of reasoning, is obtained by purely elementary means: as already discussed in IV, §3, the axiom of choice, AC, is *not* used. To summarize: a proof in which 'transfinite' induction is replaced by generalized induction is preferable, because now the proof does not use the AC.

Defn. 8 of $\mathcal{X}$ actually 'gives' the polynomials in $\prec$ –order: $\mathbf{1}, \mathbf{2}, \mathbf{3}, \ldots$ and after these, $\omega, \omega^+, \omega + \mathbf{2}, \ldots$, and so on. Of course endless constants must be passed, before ω is given. In terms of time units, therefore, ω will never be reached. But the formulation $\mathcal{X}$ gives an application of the human learning process: after the first few constants, the progression becomes evident, and ω may be given. In the same way, after a few terms of the cfs $\omega, 2\omega, 3\omega, \ldots$, the whole cfs becomes evident, and ω^2 may be given. In this way the whole of the linear hierarchy **EP** may readily be conceptualized. By contrast, the 'minimal' defn. 3 of Ch. II, which precedes the canonical assignment Ω of Ch. III, suggests no particular order. It may be noted that the notion 'a few terms' may be made strikingly precise. Suppose it is agreed that 3 terms suffice to exemplify a cfs. Then altogether $D_\varepsilon(3)$ polynomials must be written down, in order to conceptualize all of **EP**, where D is the 'descent function' III·25.

Note.

The theorem of GEN IND may be curtailed by replacing '**EP**' in thm. 12 by any proper segment.

6.2.7 Generalized Induction Implies the DCC

A Cumulative Proof of GEN IND

Notwithstanding the remarks of §2.6, GEN IND

1. may be proved independently of the DCC;
2. implies the DCC.

Now to prove thm. 12, as before the result reduces to the proof of thm. 11, $\mathcal{X} = \mathbf{EP}$. For this purpose, let GEN IND$_h$ denote axiom 2.1 with '**EP**' replaced by '$\mathbf{EP}^{(h)}$' (thus GEN IND$_0$ denotes mathematical induction). The overall proof is by induction on $h = 0, 1, 2, \ldots$. For $h = 0$: $\mathbf{1} \in \mathcal{X}$ by defn. 8·1, therefore by induction, $\mathbf{EP}^{(0)} \subset \mathcal{X}$. Now let $h \geq 0$, and suppose it has been proved $\mathbf{EP}^{(h)} \subset \mathcal{X}$. Note that this implies the validity of GEN IND$_h$. Well,

$$\omega \star (h+1) = \operatorname{diag}_\Omega \omega \star (k, n),$$

and by i.h.

$$\omega \star (h, n) \in \mathcal{X}, \; n = 1, 2, 3, \ldots .$$

Therefore by defn. 8·3, $\omega \star (h+1) \in \mathcal{X}$. It now follows that all polynomials

$$\omega \star (h+1) + \beta \in \mathcal{X},$$

by GEN IND$_h$ on $\beta \in \mathbf{EP}^{(h)}$. Therefore all polynomials

$$a\omega \star (h+1) + \beta \in \mathcal{X},$$

$\beta \in \mathbf{EP}^{(h)}$, by induction on $a \in \mathbf{N}$. Now let $e \in \mathbf{EP}^{(h)}$. Let $\mathbf{EP}^{(h+1,e)}$ denote the subclass of polynomials of $\mathbf{EP}^{(h+1)}$ of degree $\preceq e$. It is already assumed $\mathbf{EP}^{(h)} \subset \mathcal{X}$. Now let $e \succ \omega \star h$, and assume also for generalized induction hypothesis (g.i.h.).

1. If Suc(e), then $\mathbf{EP}^{(h+1,e^-)} \subset \mathcal{X}$.
2. If Diag(e), then $\mathbf{EP}^{\{h+1,(e,n)\}} \subset \mathcal{X}, \; n = 1, 2, 3, \ldots$.

Now $\alpha \prec \beta$ in $\mathbf{EP}^{(h)} \Rightarrow \mathbf{EP}^{(h+1,\alpha)} \subset \mathbf{EP}^{(h+1,\beta)}$, therefore by g.i.h. (and by III, thm. 12 if Diag(e)) the segment $\mathcal{X}'$ of polynomials of $\mathbf{EP}^{(h+1)}$ of degree $\prec e$ (note *strict*) satisfies

$$\mathcal{X}' \subset \mathcal{X}, \tag{I}$$

which implies the validity of $\mathcal{X}'$-GEN IND.[13] Now the ct's $(\omega^e, b) \prec \omega^e$ of ω^e belong to $\mathcal{X}'$, so (I) and defn. 8·3 give

$$\omega^e \in \mathcal{X}.$$

[13]Viz. in axiom 2.1, '**EP**' is replaced by '$\mathcal{X}'$'.

Now for $\beta \in \mathcal{X}'_0$, $(\omega^e + \beta)^+ = \omega^e + (\beta^+)$, and for diagonal β, $(\omega^e + \beta, n) = \omega^e + (\beta, n)$. Therefore by $\mathcal{X}'$-GEN IND, all polynomials $\omega^e + \beta \in \mathcal{X}$. So by induction on $a \geq 1$, all polynomials $a\omega^e + \beta \in \mathcal{X}$, i.e.

$$\mathbf{EP}^{(h+1,e)} \subset \mathcal{X}.$$

Therefore by GEN IND$_h$ on the variable polynomial e, for all $e \in \mathbf{EP}^{(h)}$,

$$a\omega^e + \beta \in \mathcal{X},$$

$\beta \in \mathbf{EP}^{(h+1)}$, $\deg(\beta) \prec e$ (deg$\equiv$ degree). But these are exactly the polynomials of $\mathbf{EP}^{(h+1)}$. Therefore by induction,

$$(\forall h)\, \mathbf{EP}^{(h)} \subset \mathcal{X}.$$

So by the 2nd rem. after Ch. II, defn. 5, and by the rem. after defn. 8,

$$\mathcal{X} = \mathbf{EP}.$$

Q.E.D.

Notes

1. For each value of $h \geq 0$, in order to prove

$$a\omega^e + \beta \in \mathcal{X}$$

as above, 3 techniques of induction are applied at once: (i) the induction on $a \geq 1$; (ii) on $e \in \mathbf{EP}^{(h)}$; (iii) on $\beta \in \mathbf{EP}^{(h+1)}$, $\deg(\beta) \prec e$.

2. The above proof is the route taken by GENTZEN [40], 1936, GOODSTEIN [45], 1944 and TAKEUTI [147], 1975, in every case to found the proof of consistency for formalized number theory. The proof itself (of GEN IND), despite the multiplicity of steps, is essentially elementary and free of difficulty. Also, this particular proof does not use the axiom of choice.

GEN IND Implies DCC

It remains to establish the logical equivalence of GEN IND and the DCC for **EP**. Thus, the converse to thm. 12 is

THEOREM 6.16 [14] *Given axiom 2.1, the system* $(\mathbf{EP}, \prec)$ *satisfies the DCC.*

Proof. Let $\mathcal{S} = (e_1 \succ e_2 \succ e_3 \succ \ldots)$ in **EP**. The theorem is equivalent to: $\mathcal{S}$ must contain a smallest term (with respect to $\prec$), or must be the null sequence. Suppose, therefore that $\mathcal{S}$ does not contain a smallest term. Let S be the unary relation defined (on **EP**) by

$$(\forall e)\, \mathrm{S}(e) \Leftrightarrow (\forall n)\, e \prec e_n.$$

[14] [96], thm. 3·6.

Evidently, S(**1**), since $\mathbf{1} \in \mathcal{S}$ would be the smallest term of $\mathcal{S}$. Now for arbitrary e, suppose S(e). Then S(e^+), since otherwise e^+ would be the smallest term of $\mathcal{S}$. Now suppose e is diagonal, and $(\forall n)\ \mathrm{S}\{(e,n)\}$. Then by the density result III·12,

$$(\forall \alpha)\ \alpha \prec e \Rightarrow (\exists n)\ \alpha \prec (e,n) \Rightarrow \mathrm{S}(\alpha).$$

Therefore S(e), since otherwise e would be the smallest term of $\mathcal{S}$. Therefore by generalized induction it follows $(\forall e \in \mathbf{EP})\ \mathrm{S}(e)$, so $\mathcal{S}$ must be the null sequence, and the theorem is proved.

Second proof. Suppose as above that

$$\mathcal{S} \qquad\qquad \ldots \prec e_3 \prec e_2 \prec e_1$$

is an arbitrary descending sequence in **EP**. Certainly $\mathcal{S}$ is finite if $e_1 = \mathbf{1}$. Now suppose $e_1 \succ \mathbf{1}$, and for g.i.h. that every $\prec$ –descending sequence in **EP** whose initial term is

$$\preceq \begin{cases} e_1^- & \text{if Suc}(e_1), \\ (e_1, N) & \text{for some } N, \text{ if Diag}(e_1) \end{cases}$$

is finite. If Suc(e_1), then $e_2 \preceq e_1^-$. So by g.i.h., $\ldots \prec e_3 \prec e_2$ is finite, therefore so is the old sequence. If Diag(e_1), with $e_1 = \mathrm{diag}_n\ (e_1, n)$, then $(\exists N)\ e_2 \prec (e_1, N)$, and, using the g.i.h. again a finite sequence is obtained by prefixing e_1 to a finite sequence.

□

Thus thm. 12 and the above give

COROLLARY 6.17 *For the system* $\mathbf{EP} = (\mathbf{EP}, \prec, \Omega)$, GEN IND *and the* DCC *are logically equivalent.*

□

6.2.8 Non-Constructive Proof of the DCC

The type hierarchy of Ch. II, §4 has been used to provide a proof of well-foundedness of $(\mathbf{EP}, \prec)$. Let $\mathcal{W}_{(1)} \subseteq \mathbf{EP}$ be the largest segment for which $(\mathcal{W}_{(1)}, \prec)$ is well-founded. For $h = 1, 2, 3, \ldots$, let

$$\mathcal{W}_{(h+1)} = \{F \in \mathcal{N}_{h+1} \mid f \in \mathcal{W}_{(h)} \Rightarrow F(f) \in \mathcal{W}_{(h)}\}.$$

Now, by induction on h, $E_h \in \mathcal{W}_{(h)}$. Therefore by Ch. II, thm. 52, it follows that $\mathcal{W}_{(1)} =$ **EP**. The impredicativity of the above proof is indicated by the diagram, where the arrows

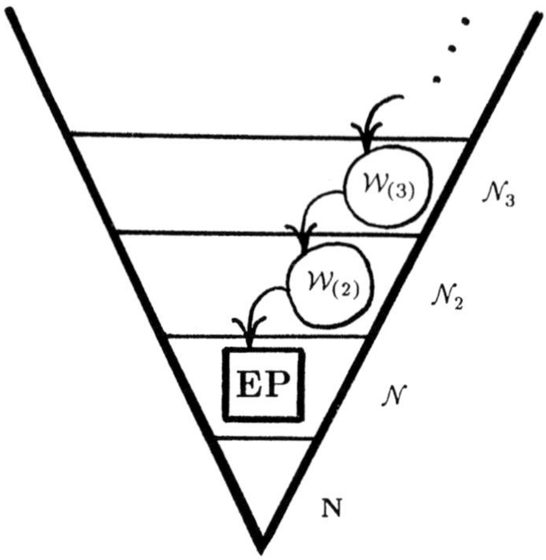

point to the corresponding domains of closure.

6.3 $\tilde{\triangleleft}$–Induction in EP

6.3.1 Induction on Segments

The GEN IND for $(\mathbf{EP}, \prec, \Omega)$ of §2 is in fact logically equivalent to an apparently stronger form:

AXIOM 6.3.1 *If* R *is a unary relation defined on* **EP**, *then from*

$$\mathrm{R}(\mathbf{1}) \,\&\, (\forall e \succ \mathbf{1}) \, [(\forall \alpha \prec e) \, \mathrm{R}(\alpha)] \Rightarrow \mathrm{R}(e)$$

may be inferred

$$(\forall e \in \mathbf{EP}) \, \mathrm{R}(e).$$

This axiom contains no reference to a canonical assignment, and may be derived by GEN IND as follows. **1** is the only polynomial in the initial case. If Suc(e) and 6.2.1,3.1 are equivalent for $(\mathbf{EP})_{e^-}$, then by theorem 12·2, so they are for $(\mathbf{EP})_e$. If Diag(e), then if the 2 forms are equivalent for $(\mathbf{EP})_{(e,n)}$, $n = 1, 2, 3, \ldots$, then by 12·3, so also for $(\mathbf{EP})_e$.

6.3.2 Well-Ordering

In the light of IV, §3·4, let R be any unary relation defined on **EP** :

THEOREM 6.18 (Well-Ordering Principle For EP) *Either*

1. $(\forall\alpha) \, \neg\mathrm{R}(\alpha)$ *or*
2. *with respect to* $\prec$, *there is a least* $e \in$ **EP** *such that* R(e).

Proof. Suppose there is no e satisfying (2). For each $\alpha \in$ **EP**, define [15]

$$\mathrm{S}(\alpha) \Leftrightarrow (\forall\beta) \, \mathrm{R}(\beta) \Rightarrow \alpha \prec \beta.$$

Then S(**1**), otherwise $e = \mathbf{1}$. Now suppose $\alpha \succ \mathbf{1}$, and $(\forall\gamma \prec \alpha) \, \mathrm{S}(\gamma)$. Then S($\alpha$), otherwise $\beta = \alpha$. Therefore by axiom 3.1 $(\forall\alpha) \, \mathrm{S}(\alpha)$, so (1), $(\forall\alpha) \, \neg\mathrm{R}(\alpha)$, holds, and thm. 18 is proved.

A second proof exposes the rôle of the AC. Let S be defined as above, and let $n_1 \geq 1$ be s.t. $S(\varepsilon_{n_1^-})$ & $\neg S(\varepsilon_{n_1})$. In order to find n_1, $\beta_1 \in$R s.t. $\varepsilon_{n_1^-} \prec \beta_1 \preceq \varepsilon_{n_1} = \beta_{(1)}$ say, had to be given. Now let n_2 be s.t. $S(\varepsilon_{n_1,n_2^-})$ & $\neg S(\varepsilon_{n_1,n_2})$. n_2 required $\beta_2 \in$R, s.t. $\varepsilon_{n_1,n_2^-} \prec \beta_2 \preceq \varepsilon_{n_1,n_2} = \beta_{(2)}$. This leads to

$$e = \beta_K = \beta_{(K)} \prec \ldots \prec \beta_{(2)} \prec \beta_{(1)}$$

and the ε–form of the least $e \in$R.

□

[15] Note that *the AC is required to secure this defn.*

Dependence of Well-Ordering on the Axiom of Choice

It may be conceded that the derivation of thm. 18 from axiom 3.1 is slightly easier than that of thm. 16 from axiom 2.1. Axiom 3.1 is less complex than axiom 2.1. The DCC for **EP**, thm. 16, is (logically) weaker than the well-ordering principle, thm. 18.

Why, then, suffer a technical penalty, to achieve less? The answer, of course, lies in the excision of the axiom of choice. The conclusion of thm. 16 and axiom 6.2.1 are logically equivalent, and as discussed in the preamble to Ch. IV, §3·4, thm. 18 cannot be obtained from the descending chain condition without the AC. The AC has, after more than a century of scrutiny, failed to establish itself as a secure foundation of reason; this is a most serious matter, and there is good reason to believe that at no time will it become rehabilitated. At the very least it is clear that the AC has no place in the study of arithmetic functions of rapid growth. It is the task of this Ch. to establish that a far-reaching generalization of mathematical induction is attainable, and applicable, by unquestionably finitary means.

6.3.3 α–Induction

Let $\alpha \in \overline{\mathbf{EP}}$, and let $\alpha_1, \alpha_2, \alpha_3, \ldots$ be a standard (that is $\alpha_1 = \mathbf{1}$) elementary enumeration of $(\mathbf{EP})_\alpha$ (II ·21). Now define the relation $<_{\text{enum}} = <_{\text{enum}(\alpha)}$ on $\mathbf{N} \times \mathbf{N}$ by

$$m <_{\text{enum}} n \Leftrightarrow e_m \prec e_n.$$

THEOREM 6.19

$$(\mathbf{N}, <_{\text{enum}}) \simeq (\mathbf{EP}, \prec)_\alpha.$$

Proof. The $1-1$ onto correspondence $n \leftrightarrow e_n$ determines the similarity.

□

The discussion of the great multiplicity (II, §4.1) of enumerations of $(\mathbf{EP})_\alpha$, for suitably $\prec$ –large α carries over to that of the relation $<_{\text{enum}}$. However, given a choice, the following generalization of mathematical induction is possible. It is best presented as a course-of-values induction:

AXIOM 6.3.2 *If* R *is a unary relation defined on* $\mathbf{N}$, *then from*

$$\mathrm{R}(\mathbf{1})\ \&\ (\forall x > 1)\ [\{(\forall i <_{\text{enum}(\alpha)} x)\ \mathrm{R}(i)\} \Rightarrow \mathrm{R}(x)]$$

may be inferred

$$(\forall x)\ \mathrm{R}(x).$$

It is a standard theorem of mathematical logic (ref. MOSTOWSKI [113], p49; GENTZEN [42], pp287-308) that for $<_{\text{enum}}$ built from primitive recursive operations, and $\alpha \in \mathbf{EP}$, axiom 3.2 is a rule of closure of the first order theory of arithmetic; while for $\alpha = \varepsilon$, it leads out of $\underline{\mathbf{N}}$.

6.4 Generalized Recursion

The scheme of (primitive) recursion

$$f(x, 1) = g(x)$$

$$f(x, y+1) = h\{x, y+1, f(x, y)\}$$

(by which f is derived from the given arithmetic functions g, h) and its usual generalizations, discussed summarily in Ch. I §3.1, has a deeper generalization, whose form is determined by the generalized induction of §2. The induction on segments[16] of §3 has no attendant generalized recursion. This is developed here; and in the following §, an equivalence is established for the generalization of recursion obtained by taking α–orderings, $\alpha \in \mathbf{EP}$.

6.4.1 Generalized Recursion Based on Ω

Generalized Iteration

The schema of pure iteration is

DEFINITION 6.20

$$f(x, \mathbf{1}) = \phi(x)$$

$$f(x, e^+) = f\{f(x, e), e\}$$

$$f(x, e) = f\{x, (e, x)\},$$

for $x \in \mathrm{N}$, $e \in \mathbf{EP}$; Diag(e); *in the last case; and* ϕ *a 1-place arithmetic function.*

An equivalent formulation is: $\phi^{(\mathbf{1})} = \phi$; $\phi^{(e^+)} = \phi^{(e)} \circ \phi^{(e)}$; with $\phi^{(e)} = \text{diag}_n\ \phi^{\{(e,n)\}}$ if Diag(e), giving $f(x, e) = \phi^{(e)}(x)$.

VI·20 has variations, generalizations and restrictions. E.g. the 2nd equation may be replaced by:

$$f(x, e^+) = \phi\{f(x, e)\}$$

(or equally by $f(x, e^+) = f\{\phi(x), e\}$). This schema reduces to: $f(x, e) = \phi^{\{e(x)\}}(x)$, the superscript denoting the usual numerical iteration.

VI·20 evidently reduces to $f(x, e) = \phi^{(K)}(x)$, where the number $K = K(e) = K(x, e)$ satisfies: $K(\mathbf{1}) = 1$; $K(e^+) = 2 \cdot K(e)$; and if Diag(e), $K(e) = K\{(e, x)\}$. Thus

$$K = 2^{D_e(x)},$$

where D is the 'descent function' of III §4·3 (defn. 25).

VI·20 may be curtailed (ref. note of §2.6) by:

---

[16]The §3 GEN IND underlies the 'transfinite' induction for the CANTOR 'below ε ordinals.' Attendant on this is 'transfinite recursion'. This is *not* the same as generalized recursion, as it is over the 'below-ε' objects, rather than over **N**.

DEFINITION 6.21 *The schema of pure generalized iteration modulo* $\alpha \in \mathbf{EP}$ *is obtained as in defn. 20, but under the restriction* $e \prec \alpha$.

□

Example. $\alpha = \omega$ gives $\mathcal{P}$–iteration.

There is little to be gained from an attempt to classify the possible variants, as they are largely accounted for by the forms of generalization to which primitive recursion is amenable. One case only, therefore, will be treated, on account of its practical applicability. Recalling the notation of rem. 3(b) after V·3, and remarks of III,§4, a si $\mathbf{N}$–sequence $\mathcal{S}$ gives a very much faster-growing descent function $D_e(\mathcal{S})$ than the constant case, as occurs in the calculation entailed by VI·20; particularly if $\mathcal{S}$ itself is rapidly-growing. In the schema GEN REC to follow, $\mathcal{S}$ is obtained by taking values of already-obtained iterations of the initial function.

Generalized Recursion

The relation between GEN IND and GEN REC is one of definedness: GEN IND may be used to show that the schema of combined arithmetic and generalized recursion always leads to a totally defined arithmetic function.

DEFINITION 6.22 (GEN REC)

$$f(\underline{x}, y, \mathbf{1}) = \varphi(\underline{x}, y)$$

$$f(\underline{x}, 1, e) = \phi(\underline{x}), \ \mathbf{1} \prec e$$

$$f(\underline{x}, y^+, e^+) = f\{\underline{x}, f(\underline{x}, y, e^+), e\}$$

$$f(\underline{x}, y, e) = f\{\underline{x}, y, (e, y)\},$$

the last case if Diag(e); *where* $\underline{x}$ *is an* $n-$*string, and* ϕ, φ *are functions of* n, $n+1$ *variables respectively.*

□

That defn. 22 leads to a total function may be seen informally as follows. For $y > 1$, $e \succ \mathbf{1}$ consider the general calculation $f(\underline{x}, y, e) = f(\underline{x}, Y, \widetilde{e})$, where

$$Y = \begin{cases} f(\underline{x}, y^-, e) \text{ and } \widetilde{e} = e^- & \text{if Suc}(e) \\ y \text{ and } \widetilde{e} = (e, y) & \text{if Diag}(e). \end{cases}$$

Writing $v^\sharp \equiv f(\underline{x}, y, e)$, $u \equiv f(\underline{x}, Y, \widetilde{e})$, $v \equiv Y$, the combinatorial unit

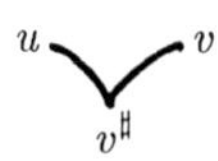

may be formed, where the edge $(v^\sharp, u)$ represents an equality; and $(v^\sharp, v)$ the necessity to evaluate the expression Y, in order to begin the calculation of u. Thus the calculation of $f(\underline{x}, y, e)$ defines a tree T, of the class $\mathfrak{D}$ (IV, defin. 16), built up from such units. Every left edge-ascent in T entails a decrement $\tilde{e} \prec e$. Therefore applying the DCC, any chain C in T has only finitely many, K say, left ascents. At the last of these, a certain value of Y has accumulated. Therefore C now has only $Y - 1$ right edge-ascents. I.e. C is finite, with eh$C = K + Y - 1$. Since C was arbitrary, it follows that all chains of T are *finite*. Since T is of finite valency, by KÖNIG's lemma KL , T is *finite*. So f is totally defined.

As discussed in Ch. IV, §2, KL is not a purely elementary principle. The following proof [17] of the result does not use KL: let $e \in \mathbf{EP}$. Within the scope of the next thm. only, let '$f(x, y, e)$' be abbreviated to '$[e, y]$' (the parameters $\underline{x}$ playing no part in the proof). Let 'Calc(e, y)' mean '$[e, y]$ can be calculated in a finite number of steps'; and let

$$\text{Calc}(e) \equiv (\forall y \geq 1)\ \text{Calc}(e, y).$$

THEOREM 6.23 $(\forall e)$ Calc(e).

Proof, by arithmetic and gen. induction. For $e = \mathbf{1}$, $(\forall y)[\mathbf{1}, y] = \varphi(x, y)$, i.e. Calc$(\mathbf{1})$. Now suppose $e = e^- + \mathbf{1}$ in **EP**, and for g.i.h. that Calc(e^-). $[e, 1] = \phi(x)$, giving Calc$(e, 1)$. Thus let $y \geq 1$, and suppose for i.h. that Calc(e, y). Well $[e, y+1] = [e^-, [e, y]]$. Therefore Calc$(e, y + 1)$ by g.i.h., and by induction, $(\forall y)$ Calc(e, y), viz. Calc(e). Now suppose Diag(e), and for g.i.h. that $(\forall n)$ Calc$\{(e, n)\}$. Then for all y, $[e, y] = [(e, y), y]$, viz. Calc(e, y), so Calc(e).

□

It is thm. 23 that provides the link between the generalized induction of axiom 6.2.1, and the generalized recursion of defn. 22.

Generalized recursion is the technique of choice for the construction, by elementary means, [18] of arithmetic functions, whose order of infinity far exceeds that of ACKERMANN's function. However the following case, in which the technique is applied to minimal effect, provides a curious example:

$\iota(\mathbf{1}) = \mathbf{1}$. Now suppose $\alpha \in \mathbf{EP}$ is non-initial. Then

$$\iota(\alpha) = \begin{cases} \iota(\alpha^-)^+ & \text{if Suc}(\alpha) \\ \text{diag}_n\ \iota\{(\alpha, n)\} & \text{if Diag}(\alpha) \end{cases}$$

THEOREM 6.24 *For every* $\alpha \in \mathbf{EP}$, $\iota(\alpha) = \alpha$.

Proof. The image of ι is the class $\mathcal{X}$ of defn. 8.

□

Note that $\iota_\Lambda = \iota$ remains invariant for all assignments Λ of fundamental sequences.

[17] Mc BETH[107], Appendix , p76.

[18] In the sense detailed in Ch. I, §3.

6.5 Enumeration and Recursion

6.5.1 α–Recursion

The correspondence which clearly links generalized induction (axiom 2.1; thm. 12) to generalized recursion (definition 22) appears again in the link between α–induction (§3.2) and the procedure of taking recursions over **N** under an un-natural order, in particular determined by $\alpha \in \mathbf{EP}$.

Thus the generalization of arithmetic recursion is obtained by taking recursion over α–sequential orderings $\prec$ of **N**, where

$$(\mathbf{N}, \prec) \simeq (\mathbf{EP}, \prec)_\alpha,$$

for some α. This procedure is known as '*α–recursion*'.

Consider F obtained by the primitive recursion

$$F(1) = k, \text{ a constant;}$$

$$F(x+1) = f\{F(x)\}.$$

In the calculation of the iteration

$$F(x) = f^{(x)}(k) = f\{f^{(x-1)}(k)\} = \dots$$

the superscript may be viewed as passing down a 1-dimensional array of values $x, x-1, \dots, 1$. Take now for example the case $\alpha = \omega^2$ (this is considered in detail in PÉTER's monograph [120], *Recursive Functions* , §12(3-9)). The 1-array is replaced by a 2-array, whose addresses are ordered lexicographically:

$$(m, n) <_{\omega^2} (M, N) \text{ iff } m < M \ \vee \ \{m = M \ \& \ n < N\}.$$

The primitive recursion is now replaced by:

$$F(1) = k; \text{ for } x > 1, \ F(x) = f\{F(y)\},$$

where the yth address $<_{\omega^2}$ the xth address. This is only determinate when y is chosen by some previously-defined 'choice-function' γ, giving the schema:

$$F(1) = k; \text{ for } x > 1, \ F(x) = f[F\{\gamma(x)\}].$$

In the new calculation of $F(x)$ as an iteration of f,

$$F(x) = f^{(X)}(k),$$

the superscript $X > x$ may be viewed as passing down a 2-array: if an address does not begin a new row, γ picks an address to the left. If an address (not $(1,1)$) begins a new row (viz. has no predecessor to the left), γ picks an address in a previous row. F may now accumulate iterations of f in downward steps in many rows, thereby reaching beyond the range of primitive recursion. In Ch. 4 of the above monograph, it is established that the ω^2–schema is equivalent to simultaneous (double) recursion, viz. is adequate for the construction of ACKERMANN's function. The method may be generalized to arbitrary $\alpha \in \mathbf{EP}$ via canonical reduction (Ch. III, §4).

It is a standard meta-theorem of mathematical logic that, given any fixed $n \in \mathbf{N}$, the thm. of generalized induction, obtained by replacing, in axiom 2.1, **EP** by $\mathbf{EP}^{(h)}$, for any given $h \in \mathbf{N}_0$, [19] is a formal theorem of $\underline{\mathbf{N}}$. GENTZEN [42] has given a direct proof that the **EP**–form 2.1 lies beyond $\underline{\mathbf{N}}$.

However, aside from the arbitrariness of the choice of the auxilliary function γ, there are 2 reservations regarding α–recursion from the finitist viewpoint.

1. In the usual formulation, the α–induction on which it is founded assumes a given *well-ordering.* However, the above sketch demonstrates that the assumption of sequential order (Ch. IV, defn. 38·2) suffices.
2. There is the question of avoiding the concept of un-natural orderings of **N**. Apart from the evident difficulties of accepting formulae such as '$3 < 2$' as true, given α, a great multiplicity of different sequential orders over **N** each correspond to α. E.g. for $\alpha = \omega^2$,

$$\begin{array}{llll} 1 & 2 & 4 & 7 & \ldots \\ 3 & 5 & 8 \\ 6 & 9 \\ 10 \\ \vdots \end{array} \qquad \begin{array}{llll} 1 & 2 & 5 & 10 & \ldots \\ 4 & 3 & 6 \\ 9 & 8 & 7 \\ \vdots \end{array}$$

each determine the same ω^2–recursion.

Since it is evident that α–recursion is equivalent to (α–)generalized recursion of the form of defn. 21, the procedure of recursion over α–orderings of **N** will not be used anywhere in the text. ROSE [131] 1984, Ch. III contains discussion of α–recursion, and recursion over many variables.

6.5.2 Fundamental Sequences and Primitive Recursive Enumerations

Suppose $e_1, e_2, e_3, \ldots$ is a primitive recursive enumeration of **EP**. Suppose $e = e_n$ is diagonal (viz. has no predecessor). Then it is possible (e.g. as in TAKEUTI, G. [147], 1976) to define a fundamental sequence for e to be any sequence

$$e_{n(1)}, e_{n(2)}, e_{n(3)}, \ldots$$

for which

1. $e_{n(i)} \prec e_{n(i+1)}$
2. $e = \lim_{i=1,2,3,\ldots} e_{n(i)}$,

where 2. has the meaning of corol. 7, viz.

$$\alpha \prec e \text{ in } \mathbf{EP} \Rightarrow (\exists I)(\forall i \geq I)\ \alpha \prec e_{n(i)}.$$

The above, however, is two steps distant from the canonical assignment of III §2.2. Not only is III·6 a specialization of the diagonal, but

$$\lim_i e_{n(i)} \neq \text{ diag}_i\ e_{n(i)}$$

is possible, as in e.g. $\psi = \lim_i\{\omega, \omega^4, \omega^9, \ldots\}$, but $\text{diag}_i\ \omega^{i^2} = \omega \star (\omega; 2) \succ \psi$.

[19] Ref. Ch. II, defn. 6.

6.6 Notes on Ch. VI

1. §2. Writing A(e) iff ¬R(e) in **EP**, the 'downwards' form of GEN IND is: from $\neg A(\mathbf{1})$ and for all $e \in \mathbf{EP}$

$$A(e^+) \Rightarrow A(e); \quad \text{and if Diag}(e) \text{ then } (\forall n)\ A\{(e,n)\};$$

may be inferred $(\forall e)\ \neg A(e)$.

2. §2·5, ex. 2. This deals with reals x_α 'added in sequential order'. The point made near the end of §5·1 is that any given sequential order, taken in composition with a permutation of the whole of **N**, gives another sequential order; for the purposes of the ex., such permutations may be ignored. However, the classical theory contains a theorem of unusual interest, for which the permutations alone are of interest, in their application to *conditionally* convergent series $\Sigma_i x_i$. Let $\rho \in \mathbf{R}$ be arbitrary:

THEOREM [REIMANN] The order of the terms of Σx_i may be altered, so that the sum takes the value ρ.

The proof is easily sketched: it may be supposed no $x_i = 0$; thus

$$\Sigma x_i = \Sigma y_i + \Sigma z_i,$$

as sums of positive and negative terms respectively, taken in order of occurrence in Σx_i. These 2 sums diverge; also their terms tend to 0. ρ may therefore be successively exceeded, then fallen short of, by alternately adding (the minimum number of) consecutive terms of Σy_i or of Σz_i. Since the terms tend to 0, the limiting value of the new sum is ρ.

□

3. §3·2. The following correspondence between integers $0, 1, 2, \ldots$ and patterns of (script) brackets '{,' '}' may be given recursively:

$$S_0 = \{\ \}; \quad \text{and for } n \geq 0,\ S_{n+1} = S_n \cup \{S_n\},$$

where '∪' is defined by $\{a, \ldots, b\} \cup \{c, \ldots, d\} = \{x, \ldots, y\}$, the patterns $x, \ldots, y$ being exactly the different patterns which occur either in $a, \ldots, b$; or in $c, \ldots, d$. E. g. $S_1 = \{\{\}\}; S_2 = \{\{\}, \{\{\}\}\}$. Let '∈' be defined by: $a \in b$ iff a is one of the patterns enclosed by the outermost brackets of b. Then it is easily seen that $m < n$ in $\mathbf{N}_0$ iff $S_m \in S_n$.

Now let '∩' be defined by: $\{a, \ldots, b\} \cap \{c, \ldots, d\} = \{u, \ldots, v\}$, where $u, \ldots, v$ are exactly the different patterns which occur both in $a, \ldots, b$ and in $c, \ldots, d$. And given any finite si $\mathbf{N}_0$−sequence $n(1) < n(2) < \ldots, < n(k)$, let $S = \{S_{n(1)}, S_{n(2)}, \ldots, S_{n(k)}\}$. Then the above recursive definition gives

$$S_{n(1)} \cap S = \{\ \}; \quad \text{and if } k > 1,\ \ S_{n(i)} \cap S \neq \{\ \}$$

for $i = 2, 3, 4, \ldots, k$. Thus, taking the normal form of the correspondence of note 4, IV, §10, the natural numbers appear as :

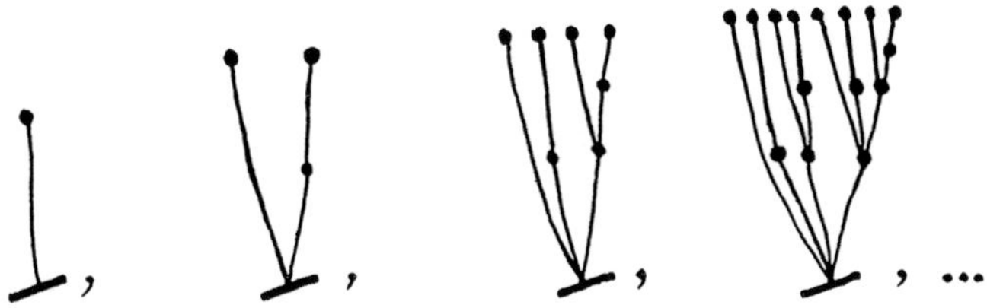

The distinguishing property of $n(1)$, described here for finite sets, is taken as the basis of the definition of well-order in the first-order theory of sets.

4. §3·2. GEN IND in 1st order set theory ZF takes the form (cf. axiom 3·2):

$$[\mathrm{P}(\{\}) \,\&\, (\forall x \neq \{\ \})\{(\forall y \in x)\ \mathrm{P}(y)\} \Rightarrow \mathrm{P}(x)] \Rightarrow (\forall z)\ \mathrm{P}(z).$$

This may be regarded as a weak form of 'the axiom of foundation', viz. every sequence $\ldots \in y \in x$ is finite.

The'axiom of infinity' of IV §10, note 4, determines an 'infinite-valency finite-chain tree'. An 'infinite Y' with the converse valency-chain properties, cannot be created in an inductive system (such as ZF). An example is $Y = \ldots\{\{\{\ \}\}\}\ldots$. Note that $Y = Y^+$, where Y^+ is obtained from Y by replacing each occurrence of '{ }' by '{{ }}'. Thus Y is inverse wrt Sc to the set $S = \{\{\{\ldots\ \ldots\}\}\} = \{S\}$ (that is, $S \in S$) which is prohibited by the 'axiom of foundation'.

5. §4. An interesting recursive schema arises with DIOPHANTINE equations and FIBONNACI numbers (see SHANKS, 'Solved and Unsolved Problems in Number Theory' (1962) §60, p178). The simplest (i.e 2-sequence) case takes the form

$$F(1) = a,\ \Phi(1) = b;\ F(x+1) = f\{F(x), \Phi(x)\};\ \Phi(x+1) = \phi\{F(x+1), \Phi(x)\}.$$

This is easily reduced to recursion over a single integer sequence, and like course-of-values recursion, is an operation of the class $\mathcal{P}$.

Chapter VII

A FINITARY DIAGONALIZATION OF ACKERMANN'S FUNCTION

Section Headings

Section Headings (Cont'd)

7.1 Introduction

The GOODSTEIN-ROBINSON $\star-$ exponentiation, Ch. II, defn. 6, arises from time to time in the literature. E.g. in logic (GOODSTEIN [45], 1944 [46], 1947; KALMAR [68], 1948; ROBINSON [129], 1952; GRZEGORCZYK [53], 1953; YESS'N-VOLPIN [163], 1970; etc.) and in combinatorics (RAMSEY [37], 1929; VAN DER WAERDEN [153], 1930; ERDÖS, RADO, HAJNAL [120],1965; etc. etc.). This is a fast-growing function, even by the standards of classical mathematics. Of course very small quantities may be inverted; but even allowing for this, large values and rapid growth such as arise in ill-conditioning problems, in transcendence theory, in arithmetic bounds, etc., are soon dominated by this function. In his 'Orders of Infinity' [59] G. H. HARDY speaks of such a function as $\star$ as not having '...yet made its appearance naturally in analysis'. However, by exchanging ordinary exponentiation for $\star$ in II $\cdot$ 6, an operation $\star\star$ is obtained which much dominates $\star$. Thus, e.g.

$$2 \star\star\, 4 = 2 \star 65536,$$

a number of great size. Finite iteration (of this process of exchange) yields a function of 3 variables (2 belonging to the binary operation, the third being the level of iteration); the diagonalization with respect to level is the classical HILBERT-ACKERMANN [1] function, now commonly called the ACKERMANN function. It is clear that the rapidly growing functions build their own dominants. Where does this hierarchy lead? It has in fact not been constructively investigated. Great effort has been spent on measuring arithmetic growth using 'ordinals' (from before HARDY, 1904 [57]). But the flaw with 'ordinals' is they are not constructively defined. And with a construction such as this, the foundations have to be sound. The main work of this monograph is to reveal the initial structure of arithmetic growth, especially that part described by normal trees. [2] It is emphasized that normal trees appear to describe only the beginnings of this hierarchy.

[1] Defined by David HILBERT 1926; proved beyond primitive recursion by Wilhelm ACKERMANN, 1928.

[2] In particular, much of §§3,4 of the present Ch. is based on McBETH [97], 1980 and [107], 1985.

7.2 The Classical ACKERMANN Function

The following double recursion leads to the standard variant of the function investigated by W. ACKERMANN in 1928 [1]. As mentioned in Ch. I, the class $\mathcal{P}$, which bears a close relationship, was known to DEDEKIND. [3] The classical function of 3 arithmetic variables is examined in the present section in some detail, prior to a treatment of the diagonalized form.

DEFINITION 7.1 [4] $(\forall n, x, y \in \mathrm{N})$

$$\mathfrak{A}_1(x,y) = x + y$$

$$\mathfrak{A}_{n+1}(x,1) = x$$

$$\mathfrak{A}_{n+1}(x,y+1) = \mathfrak{A}_n\{x, \mathfrak{A}_{n+1}(x,y)\}.$$

□

Examples

1. $\mathfrak{A}_1(x,1) = S(x)$, the successor.
2. $\mathfrak{A}_1(x,y) = x + y$; $\mathfrak{A}_2(x,y) = x \cdot y$; $\mathfrak{A}_3(x,y) = x^y$; $\mathfrak{A}_4(x,y) = x \star y$.
3. $\mathfrak{A}_5(2,3) = 2 \star 4 = 65536$, $\mathfrak{A}_5(2,4) = 2 \star 65536 = \mathfrak{A}_6(2,3)$.
4. $\mathfrak{A}_7(2,3) = \mathfrak{A}_6(2,4) = \mathfrak{A}_5(2, 2 \star 65536)$.

Corresponding to the defn. are the tables of initial cases:

$x = 2$	$y = 1$	$y = 2$	$y = 3$	...
$n = 1$	$2 + 1$	$2 + 2$	$2 + 3$	
$n = 2$	$2 \cdot 1$	$2 \cdot 2$	$2 \cdot 3$	
$n = 3$	2^1	2^2	2^3	
$n = 4$	$2 \star 1$	$2 \star 2$	$2 \star 3$	
$\vdots$				

and

$y = 2$	$x = 2$	$x = 3$	$x = 4$	...
$n = 1$	$2 + 2$	$3 + 2$	$4 + 2$	
$n = 2$	$2 \cdot 2$	$3 \cdot 2$	$4 \cdot 2$	
$n = 3$	2^2	3^2	4^2	
$n = 4$	$2 \star 2$	$3 \star 2$	$4 \star 2$	
$\vdots$				

[3] See S. S. MUCHNIC, *Zeitschr. math. Logik Grundl. Math.* (1976), 441-480.
[4] [95], §7, p509.

Remarks

1. Note that in the 2-fold recursion of the defn., the variable x plays no part than as parameter. The lower operations do satisfy the recursions

$$\mathfrak{A}_1(x, y+1) = \mathfrak{A}_1\{\mathfrak{A}_1(x,y), 1\} \quad \text{and} \quad \mathfrak{A}_2(x, y+1) = \mathfrak{A}_1\{\mathfrak{A}_2(x,y), x\}.$$

2. Using the usual superscript notation for iteration (II, §4·3), defn. 1 may be replaced by:

$$\mathfrak{A}_1(x,y) = x + y;$$
$$\mathfrak{A}_2(x,y) = x \cdot y; \quad \text{and for } n \geq 2$$
$$\mathfrak{A}_{n+1}(x,y) = \mathfrak{A}_n^{(y)}(x, \hat{1}),$$ [5]

where the circumflex denotes the place of iteration. It may be noted that

$$\mathfrak{A}_1(x,y) = \mathfrak{A}_1^{(y)}(\hat{x}, 1); \quad \text{and } \mathfrak{A}_2(x, y+1) = \mathfrak{A}_1^{(x\cdot y)}\{\hat{x}, 1\}.$$

A may be visualized [6] as essentially a progression of N−sequences, comprising the rows of an array. The first row,

$$a < b < \ldots, \quad \text{where } 1 < a,$$

is given; while for $n = 1, 2, 3, \ldots$, the nth row $k < l < \ldots$ is related to the $(n+1)$th row

$$K < L < \ldots$$

by $K = k$th term, $L = K$th term, ... of nth row.

3. $\mathfrak{A}$ as the Canonical Variant

Definition 1 has very many variants, e.g.

$$f_n(y) = \mathfrak{A}_n(2, y), \quad h_n(x) = \mathfrak{A}_n(x, 2).$$

Also $F_n(x,y) = \mathfrak{A}_n(x,y)$, for $n \leq 2$, $F_{n+1}(x,1) = x$, and

$$F_{n+1}(x, y+1) = F_n(X, X), \quad X = F_{n+1}(x,y),$$

[5]This application of iteration serves to endorse the comment of paragraph 3 of II, §4·1. In II, §4 iteration is applied to the least of all means of increasing rate-of-growth, the operation E_2 of succession. Iteration of E_2 (from $E_1 = 0$) *through all finite types* leads only to functions bounded by $\varepsilon = \omega \star \omega$. Whereas iteration as it occurs here requires no more than type 2 to obtain the fierce growth of $\mathfrak{A}_n(x,y)$.

[6]C.f. J. W. GROSSMAN and R. S. ZEITMAN, [52], 'An inherently iterative computation of ACKERMANN's function', *Theor. Comp. Sci.* **57** (1988) 327-30, p329.

giving e.g. $F_3(x, y+1) = x^{2^y}$ [7]. Detailed variations are best left unclassified, their common property being the governing interest. Different variants occur in GRZEGORCZYK [53] and in many other sources [89] [131], [23], [61], [52], [87]. See particularly Rosza PÉTER, [121], [122], and [123]. Note that the seeming advantage in growth of e.g. F_n over $\mathfrak{A}_n$ is soon all but erased as n becomes large: $\mathfrak{A}_{n+1}(2,3) > F_n(2,3) > \mathfrak{A}_n(2,3)$ for $n > 4$; a similar relation (to within a small constant) holding betwen all variants. Furthermore, it seems plain that the striking rate of growth of $\mathfrak{A}_n(x, y)$ is best appreciated when the definition is chosen so as to include the classical binary arithmetic operations as its initial cases. The remainder of this §, therefore, deals only with the variant $\mathfrak{A}_n(x, y)$, the results all applying to others.

4. The function $\mathfrak{A}_n(x, y+1)$ of the defn. is one and the same variant as $g_n(x, y)$ of CLEAVE and ROSE [23], 1966, and as the function $\psi_{n-1}(y, x)$ detailed in Ch. 9, §1 of PÉTER's monograph [122].

5. For e.g. $n = 5$, in the notation of §1,

$$x \star\star (y+1) = x \star \{x \star\star y\} = \overbrace{x^{x^{\cdot^{\cdot^{\cdot^{x}}}}}}^{x \star\star y} .$$

7.2.1 Basic Properties

Initial Values

$\mathfrak{A}_1(2,2) = 4$. And $\mathfrak{A}_{n+1}(2,1) = 2$ gives $\mathfrak{A}_{n+1}(2,2) = \mathfrak{A}_n(2,2)$, so

THEOREM 7.2 $(\forall n)\mathfrak{A}_n(2,2) = 4$.

Now consider the case $x = 1$: $\mathfrak{A}_1(1, y) = 1 + y$; $\mathfrak{A}_2(1, y) = y$.

THEOREM 7.3 $(\forall n, y)\mathfrak{A}_{n+2}(1, y) = 1$.

Proof, by induction on n. $\mathfrak{A}_3(1, y) = 1^y = 1$. Now suppose proved for $n \geq 1$.

$$\mathfrak{A}_{n+3}(1, y) = \begin{cases} x(=1) & \text{if } y = 1 \\ \mathfrak{A}_{n+2}\{1, \mathfrak{A}_{n+3}(1, y-1)\} & \text{if } y > 1. \end{cases}$$

□

LEMMA 7.4 *1.* $(\forall n, x)\mathfrak{A}_{n+1}(x, 2) = \mathfrak{A}_n(x, x)$.

2. For $n = 2, 3, 4, \ldots,$ *and* $y = 3, 4, 5, \ldots,$ $2 + y = \mathfrak{A}_1(2, y) < \mathfrak{A}_n(2, y)$.

3. $\mathfrak{A}_n(2,3) > n + 1,\ n = 1, 2, 3, \ldots$.

---

[7] Cf. the variant of $\mathfrak{A}$ attributed by C. CALUDE, *Theories of Computational Complexity* Ann. Disc. Math., **35** (1988), Ch. I, §6, to G. SUDAN.

Proof.

1. The defn. gives

$$\mathfrak{A}_{n+1}(x,2) = \mathfrak{A}_n\{x, \mathfrak{A}_{n+1}(x,1)\} = \mathfrak{A}_n(x,x).$$

2. $2+y < 2y$, $2 < y$. Thus let $2 \leq n$, $\mathfrak{A}_1 < \mathfrak{A}_n$, and consider

$$\mathfrak{A}_{n+1}(2,y) = \mathfrak{A}_n^{(y)}(2,\hat{1}) = \mathfrak{A}_n\{\overbrace{2,\ldots,\mathfrak{A}_n}^{y-3}\{2,\mathfrak{A}_{n+1}(2,3)\}\ldots\}.$$

   Now

$$\mathfrak{A}_{n+1}(2,3) = \mathfrak{A}_n(2,4) \text{ by thm. 2, } > 6 \text{ by i.h.}$$

   Therefore by repeated use of the i.h., $\mathfrak{A}_{n+1}(2,y) > y+3 > \mathfrak{A}_1(2,y)$.

3. True for $\mathfrak{A}_i$, $i = 1,2,3$; and generally , for $3 \leq n$,

$$\begin{aligned}\mathfrak{A}_{n+1}(2,3) = \mathfrak{A}_n(2,4) &= \mathfrak{A}_{n-1}\{2,\mathfrak{A}_n(2,3)\} = \mathfrak{A}_{n-1}\{2,\mathfrak{A}_{n-1}(2,4)\\ &= \mathfrak{A}_{n-1}\{2,\mathfrak{A}_{n-2}\{2,\ldots\mathfrak{A}_2\{2,\mathfrak{A}_2(2,4)\}\ldots\}\}\\ &> (n-2)+8 \text{ by part 2.}\end{aligned}$$

□

Examples

1. As already noted, $\mathfrak{A}_4(x,y) = x \star y$. Therefore II, notation 2.1 gives $\varepsilon(x) = \mathfrak{A}_5(x,2)$.

2. The expansion used in 4·3 is: $\mathfrak{A}_{n+1}(2,4) = \mathfrak{A}_n\{2,\mathfrak{A}_n(2,4)\}$.

To prove how great are the values built by ACKERMANN's function, it is necessary to begin by proving that it immediately dominates the successor function:

LEMMA 7.5

$$(\forall n \geq 2, x \geq 2, y > 2)\ 1 + \max(n,x,y) < \mathfrak{A}_n(x,y).$$

Proof, by induction on y and n. For $n = 2,3$, the lemma is clearly true. Now let $n \geq 3$, and suppose for all $x \geq 2, y \geq 3$,

$$\mathfrak{A}_n(x,y) > 1 + \max(n,x,y),$$

and consider the value $n+1$, where in the first instance, $y = 3$. Now if $x = 2$, $\mathfrak{A}_{n+1}(2,3) > n+2$ by lemma 4·3. Otherwise

$$\begin{aligned}\mathfrak{A}_{n+1}(x,3) &= \mathfrak{A}_n\{x,\mathfrak{A}_{n+1}(x,2)\}\\ &= \mathfrak{A}_n\{x,\mathfrak{A}_n(x,x)\} \text{ by lemma } 4\cdot 1\\ &= \mathfrak{A}_n(x,K),\end{aligned}$$

where $K > 1 + \max(n, x)$ by i.h. Therefore

$$\begin{aligned}\mathfrak{A}_{n+1}(x, 3) &> 1 + \max(n, x, K) \text{ again by i.h.,} \\ &> 1 + \max(n + 1, x, 3).\end{aligned}$$

Now suppose for inner inductive hypothesis (i.i.h.) that $y \geq 3$, and for all u such that $3 \leq u \leq y$,

$$\mathfrak{A}_{n+1}(x, u) \geq 1 + \max(n + 1, x, u).$$

Then

$$\begin{aligned}\mathfrak{A}_{n+1}(x, y + 1) &= \mathfrak{A}_n\{x, \mathfrak{A}_{n+1}(x, y)\} \\ &= \mathfrak{A}_n(x, L)\end{aligned}$$

where $L > 1 + \max(n + 1, x, y)$ by i.i.h. Therefore

$$\begin{aligned}\mathfrak{A}_{n+1}(x, y + 1) &> 1 + \max(n, x, L) \text{ by i.h.} \\ &> 1 + \max(n + 1, x, y + 1).\end{aligned}$$

So by inner induction,

$$(\forall y)\mathfrak{A}_{n+1}(x, y) > 1 + \max(n + 1, x, y),$$

and the result is proved by induction on n.

□

Suppose $f(x, y)$ is an arithmetic function of 2 variables. Then f is said to be *strictly increasing (si) in both variables* iff

$$(\forall x)\ \lambda y \cdot f(x, y) \text{ is si; and } (\forall y)\ \lambda x \cdot f(x, y) \text{ is si.}$$

It may be that f is si only for a parameter fixed at or above a certain value k. E.g. $x \geq k$. In such a case, f is said [8] to be si in both variables, for $x \geq k$:

THEOREM 7.6 *The ACKERMANN function $\mathfrak{A}_n(x, y)$ is si in every variable, for $x \geq 2$, $y \geq 3$.*

This basic theorem is also proved by double induction on n and y. For $n = 1, 2$ the binary operations $x + y$, $x \cdot y$ are si. Now suppose $n \geq 2$ and for i.h. suppose

1. $\mathfrak{A}_n(x, y)$ is si in x and y as above.
2. $(\forall x \geq 2, y > 2)\mathfrak{A}_1(x, y) < \mathfrak{A}_2(x, y) < \ldots < \mathfrak{A}_n(x, y)$.

[8] This definition may be generalized to functions of > 2 variables

Then since $3 \leq y < \mathfrak{A}_{n+1}(x,y)$ by lemma 5,

$$\mathfrak{A}_{n+1}(x,y+1) = \mathfrak{A}_n\{x, \mathfrak{A}_{n+1}(x,y)\} > \mathfrak{A}_{n+1}(x,y), \tag{I}$$

also by lemma 5. Therefore $\mathfrak{A}_{n+1}$ is si in the argument y, extending one of the parts of case 1. to the case $n+1$. What is more, again by lemma 5 and the fact $\mathfrak{A}_n$ is si by i.h., 1., $\mathfrak{A}_{n+1}(x,3) \geq \mathfrak{A}_n(x,4) > \mathfrak{A}_n(x,3)$; and $\mathfrak{A}_{n+1}(x,y) > y+1$, therefore by (I)

$$\mathfrak{A}_{n+1}(x,y+1) > \mathfrak{A}_n(x,y+1),$$

thus extending property (2) to the case $n+1$. It remains to consider the variable x. Well, $\mathfrak{A}_{n+1}(x,2) = \mathfrak{A}_n(x,x)$ by lemma 4; this function is si in x (by lemma 5 and) by i.h. Similarly for $\mathfrak{A}_{n+1}(x,3) = \mathfrak{A}_n\{x, \mathfrak{A}_n(x,x)\}$. Now let $y \geq 3$ and suppose for inner i.h. that for $i = 2, 3, \ldots y$, $\lambda x \cdot \mathfrak{A}_{n+1}(x,i)$ is si. Then

$$\begin{aligned}
\mathfrak{A}_{n+1}(x+1,y+1) &= \mathfrak{A}_n\{x+1, \mathfrak{A}_{n+1}(x+1,y)\} \\
&> \mathfrak{A}_n\{x+1, \mathfrak{A}_{n+1}(x,y)\} \\
&\quad \text{by i.i.h., since } \mathfrak{A}_n \text{ is si in } x \text{ and } y \text{ by i.h., (1),} \\
&> \mathfrak{A}_n\{x, \mathfrak{A}_{n+1}(x,y)\}, \text{ as } \mathfrak{A}_n \text{ is si} \\
&= \mathfrak{A}_{n+1}(x,y+1).
\end{aligned}$$

So it follows by induction that for all $y > 1$, $\mathfrak{A}_{n+1}$ is si in x, thus finishing the extension of part (1) to the case n+1. Therefore all of what was assumed for case (1) has been extended to the case $n+1$. Therefore, by the induction on n, the thm. is proved.

□

Note that the conditions of the thm. exclude the retrogression

$$\mathfrak{A}_{n+1}(x,1) < \mathfrak{A}_1(x,1).$$

Order of Potency of Argument Places

Examination of the ACKERMANN function reveals a property which only appears *after* exponentiation in the binary-operational progression.

LEMMA 7.7 ($\forall x, k \in$ N) *provided y is sufficiently large,*

$$x^{y+k} < (x+1)^y.$$

Proof. Consider the values $y = 1, 2, 3, \ldots$. The left, x^k, is successively multiplied by x; the right, 1, by $x+1$. So for $y > \frac{k \ln x}{B}$, where $B = \ln \frac{x+1}{x}$, the right is greater.

□

Contrary to this initial property:

THEOREM 7.8

$$(\forall n \geq 4)(\forall x \geq 3, y)\mathfrak{A}_n(x+1, y) < \mathfrak{A}_n(x, y+1).$$

Proof. Take $n = 4$. Note $(x+1)^a < x^b$ for $b \geq 2a$. Indeed $\log_x(x+1) \leq \log_3 4 = 1 \cdot 2618595\ldots, x \geq 3$, therefore $4^a \leq 3^{1.3a}$, giving $2(x+1)^a < x^b$, for $b \geq 2a, a \geq 3$. In the initial case $2(x+1) \star 1 < x^x = x \star 2$ for $x \geq 3$. Now let $y \geq 1$, and suppose $2(x+1) \star y < x \star (y+1)$. Then

$$\begin{aligned} x \star (y+2) = x^{x \star (y+1)} &> 2(x+1)^{(x+1)\star y} \text{ by note and by i.h.} \\ &= 2(x+1) \star (y+1). \end{aligned}$$

Thus the result is obtained by induction on y in the case $n = 4$; and the theorem as a whole may be proved by a further induction on n.

□

THEOREM 7.9

$$(\forall n, x, y)_{n, x>1, y>2}\mathfrak{A}_n(x, y+1) < \mathfrak{A}_{n+1}(x, y).$$

Proof. Using thm. 6, $\mathfrak{A}_{n+1}(x, y-1) > \mathfrak{A}_1(x, y-1) = x + y - 1 \geq y + 1$, therefore $\mathfrak{A}_{n+1}(x, y) > \mathfrak{A}_n(x, y+1)$.

□

Therefore by the 2 above results, for $n \geq 4,\ x \geq 3,\ y \geq 2$:

THEOREM 7.10

$$\mathfrak{A}_n(x+1, y) < \mathfrak{A}_n(x, y+1) < \mathfrak{A}_{n+1}(x, y).$$

□

End-Term Notation

The function of 3 variables $\mathfrak{A}_n(x, y)$ defines a binary operation : $\mathbf{N}^2 \to \mathbf{N}$, $n = 1, 2, 3, \ldots$. The assumption of brackets convention adopted after Ch. II, defn. 6, and the end-term notation introduced in II·7, may be extended to include each operation $\mathfrak{A}_n, n \geq 4$.

DEFINITION 7.11 *For* $n \geq 3$,

$$\mathfrak{A}_{n+1}(x, 1; z) = \mathfrak{A}_n(x, z)$$

$$\mathfrak{A}_{n+1}(x, y+1; z) = \mathfrak{A}_n\{x, \mathfrak{A}_{n+1}(x, y; z)\}.$$

□

Examples.

1. Ex. 1. following II·7 is $\mathfrak{A}_4(2,3;4) = 2^{65536}$.
2. $\mathfrak{A}_5(7,5;2) = 7 \star 7 \star 7 \star 7 \star 823543$.

Remark. The notation for $n \leq 2$ would be redundant because e.g.

$$\mathfrak{A}_3(x,y;z) = xx\ldots xz = x^y z = \mathfrak{A}_2\{\mathfrak{A}_3(x,y),z\}$$

would be expressible without recurrence via defn. 1.

7.2.2 Primitive Recursiveness

THEOREM 7.12 $(\forall n)\mathfrak{A}_n \in \mathcal{P}$.

Proof. The functions $x+y$, xy, x^y are primitive recursive. Thus, suppose $\lambda x, y \cdot \mathfrak{A}_k(x,y) \in \mathcal{P}$, for some $k \geq 3$. Then $F = \lambda x, y \cdot \mathfrak{A}_{k+1}(x,y)$, defined by the primitive recursive schema

$$F(x,1) = x$$

$$F(x,y+1) = \mathfrak{A}_k\{x, F(x,y)\}$$

is primitive recursive, and the thm. follows.

□

THEOREM 7.13 *Every primitive recursive function is eventually dominated by some function* $\mathfrak{A}_n$.

Proof, by double induction on order-of-definition and argument value. Thus, let $\mathcal{P}_1$ denote the initial functions of $\mathcal{P}$; and for $k \geq 1$, let $\mathcal{P}_{k+1}$ denote the functions either of $\mathcal{P}_k$, or obtainable from $\mathcal{P}_k$ by the operations of $\mathcal{P}$. For the purposes of the proof, let the relation $\ll$ of Ch. II, definition 35 be modified slightly, so that for $\alpha, \beta \in \mathcal{N}$, $\alpha \ll \beta$ iff $(\forall x \geq 2)\ \alpha(x) < \beta(x)$. By rem. 1 following defn. 1, $x \geq 3$ may be taken as constant, and $\mathfrak{A}_n(x,y)$ written $\mathfrak{A}_n(y)$. If f is an initial function of $\mathcal{P}$, [9] then $f \ll \mathfrak{A}_2$. Now let $k \geq 1$, suppose for induction hypothesis that thm. 13 has been proved for the case $\mathcal{P}_k$, and let $f \in \mathcal{P}_{k+1}$. Then $(\exists g \in \mathcal{P}_k)$ such that

$$f(1) = h, \text{ a constant, and}$$

$$f(y+1) = g\{f(y)\},\ y = 1,2,3,\ldots\ .$$

By i.h., $(\exists n \geq 1)\ g \ll \mathfrak{A}_n$. Let $N \geq n$ be the smallest integer for which

$$\max\{f(2), g(1)\} < \mathfrak{A}_N(2).$$

[9] A primitive recursive function of a vector $\underline{x} = (x_i)$ of integer variables may be accounted for by taking $x = \sum x_i$.

From this follows $f(2) < \mathfrak{A}_{N+1}(2)$. Now let $y \geq 2$, and suppose for inner i.h. that

$$f(y) < \mathfrak{A}_{N+1}(y),$$

and consider $f(y+1) = g\{f(y)\}$. If $f(y) = 1$, $f(y+1) = g(1) < \mathfrak{A}_N(2) < \mathfrak{A}_{N+1}(y+1)$. If $f(y) \geq 2$,

$$\begin{aligned} f(y+1) &< \mathfrak{A}_n\{f(y)\} \text{ as } g \ll \mathfrak{A}_n \text{ by i.h.} \\ &\leq \mathfrak{A}_N\{f(y)\} \text{ as } n \leq N \Rightarrow \mathfrak{A}_n \underline{\ll} \mathfrak{A}_N \\ &< \mathfrak{A}_N\{\mathfrak{A}_{N+1}(y)\} \text{ by i.i.h.} \\ &= \mathfrak{A}_{N+1}(y+1). \end{aligned}$$

Therefore by the induction on y, $f \ll \mathfrak{A}_{N+1}$. The thm. thus follows by induction on order-of-definition.

□

These preliminaries yield a classic property of the ACKERMANN function:

THEOREM 7.14 (ACKERMANN) $\lambda n \cdot \mathfrak{A}_n(3,2) \notin \mathcal{P}$.

Proof. Let $F_n = \lambda y \cdot \mathfrak{A}_n(3,y)$. Then by thm. 6, $F_1 \ll F_2 \ll \ldots$. By the above two results, this sequence is $\prec$ –cofinal in $\mathcal{P}$. But the diagonal $F(x) = F_x(x)$ satisfies $F_n(x) < F_x(x)$ for $x > n$. In other words, for all $n, F_n \prec F$, therefore $F \notin \mathcal{P}$. The minimal form above may now be shown to lie outside $\mathcal{P}$ by noting the property

$$(\forall x > 1)\ F(x) < \mathfrak{A}_{x+2}(3,2) = E(x),$$

say. Thus, $\lambda n \cdot \mathfrak{A}_n(3,2) \in \mathcal{P} \Rightarrow E \in \mathcal{P}$, contrary to $F \notin \mathcal{P}$.

□

Remarks

1. Putting $\mathfrak{A}(x) = \mathfrak{A}_x(x,x)$, $\mathfrak{A} \succ F$ is not primitive recursive by thm. 14 (the eventual domination following from thm. 6). Some idea of the rate of growth of this function may be gained by noting that $\mathfrak{A}(1) = 2$, $\mathfrak{A}(2) = 4$; $\mathfrak{A}(3) = 27$; and that at $x = 4$ it takes the value $\varepsilon(4)$, the number discussed in II, §2.4 and V, §5.

2. It is a variant of the ACKERMANN function which is used to define the subclasses $\mathcal{E}^n$ of $\mathcal{P}$ (ref. Ch. I, §3). Thus, for $n = 1, 2, 3, \ldots$, $\mathfrak{A}_n \in \mathcal{E}^n$ but $\mathfrak{A}_{n+1} \notin \mathcal{E}^n$.

Note on Higher Inverses

The operation μ of Ch. I, §2, used to obtain the class $\mathcal{R}$ without KLEENE's equational schemata, gives from ACKERMANN's function a hierarchy of inverse operations. Let $r(x) = (\mu\ y)[x \leq \varepsilon(y)]$. Then $(\forall k)\ r \prec l^{(k)}$, where the function on the right is the k–iterated logarithm (ref. ex. 3 after II·2). Similarly, r_n obtained by replacing $\varepsilon = \mathfrak{A}_5(x,2)$ by $\mathfrak{A}_n(x,2)$, gives a descending hierarchy

$$\ldots \prec r_7 \prec r_6 \prec r_5 = r,$$

where at every stage, a k–fold relationship, as for r itself, holds.

7.2.3 Note on Symmetric Forms

An interesting occurrence of *double induction* establishes that a symmetric polynomial $p(x_1, x_2, \ldots, x_n)$ of degree k may be expressed as a polynomial $\varphi({}^n\sigma_1, {}^n\sigma_2, \ldots, {}^n\sigma_n)$, of the elementary symmetric forms (esf's) ${}^n\sigma_i$, of weight k.

Let the notation and defns of Appx. C, §3·2 be assumed. For $n = 1$, every p of degree k has $p(x) = \Sigma_{j \leq k} a_j \cdot x^j = \Sigma a_j \cdot ({}^1\sigma_1)^j$. Thus for main i.h. let $n \geq 1$ and suppose the fundamental thm. on symmetric polynomals has been proved for $1, 2, \ldots, n$ variables. If $p(x_1, \ldots, x_n, x_{n+1})$ is of degree 0, $p \equiv c \cdot {}^{n+1}\sigma_0$ is well-represented. Thus let $k \geq 0$, and for inner i.h. suppose that the result has been proved for all cases $\deg(p) = 0, 1, \ldots, k$; and consider the case $\deg(p) = k + 1$. Let $x_{n+1} = 0$. Then by i.h. $(\exists \varphi)\ p(x_1, \ldots, x_n, 0) = \varphi({}^n\sigma_1, \ldots, {}^n\sigma_n)$, where as a polynomial of the esf's ${}^n\sigma_i$, φ is of weight $\leq k + 1$. Now (writing σ_i for ${}^{n+1}\sigma_i$) define $q \equiv p - \varphi(\sigma_1, \ldots, \sigma_n)$ (I). As a difference, q is symmetric, with $\deg(q) \leq k+1$. But $(\sigma_i)_{x_{n+1}=0} = {}^n\sigma_i$, $i \leq n$. Therefore $q(x_1, \ldots, x_n, 0) \equiv 0$. From this follows the key step of the induction: x_{n+1} divides each term of q; q is symmetric, therefore also every $x_1, \ldots, x_n$ divides each term, so the factor $x_1 \cdots x_n \cdot x_{n+1} = \sigma_{n+1}$ may be removed: $q = \sigma_{n+1} \cdot q'(x_1, \ldots, x_{n+1})$ (II) say. But $\deg(q') \leq k + 1 - n + 1 < k + 1$, so by i.i.h., $(\exists \psi)\ q' = \psi(\sigma_1, \ldots, \sigma_{n+1})$. Therefore (I), (II) give $p(x_1, \ldots, x_{n+1}) = \varphi(\sigma_1, \ldots, \sigma_n) + \sigma_{n+1} \cdot \psi(\sigma_1, \ldots, \sigma_{n+1})$. By the forgoing, the weight of the RHS is $\leq k + 1$. But $\deg(p) = k + 1$, therefore the weight is exactly $\deg(p)$. This proves the case $(n + 1, k + 1)$. By the inner induction, the whole case for $n + 1$ variables is thus proved; and the fundamental thm. now follows by the main induction.

Recursion is but definition by induction, and VII·1 may be closely compared with the above proof. In particular, compare the case $(n + 1, k + 1)$ of the proof with the defn. of $\mathfrak{A} = \mathfrak{A}_{n+1}(k + 1)$. The proof requires only the case $k + 1$ for n variables, viz. $(n, k + 1)$; whereas to calculate $\mathfrak{A}$, the case $K = \mathfrak{A}_{n+1}(k)$ of the outer induction (viz. $\mathfrak{A}_n(K)$) is required by VII·1. The case $(n + 1, k')$, $k' < k + 1$, required in the proof, is not used to determine a comparable case (n, K'), say. On the other hand, from $\mathfrak{A}_n(K)$, $\mathfrak{A}$ is obtained immediately, whereas from $(n, k + 1)$ the proof leads to $(n + 1, k + 1)$, but only by a route involving division and symmetry.

7.3 Diagonalization of $\mathfrak{A}$

The diagonalization is effected by extending the domain of the variable n of ACKERMANN's function from N to **EP**, as in McBETH [87], 1980. [10] Thus $\mathfrak{A}$: $\mathrm{N}^3 \to \mathrm{N}$ is extended to $\Delta : \mathbf{EP} \times \mathrm{N}^2 \to \mathrm{N}$. The technique is the application of the generalized recursion of the previous chapter, defn. 22. The diagonalization is therefore built on the assignment Ω of chapter III, §2.2; by the remarks of Ch. VI, §2.6, this is done by elementary means alone.

Notation

Throughout §§3·1 - 3·4 it is to be taken for granted that 'e' ranges over the domain EP.

7.3.1 The Binary Operations Δ_e.

Using the assignment of III, defn. 6, it is possible to define a class of binary operations $\Delta_e : \mathrm{N}^2 \to \mathrm{N}$.

DEFINITION 7.15 *1.* $\Delta_1(x, y) = x + y$.

2. Suppose e is such that $\mathbf{1} \prec e$. *Then*

(a) $\Delta_e(x, 1) = x$.

(b) If $\mathrm{Suc}(e)$, $\Delta_e(x, y+1) = \Delta_{e^-}\{x, \Delta_e(x, y)\}$.

(c) If $\mathrm{Diag}(e)$, *and* $y > 1$, $\Delta_e(x, y) = \Delta_{(e,y)}(x, y)$.

3. It is convenient to extend the y–domain of Δ_e by one point, by the definition $\Delta_1(x, 0) = x$; $\Delta_2(x, 0) = 0$; $\Delta_e(x, 0) = 1, \mathbf{2} \prec e$.

□

In particular, $\Delta_\omega(x, y) = \mathfrak{A}_y(x, y)$; and generally every value $\Delta_e(x, y)$ is one of the numbers $x + 1$, $2x$, x^3, $x \star 4, \ldots$. Definition 15 is an instance of the generalized recursion schemata of Ch. VI, defn. 22. Therefore, by VI, thm. 23, given $x \in \mathrm{N}, y \in \mathrm{N}_0$ and e, after a finite number of steps, the symbol $\Delta_e(x, y)$, calculated by defn. 15, becomes equated to an explicit expression, containing only the number x and the operation $+$. The length of the expression will depend on x, y and e.

[10] 'A generalization of ACKERMANN's function', *Zeitschr. math. Logik Grundl. Math.*, **26**, 509-516, §3.

Remarks.

1. Note that, as with rem. 1. after defn. 1, the variable x plays no part in the recursion or in the diagonalization, within defn. 15; except as a parameter.

2. As with rem. 2 after defn.1, $\Delta_{e^+}(x,y) = \Delta_e^{(y)}(x,\hat{1})$.

3. Note that the *values* of e and of its canonical components (such as $e(x)$) play no part in the calculation of $\Delta_e(x,y)$.

Examples.

1. For $y = 2, e \succ \mathbf{1}$,

$$\Delta_e(x,2) = \begin{cases} \Delta_{e^-}\{x, \Delta_e(x,1)\} = \Delta_{e^-}(x,x), & \text{if Suc}(e) \\ \Delta_{(e,2)}(x,2) & \text{if Diag}(e). \end{cases}$$

From this it is clear that, in the notation of III, rem. 4 after corol. 22 (taking $r = 2$), $e \succ \mathbf{1} \Rightarrow$

$$\Delta_e(x,2) = \begin{cases} \Delta_{e^-}(x,x), & \text{if Suc}(e), \\ \Delta_{(e(-),2)^-}(x,x) & \text{if Diag}(e). \end{cases}$$

2. For $e = \mathbf{n}$, a constant function, $\Delta_{\mathbf{n}}(x,y) = \mathfrak{A}_n(x,y)$, the ACKERMANN function.

3. For $e = \omega$, $\Delta_\omega(x,x) = \mathfrak{A}(x)$.

4. $\Delta_e(1,y) = 1$, except for

   (a) $\Delta_{\mathbf{1}}(1,y) = 1 + y$;

   (b) $\Delta_{\mathbf{2}}(1,y) = y$ (i.e. for $y > 1$);

   (c) $\Delta_\omega(1,2) = 2$.

   (The cases (a), (b) show that for $y > 1$ and Diag(e), $\Delta_e(1,y) \neq 1$ iff $(e,y) = \mathbf{1},\mathbf{2}$. (4) now follows by a generalized induction, using III, ex. 6 after defn. 6).

5. $\Delta_e(2,2) = 4$, as e takes each value in **EP**.

6. $\Delta_{\omega^+}(2,2) = 4$, $\Delta_{\omega^+}(2,3) = 2^{2^{2^2}} = 65536 = k$, say. Then $\Delta_{\omega^+}(2,4) = \Delta_{\mathbf{k}}(2,k) = \mathfrak{A}_k(2,k)$, so the function of x, $\Delta_{\omega^+}(2,x)$ is rapidly growing.

Δ_e as a Type-2 Iteration of A

Consider e. g. the calculation of $\Delta = \Delta_\psi(2, x^+)$, $\psi = \omega\star 2$: by defn. 15, $\Delta = \Delta_{e+\omega^+}(2, x^+)$, where e is diagonal, so that

$$\Delta = \Delta_{e+\omega}\{2, \Delta_{e+\omega^+}(2, x)\} = \Delta_{e+\omega}\{2, \alpha(x)\}$$

say. Now it may be taken without further proof that $\psi \prec\prec \alpha$. But

$$\begin{aligned}\Delta &= \Delta_{\omega^+}\{2, \beta(x)\}, \text{ where } \alpha \prec \beta \\ &\approx \mathfrak{A}^{\{\beta(x)\}}(x),\end{aligned}$$

in the II, §4·3 notation. It therefore follows that Δ_ψ is a far more elaborate iteration of $\mathfrak{A}$ than is $\lambda x \cdot \mathfrak{A}^{\{\psi(x)\}}(x)$.

7.3.2 Basic Properties

LEMMA 7.16 $(\forall e)(\forall y \in \mathrm{N})$

$$y < \Delta_e(2, y) \;\&\; [y > 1 \Rightarrow y + 1 < \Delta_e(2, y)].$$

Proof, by generalized induction on e, and by induction on y. Thus, suppose $e = \mathbf{1}$. Then $y+1 < \mathfrak{A}_1(x, y), y \in \mathrm{N}$. Thus, suppose $\mathbf{1} \prec e$, and for g.i.h. suppose it has been established that VII·16 holds in **EP** for all polynomials $\prec e$. For $y = 1, \Delta_e(2, 1) = 2$ by defn. 15· 2 (a). For $y = 2, 3 < \Delta_e(2, 2) = 4$. For inductive hypothesis, let $y \geq 2$ and suppose it has been shown $y + 1 < \Delta_e(2, y)$. Suppose Suc(e). Then

$$\begin{aligned}\Delta_e(2, y+1) &= \Delta_{e^-}\{2, \Delta_e(2, y)\} \\ &> \Delta_e(2, y) + 1 \text{ by g.i.h.} \\ &> (y + 1) + 1 \text{ by i.h.}\end{aligned}$$

Now suppose Diag(e). Then

$$\Delta_e(2, y) = \Delta_{(e,y)}(2, y) > y + 1$$

by g.i.h. The lemma is thus proved by generalized induction.

□

LEMMA 7.17

$$(\forall e)(\forall x, y \in \mathrm{N})_{x,y>2}\, \Delta_e(x, y) > \max(x, y) + 2.$$

Proof, by generalized induction on e, and induction on $y = 3, 4, 5, \ldots$. The lemma holds in the initial cases of addition and multiplication. For g.i.h., let $e \succ \mathbf{2}$ and suppose $\Delta_{e^-}(x, y)$ has the property if Suc(e); or that all $\Delta_{(e,n)}$, $n = 1, 2, 3, \ldots$ have the property, if Diag(e). If Suc(e),

$$\Delta_e(x, 3) = \Delta_{e^-}\{x, \Delta_e(x, 2)\} > \max\{x, \Delta_e(x, 2)\} + 2$$

by g.i.h., so $\Delta_e(x,3) \geq x+2 = \max(x,3)+2$. Now let $y \geq 3$, and suppose for i.h. that the property holds for all u in the range $3 \leq u \leq y$. Then

$$\begin{aligned} \Delta_e(x,y+1) &= \Delta_{e^-}\{x, \Delta_e(x,y)\} \\ &> \max\{x, \Delta_e(x,y)\} + 2 \text{ by g.i.h.} \\ &> \max(x, y+2) + 2 \text{ by i.h.} \\ &\geq \max(x, y+1) + 2. \end{aligned}$$

Therefore, by induction,

$$(\forall x, y > 2)\ \Delta_e(x,y) > \max(x,y) + 2.$$

Now suppose Diag(e). Then

$$\Delta_e(x,y) = \Delta_{(e,y)}(x,y) > \max(x,y) + 2$$

by g.i.h., since $(e,y) \prec e$. Therefore the lemma is proved by generalized induction.

□

LEMMA 7.18

$$(\forall e)(\forall x \geq 2)\ x+1 < \Delta_e(x,2).$$

Proof, by generalized induction on e. The result holds for addition. Now let $\mathbf{1} \prec e$, and suppose for g.i.h. that VII·18 has been established for all polynomials $\prec e$. If $x=2$, $\Delta_e(x,2) = 4 > x+1$. Thus, suppose $x > 2$. If Suc(e), $\Delta_e(x,2) = \Delta_{e^-}(x,x) > x+1$ by lemma 17. If Diag(e), $\Delta_e(x,2) = \Delta_{(e,2)}(x,2) > x+1$ by g.i.h. Therefore the lemma is proved by generalized induction.

7.3.3 Calculation of $\Delta_e(x,y)$

Consider the calculation of $\Delta_e(x,y)$; where it may be assumed that $\mathbf{1} \prec e$, and $1 < x, y$; as discussed after defn. 15. Put $e = e_1$, $y = y_1$. Then successive applications of the defn. may be expressed

$$\Delta_{e_1}(x,y_1) = \Delta_{e_2}(x,y_2) = \ldots = \Delta_{e_T}(x,y_T), \tag{I}$$

for some sequence of numbers $y = y_1, y_2, \ldots, y_T$, $1 < T$, and sequence of functions

$$e = e_1 \succ e_2 \succ \ldots \succ e_{T-1} \succ e_T = \mathbf{1}. \tag{II}$$

Now suppose e is a diagonal, and consider the calculation of

$$\Delta_{(e,n+1)}(x,y) \tag{III}$$

for some $n \geq 1$. Put $(e,n+1) = e_1$, and let (II) be the sequence of functions obtained in the calculation of the expression (III). Then

LEMMA 7.19 [11] *$\tilde{e} = (e, n)$ belongs to* (II).

Proof. As in Ch. III, thm. 23. Given the DCC, the lemma is a corollary to this result. Since (II) is a $\prec$ – descending sequence whose lowest term is $e_T = \mathbf{1}$, and by III, thm. 7·2, $\tilde{e} \prec e_1$, there is a $K \in \mathbf{N}$ such that $1 \leq K < T$ and

$$(A) \qquad e_{K+1} \preceq \tilde{e} \prec e_K.$$

Now suppose

$$(B) \qquad e_{K+1} \prec \tilde{e} \prec e_K.$$

Then by defn. 15, $e_{K+1} = e_K^-$ or $e_{K+1} = (e_K, y_K)$. If $e_{K+1} = e_K^-$, then (B) is impossible. If $e_{K+1} = (e_K, y_K)$, then $(e, n) = \tilde{e} \prec e_K \preceq (e, n+1)$ gives $\tilde{e} \preceq e_{K+1}$ by the BACHMANN starting property, Ch. III, thm. 14, so in this remaining case (B) is also impossible. (A) therefore gives $\tilde{e} = e_{K+1}$.

□

7.3.4 Strict Ascent

Just as with the classical ACKERMANN function, the expression $\Delta_e(x, y)$ of §3.1 defines an arithmetic function si in 3 arguments.

THEOREM 7.20 [12] *For every exponential polynomial e,*

1. $(\forall x, y)_{x>1, y>2}\, \mathrm{Diag}(e) \Rightarrow \lambda n \cdot \Delta_{(e,n)}(x, y)$ *is si.*

2. $\lambda x, y \cdot \Delta_e(x, y)$ *is si in x for $y = 1, 2, 3, \ldots,$ and in y for $x = 2, 3, 4, \ldots$.*

Proof, by generalized induction on e and induction on y. For the initial step, let $e = \mathbf{1}, \mathbf{2}, \mathbf{3}, \ldots$. Then $\Delta_e(x, y)$ is the classical ACKERMANN function, si in every argument by thm. 6. Note that since $(\omega, n) = \mathbf{n}, n \in \mathbf{N}$, from this initial case follows: $\lambda n \cdot \Delta_{(\omega,n)}(x, y)$ (x, y as in 20·1) is si. Now let the polynomial $e \succeq \omega$, and for g.i.h. suppose that for all $\alpha \prec e$ the properties (1) and (2) are satisfied. In the successor case, $e = e^- + \mathbf{1}, e^- \in \mathbf{EP}$. For $y = 1$, $\Delta_e(x, 1) = x$, si in x. For $y = 2$,

$$\begin{aligned} \Delta_e(x, 2) &= \Delta_{e^-}(x, x), \text{ si in } x \text{ by g.i.h.} \\ &> \Delta_{e^-}(x, 1), \text{ by g.i.h.,} \quad \text{as} \quad x > 1 \\ &= x = \Delta_e(x, 1), \end{aligned}$$

therefore the function is si in y for $y = 1, 2$. Now let $y \geq 2$, and suppose for induction hypothesis it has been proved that $\Delta_e(x, y)$ is si in x, and

$$(\forall x)\, \Delta_e(x, u) < \Delta_e(x, u + 1)$$

[11] [97], §3, lemma 3.
[12] Ibid., thm. 4.

for $u < y$, $1 < x$. Now

$$\Delta_e(x, y+1) = \Delta_{e^-}\{x, \Delta_e(x, y)\}. \tag{I}$$

But $\Delta_e(x, y)$ is si in x by i.h., and Δ_{e^-} is si in each argument by g.i.h. Therefore $\Delta_e(x, y+1)$ is si in x. Furthermore, by lemmas 16, 17 and 18, (I) gives

$$\Delta_e(x, y) < \Delta_e(x, y+1).$$

It therefore follows by induction on y that

$$(\forall y)\ \lambda x \cdot \Delta_e(x, y) \text{ is si and } (\forall x)_{x>1}\ \lambda y \cdot \Delta_e(x, y) \text{ is si.}$$

In the case in which e is diagonal, first of all property (1) is established; and this is used to establish property (2). Thus for $y = 3, 4, 5, \ldots$, it is to be shown

$$\Delta_{(e,n)}(x, y) < \Delta_{(e,n+1)}(x, y)$$

for $x \geq 2$. Put $(e, n+1) = e_1$, and as in lemma 19 consider the step

$$\Delta_{e_k}(x, y_k) = \Delta_{e_{k+1}}(x, y_{k+1})$$

in the calculation of $\Delta_{e_1}(x, y_1)$. Now

$$y_{k+1} = \begin{cases} \Delta_{e_k}(x, y_k - 1) & \text{if Suc}(e_k) \\ y_k & \text{if Diag}(e_k). \end{cases}$$

Therefore $y_{k+1} \geq y_k$, with $y_{k+1} > y_k$ if Suc(e_k), by lemmas 16, 17. Now let $e_{K+1} = (e, n)$, as in lemma 19.

Claim. $e_K = e^+_{K+1}$. (This was proved in note 1. after III·23).

Proof. Suppose Diag(e_K). Then $e_{K+1} = (e_K, y_K)$. But

$$\begin{aligned} (e_K, y_K) &\succ (e_K, 1), && \text{as} \quad y = y_1 > 1 \\ &\succeq (e, n) && \text{by BACHMANN's property} \\ &= e_{K+1}, && \text{a} \quad \text{contradiction.} \end{aligned}$$

Thus Suc(e_K), and therefore $e_{K+1} = e^-_K$. Returning to the proof of the thm.,

$$\begin{aligned} \Delta_{(e,n+1)}(x, y) &= \Delta_{(e,n)}(x, Y), \ y < Y \quad \text{by claim} \\ &> \Delta_{(e,n)}(x, y), \end{aligned}$$

by g.i.h., as $x \geq 2$. Thus property (1) is established.

In respect of property (2): for $y = 1, \Delta_e(x, 1) = x$, si in x. For $y = 2$,

$$\Delta_e(x, 2) = \Delta_{(e,2)}(x, 2),$$

si in x by g.i.h. Furthermore,$\Delta_e(x,2) > x$ by lemma 18, therefore for all $x > 1, \Delta_e(x,y)$ is si in y for $y = 1,2$. Now let $y \geq 2$. Then $\Delta_e(x,y+1) = \Delta_{(e,y+1)}(x,y+1)$, si in x by g.i.h.; and for all $x > 1$,

$$\begin{aligned} \Delta_{(e,y+1)}(x,y+1) &> \Delta_{(e,y)}(x,y+1) \text{ by property (1)} \\ &> \Delta_{(e,y)}(x,y) \text{ by g.i.h.} \\ &= \Delta_e(x,y). \end{aligned}$$

It therefore follows that $\lambda x, y \cdot \Delta_e(x,y)$ is si in both arguments.

□

Potency of Argument Places

Exactly the order of potency of argument places of the classical ACKERMANN function is followed by the Δ_e. Thus, applying the argument of thm. 8 to defn. 15

THEOREM 7.21 $(\forall x,y)_{x>1}(\forall e)_{\succ \mathbf{3}}\ \Delta_e(x+1,y) < \Delta_e(x,y+1)$.

□

Furthermore, for Diag(e), consider

$$\Delta_{(e,n+1)}(x,y) = \Delta_{(e,n+1,y)}(x,y) = \Delta_{(e,n+1,y,y)}(x,y) = \dots.$$

Since by Ch. III, rem. 4. after corol. 22, successive reductions determined by $y \geq 2$ will lead to a successor index *before* reaching (e,n), viz.

$$\begin{aligned} \Delta_{(e,n+1)}(x,y) &= \Delta_{e(-)}(x,y) \\ &> \Delta_{\alpha+y}(x,y) \text{ for some diagonal } \alpha \succeq (e,n), \\ &> \Delta_\alpha(x,y+1), \text{since } y \geq 2 \\ &\geq \Delta_{(e,n)}(x,y+1), \end{aligned}$$

the occurrence of the subscript (e,n) in the calculation being assured by lemma 19. Therefore for Diag(e) :

THEOREM 7.22 $\Delta_{(e,n)}(x+1,y) < \Delta_{(e,n)}(x,y+1) < \Delta_{(e,n+1)}(x,y)$.

□

Assumed Variable

As noted in earlier remarks, for the present context the variable x in the expression '$\Delta_e(x, y)$' plays no important rôle, save that it should exceed 1. Thus the expression may be reduced to '$\Delta_e(y)$,' where it is assumed x is identified with y.

Examples

1. For Diag(α),
$$\Delta_\alpha(y) = \Delta_\alpha(y, y) = \Delta_{(\alpha,y)}(y, y) = \Delta_{(\alpha,y)}(y).$$

2. Rem. 2 after defn. 15 may be developed, as in e.g.
$$\Delta_{e+3}(y) = \Delta^{(y)}_{e+2}(y) = \Delta_{e+}^{\overbrace{\Delta_{e+}^{\cdot^{\cdot^{\cdot^{\Delta_{e+}}}}}}^{y}} = \Delta_{e}^{\overbrace{\Delta_{e}^{\cdot^{\cdot^{\cdot^{\Delta_{e}}}}}}^{\overbrace{\Delta_{e+}^{\cdot^{\cdot^{\cdot^{\Delta_{e+}}}}}}^{y-1}}}$$
(cf. ex. 3 after lemma 4).

Note. It may now be convenient to replace 'y' by 'x'.

COROLLARY 7.23 $(\forall \alpha, \beta \in \mathbf{EP})\ \alpha \prec \beta \Rightarrow \Delta_\alpha \prec \Delta_\beta$

Proof. Suppose $\alpha \prec \beta$, and compare Δ_α, Δ_β. If Suc(β), by ex. 1. after defn. 15, and by thm. 20·1,
$$\Delta_\beta(2) > \Delta_{\beta^-}(2); \text{ and } \Delta_\beta(3) = \Delta_{\beta^-}(4) > \Delta_{\beta^-}(3).$$
Now let $y > 2$, and consider the calculation of $\Delta_\beta(y+1)$:
$$\Delta_\beta(y+1) = \Delta_{\beta^-}\{\Delta_\beta(y)\} > \Delta_{\beta^-}(y+1)$$
by lemma 17 and thm. 20. Therefore $\Delta_{\beta^-} \ll \Delta_\beta$, and *a fortiori* $\Delta_{\beta^-} \prec \Delta_\beta$.

Now suppose Diag(β). Let $n \geq 1$, $y > n+1$. Then
$$\Delta_\beta(y) = \Delta_{(\beta,y)}(y) > \Delta_{(\beta,n)}(y)$$
by thm. 20·1. Therefore $(\forall n)\ \Delta_{(\beta,n)} \prec \Delta_\beta$. Therefore by thm. 19 of Ch. V, provided $n^+ \geq N(\alpha)$, α and β are linked by a finite sequence
$$\alpha = \beta_t \ \triangleleft_n \ \ldots \ \triangleleft_n \ \beta_2 \ \triangleleft_n \ \beta_1 = \beta \ \ (1 < t).$$
But it has just been proved that at every step $\beta_{i+1} \ \triangleleft_n \ \beta_i$, $i = 1, 2, \ldots, t-1$, the relation $\Delta_{\beta_{i+1}} \prec \Delta_{\beta_i}$ holds. Therefore (by the transitivity of $\prec$), $\Delta_\alpha \prec \Delta_\beta$.

Second proof.

In the notation of the calculation (I) at the head of §3·3, $y = y_1 \leq y_2 \leq \ldots \leq y_t$ $(1 \leq t \leq T)$. Now let $y^+ \geq N(\alpha)$. These numbers may be taken as a segment of the $\mathcal{S}$ of the rem. foll. V·20. Viz. $\alpha \ \tilde{\triangleleft}_{\mathcal{S}} \ \beta$, which may be used in place of the above reduction. But now, $\Delta_\alpha(Y) = \Delta_\beta(y)$ actually occurs in the calculation, where $Y \geq y$; and in fact $Y > y$ by III·7·5. Therefore it is also proved
$$\Delta_\alpha \prec \Delta_\beta @ M \leq N(\alpha).$$

□

Remark

By the above, $\Delta_{(\beta,n)} \prec \Delta_\beta @ n+1,\ n > 1$. According to defn. 15, in the reduction of Δ_{β_i} to $\Delta_{\beta_{i+1}}$, if Diag(β_i) then $\beta_{i+1} = (\beta_i, Y)$, where $Y \geq y$ has accumulated over the previous steps. It therefore follows that as i increases, the indices β_i will contain very large ec's, up until the final descent, by units, to α.

The Function Δ.

Using the fundamental sequence $(\varepsilon, n) = \varepsilon_n$ of III·6·3, the diagonal

$$\Delta = \Delta_\varepsilon = \text{diag}\Delta_{(\varepsilon,n)}$$

is thus given by $\Delta(x) = \Delta_{\omega\star x}(x),\ x = 1,2,3,\ldots$. Note that $\Delta(1) = \Delta_\omega(1,1) = 1;\ \Delta(2) = 4;\ \ \Delta(3) = \Delta_e(3)$, where $e = 2\omega^{2\omega^2+2\omega+\mathbf{2}} + \ldots + 2\omega^2 + 2\omega + \mathbf{3}$.

LEMMA 7.24

$$(\forall n)\ \Delta_{(\varepsilon,n)} \prec \Delta.$$

Proof. For a given n, $(\varepsilon, n) \prec \varepsilon @ n+1$. Now for $x > n$, consider the calculation of $\Delta(x) = \Delta_{(\varepsilon,x)}(x)$. By lemma 19 repeated $(x-n)$ times, (ε, n) belongs to the sequence (II), before lemma 19. By the stage when the calculation has reached (ε, n), the $Y-$ parameter will have accumulated a value $Y > y$. Therefore, by theorem 20·2, $(\forall y)_{>2}\ \Delta_{(\varepsilon,n)}(y) < \Delta_{(\varepsilon,x)}(y)$. I.e. aside from the exceptional values $y = 1,2$, $\Delta_{(\varepsilon,n)} \ll \Delta_{(\varepsilon,x)}$. Therefore $\Delta_{(\varepsilon,n)} \prec \Delta @ n+1$.

□

THEOREM 7.25 $(\forall e)\ \Delta_e \prec \Delta$.

Proof. By II·14, the above fundamental sequence is cofinal in $(\mathbf{EP}, \prec)$. Thus given e, let N be the smallest index for which $e \prec (\varepsilon, N)$. By corol. 23, $\Delta_e \prec \Delta_{(\varepsilon,N)}$. The linear relation $\prec$ therefore gives, by the lemma, $\Delta_e \prec \Delta$.

□

7.3.5 $k-$Fold Recursion

Definition 1 of §2 is an instance of simultaneous or 2-fold recursion, viz. of recursion with respect to 2 arithmetic variables. Let **NL** be the class of linear forms of II, §2·1. Due to fundamental work of R. PÉTER [121] (see also [122] §§11-12), it is known that the restriction

$$(\Delta_e)_{e\in\mathbf{NL}}$$

of defn. 15 is cofinal in the class $\mathcal{X}$ obtained by closing $\mathcal{P}$ under $k-$fold recursion, $k = 1, 2, 3, \ldots$. That this is so may be seen by considering the $k-$fold generalization of ACKERMANN's function. The cases [13] $k = 1, 2$ are accounted for by defn. 1. For $k = 2, 3, 4, 5, 6, \ldots$, taking $x = x_0, y = x_1$, the schema of $k+1-$fold recursion is:

$$\mathfrak{A}_1(x, y, x_2, \ldots, x_k^+) = \mathfrak{A}_K(x, y, x_2, \ldots, x_{k-2}, K); \quad K = \mathfrak{A}_1(x, y, x_2, \ldots, x_k)$$
$$\mathfrak{A}_{n+1}(x, y, x_2, \ldots, x_{k-1}, 1_k) = x;$$
$$\mathfrak{A}_{n+1}(x, y, x_2, \ldots, x_{k-1}, x_k + 1) = \mathfrak{A}_n\{x, y, x_2, \ldots, x_{k-1}, \mathfrak{A}_{n+1}(x, y, \ldots, x_{k-1}, x_k)\}.$$

Remarks

1. In the case $k = 2$, $\mathfrak{A}_1(x, y, z) = \mathfrak{A}_z(x, y)$, viz. the schema presumes defn. 1.

2. If '$n+1$' were to be replaced by 'n' in the second clause, then $n = 1, k = 2, x_2 = 1$ would give
$$\mathfrak{A}_1(x, y, 1) = \begin{cases} \mathfrak{A}_1(x, y) = x + y \; (\neq x) & \text{by clause (1)} \\ x & \text{by clause (2).} \end{cases}$$

3. Far more highly interlocked $k-$fold recursion is possible. But the 'growing-power' is contained in the most potent variable x_k of the known function $\mathfrak{A}_n$. Therefore $\mathfrak{A}_{n+1}$ is constructed by recursion on x_k. The extra growth achievable by the other recursions is swamped by that of the subsequent step of the x_k-recursion.

4. Every function of $\mathcal{X}$ is eventually dominated by the diagonal
$$\mathfrak{d}(x) = \mathfrak{A}_{x+3}(2, 3, 4, \ldots, x+2).$$
Thus, e.g. $\mathfrak{d}(1) = 2 \star 3 = \mathfrak{A}_4(2, 3) = 16$; and $\mathfrak{d}(2) = \mathfrak{A}_5(2, 3, 4)$.

5. Given the above schemata, the cofinality property of (Δ_e), $e \in \mathbf{NL}$ stated at the head of §3.5 is immediate by simultaneous induction: for $k \geq 2$ and $x, y, n \geq 1$:
$$\mathfrak{A}_n\{(x)_1, (x)_2, \ldots (x)_{k-1}, y\} = \Delta_{(k-1)\omega+n}(x, y).$$

6. From the previous remark, it is clear that the function $\mathfrak{d}$ of rem. 4 satisfies
$$\Delta_{k\omega} \prec \mathfrak{d} \prec \Phi = \Delta_{\omega^2},$$
for $k = 1, 2, 3, \ldots$. Note that, in fact $\mathfrak{d} \approx \Phi$. Now consider $\lambda x \cdot \mathfrak{d}\{\mathfrak{d}(x)\}$. This function is fast-growing, because the variable of substitution is both the argument value *and* (the successor of) the # of argument places. I.e. it occurs as a variable coefficient in the $\Delta-$expression:
$$\mathfrak{d} \approx \lambda x \cdot \Delta_{x\omega}(x); \;\; \mathfrak{d} \circ \mathfrak{d} \approx \Delta_{\omega^2+1}.$$

[13]'x' may account for '0–fold recursion'.

7.3.6 Fine Structure Within (Δ_e)

By the remarks of the previous §, the functions $\mathfrak{A}$ and $\mathfrak{d}$ satisfy

1. $\lambda x \cdot \mathfrak{A}_x(x,x) \approx \Delta_\omega$;
2. $\mathfrak{A}_{(x)_{k+2}}\{(x)_1, (x)_2, \ldots, (x)_{k+1}\} \approx \Delta_{k\omega}$; and
3. $\mathfrak{d} \approx \Delta_{\omega^2}$. By rem. 6 of the previous §,
4. $\lambda x \cdot \mathfrak{d}\{e(x)\} =_{\mathrm{df}} \mathfrak{d}_e \prec \Delta_{\omega^2+1}$, $e \in$ **EP**, while on the other hand, it is clear that the upward step in growth from, e.g. $\mathfrak{d}_e$ to $\mathfrak{d}_{e^+}$ is very considerable.

In particular, the sequence $\{\mathfrak{d}_{(\varepsilon,n)}\}$, $n = 1, 2, 3, \ldots$, is cofinal in $(\mathfrak{d}_e)_{e \prec \varepsilon}$.

General Remark.

Just as with the cases considered in the subsection at the end of §2.1, all the functions of rapid growth, considered in the present section, posess higher inverses.

7.3.7 Diagonalization and Canonical Diagonalization

It should be noted that in order for defn. 15 to work, it is *not* necessary that a diagonal $\alpha \in$ EP be correlated with its canonical terms. [14] What is more, α need not be correlated with diagonal terms at all. The only requirements that must be satisfied

1. by the sequence
$$\{\alpha, 1\} \prec \{\alpha, 2\} \prec \{\alpha, 3\} \prec \ldots$$
correlated with α are
   (a) $(\forall n)\ \{\alpha, n\} \prec \alpha$
   (b) $(\forall \beta \prec \alpha)(\exists N)\ n \geq N \Rightarrow \beta \prec \{\alpha, n\}$,
   (c) $(\forall n)\ \{\alpha, n\} \preceq \{\{\alpha, n+1\}, 1\}$; [15] and
2. by the defn.: this has to be such that at no step corresponding to canonical reduction is rate-of-growth lost; and at certain steps, this is gained.

It is Δ_α which is by definition the diagonal of the $\Delta_{\{\alpha,n\}}$; as remarked, α need not be the diagonal of the $\{\alpha, n\}$, as it is in the case ($\alpha = e$) of defn. 15. Without 1(a), the DCC, which assures that the expression is evaluated in a finite number of steps, could not be used. However, note (III, §2.2) that the canonical assignment is $\prec$ –maximal: any increase in the descent 1(a) will diminish the rate of growth of

[14] E.g. given any fixed si $\mu \in \mathcal{N}$, 'y' of the subscript '(e, y)' of 15·2(a) may be replaced by '$\mu(y)$'.

[15] 1·b,c show, by exs. 2,3 of Ch. III foll. thms 12, 14 respectively, that the assignment Λ of III · 11 may be expected to fail, as a replacement for Ω in VII · 15.

the Δ_α. Without 1(b), the main thm. of Ch. V, §2.2 could not be proved, making impossible the diagonalization of induction, corresponding to the new correlation, which uses VI·10. Without 1(c), the hierarchy may be expected to collapse. Also without 2., as in the case

$$A_1(x, y) = x + y; \quad A_{\alpha^+}(x, 1) = x;$$

$$A_\alpha(x, y) = \begin{cases} A_{\alpha^-}(x, y) & \text{if } \mathrm{Suc}(\alpha) \\ A_{(\alpha, y)}(x, y) & \text{if } \mathrm{Diag}(\alpha). \end{cases}$$

It may be noted how little the definition of the $A's$ differs from defn. 15: aside from initial cases, the $A-$hierarchy collapses to the identity.

7.3.8 Diagonalization and Convergence

The relationship of dependence, of defn. 15 and the theorems that flow from it, on the DCC thm. for $(\mathfrak{T}, \prec)$ of IV, §§3, 4 is very clear. Now that functions of rapid growth are at hand, it is possible to test the argument proving finite descent in the linear hierarchy $(\mathbf{EP}, \prec)$. As observed in remarks on the Intuitionist DCC, after IV·39, it is sufficient to consider **XP**, and the restricted system $(\Delta_\alpha)_{\alpha \prec \psi}$ founded on it. The issue in question is the difference between

1. taking consecutive steps, knowing there can only be finitely many (fm), and

2. taking consecutive steps, without knowing that after fm of them, a condition is fulfilled.

This is best illuminated by the case of GOLDBACH's [16] conjecture, often cited by GOODSTEIN[17].

DEFINITION 7.26 *For $n \geq 1$, let*

$$GOLD(n) = \begin{cases} 1 & \textit{if } 4 < x \leq n \;\&\; x \equiv 0 (\mathrm{mod}\ 2) \Rightarrow \\ & (\exists \textit{ odd primes } p, q)\ x = p + q, \\ 0 & \textit{otherwise.} \end{cases}$$

□

[16] For an account of research in this problem, and of proofs of related theorems, see I. M. VINOGRADOV (1937) and CHENG JING RUN (1966) in: WANG HUAN (Ed.), *GOLDBACH's Conjecture* (1984), ISBN 9971-966-08-5.

[17] See e. g. ROSE [132], 1988.

Take now a 'difficult' case, where the terms in an **XP**–descent have deliberately been chosen for maximum delay. Thus, the coefficents may be chosen using the ACKERMANN function as follows:

$$f_1 = \omega^n;\ f_2 = \mathfrak{A}(2) \cdot \omega^{n-1};\ f_3 = 3\omega^{n-1} + \mathfrak{A}(3) \cdot \omega^{n-2};$$

$$f_4 = 3\omega^{n-1} + 26 \cdot \omega^{n-2} + \mathfrak{A}(4) \cdot \omega^{n-3}; \ldots,$$

for some suitably large n. It is plain that a sequence with initial terms such as these can be chosen to descend in **XP** very slowly indeed. Nonetheless, no matter how much ingenuity may be expended in the choice of coefficients, provided

$$f_1 \succ f_2 \succ f_3 \succ f_4 \succ \ldots,$$

then by the DCC, the index say $N(1.)$ of the final polynomial is reached. Now let

$$N(2.) = \begin{cases} \text{the smallest } n > 4 \text{ such that } GOLD(n) = 0, & \text{or} \\ 4 \text{ if every greater even integer can be well - expressed.} \end{cases}$$

These two cases exemplify the itemized possibilities, and comparison of the calculation of the numbers $N(1.), N(2.)$ proves that there is an awesome difference between them.

7.3.9 The Finite Domain

The several preceding theorems round off the project set out in §1. The progression $x + y,\ xy,\ x^y,\ x \star y, \ldots$ culminating in ACKERMANN's function is replaced by a linear hierarchy of binary operations, defined from and similar to, $(\mathbf{EP}, \prec)$. Not only is $\mathfrak{A}_n(x, y)$ extended by diagonalization, but this has been done using explicitly finite means. By the correspondence of Ch. IV, §4, other objects, such as finite trees, or finite patterns of parentheses, might have been used in place of exponential polynomials.

Because these operations have been recovered by finitary means, from generally accepted foundations, there seems no reason why the great values defined by them should not be regarded as realities of our world; just as e.g. the spiral galaxies are regarded as neighbours, millions of light-years or more distant. Certainly, therefore, $N = 2 \star 1000$ should be taken as even, and $3 \star N$ as odd. The arithmetic properties of the higher operations are virtually untouched. This implies less of a daunting burden of work awaiting future generations of mathematicians, than the prospect of a fundamental advance in mathematical knowledge, in which, surely, computers will play a part.

At this stage another matter, pointed out in §1, becomes apparent: although the replacement of the ACKERMANN function by the very much more complex **EP**–progression builds correspondingly further into the arithmetic growth hierarchy, $\mathcal{N}$–GH, the final thm. of the previous § establishes that the entire construction is only an entrance, perhaps the natural one, to a profoundly vaster hierarchy. In a sense, which is not easy to make precise, 'most' of the $\mathcal{N}-$GH is missed by such a construction; and it has to be said that it probably takes a life's study to grasp just how much weight is to be attached to this 'most'. This is not to preclude the possibility that a closure of diagonalization may be found, in which presumably objects of higher type would play a part. Such a success may turn out not unconnected with the above fundamental advance.

7.3.10 A Remark on LIOUVILLE's Theorem

Algebraic Numbers

According to a classical theorem of LIOUVILLE, if a real number α is of degree $n > 1$, then a constant C may be found such that for all integers $p, q,\ q \neq 0$,

$$| \alpha - \frac{p}{q} | > \frac{C}{q^n}$$

(ref. KHINTCHINE [69], §9, p45). The degree thus sets an upper bound on the quality of a possible rational approximation to α. This suggests that, with the variable x now ranging over the real numbers, the solutions x to the equations

$$e(x) = f(x),$$

$e, f \in \mathbf{EP}^{(h)}$, for some height h, may in some cases admit rational approximations for which the above thm. would only remain true if '$\frac{C}{q^n}$' were replaced by '$\frac{C}{q \star h}$'.

Very Small Numbers

PLANCK's constant, $\approx 6.6262\ldots \times 10^{-34} Js$ (in the SI convention), as a generally accepted constant of pure experimental science, is certainly a very small number. But a single insertion of '$-$' in the right place in a typical **EP**- expression will generally produce a much smaller number. Accepting the remarks of §3.7, there is thus a microcosmic hierarchy, differing from the hierarchies of inverses discussed in II, §6.2 but also directly related to **EP** and its extensions. As examples of convergent series of very small terms, the classical ACKERMANN function $\mathfrak{A}$, as a dominant of ω^2, defines a particular real number

$$\alpha = \sum_{n=\infty}^{1} \frac{1}{\mathfrak{A}(n)}.$$

The special diagonal Δ of §3·4 satisfies $\mathfrak{A} \ll \Delta$, and so gives

$$\beta = \sum_{n=\infty}^{1} \frac{1}{\Delta(n)},$$

where $\beta < \alpha$.

7.4 The Tree Structure of Exponential Calculations

The classical ACKERMANN function is a generalization of exponentiation $\mathfrak{A}_3(x, y)$, and so too, therefore are the binary operations $\Delta_e(x, y), e \in \mathbf{EP}$, of §3.

Consider e.g. the calculation [18] of

$$7 \star 3 = 7^{\{7\star 2\}} = 7 + 6 \cdot \sum_{i=823542}^{1} 7^i$$

(II, lemma 46). Now let '7' be taken as a string of 7 tally marks. The RHS of the above should lead to an expression consisting only of such tallies, viz. the completed calculation of $7 \star 3$. But to accomplish this, the calculation of $6 \cdot \sum 7^i$ has *first* to be effected. Thus the entire calculation determines a dyadic tree. The first step is as indicated.

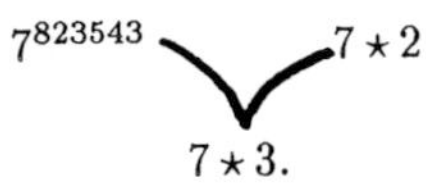

The process is formalized here, in order to expose the tree structure of generalized exponential calculations. The notation and terminology for trees will be taken from Ch. IV, §§1-2, with particlar reference to the class $\mathfrak{D}$ of defn. IV·16. The content of §4 is taken from McBETH [107][19], 1985.

7.4.1 Notation for the Numbers and Expressions $\Delta_e(x, y)$.

2 and $1 + 1$ are different expressions for the same number. The calculation of the number $N = \Delta_e(x, y)$ involves a sequence of different expressions for N, the last of which is $N = 1 + 1 + \ldots + 1$. This last will be called the *tally form* of expression of N. As noted after thm. 22, the variable x plays no part in the inductive steps of the calculation, and may be assumed to take a fixed value ≥ 3, and to be already in tally form. Therefore the notation '$\Delta_e(x, y)$' will be reduced to '$[e, y]$', (as in VI·23).

7.4.2 Dyadic Trees

Some specialized properties of the class $\mathfrak{D} \subset \mathfrak{S}$ of (finite) dyadic trees of are introduced. The notation e_1, e_L, etc. used in §4 to denote edges will not be confused with that used to denote an $\mathbf{EP}$−polynomial. Thus, let $T \in \mathfrak{D}$.

LEMMA 7.27 *The non-principal (viz. $\neq v^{\sharp}$) vertices of T fall into 2 classes. The edges of T fall into 2 classes.*

[18] Formally, a *calculation* is a proof in first-order arithmetic $\underline{\mathbf{N}}$, which uses only rules of inference of zero order logic (see Appendix A). GENTZEN [143], p114 has shown that for the system $\underline{\mathbf{N}}'$, obtained from $\underline{\mathbf{N}}$ by deleting the induction schemata, the following meta-theorem holds: if $\underline{\mathbf{N}}'$ is inconsistent, a contradiction may be derived by calculation alone.

[19] 'The tree structure of exponential calculations', *SIGACT News*, **17**(1).

Proof. Every $v \neq$ an endv in T has exactly 2 successors, v_L, v_R, say, where

$$v_L \text{ L } v_R.$$

Edges may be similarly designated, using the 1-1 correspondence between edges and non-principal vertices.

□

LEMMA 7.28 *For every endv v of T, there is a unique vertex of T, the* start *vertex.*

Proof. Let v be an endv of T. Now consider the unique chain

$$v = v_1 \text{ Sc } v_2 \text{ Sc } \dots \text{ Sc } v_n = v^\sharp$$

linking v to $v^\sharp$. Using lemma 27, let k be the smallest i for which $v_{i-1} = (v_i)_R$ if there is such an i, otherwise set $k = n$. The vertex v_k is taken as the start vertex of v.

□

LEMMA 7.29 *For every $T \in \mathfrak{D}$, there is a unique natural number, the* left semi-width *$L(T)$ of T.*

Proof. Let C be a (complete, i.e. $v^\sharp$–to-endv) chain of T, with edges $e_1, e, \dots, e_n$, where e_n is incident on v. Now, for $i = 1, 2, \dots, n-1, e_i = (e_{i+1})_L$ or $(e_{i+1})_R$ using lemma 27. Let $d = d(C) =$ the number of e_L in C *minus* the number of e_R in C. Of the fm C in T, one has a greatest value of d :

$$L(T) =_{df} \max_{C in T} d(C).$$

□

Notation. The endv of C_L is a distinguished vertex, in general different from $v^\sharp$. It is denoted by $v^\star$.

Remarks.

1. Writing C_L for the leftmost chain of T (IV, defn. 12) $d(C_L) < L(T)$ is possible.

2. If v is an endv, $v \neq v^\sharp$, $v \notin C_L$, the start vertex of v is joined by an edge to the pv of a $\mathfrak{D}$-tree whose leftmost chain has endv v.

7.4.3 Exponential Calculation Trees

First, the steps in the calculation of an expression [20] $[e, y]$ are discussed; then the calculation tree T of $[e, y]$ is introduced. Finally, the calculation sequence $\mathcal{C}$ and the reduction sequence $\mathcal{R}$ are obtained from T. Given y, as mentioned in the Introduction, the T of $[e, y]$ and the normal tree of e are independent occurrences of finite trees in the theory of **EP**.

7.4.4 The Procedure of Calculation

From defn. 15:

1. $[\mathbf{1}, y] = x + y$.

2. $[e, 1] = x, \ \mathbf{1} \prec e$.

3. $[e, y] = [\tilde{e}, Y], \ 1 < y$ and $\mathbf{1} \prec e$, where

$$Y = \begin{cases} [e, y-1] & \text{if Suc}(e) \\ y & \text{if Diag}(e), \text{ and} \end{cases}$$

$$\tilde{e} = \begin{cases} e^- & \text{if Suc}(e) \\ (e, y) & \text{if Diag}(e). \end{cases}$$

In case (3), the equation $[e, y] = [\tilde{e}, Y]$ shows that there are 2 aspects to the calculation: the reduction of e to $\tilde{e}$; and the replacement of y by Y. But to proceed with the calculation of $[\tilde{e}, Y]$, Y has to be calculated first. Thus (as seen after VI·22) the combinatorial unit

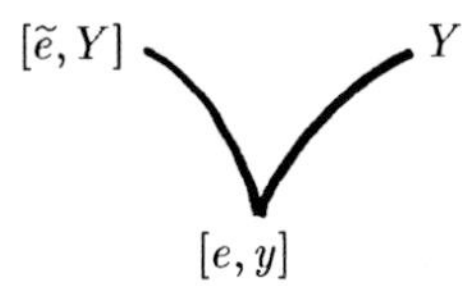

gives rise to a $\mathfrak{D}$–tree T.

Remark.

On the right, 'Y' is an expression determined by defn. 15; on the left Y is the number obtained by calculation of this expression, viz. 'Y' in tally form. Therefore the calculation of $[e, y]$ proceeds by right-to-left pass of T.

[20] The use of square brackets '[,]' in IV,§2.1 is not required; allowing the entirely different use in VII, §4.

7.4.5 Calculation Tree

DEFINITION 7.30 *The calculation tree $T = T(e, y)$ of $[e, y]$ is a unique $\mathfrak{D}$-tree whose vertices v are associated with expressions $\phi(v)$ by the rules:*

1. $\phi(v^{\sharp}) = [e, y]$.
2. *Now suppose $v \in T$, and $\phi(v) = [e_1, y_1]$. Then*
   (a) *if $e_1 = \mathbf{1}$, or if $y_1 = 1, v$ is an* endv *of T.*
   (b) *if $\mathbf{1} \prec e_1$ and $1 < y_1, v$ has successors v_L, v_R for which (in the notation of §4.4)*

$$\phi(v_L) = [\tilde{e}_1, Y_1], \ \ \phi(v_R) = Y_1.$$

Notes.

1. $T(e, y)$ is defined inductively, so that conditions 2., 3. of IV, defn. 16 are satisfied. As to condition 1., T has plane structure by the designations of lemma 27; and that T is finite, is a corol. of VI, thm. 23.
2. By the defn., if $e = \mathbf{1}$ or $y = 1$, T is a point (isolated vertex)
3. It is possible to find $\mathfrak{D}$-trees which are not calculation trees. Therefore the calculation trees determine a specialized subclass of $\mathfrak{D}$.

Examples.

1. Classical ACKERMANN function. Let $x = 2$, so $[\mathbf{n}, y] = \mathfrak{A}_n(2, y)$. Then for $\mathbf{n} = \mathbf{3}$ (exponentiation):

   $[\mathbf{3}, 1] = 2^1$, $T(\mathbf{3}, 1) =$ 

   $[\mathbf{3}, 2] = 2^2$, $T(\mathbf{3}, 2) =$ 

   $[\mathbf{3}, 3] = 2^3$, $T(\mathbf{3}, 3) =$ 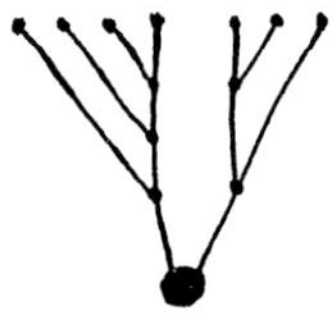

   (Thus $T(\mathbf{3}, 3) \notin \mathfrak{T}$).

2. For $n = 4$ (repeated exponentiation):

   $[\mathbf{4}, 2] = 4$, $T(\mathbf{4}, 2) =$ 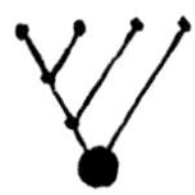

   Although $[\mathbf{3}, 2] = [\mathbf{4}, 2]$, the trees for the 2 expressions are different.

3. the combinatorial unit for $T = (\mathbf{n}, y)$ takes the form

$$[\mathbf{n}, Y] \qquad [\mathbf{n}+\mathbf{1}, y]$$

$$[\mathbf{n}+\mathbf{1}, y+1],$$

   where $Y = [\mathbf{n}+\mathbf{1}, y]$ is in tally form. Therefore an edge e_L requires a decrement of n by 1, while an edge e_R leaves this variable unchanged. Therefore $L(T) = n - 1$. For example, let $n = 4$. For any $x > 1$, the tree $T(\mathbf{4}, 5)$ branches to the right by more than 10^{10000} edges, yet $L(T)$ is only 3.

4. Consider the diagonalization of §3. Suppose Diag(e) in **EP**. Then the combinatorial unit for $T(e, y)$ takes the form

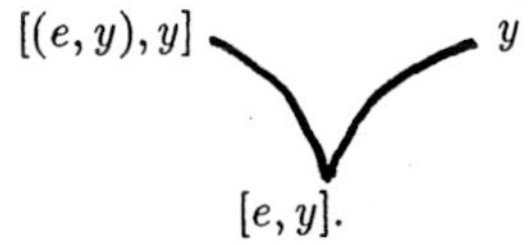

Now the descending sequence $e, (e, y), (e, y, y), \ldots$ ends at a successor or at **1**. The sequence, however, may be very long. [21] For $n \geq 1$, $e =$ the function x^n, the sequence contains $n + 1$ terms, so $L(T) \geq n$, so $L(T)$ may be arbitrarily large. In fact, taking $n = 3$, it may be verified using defn. 15 that $T(\omega^3, 2)$ is

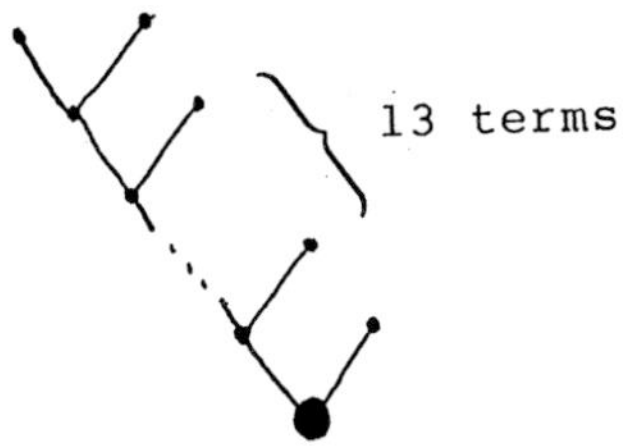

7.4.6 Calculation Sequence

A sequence $\mathcal{R}$ of *reductions* starting with $[e, y]$ may now be obtained. Each is of the kind which occurred in §3, and results from an application of defn. 15. $\mathcal{R}$ is embedded in a larger sequence $\mathcal{C}$, the *calculation sequence* of $[e, y]$. $\mathcal{C}$ is obtained by edge-traversal of $T(e, y)$, determined by the following algorithm: it may be supposed that a *cursor* covers a vertex of $T = T(e, y)$; and that the cursor may be moved to another vertex of T according to

ALGORITHM 7.4.1 *1. The cursor is placed at $v^\natural$, and is moved to $v^\natural_R$.* [22]

2. Now suppose $v \in T$ is interior, and suppose that the cursor has been moved to v.

   (a) If the move was *from v' to v, where v Sc v', then the cursor is moved to v_R.*

   (b) If the move was *from v_R to v, ('Y has been calculated') then the cursor is moved to v_L.*

3. (a) If $v \in T$ is an endv $\neq v^\star$ (and the cursor was moved from v' to v, where v Sc v'), then the cursor is moved to the start vertex (ref. lemma 28) of v.

[21] The length is the value $D(y)$ of the 'descent function' of Ch. III, §5.

[22] It may be assumed $1 \prec e$, viz. $v^\natural \neq v^\star$.

(b) If $v = v^\star$, the cursor halts.

□

THEOREM 7.31 *Algorithm 4.1 halts in a finite number of steps, in which every vertex of T is covered at least once.*

Proof, by induction on right-to-left chain order (ref. Ch. IV, lemma 11). By rule 1., the cursor K covers $v^\sharp$. By rules 1. and 2(a), K covers every v of the rightmost chain of T. Now suppose K covers the endv, v, say of the mth chain of T, where $m < M$, the number of chains of T. Then by the return to the start vertex of v (rule 3(a)), rule 2(b) followed by repeated applications of rule 2(a), K covers every vertex of the $(m+1)$th chain of T (viz. the vertices from v to the start vertex are already covered, and by these rules, the remaining vertices of the chain are covered). Therefore by induction K covers the v's of every chain. But T is finite, with M chains, so K reaches the endvertex $v^\star$ of the Mth chain and halts, by rule 3(b).

□

Remarks.

1. $v^\sharp$ is covered twice, by rules 1. and 3(a).
2. Interior vertices v are covered exactly twice, as each such v is a successor (rules 2(a)-(b)); and each is the start vertex of a unique endv (rule 3(a)).
3. Endv's v are covered exactly once by K, since $v^\star$ is covered exactly once before K halts at $v^\sharp$; otherwise the start vertex of v leads K to vertices in chains to the left of v.

DEFINITION 7.32 [23] *The calculation sequence C of $[e, y]$ is*

$$\phi(v_1),\ \phi(v_2), \ldots, \phi(v_N),$$

where v_i is the ith vertex covered by K, excluding repetitions.

Broadly speaking, C is the sequence of expressions which have to be 'remembered' during the calculation of $[e, y]$. In the case Suc(e), and in particular $e = \mathbf{n}^+, n \geq 1$, the calculation gives

[23] [107] , 3.4.

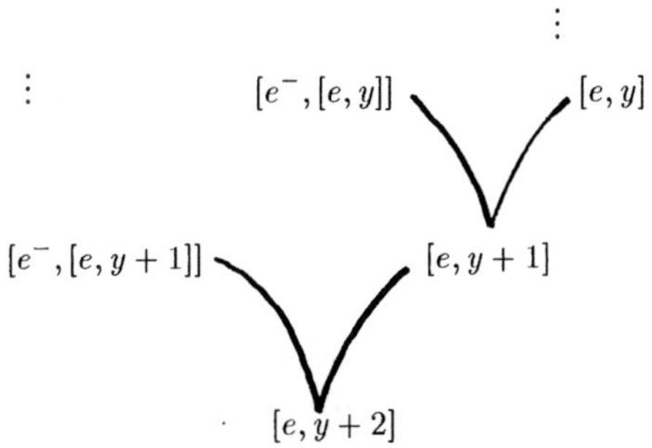

so that in the calculation of $[e^-, [e, y+1]]$, the expression $[e^-, [e, y]]$ arises again. But for $2 < y$, $[e, y]$ is very much less than $[e, y+1]$, so that the reduction in the number of steps obtained by removing the repeated expressions would be negligible.
Remark. $v_N = v^\star$. $\phi(v^\star)$ is the value of $[e, y]$ expressed in tally form.

7.4.7 Calculation is Executed in Pre-Order

The pre- and post-orders for a tree of $\mathfrak{S}$ (IV, §2.5) are respectively pv-first and pv-last (see also KNUTH [77] p316). Often the supporting vertices for the 2 orders are taken in L to R order. The syntax $\Delta_e(x, y)$ suggests the unit given in the figure of §4.4, which gives the R to L order. Taking these conventions as equivalent,

THEOREM 7.33 *The vertex order determined by C is the pre-ordering of $T(e, y)$.*

Proof, by considering the rules of algorithm 4.1 By rule 2. and by the supposition of 2., the order is pv-first. By 2(a), 3(a) and 2(b), the order of succession is R to L.

□

Reduction Sequence

DEFINITION 7.34 *The* reduction sequence $\mathcal{R}$ *of the expression* $[e, y]$ *is the sequence*

$$\phi(v_1),\ \phi(v_2), \ldots, \phi(v_t),$$

where $v_1 = v^\sharp, v_t = v^\star$, *and the* v_i *are the vertices of* C_L *in chain* (Sc−) *order.*

This in the sequence (I) given before lemma 19; it is the sequence which is treated in the proof of part 1 of thm. 20.

Remarks.

1. The sequence $\mathcal{R}$ for $[n, y] = \mathfrak{A}_n(x, y)$ has $t - n$ terms.
2. All but reduction sequence edges of T are traversed twice. The edges determined by $\mathcal{R}$ are traversed once.

7.4.8 A Remark on Complete Trees

DEFINITION 7.35 *T_n, $1 \leq n$ is the unique $\mathfrak{D}$-tree for which all (maximal) chains are of edge-length n.*

□

T_n is of edge-height n; the sequence $T_1, T_2, \ldots$ may be used to reveal the evolution of the higher arithmetic operations. Let $\tau(1) = 2$, and for $n \geq 1$, $\tau(n+1) =$ number of endv's of $T_{\tau(n)}$. Thus,

$$\begin{aligned}\tau(n) &= \text{exponential 'ladder' of } n \text{ 2's} \\ &= 2 \star n \text{ in ROBINSON's notation (II, §2)} \\ &= \mathfrak{A}_4(2, n) \text{ as ACKERMANN's function.}\end{aligned}$$

Now putting $U_n = T_{\tau(n)}$, let $\xi(1) = 2$, $\xi(n+1) =$ edge height of $U_{\xi(n)}$. Therefore $\xi(1) = \tau(1)$ $\xi(n+1) = \tau\{\xi(n)\}$. And therefore $\xi(n) = \mathfrak{A}_5(2, n)$.

7.5 Variants of the Minimal Diagonalization

The diagonalization of defn. 15 aims for: (i) the inclusion of the usual arithmetic operations, and (ii) greatest simplicity. In addition to the variants of the classical ACKERMANN function discussed in §2, each giving a unique general form, there are variants arising from changes in the technique of diagonalization. Some variants are itemized:

1. Defn. 15·2(c) may be amended e.g. by substituting '$(e, \Delta_e(x, y-1))$' in place of the subscript '(e, y)', to give a new operation Δ' say. Thus, instead of progressing without gain from $\Delta_e(x, y)$ to $\Delta_{(e,y)}(x, y)$, the calculation is lengthened after every diagonal step, by using Δ' to pick out a much later term in the fundamental sequence. But little is in fact gained. For $x \geq 2$:

   THEOREM 7.36 $(\forall e)(\exists \alpha \succ e)$ in **EP** such that $\Delta'_e \prec \Delta_\alpha$.

   Remark. A slight gain in growth may be achieved by a similar substitution in the less potent $y-$ and $x-$argument places.

2. In 1903, HARDY gave an infinitary formulation of what is essentially defn. 15, as follows. In the notation of CICHON and WAINER [24]:

$$\begin{cases} H_1(y) = 1 & \\ H_e(y) = H_{e^-}(y+1) & \text{if } \mathrm{Suc}(e) \\ H_e(y) = H_{(e,y)}(y) & \text{if } \mathrm{Diag}(e). \end{cases}$$

   For this schema, a result of converse form to that of thm. 36 holds: every Δ_e of defn. 15 is eventually dominated by some H_α. Note that by a non-commutation of unary operators, taking instead $H^{\cdot}_e(y) = H^{\cdot}_{e^-}(y) + 1$ causes HARDY's hierarchy to collapse: $H^{\cdot}_e(y) = e(y)$ (cf. VI·24).

   Remarks.

   (a) The above schema relies on the assignment Ω, as discussed in III §1.

   (b) It is a remarkable fact that HARDY obtained in 1904, by infinitary means, what was essentially a diagonalization of a function claimed by ACKERMANN, via 2-fold recursion, in 1928.

   (c) Notwithstanding the simplicity of the present variant, it misses the usual binary arithmetic operations.

   (d) Unlike the numbers $\Delta_e(1), \Delta_e(2) \leq 4$, $H_e(1)$ is unbounded, and is better compared with, e. g. $\Delta_e(3)$. Ref. II, §2·14.

[24] *J. Symb. Logic*, **48**, 2, June 1983, p400. See also WAINER [157], ibid. **37**(2), June 1972, p283.

3. The sequence $1, 2, 3, 4, \ldots$ on which the previous case is based may be replaced by any strictly increasing arithmetic sequence $n_1, n_2, n_3, \ldots,\ n_1 > 1$, e.g. that of the prime numbers. Thus $H_e(y)$ takes the form

$$n_n{}^{\cdot^{\cdot^{\cdot^{n_y}}}}$$

(where the number of occurrences of 'n' has of the order of the value determined by this pattern).

7.5.1 A Result of SOLOVAY and KETONEN

Consider the variants Δ', H of defn. 15, given in the previous section. Suppose $\alpha \succ \mathbf{1}$ in **EP**, and let k be the smallest natural number such that, by means of the sequence $1, 2, 3, 4, 5, \ldots$ on which H is based, α is canonically reduced to $\mathbf{1}$. Thus, as in III·24, k is the smallest number for which $\alpha_{(k)} = \mathbf{1}$.

The following is, in effect, the result credited by PARIS 1979 (Springer Lecture Notes 834, KARPACZ Coll.) to SOLOVAY and KETONEN 1981 (Annals. Math., 113, pp267-314).

THEOREM 7.37 $H_\alpha(1) - 1 \leq k \leq \Delta'_\alpha(3)$.

Proof. Since $\Delta'_\alpha(2) = \Delta_\alpha(2) = 4$, an argument > 2 is required. Consider the calculation of $\Delta'_\alpha(3)$. This entails the stepwise canonical reduction of α to $\mathbf{1}$, as in (II) of §3.3 (before lemma 19). Since for $y > 1$, $y < \Delta'_\alpha(y-1)$; and if Diag(α),

$$N = \Delta'_\alpha(y) < \Delta'_{(\alpha,N)}(y),$$

it follows that the canonical index is si at each step. Since, as noted in III, after defn. 24, the greater the index, the less the $\prec$ −descent, it therefore follows

$$k \leq \Delta'_\alpha(3).$$

On the other hand, consider the value of $H_\alpha(1)$. This consists only of 1 plus unit contributions at reductions of a successor subscript. After each such reduction, the canonical index undergoes unit increase, while at diagonal reductions it remains fixed. Therefore again by the above-cited remark of Ch. III,

$$H_\alpha(1) - 1 \leq k,$$

and the thm. is proved.

□

Remarks.

1. By rem. (d) and the comment after the defn. of (H_e), thm. 37 establishes that **the # of steps required to complete a canonical reduction of e (via $1, 2, 3, 4, \ldots$) is of the order of $\Delta_e(m)$, $m \geq 3$ a constant.**

2. Since the clause $\Delta_{\alpha^+}(y+1) = \Delta_\alpha\{\Delta_{\alpha^+}(y)\}$ of defn. 15 generally gives a great gain in canonical index, and since diagonal steps are 'less common', Δ could in fact be substituted for Δ' in the thm.

3. By the above argument, the function $k = k(y) = k(y; 1, 2, 3, \ldots)$ satisfies $k \prec \Delta_\alpha$; and the alteration to $K(y) = K(y; \Delta_\alpha(1), \Delta_\alpha(2), \ldots)$, while giving $k \prec K$, nonetheless yields only $K \approx \Delta'_\alpha$. Thus note that the large indices of §3.7 are significant in retardation of descent only on account of the restriction to **XP**.

4. Consider the canonical reduction (cr) of ε (III·24) by $n, n+1, n+2, \ldots;\ 2 < n$. It will be noted that prior to reaching a successor, a form

$$\omega^{\alpha+e}, \quad \text{where } \alpha = \omega \star (n-2),\ \text{ht}(e) = n-2,\ \alpha \prec e,$$

is encountered. Now, before the reduction can reach a successor, e must be reduced to 1 : $(\omega^{\alpha+1}, N) = N \cdot \omega^\alpha$. It therefore follows that, while $H_{(\varepsilon,n)}(1) - 1 \leq k_n$, the function $k_{n,\text{Suc}}$, obtained by requiring only that the reduction reaches a successor, satisfies

$$H_{(\varepsilon,n-2)}(1) - 1 \leq k_{n,\text{Suc}}.$$

5. **SOLOVAY's Concept of α-Large**
In the notation of III, §4·3, let $\alpha \in$ **EP** be canonically reduced via $n(1) < n(2) < \ldots$ to give $\alpha_{(1)} \succ \alpha_{(2)} \succ \ldots$. Then the finite set $X = \{n(1), n(2), \ldots, n(K)\}$ is *α-large* if $(\exists k \leq K)\ \alpha_{(k)} = 1$. By the preceeding remarks, X may effectively be taken to be $\{1, 2, 3, \ldots, K\}$; in which case 'α-large' merely means '$\geq \Delta_\alpha(3)$'.

A Theorem of LOEBL and MATOUŠEK

These authors [25] have applied thm. 37 to obtain a contravariant form of KRUSKAL's result, IV·76: there is a 'long' sequence of trees without a $\subseteq_{\text{homeo}}$ –embedding. A fast proof is obtained via $\mathfrak{T}$:

THEOREM 7.38 *Let T_1 = the chain on n vertices; and $T_{n+1} = (T_n, |\ T_n\ |)$ for $n = 1, 2, 3, \ldots$. Then by the first step at which* Suc(T_N), *for no $i \neq j$ is $T_i \subseteq_{\text{homeo}} T_j$.*

Proof. Note first that, by the thm. of IV, §7·1, the $\mathfrak{T}$-sequence (T_n) is finite. At each cr, $|\ T\ |$ increases, so by note 3 after IV·72, $T_i \not\subseteq_{\text{homeo}} T_j$ for $j < i$. On the other hand by III·7·3 and IV·62, $T_{n+1} \prec T_n$, $1 \leq n < N$. But by IV·73, $T_n \subseteq_{\text{homeo}} T_{n+1}$ would imply $T_n \preceq T_{n+1}$, so $T_i \not\subseteq_{\text{homeo}} T_j$ for $i < j$.

□

As noted in the proof, cr's in **EP** and in $\mathfrak{T}$ correspond. The thm. therefore provides, via IV, §6, VII·37 and rem. 4 above, a 'long' sequence which contains no homeomorphic embedding.

[25] 'On the undecidability of the weaker KRUSKAL theorem', in *'Logic and Combinatorics'*, *A.M.S. Contemporary Math. Ser.* **65** (1987).

7.6 Finite Persistence Property for EP

7.6.1 Introduction

The BACHMANN Ω–property, III·14, arises in the context of G. H. HARDY's notion of persistent values: the following finiteness property for the hierarchy $\{\Delta_\alpha\}_{\alpha \prec \varepsilon}$ based on **EP** is shared by all variants of the kind considered in §5.

This section is based on the treatment of persistent values in: Mc BETH 1981, 'A note on exponential polynomials and prime factors', *Zeitschr. math. Logik Grundl. Math.*, **27**, 213-214; and Mc BETH 1979, 'A note on HARDY's persistent numbers', ibid., **25**, 375-378.

7.6.2 $\mid \alpha, x : \Delta_\alpha(x) = N \mid < \infty$

Folowing ex. 5 after defn. 15, take $\Delta_\alpha(x) =_{df} \Delta_\alpha(3, x)$. Let

$$\mathcal{S} : \qquad \alpha(1) \prec \alpha(2) \prec \alpha(3) \prec \ldots$$ [26]

be a given si sequence in **EP**.

DEFINITION 7.39 *1. $N \geq 4$ in* **N** *is* finitely persistent *if for any $\mathcal{S}$ as above,*

$$(\exists k)(\forall y \geq k)(\forall x > 2)\ N \neq \Delta_{\alpha(y)}(x).$$

2. $\alpha \to \beta$ in **EP** *iff $(\forall x)(\exists y)\ \Delta_\alpha(y) = \Delta_\beta(x)$. Viz. $\alpha \to \beta$ iff every value of Δ_β is a value of Δ_α.*

□

Thus

LEMMA 7.40 *The relation $\to$ is transitive.*

□

Remark. Suppose $\mid (\alpha, x) : \Delta_\alpha(x) = N \mid = \infty$. Then by the DCC (and thm. 20·2 of §3·4), 39·1 cannot be satisfied. Viz. the condition implies that N is a value of only fm Δ_α.

Now let e, α Diag(e); and $n \geq 1$ be such that

$$(e, n) \preceq \alpha \preceq (e, n+1);$$

with the single exception $e = \omega, n = 1$.

[26] In order to avoid recurrent subscripts, the i-th polynomial in $\mathcal{S}$ is written $\alpha(i)$ (the value $\alpha(i)(x)$ at x is not required).

LEMMA 7.41 $(e, n) \to \alpha$.

Proof. Let x, y be as in defn. 39. In all cases, if $x = 1$,

$$\Delta_\alpha(1) = 3 = \Delta_{(e,n)}(1),$$

so $y = 1$ may be taken. Thus assume $x > 1$. The proof proceeds by generalized induction on α. In the first instance, $(e, n) \to (e, n)$. Now let $\alpha \succ (e, n)$, and suppose for g.i.h. that the lemma holds for α^- if Suc(α); or for all (α, m), $m = 1, 2, 3, \ldots$ if Diag(α). If Suc(α),

$$\begin{aligned} \Delta_\alpha(x) &= \Delta_{\alpha^-}\{\Delta_\alpha(x-1)\}, \text{ as } x > 1 \\ &= \Delta_{\alpha^-}(u), \text{ say.} \end{aligned}$$

But by g.i.h.

$$(\exists y)\, \Delta_{(e,n)}(y) = \Delta_{\alpha^-}(u) = \Delta_\alpha(x).$$

Now suppose Diag(α). By III·14, $(e, n) \preceq (\alpha, m) \prec \alpha$, $m \geq 1$, so applying the g.i.h. in the case of the particular value x,

$$(\exists y)\, \Delta_{(e,n)}(y) = \Delta_{(\alpha,x)}(x) = \Delta_\alpha(x),$$

and the lemma is proved.

□

Remark. The lemma takes the form of the starting theorem on which it rests.

COROLLARY 7.42 $(\forall e \in \mathbf{EP})_{\text{Diag}(e)}(\forall m, n)\, m < n \Rightarrow (e, m) \to (e, n)$.

Proof. Let $d = n - m$, $i = 0, 1, 2, \ldots, d - 1$. Then $(e, m + i) \to (e, m + i + 1)$ by the lemma; and the general result for homologous terms follows by lemma 40.

□

By III·12, VII·41, 42 give an analogue of III·15:

COROLLARY 7.43 $(\forall e, \alpha)_{\text{Diag}(e)}(\forall n)\, (e, n) \preceq \alpha \prec e \Rightarrow (e, n) \to \alpha$.

Remarks

1.

2. For large n, $(e, n) \not\to e$, since for $3 \leq x$; $2x < n+2$, $\Delta_{(e,n)}(2) > \Delta_{(e,2x-2)}(2) > \Delta_{(e,x)}(x) = \Delta_e(x)$ (using thm. 22).

The main theorem is obtained by first getting the result for canonical fundamental sequences:

LEMMA 7.44 $(\forall e)_{\mathrm{Diag}(e)}(\forall N)(\exists n)(\forall x > 2)\ N \neq \Delta_{(e,n)}(x)$.

Proof. Let $1 \leq N$. Then by thm. 20·1 $(\exists n)\ N < \Delta_{(e,n)}(3)$. By 20·2 $x \geq 3 \Rightarrow N < \Delta_{(e,n)}(x)$.

□

THEOREM 7.45 *Every natural number is finitely persistent.*

Proof. Let $\alpha(1) \prec \alpha(2) \prec \alpha(3) \prec \ldots$ be as in defn. 39 and using VI, corol. 7, let $\sup_i \alpha(i) = \alpha$. For the diagonal α,

$$(\forall n)\{\exists k = k(n)\}\ (\alpha, n) \prec \alpha(k).$$

Now, given N, choose n according to lemma 44, so $N \neq \Delta_{(\alpha,n)}(x)$. Then as

$$(\alpha, n) \prec \alpha(k) \preceq \alpha(u) \prec \alpha, \quad k \leq u,$$

by corol. 43 the finite persistence property follows.

□

7.7 Notes on Ch. VII

1. §1. The results of the procedure of exchange mentioned in this section seem so profound that it is perhaps worthwhile to express it in ordinary terms. Starting with the operation of multiplication, the secret is simply to observe that the new operation [27] to the n–th power gives some (large) number, N say. Then the $(n+1)$ -th power is obtained by taking the old operation to the N–th power, giving the ramus

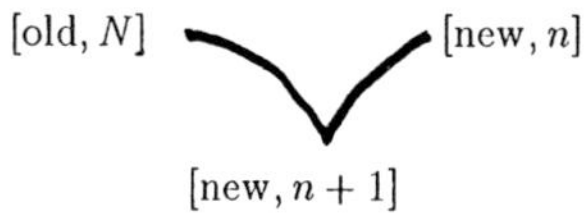

of §4. It is this procedure whch makes it possible to set, with a few strokes of the pen an explicit arithmetic problem, e.g. 'is $\varepsilon(4)+1$ a prime number?'; far beyond the capabilities of the world's largest and fastest computers.

2. §§1, 2.2. As indicated in footnote 1, the function $\mathfrak{A}_n(x, y)$ of 3 variables was known prior to ACKERMANN's 1928 paper [1]. HILBERT 'Über das Unendliche', *Math. Annalen* (1925), informally lists recursive definition schemata, for $n = 1, 2, 3, 4$ before discussing the definition of $\mathfrak{A}$ by 'simultaneous' recursion, and by the following scheme: let ι be a functional of a single function f, and 2 number arguments x, y (where f itself is of 2 number arguments) defined recursively by

$$\iota(f, x, 1) = x$$

[27] Which to the first power is the identity.

$$\iota(f, x, y+1) = f\{x, \iota(f, x, y)\}.$$

Then defn. 1 (§2) may be replaced by

$$\mathfrak{A}_1(x, y) = x + y$$

$$\mathfrak{A}_{n+1}(x, y) = \iota(\mathfrak{A}_n, x, y).$$

3. §3.4. Accepting the constructivity of the numbers $N = \Delta_\alpha(x, y)$, the arithmetic properties of the N depend on x. However the analogue of corol. 23 obtained by holding $y \geq 2$ fixed does not hold. For $k \geq y$:

$$(\forall x)\ \Delta_{\alpha+\omega}(x, y) \leq \Delta_{\alpha+\mathbf{k}}(x, y).$$

4. §3.4, first remark. Given any integer sequence $N = N(\alpha) \leq n_1 \leq n_2 \leq \ldots$, the polynomials α, β are linked by

$$\alpha \lhd_{n_t} \ldots \lhd_{n_1} \beta, \qquad \text{(I)}$$

for some $t \geq 1$. Now $\alpha = (e, n)$ for some diagonal e, only for $n < N$ by III·7·5. Therefore $n_t > n$, and (I) posesses an end-segment of K unit descents (viz. passage to the predecessor), where $K > 1$. In the case of thm. 23, the integer sequence has large increments at successor steps, thereby contributing large parts to K.

5. §§2,3. KNOEBEL [76] (1981), has put the question: can the domain of the (classical) ACKERMANN function be extended, so as to provide an analytic continuation to the complex plane? Since there is no bound on the growth of analytic functions, it seems probable that an affirmative answer would lead to such a continuation for the diagonalized forms (Δ_e) of §3.

6. §3.4 Of the 3 cases of thm. 20, it is 20·1 which is most important, the other 2 being essentially (cases of) thm. 6. The indispensible foundation for 20·1 is lemma 19, which is based on III·14. It may thus be said that (Δ_e) rests on the DCC to assure definedness; and on BACHMANN's thm. to ensure strict increase.

Chapter VIII

EXTENSIONS OF EP

Section Headings

Section Headings

8.1 Introduction

The foundation **EP**, and the structure $(\Delta_e)_{e \in \mathbf{EP}}$ are the 2 characteristic hierarchies of modern mathematical logic. Nonetheless, **EP** may be extended in various ways.

The exponentials may be extended to larger domains such as the real or complex fields. Alternatively, new functions may be included, by taking further rules of closure; or by allowing modifications of the rules **EP**.1-.3. §§2 - 6 of Ch. VIII will concentrate on the admission of new rules of closure based on ACKERMANN's function and its diagonalizations. This has the advantage of producing extensions posessing the properties which make **EP** itself interesting. The question of producing further extensions by inclusion of slow-growing unbounded terms, touched on in Ch. II, §6 is left open.

The underlying structure of the process of iteration and diagonalization of II, §5·2, is the feather. This notion (and its dual) is formulated in §7. The feather-diagram encompasses both the canonical evolution of **EP**, VI, §2·4; and the calculation of $\mathfrak{A}$. It is the key to infinite hierarchies.

Returning to the comment of the 3rd paragraph of Ch. II, §4.1, there is no opportunity to take up, in this volume, what may indeed be considered the true application of type-objects to hierarchies: higher diagonalization. This complex construction, begun by Oswald VEBLEN [155] and systematized by Heinz BACHMANN (1950), constitutes a discipline in its own right, and will not be considered here. The methods of Ch. VIII are only those thrown up as 'by-products' of Chs. II - VII.

Extension to larger domains is considered in a special section of §8.

8.2 Enlargement of EP

8.2.1 Introduction

Restricting attention now to functions which transform a natural number, consider the enlargement of **EP** by inclusion of new polynomials.

8.2.2 The Class EP*.

The ROBINSON 'Superpower'

The operation $x \star y; x, y \in \mathbf{N}$, introduced by Julia ROBINSON in 1952 [129], p438, was defined in Ch. II by

$$x \star 1 = x; \quad x \star (y+1) = x^{x \star y}, \quad y \geq 1$$ [1].

This function occurs in various places in the literature. SKOLEM [141] p59 uses the notation $f(x, y) = x \star (y+1)$, and for $x = 2$, YESSENIN-VOLPIN [163], p40 uses the notation $2^{(y)} = 2 \star y$. Even this special case is a function of shocking rate of growth. As y takes the values $1, 2, 3, \ldots, 2^{(y)}$ is

$$2, 4, 16, 65536, 20035 \ldots 56736, \ldots$$

where, to save space here, 19718 digits have been omitted in the fifth term. It is a remarkable phenomenon that such a function can so easily be defined.

The relation $\prec$ of eventual domination is taken as extended to **EP***. It will be shown that the relation is connected, solving a special case of a problem of SKOLEM [141], p59. The sign **NL** is used to denote the subclass of **XP** (Ch. II, §2.1) of linear polynomial forms $b\omega + a$; and $\mathbf{NL_0} =_{df} \mathbf{NL} \cup \mathbf{0}$. The operations $+, ., , \star$ of addition, multiplication, exponentiation and $\star$–exponentiation respectively may be regarded as forming the initial terms of a sequence, and for applications it is useful to assign them the notations $\star_1, \star_2, \star_3, \star_4$. Indeed, this observation has already been made, a number of times by GOODSTEIN e.g. [46] who calls his operation ${}^y x, = x \star y$, 'tetration'.

8.2.3 Exponential Polynomials of Linear Height

The content of this section is based on McBETH [95] [2], 1980.

[1]Extending $\star$ to include $x, y = 0$, the evaluation $0^0 = 1$ (ref. II, footnote 3) gives

$$0 \star y = \begin{cases} 1 & \text{if } 2 \mid y \\ 0 & \text{if } 2 \not\mid y. \end{cases}$$

[2]'Exponential polynomials of linear height', *Zeitschr. math. Logik Grundl. Math.*, Bd. 26, S. 399-404 (1980).

DEFINITION 8.1 *The class* $\mathbf{EP}^\star$ *is the smallest class closed under the following operations:*

rule $\mathbf{EP}^\star$*.1* $\lambda x \cdot 1 = \mathbf{1} \in \mathbf{EP}^\star$
rule $\mathbf{EP}^\star$*.2* $\mathbf{EP}^\star(e,f) \Rightarrow \mathbf{EP}^\star(e+f)$
rule $\mathbf{EP}^\star$*.3* $\mathbf{EP}^\star(e) \Rightarrow \mathbf{EP}^\star(\omega^e)$
rule $\mathbf{EP}^\star$*.4* $\mathbf{EP}^\star(e) \Rightarrow \mathbf{EP}^\star(\omega \star e)$. [3]

□

For rule $\mathbf{EP}^\star$.4, it may be assumed $e \neq \mathbf{1}$, and $e \neq f^+$, $f \in \mathbf{EP}^\star$. Note that, from the defn., every $e \in \mathbf{EP}^\star$ is cst or si. The functions of $\mathbf{EP}^\star$ are classified as follows:

1. $\mathbf{1}$ is the *initial polynomial*
2. if $e = f^+$, for some $f \in \mathbf{EP}^\star$, then e is a *successor polynomial*
3. if $e \neq \mathbf{1}$, is not a successor polynomial, then e is a *diagonal polynomial.*

These 3 classes thus partition $\mathbf{EP}^\star$.

Example.

The function $\varepsilon(x) = \omega \star \omega(x)$ belongs to $\mathbf{EP}^\star$. As in II, corol. 16, $\varepsilon \notin \mathbf{EP}$, since for each n, $\lambda x \cdot x \star n \prec \varepsilon$; and every $e \in \mathbf{EP}$ is eventually dominated by some $\omega \star n$. Thus $\mathbf{EP} \subset \mathbf{EP}^\star$.

Note.

$\mathbf{EP}^\star$ is not closed under *product,* so e.g. $\omega \cdot \varepsilon \notin \mathbf{EP}^\star$.

DEFINITION 8.2 *The* height $\mathrm{ht}(e) \in \mathbf{NL_0}$ *of* $e \in \mathbf{EP}^\star$ *is defined by : if* e *is cst,* $\mathrm{ht}(e) = 0$; *if* $e = e_1 + e_2$ *in* $\mathbf{EP}^\star$, $\mathrm{ht}(e) = \max\{\mathrm{ht}(e_1), \mathrm{ht}(e_2)\}$. *Now suppose* $f \in \mathbf{EP}^\star$ *and* $\mathrm{ht}(f) = b\omega + a$; $a, b \in \mathbf{N}_0$. *Then if* $e = \omega^f$,

$$\mathrm{ht}(e) = \begin{cases} b\omega + \mathbf{a} & \textit{if } b > 0 \\ \mathbf{a}^+, & \textit{if } b = 0, \end{cases}$$

and if $e = \omega \star f$ *(where* $\mathrm{ht}(f) \neq 0$ *by convention after defn. 1)*

$$\mathrm{ht}(e) = \begin{cases} (b+1) \cdot \omega + \mathbf{a} & \textit{if } b > 0 \\ \omega + \mathbf{a}^-, & \textit{if } b = 0. \end{cases}$$

Examples

1. $\mathrm{ht}(\varepsilon) = \mathrm{ht}(\omega^\varepsilon) = \omega$; $\mathrm{ht}(\omega \star \varepsilon) = 2\omega$; $\mathrm{ht}(\omega \star \psi) = \omega^+$.

[3] Thus if rule $\mathbf{EP}^\star$.4 is omitted, then in place of $\mathbf{EP}^\star$, the class $\mathbf{EP}$ is obtained.

2. For $e \in \mathbf{EP}$, ht(e) is equal to a constant.

The following contains an extension of a function of SKOLEM:

DEFINITION 8.3 *The* reflex φ_e *of a linear polynomial* $e \in \mathbf{NL_0}$ *is defined by recursion on* $b \in$ N. *For* $a \in \mathbf{N}_0$,

$$\varphi_{\mathbf{0}} = \mathbf{1}; \quad \varphi_{\mathbf{a}^+} = \omega^{\varphi_{\mathbf{a}}}; \quad \varphi_{\omega+\mathbf{a}} = \omega \star \varphi_{\mathbf{a}^+};$$

$$\varphi_{(b+1)\omega+\mathbf{a}} = \omega \star \varphi_{b\omega+\mathbf{a}}.$$

Thus $\varphi : \mathbf{NL_0} \to \mathbf{EP}^\star$.
Examples

1. $\varphi_{\mathbf{1}} = \omega$; $\varphi_\omega = \varepsilon$ and generally $\varphi_{b\omega} = \omega \star\star\, b^+$.
2. ht$(\varphi_e) = e$. Note that φ is 1-1, ht is many-1.
3. The reflex of a cst polynomial $\mathbf{n}$ is the function φ_n of SKOLEM given on p55 of [138].
4. $(\forall e \in \mathbf{EP}^\star)(\exists f \in \mathbf{NL})\; e \prec \varphi_f$.

Segments of EP*

The natural means to measure the rate of growth of $e \in \mathbf{EP}^\star$ is by comparison by eventual domination, $\prec$. From defn. 1 alone, it is not clear that $(\mathbf{EP}^\star, \prec)$ is either linear or well-founded. But the early terms of the system can readily be listed. Under the convention given after defn. 1, the $\prec$ –least polynomial obtained using $\mathbf{EP}^\star$.4 is ε. Since this dominates of all of **EP**, $\mathbf{EP}^\star$ contains the segment

$$\mathbf{1}, \mathbf{2}, \mathbf{3}, \ldots, \omega, \omega^+, \ldots, e, \ldots, \varepsilon, \ldots,$$

with e ranging over **EP**. Inspection of the defn. shows the segment continues as

$$\ldots, \varepsilon, \varepsilon^+, \ldots, \varepsilon + e, \ldots, 2\varepsilon, \ldots, a\varepsilon + e, \ldots,$$

where $a \in \mathbf{N}$ and e is as above. This subsystem is closed under addition, so the next term, wrt $\prec$, requires $\mathbf{EP}^\star$.3, the least possibility being

$$\kappa = \omega^\varepsilon = \omega \star \omega^+ = \lambda x \cdot x \star (x+1).$$

κ represents a considerable leap forwards, from the general term of the above segment. This insight into the initial stages of $\mathbf{EP}^\star$ may be carried somewhat further, by the same means. E.g. before $\omega \star (2; \varepsilon)$ is encountered, general terms

$$\sum a_i \omega^{m_i\varepsilon + e_i} + m\varepsilon + e$$

are passed, $a_i, m, m_i \geq 0,\; e, e_i \in \mathbf{EP_0}$.

A Spanning Sequence for EP*.

The comparatively early polynomial functions of $(\mathbf{EP}, \prec)$ do in fact take small values for low x, as in $\alpha(x) = x^{x^x+x} + x^x + x$, with $\alpha(1) = 3$, $\alpha(2) = 70$, $\alpha(3) < 10^{15}$, where α is of height 3. However, as the parameter a in the height $a\omega + b$ of $e \in \mathbf{EP}^\star$ takes the values $1, 2, 3, \ldots$, e becomes very fast growing. Such a sequence of e is given by

$$\omega,\ \omega \star \omega,\ \omega \star \omega \star \omega, \ldots,$$

(with as usual assumed brackets associating from the right). Thus every $e \in \mathbf{EP}^\star$ of height $\geq 3\omega$ satisfies $e(3) \geq 3 \star 7625597485987$. It is a remarkable fact that it is possible to reason so closely about a hierarchy of functions taking such values. The nth term of the previous sequence may be expressed

$$\omega \star\star\, n$$

in the convention of II, §2.5, and the notation of the introduction to the previous Ch. Then

THEOREM 8.4

$$(\forall e \in \mathbf{EP}^\star)(\exists n \in \mathbf{N})\ e \prec \omega \star\star\, n.$$

Proof, as in II, thm. 14.

□

The Arithmetic Function η.

The function $\eta =_{df} \omega \star\star\, \omega$ satisfies $(\forall x > n)\ \omega \star\star\, n(x) < \eta(x)$. Therefore by thm. 4,

THEOREM 8.5

$$(\forall e \in \mathbf{EP}^\star)\ e \prec \eta.$$

η thus stands in relation to $\mathbf{EP}^\star$ in much the same way as $\varepsilon \prec \eta$ does to the smaller class **EP**. Bearing in mind the remark after Ch. II, notation 2.1, it may be noted that $\eta(4) = 4 \star (4 \star\star 3) = 4 \star (4 \star (4 \star\star 2))$ is obtained by taking a ladder of $\varepsilon(4)$ occurrences of 4; then building another ladder, containing this number of occurrences.

8.2.4 EP* is Linear

The asymmetry, irreflexivity and transitivity of the relation $\prec$ on $\{\mathbf{EP}^\star\}^2$ follow by definition, as with **EP**, II, §2.9. In order to prove $\mathbf{EP}^\star$ is totally ordered, the notion of definition tree is introduced. Consider $e \in \mathbf{EP}^\star$. By defn. 1, there is an $m \geq 1$, and there are functions $e_1, e_2, \ldots, e_m \in \mathbf{EP}^\star{}_0$ and numbers $3 \le c_1, c_2, \ldots, c_m \le 4$ such that

$$\text{(I)} \qquad e = \sum_{i=0}^{m} \omega \star_{c_i} e_i.$$

Now suppose $e \in \mathbf{EP}^\star$.

DEFINITIONS 8.2.1 *1. If e has the form (I), then the relation* Sc *is defined by* e_i Sc e, $i = 1, 2, \ldots, m$. $\widetilde{\text{Sc}}$ *is the transitive closure of* Sc.

2. For $i = 1, 2, \ldots, m$, $\star_{c_i}$ *is called the* associated operation of e_i *in the expression (I).*

3. The definition tree T_e *of e is the oriented tree (without assigned plane structure) defined recursively by:* T_e *is the isolated vertex* $v^\sharp$ *if e is cst. If e is non-constant,* T_e *is defined as follows: first form the tree* T_e' *whose vertices are*

(a) e;

(b) $\alpha \in \mathbf{EP}^\star$ *for which* $\alpha \, \widetilde{\text{Sc}} \, e$.

The edges of T_e' *are edges joining each* $\alpha \, \widetilde{\text{Sc}} \, e$ *to the unique* β *for which* α Sc β. *The defn. tree* T_e *of e is the tree obtained from* T_e' *by replacing each vertex* $\alpha \ne e$ *of* T_e' *by the pair* $(\alpha, \star_i)$, *where* $\star_i$ *is the associated operation of* α.

T_e is finite, since e is built up in a finite number of steps from the initial polynomial. Thus every maximal chain of T_e contains an endv equal to $(\mathbf{0}, \star_3)$. [4]

DEFINITION 8.6 *For* $N \ge 1$, *the class* $\mathbf{EP}^\star{}_N$ *of polynomials of order (of defn.) N in* $\mathbf{EP}^\star$ *contains exactly those obtainable by not more than N applications of the rules* $\mathbf{EP}^\star$*.1,.3 and .4.*

Thus, for each N, $\mathbf{EP}^\star{}_N$ is closed under the rule $\mathbf{EP}^\star$.2 of addition; $\mathbf{EP}^\star{}_1$ is the class of cst functions.

THEOREM 8.7 [5] *The relation* $\prec$ *is connected on* $\{\mathbf{EP}^\star\}^2$

Proof, by induction on the order N of $e \in \mathbf{EP}^\star$. In the initial case, the constant polynomials are $\prec$ −comparable. Suppose for induction hypothesis that $\prec$ is connected on $\{\mathbf{EP}^\star{}_n\}^2$, for some $N > 1$ and all $n < N$. Consider any $e_{(1)}, e_{(2)}$ in $\mathbf{EP}^\star{}_N$ of the form

$$\text{(II)} \qquad e_{(1)} = \omega \star_a \alpha, \; e_{(2)} = \omega \star_b \beta, \; 3 \le a, b \le 4.$$

[4]Suppose $e \in \mathbf{EP}^\star$ is diagonal. Then $\omega \star e$ has no decomposition $\omega \star e = \omega^u$, $u \in \mathbf{EP}^\star$. Finally the expressions $\omega^e, \omega \star e, e \in \mathbf{EP}^\star$, cannot be obtained as (multiple) finite sums of terms in $\mathbf{EP}^\star$ (ref. end of proof of II·22); therefore it follows the defn. tree T_e is *unique*. See, however note 7, end of this Ch.

[5][95], thm. 2·6, p401.

Thus $\alpha,\ \beta \in \mathbf{EP}^{\star}{}_{N-1}$. If $a = b$, it follows from the induction hypothesis that the expressions (II) are $\prec -$ comparable. Now suppose $a \neq b$. The possibilities are

	$\alpha \prec \beta$	$\beta \prec \alpha$	$\alpha = \beta$
$a = 3,\ b = 4$	$\prec$		$\prec$
$a = 4,\ b = 3$		$\succ$	$\succ$.

The cases marked show the relation between the expressions (II). The remaining cases are dual, so it may be assumed $\alpha \prec \beta$ and $b < a$, i.e. $b = 3, a = 4$. Each chain in T_β starting at the $\beta-$vertex, reaches, in a finite number of steps, either a cst vertex, or a pair $(g, \star_4)$. By the convention after defn. 1, both α and g are diagonal. Thus, if no chain contains a pair $(g, \star_4)$,

$$\omega^\beta \prec \omega \star \alpha,$$

The LHS being of constant height, the RHS being of unbounded height. Thus, suppose at least one chain contains a pair $(g, \star_4)$ and let the set S contain all functions which occur in such a pair. Furthermore, for $g \in S$, let C_g be the unique chain linking $(\beta, \star_3)$ to $(g, \star_4)$. By induction hypothesis, α and g are $\prec -$comparable. To compare the expressions (II), in the first case if $(\exists g \in S)\ \alpha \preceq g$, then

$$\omega \star \alpha \preceq \omega \star g \preceq \beta \prec \omega^\beta.$$

Thus the remaining case is that in which $(\forall g \in S)\ g \prec \alpha$. Now an expression $\omega^v, v \in \mathbf{EP}^\star$ and v diagonal, satisfies $\omega \star v = \omega^u$, where $u : \mathbf{N} \to \mathbf{N}$ is si. (But $u \notin \mathbf{EP}^\star$, e.g. $v = \omega$ and $u = \omega \star \omega^-$.) Now α, g are diagonal polynomials. Therefore $g \prec \alpha$ implies, by the expressions for g, α as sums of exponential terms, that the difference $d(x) = \alpha(x) - g(x)$ becomes larger than any fixed bound, i.e.

$$(\forall N)(\exists k)(\forall x \geq k)\ N < d.$$

Let the vertices of C_g contain $n(g)$ operations $\star_3$. Let

$$g^+ = \max_{\prec} S, \quad \text{and} \quad K = 1 + \max_{g \in S} n(g).$$

Then

$$\omega^\beta \prec \omega \star (g^+ + K) \prec \omega \star \alpha.$$

Thus it is proved $e_{(1)} \prec e_{(2)}, e_{(1)} = e_{(2)}$, or $e_{(1)} \succ e_{(2)}$. But any polynomials $e, f \in \mathbf{EP}^\star{}_N$ are sums of terms of the form (II). Thus, collecting equal terms,

$$\text{(III)} \qquad e = \sum_{i=1}^{m} a_i \omega \star_{c_i} e_i + \mathbf{a} \qquad f = \sum_{i=1}^{n} b_i \omega \star_{d_i} f_i + \mathbf{b}$$

where $a_i, b_i \geq 1$; $a, b \geq 0$; $e_i, f_i \in \mathbf{EP}^\star{}_{N-1}$, $3 \leq c_i, d_i \leq 4$. By the $\prec -$comparability of the expressions (II), it follows that the terms of the expressions (III) may be assumed to be in strictly $\prec$ −decreasing order. And also that it is possible to establish $e \prec f, e = f$, or $f \prec e$ exactly as for the **EP**−result, II·18. The generalized exponential polynomial functions of $\mathbf{EP}^{\star}{}_N$ are therefore pairwise $\prec$ −comparable, and so it follows by induction that $\prec$ is connected on $\{\mathbf{EP}^{\star}\}^2$.

□

Normal Form

COROLLARY 8.8 *If* $e \in \mathbf{EP}^{\star}$, *then* e *has a unique expression*

$$e = \sum_{i=r}^{1} a_i \omega \star_{c_i} e_i + \mathbf{a},$$

where $r \in \mathrm{N}_0$, $\sum =_{df} \mathbf{0}$ *if* $r = 0$, *and where* $3 \le c_1, c_2, \ldots, c_r \le 4, e_1, e_2, \ldots, e_r \in \mathbf{EP}^{\star}$,

$$\omega \star_{c_1} e_1 \prec \omega \star_{c_2} e_2 \prec \ldots \prec \omega \star_{c_r} e_r,$$

$a_1, a_2, \ldots, a_r \in \mathrm{N}$, *and where*

$$a \in \begin{cases} \mathrm{N}_0 & \text{if } r > 0, \\ \mathrm{N} & \text{if } r = 0. \end{cases}$$

□

EP* is a more complex class than **EP**, and e may be given the 2nd canonical expression

$$e = \sum_{i=s}^{1} b_i \omega \star \theta_i + a_i \omega^{\theta_i} + \mathbf{a},$$

where $\theta_1 \prec \theta_2 \ldots \theta_s$. The terms of the 2nd expression form a permutation of the 1st.

8.2.5 Unique Representation in EP*

Given a base $m \ge 2$, $N \in \mathrm{N}$ has a unique representation in **EP***, relative to m. Thus, let the explicit coefficients of $e \in \mathbf{EP}^{\star}$, and mult($e$) be defined as in Ch. II, §2 for $e \in \mathbf{EP}$.

LEMMA 8.9 *For fixed* $A, B, m \ge 2$, *there are only* finitely many $e \in \mathbf{EP}^{\star}$ *such that mult*(e) $< m$ *and* $a < A$, $b < B$, *where ht*(e) $= b\omega + a$.

Proof, by simultaneous induction on a, b.

□

Now let the $m-$height of N be $b\omega + a$, viz.

$$m \star\star (b, m \star a) \le N < m \star\star \{b, m \star (a+1)\}.$$

N may now be successively approximated by constructing polynomials using the closure rules of **EP***, under the restriction that no coefficient exceeds $m - 1$. By the lemma, this process reaches N after fm steps.

8.2.6 The Sub-class $\mathbf{L}^{3,4}$ of $\mathbf{EP}^\star$

The class $\mathbf{L}^{3,4}$ of 'ladder functions' is the class $\mathbf{EP}^\star$ with the rule of closure $\mathbf{EP}^\star.2$ deleted. Thus the functions of $\mathbf{L}^{3,4}$ can conveniently be represented by expressions

$$e = a_0 b_1 a_1 b_2 a_2 \ldots b_{n-1} a_{n-1} b_n a_n,$$

where a_0, n are non-negative integers, and where the remaining a's and b's are positive integers. e denotes the polynomial in $\mathbf{EP}^\star$ obtained from the initial polynomial by successive applications of rule $\mathbf{EP}^\star.3$ (for a's) and rule $\mathbf{EP}^\star.4$ (for b's), passing from right to left. In the special case $n = 0, a_0 = 0, e$ is set equal to the initial polynomial. Now let **XP** be the class of natural polynomials of II, §2.1. A 1-1 correlation $\Phi : \mathbf{L}^{3,4} \to \mathbf{XP}$ can be set up by:

$$\Phi(a_0 b_1 a_1 b_2 a_2 \ldots b_n a_n) = \lambda x \cdot a_n x^b + \ldots + a_2 x^{b_2+b_1} + a_1 b^{b_1} + a_0,$$

where $b = \sum_{i=n}^{1} b_i$. Thus, e.g. $\Phi(01131) = \omega^4 + \omega$. Now consider an expression

$$e = a_0 b_1 \ldots a_n$$

as a ladder

(I) $$\overbrace{\omega \star \omega \star \ldots \star \omega}^{b}$$

containing $b\star$'s into which, in certain places above, between or below the $\star$'s, have been inserted ladders of the form

(II) $$\overbrace{\omega^{\omega^{\cdots^{\omega^\omega}}}}^{a_i}.$$

A single insertion of ω (i.e. a (II)-ladder, with $a_i = 1$) into (I) at a certain place P gives a function which eventually dominates any obtained by insertion of a (II)-ladder of a_i terms, $a_i = 1, 2, 3 \ldots$, at a place below P. Thus e.g. let (I) be $\omega \star \omega$. A 1-term (II)-ladder can be inserted in 2 different places:

$$\omega^{\omega \star \omega}, \omega \star (\omega^\omega).$$

From these remarks, it is evident that $e \prec e' \Leftrightarrow \Phi(e) \prec \Phi(e')$. Thus $\Phi : (\mathbf{L}^{3,4}, \prec) \to (\mathbf{XP}, \prec)$ conserves order, viz. $\mathbf{L}^{3,4} \simeq \mathbf{XP}$. Note that for e as given above, $\mathrm{ht}(e) = b\omega + a_n$.

8.2.7 Other Subclasses of $\mathbf{EP}^\star$

Let $\mathbf{EP}^{\star(b)}$ be the subclass obtained by restricting the rule of closure $\mathbf{EP}^\star.4$ to at most b applications. Then using the normal form of corol. 8,

$$\mathbf{EP}^{\star(b)} = \{e \in \mathbf{EP}^\star \mid e \prec \omega \star \star (b^+)\}.$$

The subclass $\mathbf{EP}_\star$ obtained by omitting the rule of closure $\mathbf{EP}^\star.3$, is $\prec$ −similar to **EP**.

8.3 Fundamental Sequences in EP*

A further operation of closure entails more than is contained in Ch. III, defn. 6 to assign fundamental sequences in **EP*** to the diagonals of this class. Besides, there is a serious difficulty: the informal examination of the early terms of **EP*** showed the sequence $(n\varepsilon)_{n=1,2,3,...}$ to be cofinal in the segment defined by ω^ε. The assignment suggested by III·6 is

$$\omega^\varepsilon = \text{diag}_n \omega^{\omega\star n}, \quad \text{so that } (\omega^\varepsilon, n) = \omega \star (n+1) \prec \varepsilon.$$

There is no hope [6], therefore, of proving an extension of the density thm., III·12, applicable to **EP***, if an assignment is made on this basis. This is an opportune place to trace dependencies on the density result, which is a foundation of the hierarchy $\{\Delta_e\}$. Together with the DCC, it is needed to prove V·3: the equivalence of $\prec$ and $\tilde{\triangleleft}$. This equivalence is required to prove the validity of diagonalized induction (ref. VI·10,12); and the diagonalization of induction is required, to support that of recursion (VI·23).

The motivation to assign the diagonals of **EP*** is to build an extension of the Δ_e–hierarchy by allowing e to range over the vastly greater system **EP***, at the same time initiating a general procedure for extension of the hierarchy, under which e will range still further. Such extensions, as here envisaged, use generalized recursion.

A solution to the above obstacle is to assign to a diagonal $\alpha \in \mathbf{EP}^\star$ a sequence $(\alpha,1),(\alpha,2),\ldots$ which *has* the property

$$(\forall e \in \mathbf{EP}^\star)\, e \prec \alpha \Rightarrow (\exists N)\, e \prec (\alpha, N),$$

(viz. that of III·12 for $(\mathbf{EP},\Omega)$), but for which α is not necessarily the diagonal. As noted in VII, §3.6 after the defn. of the hierarchy $\{\Delta_e\}$, the diagonal chosen for the construction of a non-collapsing hierarchy does not have to be the index. It turns out that larger classes than **EP*** may be assigned, using this expedient; so first of all **EP*** should be enlarged.

8.4 Enlargements Beyond EP*

8.4.1 Further Operations of Closure

It is evident from §2 that the inclusion of $x \star y$ as an operation of closure greatly enlarges the class **EP**. Certainly the question 'is it possible to go further?' is answered in the affirmative. **EP** lies at the entrance to a vast and complex system of generalized exponential polynomial functions, sequentially ordered by eventual domination. E.g. $\mathcal{X}$ may be taken as the class obtained from **EP*** by closure under the operation

$$e \in \mathcal{X} \Rightarrow \omega \star\star\, e \in \mathcal{X}.$$

[6]Note that the assignment $\omega^\varepsilon = \text{diag}_n\, n^\varepsilon$ is 'ex-**EP***'.

The new operation, the 5th term of the classical ACKERMANN function gives $\mathbf{EP}^\star \subset \mathcal{X}$. This approach leads to the definition of the class **E2**, obtained by closure, for each $\mathbf{3} \preceq e$ in **EP**, of the class $\mathbf{E1} = \mathbf{EP}$ under the binary operations Δ_e :

$$e \in \mathbf{E2} \Rightarrow \lambda x \cdot \Delta_e\{x, e(x)\} \in \mathbf{E2}.$$

It is at this point that the difficulty of proceeding without use of the axiom of choice becomes serious. The door to the extension **E2** of the class $\mathbf{EP}^\star$ remains locked, until it is shown (i) that $(\mathbf{E2}, \prec)$ is linear, and (ii) that $(\mathbf{E2}, \prec)$ satisfies the DCC. Instead of speaking of the 'class of all structures which have the required property', one such structure has to be constructed.

8.4.2 Construction vs Deduction

The classical non-constructive approach to hierarchy theory, viz. to well-order (Ch. VI, §3.1) is via a theory of (well-)order types. α is the type of all well-ordered systems equivalent under similarity. $\alpha < \beta$ iff α is similar to a proper segment (VI, §2.2) of β. This involves the collection of all entities with a given property, and leads to the well-known BURALI-FORTI paradox. The classical resolution of the paradox, the first-order axiomatization of the theory of sets, ZFC [7] necessitates a restriction. This is eventually a very important restriction: it amounts to the cutting off of a tail of a convergent sequence $\mathcal{S}$. [8] The purpose of the axiomatization is to make the tail as small as possible. In the ex. of the footnote, the tail may soon be made invisible: but still only fm points lie outside it, while the tail contains infinitely many. From this perspective, the building of ever 'stronger' axiomatizations is absurd, because the part omitted is always the greater. [9]

It is with the presumption of well-order, above, that the AC enters. For the earlier treatment here of fast-growing arithmetic functions this intervention is unnecessary, since the diagonalization of Ch. VII uses only those generalized devices of Ch. VI which may be built from the logically weaker principle of sequential order (IV, §3.8, defn. 38·2). It should be pointed out, however, that the restriction discussed above applies equally to an axiomatization (such as ZF) of the theory of sequential order.

In particular, in the case of (segments of) CANTOR's 'second number class', there is a 1-1 correspondence between deductions (as existence proofs of fixed points [ref. the final section of Ch. II, §2.18]) and descriptions (of formal 'ordinal' notations) for any given first-order theory $\underline{\mathbf{T}}$. The descriptive power of $\underline{\mathbf{T}}$ fails at the 'proof-theoretic ordinal' of $\underline{\mathbf{T}}$. $\underline{\mathbf{T}}$ need not be a theory of sets, and may e. g. be $\underline{\mathbf{N}}$. Thus the existence of the initial polynomial corresponds to the sentence '$\mathbf{1} \in \mathbf{N}$'; given the existence of $\alpha \in \mathbf{EP}$, that of α^+ corresponds to an inference by modus ponens; and given a cfs $\alpha_1, \prec \alpha_2 \prec \ldots$ in **EP**, that of $\alpha = \text{diag}_i \alpha_i$ corresponds to an application of the induction schemata. So by the cumulative proof of VI, §2.7, $\underline{\mathbf{N}}$ describes every $e \in \mathbf{EP}$. But $\underline{\mathbf{N}}$ cannot describe ε. Using the function symbol '$\cdot$', x^n may be defined from the induction schemata; but in the absense [10] of a function symbol

---

[7] Viz. the axioms of ZERMELO-FRAENKEL set theory, plus the AC.

[8] $\mathcal{S}$ may be thought of as $x_1 < x_2 < \ldots$ in $\mathbf{R}$, with $\sup \mathcal{S} = x < \infty$, so that a tail of $\mathcal{S}$ may be identified with a remainder.

[9] This discussion is independent of the '$\aleph_1 = \mathbf{c}$' problem which arises from CANTOR's conception of the linear continuum as a set of points.

[10] As with classical Greek mathematics.

for exponentiation, $x \star n$ is not definable in $\underline{\mathbf{N}}$, while this operation is required, to provide a fundamental sequence for ε.

The GOODSTEIN-SKOLEM hierarchy is, however, at the 'low' end, and the preceding objections are not the seat of the problem. The descriptions fail to produce an explicit member of the described type. In particular, the $T \in \mathfrak{T}$ have profound combinatorial properties quite omitted by the description; similarly, $e \in \mathbf{EP}$ is more than merely a featureless object determining a certain type: it is an arithmetic object which, e.g. could possibly posess infinitely many prime values

$$e(n_1),\ e(n_2),\ e(n_3),\ \ldots\ .$$

The hierarchy is introduced here in the first place as a domain of arithmetic and combinatorics, and in the second place as a means of possibly simplifying or getting rid of some of the obscurities which devalue modern logic.

8.4.3 Note on Functions and Functionals

Let $f : \mathrm{N} \to \mathrm{N}$ be a function on the natural numbers. Then taking as usual, functions as functionals of type 1, f defines an object $\langle 1, f\rangle$ of type 1:

$$(\forall x \in \mathrm{N})\langle 1, f\rangle(x) = f(x),$$

and an object $\langle 2, f\rangle$ of type 2:

$$(\forall g : \mathrm{N} \to \mathrm{N})\langle 2, f\rangle(g) = f,$$

and generally, for all objects ϕ of type n :

$$\langle n+1, f\rangle(\phi) = \langle n, f\rangle.$$

Indeed it is evident that a definition may be found for any constant functional $\langle e, f\rangle$ where e is a polynomial in such a class as **EP**. Thus, with regard to type, f may be viewed as any object from a function upwards. Furthermore, any $n \geq 1$ defines a function $\mathbf{n}$ on N, the constant function identically equal to n. Thus, while a function cannot be a number, a number can be a function, and generally it may be nearer the truth to speak of the 'minimum type' rather than the 'type' of an object, or to understand 'type' as meaning 'minimum type'.

Notation.

Many notations adopted for **EP** may be taken for larger classes without difficulty. E.g. if $e \in \mathbf{E2}$, and $f = e^+$, then e is written f^-.

8.4.4 The Hierarchy E2

By VII, §3.4, the binary operations $\Delta_e(x, y)$ define an increasing unary operator on functions f on the natural numbers: the transform of f is taken to be $\lambda x\ \cdot$

$\Delta_e\{x, f(x)\}$.[11] Thus, for example, $\Delta_\omega\omega = \mathfrak{A}$, ACKERMANN's function. A formal definition of the class **E2** follows:

DEFINITION 8.10 *1.*

2. **E2**.1 $\quad \mathbf{1} \in \mathbf{E2}$

3. **E2**.2 $\quad e, f \in \mathbf{E2} \Rightarrow e + f \in \mathbf{E2}$

4. **E2**.e $\quad e \in \mathbf{EP}\ \&\ \mathbf{3} \preceq e\ \&\ f \in \mathbf{E2} \Rightarrow \Delta_e f \in \mathbf{E2}$.

Remark. The problem discussed in §4.1 prevents the substitution of '**E2**' for '**EP**' in 10·3: the defn. of the Δ_e succeeds only if it is given that e varies over a sequentially ordered system.

Examples.

1. For $e = \mathbf{3}$, $f \in \mathbf{E2} \Rightarrow \lambda x \cdot x^{f(x)} \in \mathbf{E2}$, so $\mathbf{EP} \subseteq \mathbf{E2}$, and the higher operations of closure establish $\mathbf{EP} \subset \mathbf{E2}$ (and $\mathbf{EP}^\star \subset \mathbf{E2}$).
2. The rule **E2**.ω ensures $\mathfrak{A} \in \mathbf{E2}$.
3. By VII·20, the polynomial functions of **E2** are all si.
4. The td's (II, §2.15) determine a proper subclass $\mathcal{X}$ of **E2** : let $e \in \mathbf{EP}$ be the outline of a td. Suppose the occurrence of x in $e(x)$ pinpointed by

$$e(x) = \quad \Sigma_1 + \underset{\therefore}{x^\alpha} + \Sigma_2$$

has inscribed on the right, the numeral i. Then using the classical ACKERMANN function, e may be transformed to the higher polynomial

$$\widetilde{e}(x) = \quad \Sigma_1 + \underset{\therefore}{\Delta_{\mathbf{i}}\{\alpha\}} + \Sigma_2.$$

If such a transform is now effected, for all remaining occurrences of x in $e(x)$, the result is some F in **E2**, and $\mathcal{X}$ may be taken as the class of all such F, $e \in \mathbf{EP}$.

8.4.5 Classification in E2

As in Ch. II, §2, the polynomials of **E2** may be classified into initial, successors and diagonals. Furthermore if $e = \alpha + \beta$; $\alpha, \beta \in \mathbf{E2}$, and α is diagonal and β is constant, then α is the *diagonal part* and β the *constant part* of e. The identity polynomial ω is the *initial diagonal* of **E2**. As a class, the diagonals of **E2** are denoted by Diag**E2**.

[11] Ref. the remark of VII, after thm. 22. The transform might as appropriately be taken to be $\lambda x \cdot \Delta_e\{n, f(x)\}$, for any fixed integer $n \geq 2$. The §4·4 convention ensures standard inclusions.

8.4.6 Reductions

Let $\mathbf{3} \preceq e$ in **EP**. Then

1. $\Delta_e \mathbf{1} = \omega$.

2. For Suc(e) and $f \in \mathbf{E2}$, Suc(f) $\Rightarrow \Delta_e f = \Delta_{e^-} \Delta_e f^-$. And by defn. 10, $\Delta_e f^- \in \mathbf{E2}$.

3. For Diag(e), $\Delta_e \mathbf{n} = \Delta_{(e,n)} \mathbf{n}$, $n \in \mathrm{N}$.

Therefore, in defn. 10, an operator of successor index $\succ \mathbf{3}$ need only act on a diagonal argument; and any operator of index $\succ \mathbf{3}$ need only act on polynomials $f \succeq \omega$. In the following, these conventions will be adhered to strictly. The '$\star$' notation will be dropped, in favour of square brackets [12]:

NOTATION 8.4.1 $(\forall e \in \mathbf{EP})(\forall f : \mathrm{N} \to \mathrm{N})\ [e, f] = \lambda x \cdot \Delta_e\{x, f(x)\} = \Delta_e f.$

□

Thus, for example, the number $[\mathbf{4}, \varepsilon](7) = 7 \star \varepsilon(7)$. Note that $[\mathbf{3}, \mathbf{0}] = \mathbf{1}$; $[e, \mathbf{1}] = \omega$, $e \succ \mathbf{1}$.

THEOREM 8.11 *Every polynomial of* **E2** *is either initial, the sum of* ≥ 2 *polynomials of* **E2**, *or of the form* $[e, f]$, *where* $e, f \in \mathbf{E2}$, $\mathbf{3} \preceq e$.

□

8.4.7 Calculation Sequence

Let $\mathbf{3} \prec e$ in **EP**, and let f be any si function. As in McBETH [13] §2, consider the calculation of $[e, f](x)$. Let

$$\mathcal{S} = (e_i)_{i=1,2,\ldots,T=T(e,x)},\quad e_T = \mathbf{1},$$

be the sequence of functions and

$$(y_i)_{i=1,2,\ldots,T}$$

the sequence of numbers obtained in the calculation of $\Delta_e f(x)$, of VII, §3.3, (I) and (II). Note that these depend on x; in particular, T is si in x. In the proof of VII, thm. 20, it is established that (y_i) is non-decreasing.

[12] There will be no clash with the entirely different usage of Ch. IV.

[13] 'The class **E2**', manuscript, accepted for publication under title 'The class **GEP**-2', *Zeitschr. math. Logik Grundl. Math.*, 1980.

LEMMA 8.12

$$\alpha \prec e \text{ in } \mathbf{EP} \Rightarrow (\exists N > 1)(\forall x \geq N)\, \alpha \in \mathcal{S}.$$

Proof. By V·3, $\alpha \tilde{\lhd} e$, i.e.

$$\alpha = e_{(K)} \lhd e_{(K-1)} \lhd \ldots \lhd e_{(2)} \lhd e_{(1)} = e, \tag{I}$$

for certain $e_{(K-1)}, \ldots, e_{(2)}$ in **EP**. Let $N \geq 1$ be the largest index of a canonical term occurring in a $\lhd -$ step of (I). Thus for some k in the range $1 \leq k < K$,

$$e_{k+1} = (e_k, N).$$

Now since (y_i) is non-decreasing, it follows by VII, lemma 19, that

$$\beta \in \mathcal{S} \Rightarrow \begin{cases} \beta^- \in \mathcal{S} & \text{if } \mathrm{Suc}(\beta) \\ (\beta, j) \in \mathcal{S},\ j = 1, 2, \ldots, y_1 & \text{if } \mathrm{Diag}(\beta). \end{cases}$$

Thus, for $x \geq N$, i.e. $f(x) = y_1 \geq N$, $\{e_{(k)}\} \subset \mathcal{S}$. Therefore (I) gives $\alpha \in (e_i)$.

□

Now let $\mathbf{3} \preceq \alpha \prec e$ in **EP**, and consider $[e, f]$, where $\omega \preceq f$ in **E2**.

LEMMA 8.13 $(\exists N \geq 1)(\exists \beta \in \mathcal{N})_{\beta \text{ si}}(\forall x \geq N)\, [e, f](x) = [\alpha, \beta](x)$.

Proof. By the previous lemma, for $f(x)$ suitably large, α occurs in the sequence $(e_t)_{t=1,2,\ldots,T}$ of functions obtained in the calculation of $[e, f](x)$. But f is si, thus for x suitably large, say $x \geq N > 1$, α occurs. Let $f^{(1)} = f$; and for $t = 1, 2, \ldots, T-1$,

$$f^{(t+1)} = \begin{cases} [e_t, f^{(t)-}] & \text{if } \mathrm{Suc}(e_t) \\ f^{(t)} & \text{if } \mathrm{Diag}(e_t). \end{cases}$$

The upper line following the bracket is to be interpreted:

$$(\forall x \geq N)\, f^{(t+1)}(x) = [e_t, f^{(t)-}](x)$$

if $\mathrm{Suc}(e_t)$. Thus by notation 4.1

$$[e, f](x) = [e_t, f^{(t)}](x), \quad t = 1, 2, \ldots, T.$$

As in the previous lemma, $T = T(x)$ is a si function of x. Now let $x \geq N$. Then for a certain value $K = K(x)$, which increases strictly with x, and which lies in the range $1 \leq K \leq T$, $e_K = \alpha$. $1 < K$ since $\alpha \prec e$; $K < T$ since $\alpha \succeq \mathbf{3}$. Now put $\beta(x) = f^{(K)}(x)$. Then

$$[e, f](x) = [\alpha, \beta](x). \tag{I}$$

$[e, f] \in \mathbf{E2}$ is si. By VII·20·2, $\lambda y \cdot \Delta_\alpha(x, y)$ is si. Therefore by equation (I), using VII·23, it follows β is (eventually) si.

□

Remark. Since $\alpha \prec e$, and $[\alpha, \beta]$ is eventually equal to $[e, f]$, it follows by Ch. VII, corol. 23 that $f \prec \beta$.

8.4.8 Connexity of the Relation $\prec$ on $\{\mathbf{E2}\}^2$

Rate of Growth in E2

E2 is closed under an **EP**–system of operations. By II·17, followed by VIII·23, in terms of growth-amplification power, these are spanned by

$$\Delta_{\omega},\ \Delta_{\omega^{\omega}},\ \Delta_{\omega^{\omega^{\omega}}},\ \ldots\ .$$

Now it is evident that a single application of $\Delta_{\omega\star(n+1)}$ amplifies very much more than than fm applications of $\Delta_{\omega\star n}$; $n = 1, 2, 3, \ldots$. From this alone it is clear that $(\mathbf{E2}, \prec)$ is spanned by the sequence of polynomials

$$\Delta_{\omega}\omega,\ \Delta_{\psi}\omega,\ \Delta_{\omega\star 3}\omega, \ldots\ .$$

While this gives an idea of the place of **E2** in terms of growth, there remains the problem of proving that the very many polynomials generated at the various levels of growth are in fact linear with respect to $\prec$.

Indexed Definition Tree

Some preliminary notions are needed. Consider the definition of the polynomials f in **E2**. By defn. 10 ($\exists R \geq 1$) and functions $e_R, \ldots, e_2, e_1 \succeq \mathbf{3}$ in **EP**, and $f_R, \ldots, f_2, f_1$ in $\mathbf{E2_0}$ such that

$$f = \sum_{i=R}^{1} [e_i, f_i]. \tag{I}$$

In case $f_i = \mathbf{0}$, it may be assumed $e_i = \mathbf{3}$. Returning now to the usage of defns. 2·1,

DEFINITION 8.14 *1. The relation* Sc *on* $\{\mathbf{E2}\}^2$ *is defined by*

$$f_i \text{ Sc } f,\ \ i = 1, 2, \ldots, R.$$

The transitive closure of Sc *is denoted by* $\widetilde{\text{Sc}}$.

2. *For* $r = 1, 2, \ldots, R, \Delta_{e_i}$ *is called the* associated operation *of* f_i *in the expression* (I).

3. *The* non-indexed defn. tree T'_f *of* $f \in \mathbf{E2_0}$ *is defined inductively on order:* $T_0 = \{\ \}$; *if* f *takes the form (I), and* $T_i = T_{f_i}$ *have been defined,* $1 \leq i \leq R$, *then*

$$T_f = \{T_1, \ldots, T_R\}$$

(Where the script brackets $\{,\}$ *are used to indicate that the left-to-right order of the branches (contrary to Ch. IV, §2) is* not *taken into consideration).*

4. *The* (indexed) defn. tree T_f *of* $f \in \mathbf{E2}$ *is obtained from* T'_f *by replacing each vertex* $v(\alpha)$ *for which* $\alpha\ \widetilde{\text{Sc}}\ f$ *by the pair* $[e, \alpha]$, *where* Δ_e *is the associated operation of* α.

□

T_f is finite, since f is built up in a finite number of steps from the initial polynomial. A part of T_f may thus take the form

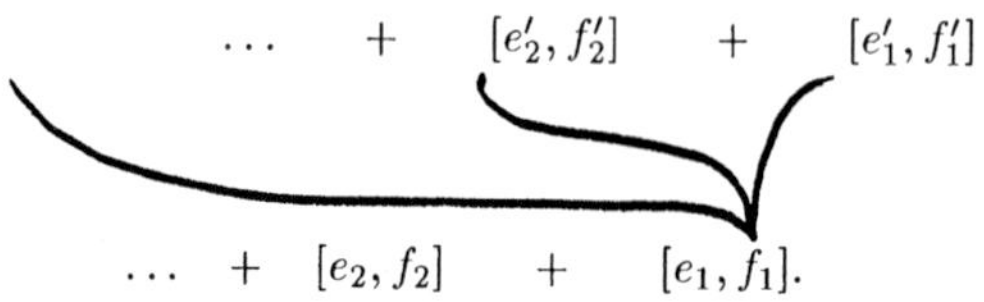

[14]

DEFINITION 8.15 *Let* $\mathbf{E2}_0 = \{\mathbf{1}, \mathbf{2}, \mathbf{3}, \ldots\}$.[15] *For* $N \geq 0$ *let* $\mathbf{E2}_{N+1}$ *contain exactly those forms*

$$f = \sum_{i=R}^{1} [e_i, f_i],$$

for which $\mathbf{3} \preceq e_i$ *in* **EP**, *and* $f_i \in \mathbf{E2}_N \cup \{\mathbf{0}\}$, $i = R, \ldots, 2, 1$. *The smallest* N *for which* $f \in \mathbf{E2}_N$ *is the* order (of defn.) of f in **E2**.

□

Example. $\mathbf{E2}_1$ contains all terms $[\varepsilon_m, n]$, $m, n \geq 1$.

LEMMA 8.16 $(\forall e, \alpha \succeq \mathbf{3} \text{ in } \mathbf{EP})(\forall f, \beta \in \mathbf{E2})$

$$\{\text{Diag}(f) \vee \text{Diag}(e)\} \ \& \ \alpha \prec e \Rightarrow \{\beta \prec [e, f] \Rightarrow [\alpha, \beta] \prec [e, f]\}.$$

Proof. This relies on the $\prec$ – comparability of polynomials in **EP**. Consider $T = T_\gamma$, where $\gamma = [\alpha, \beta]$. If T contains no vertex of index $\mu \succeq e$, then $\gamma \prec [e, f]$. Thus, suppose T contains such a vertex. Each chain, starting at the γ–vertex, reaches, in a finite number of steps, either a zero function vertex, or a pair $[\mu, h]$, where $e \preceq \mu \in \mathbf{EP}$, and $h \in \mathbf{E2}$. Let $[\mu_1, h_1], \ldots, [\mu_s, h_s]$ be the set of such pairs. Let $[\mu, h] = \max_{i,\prec}[\mu_i, h_i]$. Let $C(h_i)$ be the unique chain connecting the γ– vertex

---

[14] Furthermore, suppose $f \in \mathbf{E2}$, $e \in \mathbf{EP}$, $\mathbf{3} \prec e$, with either Diag(f) or Diag(e). Then $[e, f]$ has no decomposition

$$[e, f] = [\alpha, \beta],$$

where $\alpha \prec e$ is in **EP**, and $\beta \in \mathbf{E2}$. And the expression $[e, f]$, cannot be obtained as a (multiple) finite sum of terms in **E2**; it therefore follows that the defn. tree T_f of $f \in \mathbf{E2}$ is *unique* (c.f. footnote 4).

[15] The notation $\mathbf{E2}_0 \simeq \mathbf{N}$ should not be confused with the (boldface subscript) notation, used more frequently in this Ch., viz. $\mathbf{E2}_{\mathbf{0}} = \mathbf{E2} \cup \{\mathbf{0}\}$.

to $[\mu_i, h_i]$. Let ξ be the highest index [16] occurring in the vertices in $C(h_i)$ below $[\mu_i, h_i]$, $i = 1, 2, \ldots, s$, as indicated in the figure.

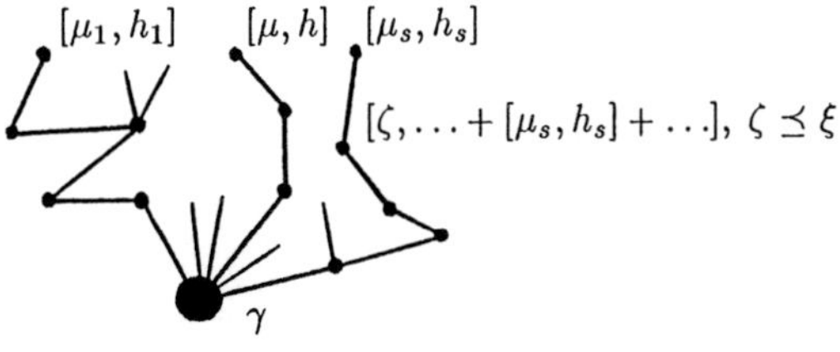

The growth of γ will be roughly approximated, using the function $[\mu, h]$ and the operation defined by ξ. Let $\gamma_0 = [\mu, h]$, and $\gamma_{t+1} = [\xi, \gamma_t]$, $t \geq 0$. Then $(\exists t > 0)\ \gamma_{t-1} \preceq \gamma \prec \gamma_t$. Now suppose $\mu = e$. Then $\gamma_t \preceq [e, h + \mathbf{t}]$. If Suc$(e)$, then Diag$(f)$: now $h \prec f$, and since $f, h \in \mathbf{E2}$, $h + \mathbf{t} \prec f$, and

$$[\alpha, \beta] \prec \gamma_t \prec [e, f].$$

If Diag(e),

$$[\alpha, \beta] \prec \gamma_t \prec [e, h^+] \preceq [e, f].$$

Now suppose $e \prec \mu$. Then by lemma 13, $(\exists \text{si}\ \widetilde{h})\ [\mu, h] = [e, \widetilde{h}]$, where $\widetilde{h} \prec f$. Then $\gamma_t \preceq [e, \widetilde{h} + \mathbf{t}]$. If Suc$(e)$, then Diag$(f)$. Furthermore, from the form of $\widetilde{h}$ given by lemma 13, no function $\widetilde{h} + \mathbf{m}$, $m \geq 1$, is a diagonal of $\mathbf{E2}$. Therefore, $\widetilde{h} + \mathbf{t} \prec f$, and the result follows as in the case $e = \mu$, Suc(e). If Diag(e),

$$[\alpha, \beta] \prec \gamma_t \prec [e, \widetilde{h}^+] \preceq [e, f]\ ^{17}$$

Thus the property is established in all cases.

□

Examples. $\beta \prec \omega \star f \Rightarrow \omega^\beta \prec \omega \star f$, since Diag$(f)$. $[\omega, \omega] \prec [\omega, \omega^+]$; and for all $n \geq 1$, $\Delta_n[\omega, \omega] \prec [\omega, \omega^+]$.

THEOREM 8.17 *The relation $\prec$ is connected on* $\{\mathbf{E2}\}^2$.

Proof, by induction on the order N of $f \in \mathbf{E2}$. The proof follows the lines of that of thm. 7, which is referred to as 'the proof for **EP***'. In the initial case, the constant polynomals are $\prec$ – comparable. Let $N \geq 1$, and suppose for induction hypothesis that $\prec$ is connected on $\{\mathbf{E2}_N\}^2$. Consider terms $e^{(1)}, e^{(2)} \in \mathbf{E2}_{N+1}$

$$e^{(1)} = [e, f]; \qquad e^{(2)} = [\alpha, \beta] \tag{I}$$

[16] Note that $\xi = \alpha$ is possible.

[17] The last inequality is in fact strict, as $\widetilde{h} \notin \mathbf{E2}$.

where $e, \alpha \succeq \mathbf{3}$ are polynomials of **EP**, and f, β are polynomials of $\mathbf{E2}_N$.

If $e = \alpha$, by the i.h. the expressions (I) are $\prec$ −comparable. Thus suppose $e \neq \alpha$. By II·18 $\prec$ is connected on $\{\mathbf{EP}\}^2$, so the possibilities are

	$f \prec \beta$	$\beta \prec f$	$f = \beta$
$e \prec \alpha$	$\prec$		$\prec$
$\alpha \prec e$		$\succ$	$\succ$

The cases marked show the relation between the expressions (I). The remaining cases are dual, so it may be assumed $f \prec \beta$, and $\alpha \prec e$.

Each chain in T_β starting at the β−vertex reaches, in a finite number of steps, either a zero function vertex, or a pair (h, μ), where $e \preceq \mu$. By the convention of §4.5, both f and h are diagonal, unless e or μ are diagonal, in which case f and h may respectively be successors. Also by induction hypothesis, f and h are $\prec$ −comparable. Thus if no chain contains a pair $[\mu, h]$,

$$e^{(1)} \succ e^{(2)},$$

the 'height' of the RHS being less than that of the LHS. Thus, suppose at least one chain contains a pair $[\mu, h]$, and let the finite set S be the least containing all functions h which occur in such a pair. S is partitioned into

$$S_1 = \{h \in S : e = \mu\}$$

and

$$S^+ = \{h \in S : e \prec \mu\}.$$

The inductive step is established by 2 arguments, one based on each of S_1, S^+. The S_1−argument is essentially the S−argument of the proof for $\mathbf{EP}^\star$. If $e^{(1)} \succ e^{(2)}$ is proved, ignoring the functions of S^+, this is written $(\succ)_1$. If $e^{(1)} \succ e^{(2)}$ is proved, ignoring those of S_1, this is written $(\succ)^+$. To compare the expressions (I), in the first instance, consider the following argument, based on the class S_1. Suppose firstly $(\exists h \in S_1)\, f \preceq h$. Then by VII, thm. 20·2:

$$[e, f] \preceq [e, h] \preceq \beta \prec [\alpha, \beta].$$

It therefore may be assumed that

$$(\forall h \in S_1)\, h \prec f.$$

Now an expression $[\sigma, \rho]$, $\mathbf{3} \preceq \sigma$, in **EP**; $\rho \in \mathbf{E2}$, satisfies $[\sigma, \rho] = \omega^u$, where $u : \mathbf{N} \to \mathbf{N}$ is si, by lemma 13. And $h \prec f$ implies, by the expressions for h, f as sums of exponential terms ω^u, that the difference $d(x) = f(x) - h(x)$ is either constant or unbounded. Suppose therefore, secondly,

$$(\exists h \in S_1)(\exists t \in \mathbf{N})\, h = f - \mathbf{t}.$$

So f is a successor function, and it may be assumed that e is a diagonal. Let $\tilde{h} = \max_{\prec} S_1$. Then $\tilde{h}^+ \preceq f$, and using Ch. VII · 20

$$[\alpha,\beta] \prec \overbrace{[\xi[\xi,\ldots[\xi,}^{K}[e,\tilde{h}]]\ldots]]$$

for some $K \geq 1, \xi \prec e$ in **EP**, depending on the structure of T_β. But, since Diag(e), the RHS$\prec$ $[e,\tilde{h}^+] \preceq [e,f]$. Therefore $[\alpha,\beta] \prec [e,f]$. I.e. $(\succ)_1$. Suppose thirdly $(\forall h \in S_1) f - h \succeq \omega$, and let $C(h)$ be the unique chain connecting $[\alpha,\beta]$ to $[e,h]$. Let $C(h)$ contain n_h edges, and let $K = 1 + \max_{h \in S} n_h$. And of the finite set of indices occurring in the vertices of the chains $C(h)$, below each vertex $[e,h]$, let the $\prec$ − maximum be ξ. Then by VII·20,

$$[\alpha,\beta] \prec \overbrace{[\xi\ldots[\xi,}^{K}[e,\tilde{h}]]\ldots] \preceq [e,\tilde{h}+K] \prec [e,f]$$

i.e. $(\succ)_1$. This completes the argument based on S_1. Consider now the final argument, based on the class S^+. By induction hypothesis $(\forall h \in S^+)$ f and $[\mu,h]$, where μ depends on h, are $\prec$ −comparable. Suppose $(\exists h \in S^+)$ $f \prec [\mu,h]$. Then by lemma 16, since $e \prec \mu$,

$$[e,f] \prec [\mu,h] \prec [\alpha,\beta].$$

Thus suppose $(\forall h \in S^+)$ $[\mu,h] \preceq f$. Then

$$[\alpha,\beta] \prec \overbrace{[\xi,\ldots,[\xi,}^{K}[\mu,h]],\ldots,] \prec [e,f],$$

for some $\xi \prec e$ in **EP**, and some $K \geq 1$, i.e. $(\succ)^+$. This completes the argument based on S^+.

Therefore, taking together all cases, either

$$e^{(1)} \prec e^{(2)} \text{ or } \{(\succ)_1 \mathbin{\&} (\succ)^+\}.$$

Evidently the condition in brackets implies

$$e^{(1)} \succ e^{(2)}.$$

But any functions $f, \phi \in$ **E2** are sums of terms of the form (I). Thus collecting equal terms

$$f = \sum a_i[e_i,f_i], \quad \phi = \sum b_i[\alpha_i,\beta_i], \tag{II}$$

where $a_i, b_i \in \mathbf{N}; e_i, \alpha_i \in \mathbf{EP} - \{\mathbf{1},\mathbf{2}\}; f_i, \beta_i \in \mathbf{E2}_N \cup \mathbf{0}$. By the $\prec$ − comparability for (I), it follows that the terms of (II) may be taken in strictly $\prec$ −decreasing order. So $f \prec \phi, f = \phi$ or $\phi \prec f$ by lexicographic comparison. So it follows by induction that the higher exponential polynomials of **E2** are $\prec$ −pairwise comparable.

□

Remark. From the proof of thm. 17, it follows that

$$e \neq \alpha \Rightarrow [e,f] \neq [\alpha,\beta].$$

8.4.9 Normal Form of Functions in E2

COROLLARY 8.18 *$f \in \mathbf{E2}$ has a (unique) expression*

$$f = \sum_{i=r}^{1} a_i[e_i, f_i] + \mathbf{a},$$

where $e_1, e_2, \ldots, e_r \in \mathbf{EP} - \{\mathbf{1}, \mathbf{2}\}$; $f_1, f_2, \ldots, f_r \in \mathbf{E2_0}$;

$$[e_1, f_1] \prec [e_2, f_2] \prec \ldots \prec [e_r, f_r];$$

and $a_1, a_2, \ldots, a_r \geq 1$, with a, r as in corol. 8.

□

It may be noted that, switching notations, $f = \sum a_i \Delta_{e_i} f_i + \mathbf{a}$.

As with the class $\mathbf{EP}^\star$, functions of **E2** may be given the second canonical expression

$$f = \sum_{i,j} a_{ij}[e_{ij}, \phi_i] \qquad \text{(I)}$$

where the ϕ_i are the distinct f_i; where

$$\phi_1 \prec \phi_2 \prec \ldots \prec \phi_s;$$

and the indices e_i are sd in j wrt $\prec$. By thm. 17, the terms of (I) determine a permutation of those of the *term normal form* of the corol. (I) is the *exponent normal form* of f.

8.4.10 Well-Foundedness of $(\mathbf{E2}, \prec)$

THEOREM 8.19 **E2** *contains no endlessly $\prec$ –descending sequences of functions.*

Proof, by induction on $N \geq 0$, in $\mathbf{E2} = \cup_N \mathbf{E2}_N$. By thm. 17, $f, g \in \mathbf{E2}$ are $\prec$ –comparable. The argument of Ch. IV, §3.8 may now be applied: $\mathbf{E2}_0 \simeq \mathbf{N}$ is well-founded. Let $N \geq 0$, and suppose for induction hypothesis that $\mathbf{E2}_N$ has been proved well-founded. Consider a decreasing sequence $\mathcal{S}$ in $\mathbf{E2}_{N+1}$:

$$\ldots \prec f_3 \prec f_2 \prec f_1,$$

where f_i has term normal form $\sum_j a_{ij}[e_{ij}, f_{ij}]$, $i = 1, 2, 3, \ldots$.

Then corresponding to the sequence of principal branches

$$\ldots \prec T_{i(2),1} \prec T_{i(1),1}$$

in the argument of IV, §3.8, thm. 39, there is a sequence of leading terms

$$\ldots \prec [e_{i(2),1}, f_{i(2),1}] \prec [e_{i(1),1}, f_{i(1),1}].$$

For convenience let these be written

$$\ldots \prec [e_2, f_2] \prec [e_1, f_1], \tag{I}$$

and consider the sequences (e_n), (f_n), $n = 1, 2, 3, \ldots$. Suppose the sequence (e_n) contains an endless $\prec$ –increasing subsequence $e_{n(1)} \prec e_{n(2)} \prec e_{n(3)} \prec \ldots$, so that from (I),

$$\ldots \prec [e_{n(2)}, f_{n(2)}] \prec [e_{n(1)}, f_{n(1)}].$$

Now consider the first 2 expressions

$$\alpha = [e_{n(1)}, f_{n(1)}] \quad \beta = [e_{n(2)}, f_{n(2)}].$$

As in lemma 13, $f_{n(1)} \preceq f_{n(2)} \Rightarrow \alpha \prec \beta$, contrary to assumption. Therefore by thm. 17, $f_{n(2)} \prec f_{n(1)}$. And generally the argument shows that the sequence $\ldots \prec f_{n(2)} \prec f_{n(1)}$ is endless. But this is impossible, as by induction hypothesis $\mathbf{E2}_{(N)}$ is well-founded. Therefore: (e_n) contains no endless *ascending* subsequence. Furthermore, by IV·48, (e_n) contains no endlessly descending subsequence. Therefore (e_n) contains only finitely many distinct terms. By the same argument, but with (e_n) and (f_n) exchanged, (f_n) contains only fm distinct terms. But all the terms of (I) are distinct, therefore the sequence is finite. Hence it is possible to proceed with the argument of Ch. IV, and establish that $\mathcal{S}$ contains only fm terms; thus it is proved $\mathbf{E2}_{(N+1)}$ is $\prec$ –well founded. So by induction, **E2** is $\prec$ –well-founded.

□

COROLLARY 8.20 **EP*** *is* $\prec$ *–well-founded.*

□

ACKERMANN's Consistency Proof

In a series of papers from 1950 to 1953 (see KNEEBONE [75], Appendix, §2, p386) Wilhelm ACKERMANN extended the reduction method of GENTZEN (I, §3.5) to a proof of the consistency of 1st order analysis, $\underline{\mathbf{R}}$. The correlation, with derivations in $\underline{\mathbf{R}}$, uses the initial segment corresponding to the first fixed point, η, in CANTOR's ε–sequence, viz. the sequence of fixed points of similarity exponentiation (I, §2.18). It may be shown (see [94], §3, p40; and thm. 20 above) that $(\mathbf{EP}^\star, \prec)$ is similar to this segment. The sequentially ordered system **EP*** may thus be used to constructivize ACKERMANN's proof.

8.4.11 Particular Functions of E2

E2 is greatly more comprehensive than **EP** (or **EP***), since for every **EP**-polynomial, **E2** has an operation of closure. By VII, thm. 14, and by example 2. following defn. 10, $\mathbf{E2} \not\subseteq \mathcal{P}$, the class of primitive recursive functions. Let $f \in \mathbf{EP}$, with $\mathbf{3} \prec f$, and let $\mathbf{E2}(e)$ denote the subclass of **E2** obtained by restricting closure to the rules $\varepsilon \cdot \alpha$, $\alpha \prec e$. Then $\mathbf{E2(4)} = \mathbf{EP}$; $\mathbf{E2(5)} = \mathbf{EP}^\star$; $\mathbf{E2(n^+)} \subset \mathcal{E}^n$, $n = 3, 4, 5, \ldots$; and $\mathbf{E2}(\omega) \subset \mathcal{P}$; $\mathbf{E2}(\omega^+) \not\subseteq \mathcal{P}$, where the classes $\mathcal{E}^n$ are those presented by GRZEGORCZYK, discussed in Chs. I, §§3.2, 3.5; and VII, §2.2, rem. 2. The function $\mathfrak{d}$ of VIII, §3·5 and its iterations are dominated by polynomials of $\mathbf{E2}(\omega^2 + \mathbf{1})$.

The ACKERMANN-TAKEUTI Class E2(ω)

Let $F \in \mathbf{E2}(\omega)$. Then if each application of rule $\mathbf{E2} \cdot \mathbf{i}$, $3 \leq i$, is replaced by $\mathbf{E2} \cdot \mathbf{3}$, F collapses to some $e \in \mathbf{EP}$. Overlooking the left indices, if now e is taken as the outline of a td (whose right indices are those just replaced by **3**), and e is $\mathfrak{X}$–transformed as in ex. 4 after defn. 10, then F is recovered. Therefore the class of the ex. $\mathcal{X} = \mathbf{E2}(\omega)$, the class of TAKEUTI diagrams, interpreted as higher exponential polynomials. Note that the order of td of $F = \max_i$ s.t. $\{\mathbf{E2} \cdot \mathbf{i}$ is used in defn. of $F\}$.

The LÖB-WAINER Hierarchy

The operations $\Delta_e, e \in \mathbf{EP}$ closely resemble the LÖB-WAINER arithmetic functions F_α, where α is an 'ordinal' less than the proof-theoretic order type of $\underline{\mathrm{N}}$. Closure of the F_α under substitution, etc. yields an unordered class $\mathcal{X}$ properly containing, but spanned by **E2**. It should be noted, however, that **E2** does not in any way rely on an assumed well-founded system which itself may rely on the axiom of choice.

The Diagonal of E2

The diagonal Δ of VII, lemma 24 is an example of a function which dominates every polynomial of **E2**; and by the discussion following V, defn. 9, Δ may be taken as the supremum of this hierarchy.

8.4.12 KREISEL's Meta-Theorem on E2

Using induction α–induction over **N**, (Ch. VI, §5) KREISEL [80] [18] characterized the subsystem $\mathcal{K} \subset \mathcal{R}$ of I, §3.4.2, obtained by taking the closure of **E2** under the operations of primitive recursion, as containing precisely those arithmetic functions whose $\underline{\mathrm{N}}$–representing formula $\mathrm{A}(x, y)$ satisfies

$$\vdash (\forall x)(\exists y)\ \mathrm{A}(x, y).$$

[18] Akiko KINO [70], *J. Math. Soc. Japan*, **20**(3) (1968) 456-476, and ERDÖS and MILLS [34], J. Comb. Theory, Ser. A, **30** (1981) 53-70 contain discussion of this result.

The *proof* of KREISEL's result may be sketched as follows: by a standard meta-theorem of HILBERT and BERNAYS ([64], *Grundlagen der Mathematik*, Vol II) for each $n = 1, 2, 3, \ldots$, generalized induction determined by $\omega \star n$ (see II·2; and the note of VI, §2.6) has both expression and proof in $\underline{\mathbf{N}}$. On the other hand, by another result of GENTZEN [42], generalized induction (viz. determined by ε) is demonstrably unprovable in $\underline{\mathbf{N}}$. (In either case, the generalized induction may be correlated with α–induction, as shown in the critique of VI, §5.1). It therefore follows that

$$B \equiv (\forall x)(\exists y)\; F(x) = y$$

satisfies

$$\vdash B$$

if $F = \Delta_{\varepsilon_n}$; while on the other hand

$$\nvdash B \text{ and } \nvdash \neg B$$

for $F = \Delta_\varepsilon$ (before VII·24). The result now follows by taking the functions primitive recursive in the Δ_{ε_n}.

Examples

Since the length of the sequence (T_i) of VII·38 determines a function of $n = \mid T_1 \mid$ which eventually dominates every $e \in \mathbf{E2}$, by KREISEL's result it follows that (the 1st order expression of) this theorem is $\underline{\mathbf{N}}$–undecidable.

8.5 Canonical Sequences for Diagonals of E2

Remarks on Fundamental Sequences

Corollary 16 of Ch. II establishes that **EP** is *not* closed under diagonalization; and the function Δ cited at the end of §4.11 establishes a similar result for **E2**. However, the 'new' closure operations are diagonalizations in **E2** : if $e \in \mathbf{EP}$ is diagonal, then for nc $\alpha \in \mathbf{E2}$, $\beta_n = \lambda x \cdot \Delta_{(e,\alpha(n))}\{x, \alpha(x)\}$,

$$\beta = \Delta_e \alpha = \mathrm{diag}_n\, \beta_n,$$

where the fundamental sequence $\beta_1, \beta_2, \beta_3, \ldots$ lies in **E2**.

As in III, §2.2, an assignment functional

$$\Omega : \mathrm{Diag}\mathbf{E2} \times \mathrm{N} \to \mathbf{E2}$$

is obtained inductively. Ω is chosen on the basis of the remarks of §3, but so as to include III, §2.2 as a special case. As usual, the canonical term $\Omega(e, n)$ is written (e, n), $n = 1, 2, 3, \ldots$. Upper case letters $E, F, \ldots$ may be used to denote exponentials of **E2**. Let $F \in \mathbf{E2}$ be diagonal, in the normal form of corol. 18:

$$F = [e_R, f_R] + \ldots + [e_1, f_1], \quad R \geq 1,\ e_i \succeq \mathbf{3};$$

and write $e_1 = e,\ f_1 = f$, so that

$$F = \Sigma + [e, f];\ e \in \mathbf{EP},\ e \succeq \mathbf{3};\ f \in \mathbf{E2};\ \Sigma \in \mathbf{E2_0}.$$

Furthermore, let $f = [\nu_S, g_S] + \ldots + [\nu_1, g_1]$ in $\mathbf{E2_0}$, with $\nu_1 = \nu,\ g_1 = g$, so that if $S = 1$,

$$f = [\nu, g].$$

DEFINITION 8.21 *For* $n = 1, 2, 3, \ldots$

$$(F, n) = \Sigma + ([e, f], n),$$

where

$$([e,f],n) = \begin{cases} \text{(i) } \mathbf{n} & \textit{if } f = \mathbf{1}; \\ \text{(ii) } n\omega^{f^-} & \textit{if } \mathrm{Suc}(f),\ e = \mathbf{3}; \\ \text{(iii) } n \cdot f & \textit{if } \mathrm{Diag}(f),\ e = \mathbf{3},\ S = 1\ \nu \succeq \mathbf{4}, \\ & g \succeq \omega \textit{ if } \nu = \mathbf{4};\ g \succ \mathbf{1} \textit{ if } \nu \succ \mathbf{4}; \\ \text{(iv) } [(e,n),[e,f^-]] & \textit{if } \mathrm{Diag}(e),\ \mathrm{Suc}(f); \\ \text{(v) } \overbrace{[e^-, \ldots, [e^-}^{n}, f] \ldots] & \textit{if } \mathrm{Suc}(e),\ e \succ \mathbf{3},\ S = 1,\ e \prec \nu; \\ & g \succ \omega \textit{ if } \nu = e^+; \\ \text{(vi) } [(e,n), f] & \textit{if } \mathrm{Diag}(e) \textit{ and } f = \omega, \textit{ or } (S = 1\ \&\ e \prec \nu); \\ \text{(vii) } [e, (f, n)] & \textit{in all remaining cases.} \end{cases}$$

Examples. Corresponding to the cases of the defn.,

1. $F = [\psi, \mathbf{1}] = \lambda x{\cdot}\Delta_\psi\{x, \mathbf{1}(x)\} = \omega$; $(F, n) = \mathbf{n}$. $F = [\varepsilon, \omega] = \lambda x{\cdot}\Delta_\varepsilon(x, x)$; $(F, n) = [\varepsilon_n, \omega]$.

2. $F = \Sigma + \omega^{\varepsilon^+}$; $(F, n) = \Sigma + n\omega^{\varepsilon}$. Thus, $F = \mathrm{Diag}_n\ (F, n)$.

3. $(\omega^\varepsilon, n) = n\varepsilon$; $(\omega^\eta, n) = n\eta$; and generally $(\omega^{[\nu,\omega]}, n) = n[\nu, \omega]$, $\nu \in \mathbf{EP},\ \nu \succeq \mathbf{4}$. In every instance, $\mathrm{diag}_n\ (F, n) \prec F$.

4. $F = [\omega, \omega^+]$. Then $[(\omega, n), \omega^+] = \Delta_{\mathbf{n}}(\omega^+) \prec \Delta_\omega\omega \prec \Delta_{\mathbf{n}}\Delta_\omega\omega = (F, n)$. Thus $\mathrm{diag}_n\ (F, n) \prec F$.[19]

5. $F = [\mathbf{4}, [\mathbf{5}, \omega]] = \omega \star \eta$; $(F, n) = \omega \star (\eta, n)$.

6. (a) $F = [\omega, \omega] = \mathfrak{A}(x)$ (ref. VII, §2);

$$\begin{aligned} (F, n) &= [(\omega, n), \omega] = \lambda x \cdot \mathfrak{A}_n(x, x), \\ & \quad \text{the } n\text{th ct of ACKERMANN's function} \\ &\succ \lambda x \cdot \mathfrak{A}_n(x, n) = [\omega, \mathbf{n}]. \end{aligned}$$

[19]Note that the generalized polynomial $\mu(x) = \mathfrak{A}_x\{x, \mathfrak{A}(x)\}$, for which

$$(F, n) \prec \mu \prec F,\ n = 1, 2, 3, \ldots$$

is unobtainable via the rules of defn. 10.

(b) $g = 2\omega,\ F = [\omega, [\omega^2, g]]$,

$$\begin{aligned}(F, n) &= [\mathbf{n}, [\omega^2, g]] \\ &\succ [\omega, [\omega^2, \omega + \mathbf{n}]] \\ &= [\omega, [\omega^2, (g, n)]].\end{aligned}$$

7. (a) $F = \omega^{\eta+\varepsilon+\psi}$; $(F, n) = \omega^{\eta+\varepsilon+\omega^n}$.
   (b) $F = [\omega, \omega^2]$; $(F, n) = [\omega, n\omega] \succ [\mathbf{n}, \omega^2]$.
   (c) Let $\mu \in \mathbf{E2}$, $\omega \preceq \mu \prec \mathfrak{A}$. Then an $e = \nu\ (= \omega)$ case arises with $f = \Delta_\omega \mu$:

$$\begin{aligned}\mathfrak{A}_n \mathfrak{A}_\mu &\prec \mathfrak{A}_{\mathfrak{A}(\mu,n)} \mathbf{2} \text{ by VII} \cdot 9 \\ &\prec \mathfrak{A}\mathfrak{A}\{(\mu, n)\} = ([e, f], n).\end{aligned}$$

In all such instances, $F \succeq \text{diag}_n\,(F, n)$.

Notes

1. $(\omega, 1) = \mathbf{1}$, and for each diagonal e of $\mathbf{E2}$, $e \succ \mathbf{1} \Rightarrow (e, 1) \succ \mathbf{1}$.
2. The case Diag(α), $e = \mathbf{3}$ requires a separate clause, on account of the note of §2.3, before defn. 2.
3. The defn. is chosen so that

$$(F, n) \succeq \begin{cases} [(e, n), f] & \text{if Diag}(e)\text{; and} \\ [e, (f, n)] & \text{if Diag}(f). \end{cases}$$

4. The observation made in §4 is exploited in this definition. For example, in case 6, Diag(e),

$$(f, n) = [(e, f(n)), f]$$

   could have been taken in place of the minimal defn.
5. In particular, defn. 21 assigns an ascending sequence to each diagonal in $\mathbf{EP}^\star$.

Simple and *Compound* diagonals (cd's) of **E2** are defined as in II, §3. And the functions (e, n, m), $e \in \mathbf{E2}$; as in V, §1.

8.5.1 Density in E2

THEOREM 8.22

$$(\forall E, F \in \mathbf{E2})_{\mathrm{Diag}(F)}\ E \prec F \Rightarrow \{(\exists N) E \prec (F, N)\}.$$

Proof, by induction on the order-of-definition N of $e \in \mathbf{E2}$. There are no diagonals among the functions of order 0. Thus, let $N \geq 0$, and suppose for i.h. that the diagonals of $\mathbf{E2}_N$ have the property. As with the proof for **EP**, viz. that of III·12, the present argument reduces to the comparison of single terms. Thus, suppose

$$E = [\alpha, \beta] \prec [e, f] = F$$

in $\mathbf{E2}_{N+1}$, where $f \succ \mathbf{1}$, $e \succ \mathbf{3}$ may be taken.

Let $\alpha = e$, so that $\beta \prec f$.

1. If Suc(f), then Diag(e), and $\beta \preceq f^-$. Therefore

$$E \preceq [e, f^-] \prec [(e, n), [e, f^-]] = (F, n), \quad n = 1, 2, 3, \ldots.$$

2. If Diag(f), by i.h., $\beta \prec (f, k)$ for some k, and

$$E = [e, \beta] \prec \begin{cases} f & \text{if } e \prec \nu \text{ by lemma 16, and} \\ [e, (f, n)] & \text{for } n \geq k \text{ by i.h.,} \end{cases}$$

where $f = [\nu, g]$. Therefore, according to the case of defn. 21, $E \prec (F, 1)$, or $E \prec (F, n)$, for $k \leq n$. [20]

Let $e \prec \alpha$, and suppose $f \prec [\alpha, \beta]$. Then by lemma 16 $[e, f] \prec [\alpha, \beta]$, contrary to assumption. Therefore $[\alpha, \beta] \preceq f \preceq (F, 1)$.

Let $\alpha \prec e$ ('tree case'). Consider $T = T_E$, in the notation of §4.8. If T does not contain an index $\succeq e$, then $(\exists k)[\alpha, \beta] \prec [e, \mathbf{k}]$, giving

$$n \geq k \Rightarrow E \prec (F, n).$$

Now suppose T contains such a vertex. Let $[\mu, h]$, $e \preceq \mu$, be the function, and ξ the index obtained from T, as from T_γ in lemma 16. Then

$$(\exists t \geq 1)\ E \prec \widetilde{E} = \overbrace{[\xi, \ldots, [\xi}^{t}[\mu, h]] \ldots].$$

1. Suppose $e = \mu$. Then as in the case $\alpha = e$, if Suc(f), then Diag(e) and $h \preceq f^-$, and

$$[\mu, h] = [e, h] \preceq [e, f^-].$$

By III·12, $(\exists k)\ \xi \prec (e, k)$, therefore applying lemma 16,

$$E \prec \widetilde{E} \prec (F, k).$$

[20] Note that, in the case Suc(e), e.g.

$$([\mathbf{4}, [\mathbf{6}, \omega^+]], n) = \overbrace{[\mathbf{3}, \ldots [\mathbf{3}}^{n}, [\mathbf{6}, \omega^+]] \ldots] \succ [\mathbf{4}, n \cdot [\mathbf{6}, \omega]].$$

1. If Diag(f), $h \prec f$ and i.h. give $(\exists k)\ h \prec (f,k)$. Therefore, as $\xi \prec e$,

$$n \geq k \Rightarrow E \prec \widetilde{E} \prec [e,(f,n)] \preceq (F,n).^{21}$$

2. Suppose $e \prec \mu$. Then by lemma 16, $[\mu, h] \preceq f$. Now by defn. of T, $\xi \prec e$.

   (a) If Diag(e), by III·12, $(\exists k)\ \xi \prec (e,k)$, so for $n > k$,

   $$E \prec \widetilde{E} \prec [(e,n), f] \preceq (F,n).$$

   (b) If Suc(e), then since $\xi \prec e$, [22]

   $$\begin{aligned} \widetilde{E} &\prec \overbrace{[e^- \ldots [e^-}^{n}, f] \ldots] \qquad \text{for } n > t \\ &\preceq (F,n). \end{aligned}$$

□

Reduced Starting Property

Let F be a compound diagonal of **E2**.

THEOREM 8.23

$$(\forall n)(F,n) \preceq (F, n+1, 1).$$

Proof, by induction on order N of F in **E2**. There are no diagonals of order 0. Thus, let $N \geq 0$, and suppose for induction hypothesis that the cd's of $\mathbf{E2}_N$ have the property. Let $F \in \mathbf{E2}_{N+1}$. As with the previous proof, the general case follows from that for a single term, viz.

$$F = [e, f].$$

Suppose first that $f = 1$. Then $F = \omega$, the initial (simple) diagonal. Thus suppose $f \succ 1$; it may also be assumed that $e \succ e$.

1. Suc(f). If $e = 3$, the case follows as in Ch. III, §3.5. Thus for Diag(e),

$$(F, n+1) = [(e, n+1), [e, f^-]].$$

[21] Note that the last inequality uses the lemma of note 3 after definition 21.

[22] Note that, e.g.

$$\overbrace{[6, \ldots, [6}^{n}, [4, \omega^3] \ldots] \prec [7, [4, n\omega^2]] = ([7, [4, \omega^3]], n).$$

So $\mathrm{Diag}\{[e, f^-]\}$ and

$$(F, n+1, 1) = \begin{cases} [(e, n+1)^-, [e, f^-]] & \text{if } \mathrm{Suc}\{(e, n+1)\}, \text{ or} \\ [(e, n+1, 1), [e, f^-]] & \text{if } e \text{ is compound.} \end{cases}$$

In the first case $(e, n+1)^- = (e, n)$; in the second $(e, n) \preceq (e, n+1, 1)$ by III·14. So by VII·20

$$(F, n) = [(e, n), [e, f^-]] \preceq (F, n+1, 1).^{23}$$

2. Diag(f).

   (a) If, as in defn. 21, $e \prec \nu$: if Suc(e),

$$(F, n+1) = [\overbrace{e^-, \ldots [e^-}^{n+1}, f] \ldots] = [e^-, (f, n)],$$

   therefore

$$(F, n+1) = \left\{ \begin{array}{ll} n \cdot f + (f, 1) & \text{if } e = \mathbf{3} \\ [e^{--}, (F, n)] & \text{if } \mathrm{Suc}(e^-) \\ [e^-, ((F, n), 1)] & \text{if } \mathrm{Diag}(e^-) \end{array} \right\} \succeq (F, n),$$

   the inequality in the diagonal case following because

$$(F, 1, 1) = ([e^-, f], 1) = [(e^-, 1), f] \succ f.$$

   If Diag(e), $(F, n+1) = [(e, n+1), f]$. Now $(e, n+1) \prec \nu$ by III·7·3, therefore

$$(F, n+1, 1) = \begin{cases} [(e, n), f] & \text{if } e \text{ is simple,} \\ [(e, n+1, 1), f] & \text{if } e \text{ is compound.} \end{cases}$$

   The first case is $\succeq (F, n)$; also the second, by III·14 and VII·20.

   (b) If $\nu \preceq e$, then $(f, n+1) = [\mu, g]$ in **E2**, where $\mu \preceq \nu$ (or the ct is the sum of many terms), so the scheme 21(vii) applies: $(F, n+1) = [e, (f, n+1)]$ and again

$$\begin{aligned} (F, n+1, 1) &= [e, (f, n+1, 1)] \\ &\succeq [e, (f, n)] \text{ by i.h. and VII} \cdot 20 \\ &= (F, n). \end{aligned}$$

[23] Note that the last case of defn. 21 does not arise, since $(e, n+1) \prec e$.

(c) $f = u + v$ in **E2**, if f is simple, $(f, n+1) = w^+$ say, $w \in$ **E2** (so $(F, n+1) = [e, w^+]$ by 21·vii) 2 cases arise: if Suc(e),

$$\begin{aligned}(F, n+1, 1) &= ([e, w^+], 1) = ([e^-, [e, w]], 1) \\ &\succ [e, w] \text{ by } 21 \cdot \text{v, vi} \\ &= (F, n);\end{aligned}$$

while if Diag(e),

$$\begin{aligned}(F, n+1, 1) &= [(e, 1), [e, w]] \text{ by } 21 \cdot \text{iv} \\ &\succ [e, w].\end{aligned}$$

If f is compound,

$$\begin{aligned}(F, n+1, 1) &= [e, (f, n+1, 1)] \text{ by } 21 \cdot \text{vii, since } \nu \preceq e \\ &\succeq [e, (f, n)] \text{ by i.h.} \\ &= (F, n).\end{aligned}$$

Thus in all cases $F \in \mathbf{E2}_{N+1}$ has the reduced starting property, and thm. 23 follows by induction.

Now using defn. 21, let the 'step-down' $\lhd_1$ of Ch. III, §4 be extended to **E2**. Allowing a small shift in notation:

THEOREM 8.24 *Suppose f is a cd of* **E2**. *Suppose for some $n \geq 1$,*

$$(f, n) \prec (f, n+1, 1).$$

Then

$$(f, n) \lhd_1 (f, n+1, 1).$$

Proof, by induction on the order N of f in **E2**, using the DCC, thm. 19, and the arguments of the previous thm.

□

Now let e be a diagonal, and f a cd of **E2**.

THEOREM 8.25 *If an assignment Ω for* Diag**E2** *satisfies thms. 23 and 24, then Ω has the BACHMANN starting property (cf. III·14):*

$$\text{if for some } n,\ (f, n) \prec e \prec (f, n+1), \text{ then } (f, n) \preceq (e, 1).$$

Proof. Put $f = f_0, n = n_0$, so $(f, n) = (f_0, n_0)$. Put $f_1 = (f_0, n_0 + 1)$, with $e \preceq f$. Now suppose for some $S \geq 1$, f_s has been defined, with

$$e \preceq f_s.$$

If $e = f_s$, the defn. of f_i, $i = 0, 1, 2, \ldots, s$ is complete. Thus, suppose $e \prec f_s$. If Suc(f_s), put $f_{s+1} = f_s^-$; then $e \preceq f_{s+1}$. If Diag(f_s), then putting $(f_s, 0) = (f_{s-1}, n_{s-1})$, by thm. 22

$$(\exists n_s \geq 0)\,(f_s, n_s) \prec e \preceq (f_s, n_s + 1).$$

Put $f_{s+1} = (f_s, n_s + 1)$. Then since

$$\ldots \prec f_s \prec \ldots \prec f_1 \prec f_0,$$

by thm. 19 it follows that for some $t \geq 0$, $e = f_{t+1}$. But by thms 23 and 24,

$$(f, n) = (f_0, n_0) \preceq (f_1, n_1) \preceq \ldots \preceq (f_t, n_t) \preceq (e, 1).$$

Therefore $(f, n) \preceq (e, 1)$, and the thm. is proved.

□

COROLLARY 8.26 *The assignment Ω of defn. 21 has the BACHMANN starting property.*

Note on Consequences of Properties of E2

By thm. 19, **E2** is well-founded. The results of §5, together with defn. 21, are therefore sufficient to allow the replacement of **EP** by **E2** as index class, in Ch. VII, defn. 15. The extended class $(\Delta_e)_{e \in \mathbf{E2}}$ of operators obtained enables the definition of a new class, **E3**, viz. the smallest class containing **EP** and closed under the $(\Delta_e)_{\mathbf{3} \prec e \in \mathbf{E2}}$. However, to show **E3** is sequentially ordered, a generalization of VI·12 is required. And this in turn depends on the present results, thm. 22 and corol. 26.

8.6 A Closure For the ε - Δ Construction

8.6.1 The System $(\mathbf{E3}, \Omega)$

The object is to construct for the class **E3**, discussed at the end of the previous section, a DS-assignment Ω, viz. that has both the density and starting properties.

1. $(\Delta_e)_{e \in \mathbf{E2}}$ is obtained from VII, defn. 15 by replacing '$e \in \mathbf{EP}$' by '$e \in \mathbf{E2}$'. $(\Delta_e)_{e \in \mathbf{E2}}$ has the properties of VII, thms 20, 23, as **E2** has a DS-assignment by articles 22, 26.

2. The class **E3** is obtained from defn. 10, by allowing e to vary over **E2**. From $\mathbf{EP} \subset \mathbf{E2}$, follows $\mathbf{E2} \subset \mathbf{E3}$.

3. **E3** is proved $\prec$ - connected by
   (a) using thm. 17 (the $\prec$ - connexity of **E2**)
   (b) repeating identically the argument of this theorem, with '$e, \alpha \in \mathbf{E2}$'.

4. A DS-assignment Ω for the diagonals of **E3** is obtained by
   (a) taking '$e \in \mathbf{E2}$' in '$[e, f]$' of notation 4.1. This is possible, since Diag**E2** has already been assigned.
   (b) taking '$e \in \mathbf{E2}$' and '$f \in \mathbf{E3}$' in defn. 21, and repeating identically the proofs of thms 22, 26.

Notes

1. II·18 cannot be directly generalized to the present thm. 17, because $T_e, e \in \mathbf{EP}$, is non-indexed, while $T_e, e \in \mathbf{E2}$, is vertex -e- indexed, $e \in \mathbf{EP}$.

   The method of proof, however, is not different if the indices are taken from **E2** rather than **EP**. Therefore VIII·17 can be directly generalized to the proof of $\prec$ - connexity of **E3**, 3(b) above. (In fact much of the method of thm. 17 is contained in the proof for $\mathbf{EP}^*$, thm. 7).

2. Defn. 21 contains new clauses, besides those of III·6:
   (a) in the successor case, special steps having to be taken to ensure the DS-properties for the assignment of such functions as $[\mathbf{3}, [\mathbf{4}, \omega]]$.
   (b) in the diagonal case, having to take account of the effect of high-power exponential operations. E.g. for $\alpha = [\omega, [\omega^2, \omega]]$ the diagonal assignment

$$(\alpha, n) = [\omega, [n\omega, \omega]]$$

   would give canonical terms all $\prec [\omega^2, \omega]$.

No extra complications arise, however in the extension of defn. 21 from **E2** to **E3**.

3. The $\prec$ –least polynomial in **E3**–**E2** is the supremum Δ, of the diagonalization of Ch. VII, as with the supremum ε of **EP**, viz. the least term of **E2** – **EP**.

For **E2** $= \mathcal{X}$, **E3** $= \mathcal{Y}$, the passage from $\mathcal{X}$ to $\mathcal{Y}$ may be summarized by:

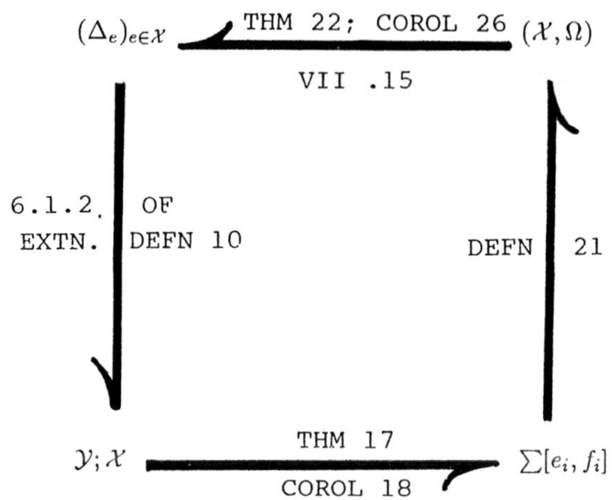

8.6.2 The Classes En

Building on **E3** let $n \geq 3$, and suppose the system $(\mathbf{En}, \Omega)$ has been defined, Ω a DS-assignment for Diag**En**. Now let $(\mathbf{E(n+1)}, \Omega)$ be obtained from $(\mathbf{En}, \Omega)$ by the procedure by which **E3** was obtained from **E2**. So (putting **E1** = **EP** ref. §4.1)

THEOREM 8.27 $(\forall n)$ *the system* $(\mathbf{En}, \Omega)$ *is defined.*

□

DEFINITION 8.28 *Let* $\theta_1 = 4$, *and for* $n \geq 1$, *let* $\theta_{n+1} = \Delta_{\theta_n}\omega$.

Thus for each $x = 1, 2, 3, \ldots$, $\theta_1(x) = 4$, $\theta_2(x) = x \star x$, and $\theta_2 = \varepsilon \in \mathbf{E2}$; $\theta_3 = \Delta_\star = \Delta \in \mathbf{E3}, \ldots, \theta_n \in \mathbf{En}, \ldots$. The rapidly growing monomial θ_3 is associated with **E2** in the same way the monomial η is associated with $\mathbf{EP}^\star$, as in thm. 5. By the generalization of VII, §3,

$$\theta_1 \prec \theta_2 \prec \ldots \prec \theta_n \prec \theta_{n+1} \prec \ldots .$$

THEOREM 8.29

$$(\forall n)(\forall e \in \mathbf{En})\, e \prec \theta_{n+1}.$$

Proof, by induction on n. By II·16, $(\forall e \in \mathbf{EP})\ e \prec \varepsilon \in \mathbf{E2}$. Now let $n > 1$, and suppose it has been shown

$$\{\forall e \in \mathbf{E(n-1)}\}\ e \prec \theta_n.$$

Let the (indexed) defn. tree T_f of $f \in \mathbf{En}$ be as in definition 14·4, but with the indices taken from $\mathbf{E(n-1)}$. Let $\mu \in \mathbf{E(n-1)}$ be the $\prec$ –greatest index occurring in the vertices of T_f. Then

$$f \prec \Delta_{\mu+1}\omega \in \mathbf{En}, \text{ with } \mu + \mathbf{1} \in \mathbf{E(n-1)}.$$

But by i.h., $\mu + 1 \prec \theta_n$, so by the generalization of VII·23,

$$f \prec \Delta_{\theta_n}\omega \in \mathbf{E(n+1)}.$$

I.e. every $f \in \mathbf{En}$ is eventually dominated by $\Delta_{\theta_n}\omega = \theta_{n+1}$, and the thm. is proved.

□

COROLLARY 8.30

$$(\forall n)\ \mathbf{En} \subset \mathbf{E(n+1)}.$$

□

Remark. $\mathbf{E(n+1)}$ is greatly more extensive than $\mathbf{En}$, since the $\prec$ –lowest new operation, Δ_{θ_n}, leads out of $\mathbf{En}$, and the excess in extent of operations of $\mathbf{E(n+1)}$ over $\mathbf{En}$ is the excess in extent of $\mathbf{En}$ over $\mathbf{E(n-1)}$, $n > 1$.

Cofinal Fundamental Sequences

THEOREM 8.31 *The fundamental sequence*

$$(\theta_{n+1}, 1), (\theta_{n+1}, 2), \ldots$$

is $\prec$ –cofinal in $\mathbf{En}$, $n = 1, 2, 3, \ldots$.

Proof, by induction on n.

□

Thus, for $n > 1$, $\mathbf{En}$ can be traversed in various ways: as e.g. by the above sequence $\mathcal{S}$; by the longer $\mathbf{E(n-1)}$–sequence $(\Delta_e\omega)$ or by the entire sequence itself.

8.6.3 The Class E

DEFINITION 8.32

$$\mathbf{E} = \cup_n \mathbf{En}$$

The class $\mathbf{E2}(\omega)$ (see end of §4) is $\prec$ −cofinal in the class of functions obtainable by primitive or single recursion. The class **E** is cofinal in the class obtainable by single recursion and single diagonalization. Primitive recursion may be regarded as diagonalization of type 0, diagonalization as of type 1, and diagonalizations over functionals as diagonalizations of higher type.

COROLLARY 8.33 *1.* **E** *is sequentially ordered by* $\prec$.

2. $e, f \in \mathbf{E} \Rightarrow \Delta_e f \in \mathbf{E}$.

□

Remarks

1. According to the corol., **E** is closed under a binary operation:

$$\alpha, \beta \in \mathbf{E} \Rightarrow [\alpha, \beta] \in \mathbf{E}.$$

This is untrue of **EP**; or indeed of any class **En**, $n = 1, 2, \ldots$. However, with the exception of the closure of **EP** wrt multiplication, none of the above classes, **E** included, is closed under any arithmetic binary operation $\Delta_\alpha(x, y)$ other than addition:

$$\alpha, \beta \in \mathcal{X} \Rightarrow \alpha + \beta \in \mathcal{X}.$$

It is just that **E** is so large that such an operation becomes tertiary: putting '+' as $\otimes_1$, etc., the 'binary' operation would be

$$\alpha, \beta, \in \mathbf{E} \Rightarrow \omega \otimes_\alpha \beta \in \mathbf{E}.$$

2. Using defn. 28, let $\theta = \mathrm{diag}_n \theta_n$. Then

THEOREM 8.34

$$(\forall e \in \mathbf{E})\, e \prec \theta.$$

□

8.6.4 The KREISEL-FEFERMAN Conjecture For $(\mathbf{E}, \prec)$

KREISEL [80] and FEFERMAN [38], [39] have conjectured that **E** represents the smallest class for which there can be any doubt that extension (in terms of rate of growth) is definable via predicative means. Well, the presentation here of **E** is predicative. As noted after defn. 28, $\theta(1) > 1$. The diagonalization of VII, §3 may thus be altered, by replacing 'Δ_1' of VII·15·1, by 'θ,' reaching by predicative means far beyond **E**. Alternatively, applying the remark after corol. 33, the purely finitary diagonal

$$\lambda x \cdot \mathfrak{d}^{(x)}(x)$$

reaches beyond **E**, without even relying on cfs's.

Sentences Undecidable in Theories Stronger Than N

In the context of the 'descriptive power' of $\underline{\mathbf{N}}$, §4.2, it may be noted that a stronger theory $\underline{\mathbf{T}}$ contains gen. ind. corresponding to a segment $\mathbf{E}_\xi$ of $\mathbf{E}$,[24] $\varepsilon \prec \xi$: if (noting rem. 2. after VII·37) the relation

$$T_{n+1} = (T_n, \Delta_\xi\{| T_n |\})$$

is substituted for that of VII·38, then a sentence true, but not provable in $\underline{\mathbf{T}}$ is obtained.

[24] Or of a larger class.

8.7 Fundamental Sequences and the Feather Diagram

8.7.1 Edge-Diagram

Diagonalization and Trees

Theorem 39 of Ch. IV, together with defn. 15 of Ch. VII together establish the fundamental link between trees and diagonalization, at least as far as is essential for the analysis of **EP**. It is a remarkable fact, however, that the pattern of repetition (or succession, or iteration, or recursion) and of diagonalization

1. is well-established within as small a segment of **EP** as **XP**;

2. the tree structure of **XP** is virtually trivial, being limited to edge-height 2.

There is, however, another (infinite) ramified structure, posessing an even closer link with diagonalization, and well-developed by the cubic or even quadratic levels of **XP**.

Construction

Let $e \in \overline{\mathbf{EP}}$ be diagonal, and denote the nth canonical term [25] (e, n) of e by e_n, $n = 1, 2, 3, \ldots$. Now, as in III, ex. 1., after defn. 3, consider the unit squares of the plane, ruled by the lines $x = 1, 2, 3, \ldots$, and $y = 1, 2, 3, \ldots$. For $m, n \geq 1$ put into the square, whose upper right corner has coordinates (m, n), the integer

$$e_m(n).$$

The verticals of the figure thus contain the images of the canonical terms (ct's) of e; the diagonal, those of e itself. If the entry of square $(m, 1)$ is now replaced, by a copy of the entry of square (m, m), the new $(y = 1)-$row may be taken as the $e-$*spine* of the figure; the verticals as the *leaves*. The leaf which contains the images of (e, n) is the *nth leaf* of the $e-$spine. This is the *construction for e*. The point or vertex on which a leaf is incident on its spine is a *contact*. The interval of a spine determined by 2 consecutive contacts is an *edge* of the figure.

[25] Recall (III, defn. 6·3) ε is assigned $\omega \star 1, \omega \star 2, \ldots = \varepsilon_1, \varepsilon_2, \ldots$.

This is as indicated. Some curvature allows the leaves to be conveniently drawn

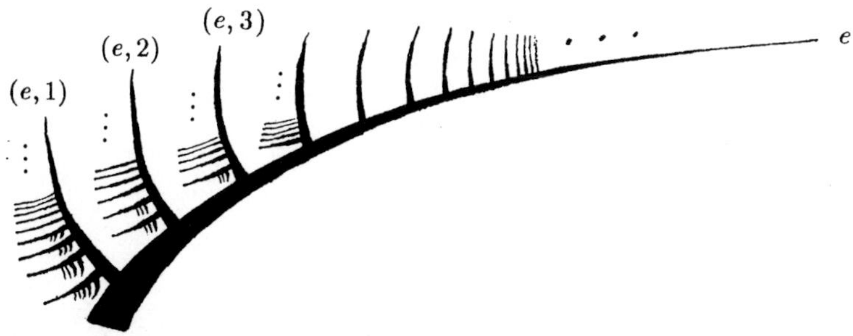

on the convex side. If e is compound (III, defn. 16, and thm. 17), all the ct's are diagonal, so the construction for the ct's may be made, as shown in the case of e_1. Note that an $e-$leaf is an e_n-spine.

Feather Diagram

Recall the class $\mathcal{X}$ of VI, defn. 8:

1. $\mathbf{1} \in \mathcal{X}$
2. $e \in \mathbf{X} \Rightarrow e^+ \in \mathcal{X}$
3. For $e \in \mathbf{EP}$, Diag(e),

$$\{(\forall n)(e,n) \in \mathcal{X}\} \Rightarrow e \in \mathcal{X}.$$

By thm. 11 of the same Chapter, $\mathcal{X}$ is **EP**. Now consider $\alpha \succeq \omega$ in **EP**. The *feather diagram* (fd) for α or $\alpha-$*feather* is obtained so as to display the build-up of α via 1. - 3. above. Thus if $\alpha = \mathbf{1}$ or Suc(α), then the $\alpha-$feather is a *leaf*. If Diag(α), and there is a non-diagonal e satisfying $e \mathrel{\tilde{\lhd}} \alpha$ via $e \lhd \alpha_K \lhd \alpha_{K-1} \lhd \ldots \lhd \alpha_1 = \alpha$ such that Diag(α_i), $i \leq K$, then *the construction for each* α_i *is made.* Note that, as a consequence, spines obtained at constructions which start at $e \prec \alpha$ become themselves leaves of the subsequent construction.

Example.

By VI, corol. 6, the feather diagram of a segment **X** of $\overline{\mathbf{EP}}$ may be taken to be that of α, where $\mathbf{X} = (\overline{\mathbf{EP}})_\alpha$. Far from being trivial, the diagram for $\mathbf{X} = \mathbf{XP}$ is too highly recurrent to sketch, being the sum of the

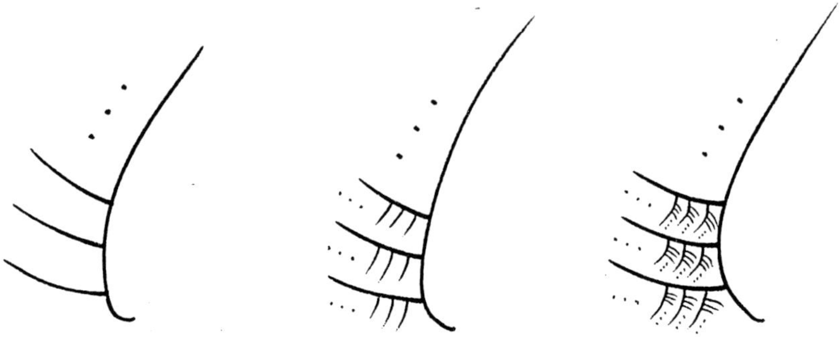

constant, linear, quadratic , ...

cases. It may be noted that, as in the **XP**−example, one endpoint of the leaf corresponding to **1** is joined to an endless hierarchy of spines. Note that for $\alpha \succeq \omega$, the feather is an infinite structure, with a well-founded, recurrent pattern.

Classification

Suppose α is a cd of **EP**, viz. of the form $\alpha = \omega^\beta$, $\mathbf{1} \prec \beta$. If α is of the first kind, $\alpha_n = n\omega^{\beta^-}$, $n = 1, 2, 3, \ldots$. Now the canonical reductions from $2\omega^{\beta^-}$ to $\omega^{\beta^-} + \mathbf{1}$ are identical to those from ω^{β^-} to **1**, save for the added term ω^{β^-}. Similarly for subsequent ct's. The leaves of the α−feather are therefore *identical.* If α is of the second kind (e.g. $\alpha = \psi$, as in the previous ex.), then a proof by gen. ind. gives

THEOREM 8.35 *The leaf corresponding to α_{n+1} in the α−feather contains endless copies of that corresponding to α_n.*

Thus e.g. ω^{ω^+} contains endless copies of the ψ-feather.

Interpretation

The fd of $e \in \mathbf{EP}$ is constructed from the ct's of e, in order to expose the build-up of e, according to VI, defn. 8, by iteration (of the passage from$(\alpha, n)$to$(\alpha, n+1)$); and the diagonalization of the (α, n), to give α; for $\alpha \mathrel{\tilde{\lhd}} e$.

The fd may be used to help understand ideas of Chs. II-VII. Thus, if the spine of an e−feather is dipped in ink, which rises to the n−th leaf, and similarly as each leaf is a spine; then the stained part of the feather corresponds to the class of $\alpha \mathrel{\tilde{\lhd}_n} e$, viz. to the class $\mathbf{EP}[e,n]$ (III, §4).

The DCC for fd's may be obtained via edge-excision: for $n \geq 1$, if the edge of fdα corresponding to α_n, α_{n+1} is removed, a tail of the diagram falls away, leaving the fd of α_n. Let this be called a 'break'. Then by IV·48

THEOREM Starting with fdα, after finitely many breaks, fd$\mathbf{1}$ is obtained. □

Every $e = \varepsilon_{n_1,n_2,\ldots,n_r} \in \mathbf{EP}$ corresponds to a particular leaf l of the ε–feather. As e is or is not a diagonal, l itself is or is not a spine. l is reached in r steps by taking the n_1st leaf l_1 of the ε–feather, the n_2nd leaf of $l_1, \ldots$. Generally, iterated canonical reduction via (n_i) determines post-order traversal of the finite subtree of the ε–feather picked out by (n_i). A sketch of the feather might contain 3-4 (serial) spine-leaf descents; thm. 35 shows that if (n_i) is si, leading ever further out along an e–leaf, $\psi \preceq e$, there are of the order of $\Delta_e(n_1)$ descents, in accordance with the SOLOVAY result, VIII·37, before reaching a leaf which is not a spine.

8.7.2 Vertex Feather Diagram

Dual FD

The spine of the fd of e corresponds to e, and the leaves to its ct's. Thus, exponential polynomials correspond with *edges* of the diagram, which is infinite 'outwards', and is best oriented so as to build up. A dual, in which exponentials correspond with vertices, is obtained by reversing the orientation of the leaves, to give an 'infinite-inwards' or 'downwards hanging' figure. Thus the construction for e now comprises a non-connected infinite linear graph $\Gamma = \Gamma(e)$, whose vertices v_n are the rational points $\frac{n}{n+1}$, and $v = 1$; and whose edges are the (open) intervals $(\frac{n}{n+1}, \frac{n+1}{n+2})$, $n = 1, 2, 3, \ldots$. [26] Then v_n corresponds with (e, n), and v with e itself.

The *dual* fd for e is then obtained by

1. effecting the construction for each diagonal α, $\tilde{\unlhd}\, e$;
2. for $\alpha = (\beta, n) \lhd \beta \mathrel{\tilde{\unlhd}} e$, identifying the α–vertex of $\Gamma(\alpha)$ with the (β, n)–vertex of $\Gamma(\beta)$.

[26] Γ is thus non-connected in the sense that there is no finite vertex-edge path from any v_n to v.

Some of the vertices and edges of the dual fd are as indicated:

dual

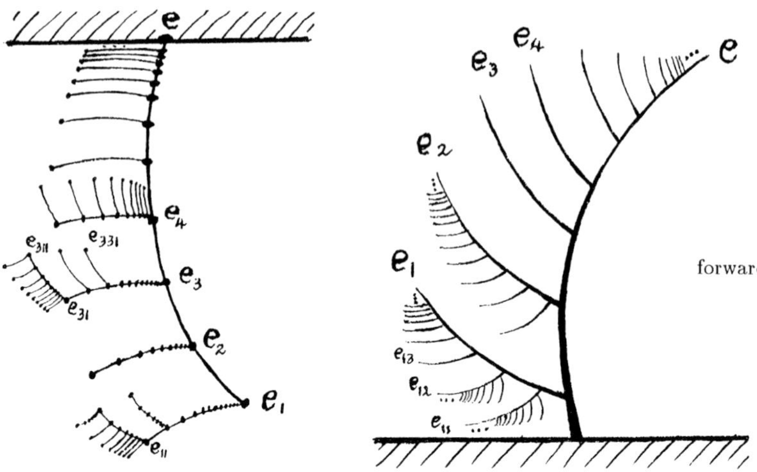

forward

Interpretation

Although a fundamental finiteness property is lost on dualizing the fd, the result in many ways more explicitly reflects the structure of e.g. the hierarchy **EP**; in particular tracing an infinite upward path from any given $e = \varepsilon_{n_1,n_2,\ldots,n_r}$, via each $\varepsilon_{n_1,\ldots,n_i}$, $i = r-1, r-2, \ldots, 1$, to ε itself.

If the α_{n+1}–th leaf is folded onto the (α_{n+1}, α_n)-interval of the α–spine, $\alpha \mathrel{\tilde{\triangleleft}} e$, then the linear structure resulting satisfies the metric version of III ·14: distance $d(\beta, \beta_1) \leq d(\beta, \alpha_n)$ for all $\alpha_n \prec \beta \preceq \alpha_{n+1}$.

The Vertex-FD of a Higher Induction

As noted in VI, §5 ·1, gen. ind. up to any given $e \in$ **EP** is provable in **N**. This can be perceived as a vertex-fd, built up from individual inductions. Thus, suppose A=A$_e$ is a sentence of **N** having a proof by e–induction: ⊢ A. Then for every $\alpha \mathrel{\underset{-}{\tilde{\triangleleft}}} e$, there corresponds a sentence A=A$_\alpha$ and, if Diag(α), the proof contains an induction tableau

$$\frac{\mathrm{A}(1)\ \&\ \mathrm{A}(n) \Rightarrow \mathrm{A}(n+1)}{\mathrm{A}},$$

where A(n) is the sentence corresponding to (α, n), $n = 1, 2, 3, \ldots$; and A=A$_\alpha \equiv (\forall n)$ A(n). Corresponding to the spine A, are the leaves A(1), A(2), A(3), ... The dual diagram is chosen, as the vertex A should be the limit of the convergent sequence A(1), A(2), ...

Example. Assuming a suitable formalization, the proof of a thm. such as that of VII, §2·3 has an ω^2-fd.

8.7.3 FD of a Calculation

MILLS Construction of $\mathfrak{A}$

MILLS [111] obtained ACKERMANN's function as follows. Given $y, n \in \mathbf{N}$, suppose the vertex-indexed tree T is constructed, as in IV, §2.1, starting from $v^\sharp$, and then successively adjoining vertex-edge units, such that

1. $\mathrm{va}(v^\sharp) = y$;
2. $\mathrm{eh}(T) = n$;
3. the vertices are adjoined in (R-to-L) pre-order;
4. the i–th vertex v adjoined, is either an endv, or $\mathrm{va}(v) = y + i$.

Note that such a T has: 5. if v is adjoined after u, then v lies in a branch lying to the left of a branch containing u; and: 6. $u \text{ Sc } v \Rightarrow \mathrm{va}(v) < \mathrm{va}(u)$, unless u is an endv.

LEMMA 8.36 *T as above is* **finite** *(viz. $T \in \mathfrak{S}$).*

Proof, by induction on edge-height n. For $n = 0$, only one T is possible. Now let $n \geq 0$, suppose that $\mathrm{eh}(T) \leq n \Rightarrow T \in \mathfrak{S}$, and consider

$$T = [T_y, T_{y-1}, \ldots, T_1], \text{ of eh} = n + 1.$$

By induction hypothesis, $| T_1 | = N(1) < \infty$, after construction via $\mathrm{va}(\mathrm{pv}T_1) = y + 1$ as in (1) - (4) above. And in general, $y \geq i > 1 \Rightarrow | T_i | = N(i) < \infty$, after construction via $\mathrm{pv}T_i = \sum_{j<i}^{1}\{N(j)\} + y$. Evidently $| T | = 1 + \sum_{i=y}^{1} N(i)$, $T \in \mathfrak{S}$, and the result follows.

□

Remark. By ex. 4 after IV·19, $T \in \mathfrak{T}$.

Note that $N = | T |$ is a si function of y, n. Also, it is clear that N depends on the pre-order construction of T. Indeed, suppose the v's of T are adjoined in another order, to give $| T | = M$. Then

LEMMA 8.37 $M < N$.

Proof, by induction on eh. $v^\sharp$ alone determines a unique order; thus suppose that for eh$= n \geq 0$, the lemma holds, and consider the construction of

$$T = [T_y, \ldots, T_1], \text{ of eh} = n + 1.$$

Let $\mathrm{pv}T_i = v_i$. Suppose T is constructed in pre-order, with the following violation: v_1 is adjoined first, then v_2; then the pre-order construction of T_1 is completed, then similarly for T_2, and each of $T_3, T_4, \ldots, T_y$. This will afford a unit increase in the starting value for T_1 ($y + 2$ in place of $y + 1$), giving a *greater* N_1; but now T_2 is constructed from $y + 2$ in place of (the generally considerably greater) $1 + N_1 + y$, more than offsetting this increase. Similarly for more gross violations of the pre-order.

□

THEOREM 8.38 (MILLS) $N \approx \mathfrak{A}_n(2, y)$.

Proof. It will be shown that in fact $N > \mathfrak{A}_n(2,y)$ (while an increase in the subscript n (on the right) will give a number greater than N). For this purpose, given n, y, consider the *recursion tree* $RT(\mathbf{n}, y)$ of $\mathfrak{A}_n(2,y)$. This is closely related to the calculation tree $T(\mathbf{n}, y)$ of VII, §4.5. Indeed, $[\mathbf{n}, y]$ is the pv of RT; and the vertex $[\mathbf{m}, k]$ is an endv if either $\mathbf{m} = \mathbf{1}$ or $k = 1$; otherwise

$$[\mathbf{m}^-, i] \text{ Sc } [\mathbf{m}, k],$$

where $i = [\mathbf{m}, k^-], [\mathbf{m}, k^-]^-, \ldots, 1$. Note that $[\mathbf{m}, k-1] = ([\mathbf{m}, k])_R \ \ (m, k > 1)$ in $T(\mathbf{n}, y)$ is replaced by

$$[\mathbf{m}, k] \text{ L } [\mathbf{m}, k-1]$$

in $RT(\mathbf{n}, y)$. Thus, eh$RT = n$; and the calculation is described (with repetition [27]), as in VII, §4.7 by the pre-order traversal of RT. The endv's joined to the vertex

$$[\mathbf{2}, [\mathbf{3}, \ldots [\mathbf{n}^-, [\mathbf{n}, y^-]^-]^- \ldots] \,]$$

represents the entire calculation of ACKERMANN's function. Viz.

$$|\, RT \,| < |\, T \,| \, .$$

It is easily shown that this inequality is reversed, if the index n is incremented for the tree RT.

Second proof.

The lower bound for $|\, T \,|$ will be found. $\mathfrak{A}$ has the property (taking $x = 2$):

$$\mathfrak{A}_{n+1}(2,y) = \mathfrak{A}_n\{2, \mathfrak{A}_{n+1}(2, y-1)\} = \ldots = \mathfrak{A}_n^{(y-1)}(2,2),$$

the superscript denoting iteration in the y-variable. Let $T = [T_y, \ldots, T_2, T_1]$ be the tree of lemma 4.1. Taking as initial case eh$T = 2$, $|\, T \,| = 3y + 4 > \mathfrak{A}_2(2,y)$. Now suppose the result is obtained for eh$= n \geq 2$, and consider the case eh$T = n+1$. The hypothesis gives va(pvT_2) $> \mathfrak{A}_n(2, y+1)$; therefore va(pvT_3) $> \mathfrak{A}_n^{(2)}(2, y+1)$; and after $y - 1$ steps,

$$\begin{aligned} |\, T \,| > \text{va}(\text{pv}T_y) \ &> \ \mathfrak{A}_n^{(y-1)}(2, y+1), \\ &> \ \mathfrak{A}_{n+1}(2,y) \end{aligned}$$

by the above property, and the result follows by induction.

□

The Feather Diagram of the Calculation of $\mathfrak{A}_n(x,y)$.

The calculation tree $T(\mathbf{n}, y)$, of which $RT(\mathbf{n}, y)$ is a combinatorial transform, is in fact a finite substructure of the (dual) fd of the subhierarchy **XP** of **EP** :

1. the (indices of) the vertices of the dual **XP**−fd are 'inwards $\prec$ - ascending', as with those of $T(\mathbf{n}, y)$.

[27] Ref. discussion after VII, def. 32.

2. The steps

$$[\mathbf{m}, [\mathbf{m}^+, i]] = ([\mathbf{m}^+, i^+])_L$$

may be put into correspondence with spine-to leaf (i.e. $\lhd-$) descents in the **XP** diagram. The $n-1$ possible v to $(v)_L$ steps correspond with the $n-1$ possible canonical reductions of ω^n to a successsor polynomial.

$T(\mathbf{n}, y)$ may now be obtained via

1. a $\prec$ –lowest canonical correspondence of vertices, so that e.g. $[\mathbf{1}, y], [\mathbf{2}, y], \ldots, [\mathbf{n}, y]$ in $T(\mathbf{n}, y)$ correspond with $(\psi, 1), (\psi, 2), \ldots, (\psi, n)$ in the dual **XP**–fd.
2. by excising the remainder (viz. an infinite hierarchy of tails) of the **XP**–Diagram.
3. introducing an edge (corresponding to the step v to $(v)_R$ in $T(\mathbf{n}, y)$) to replace every tail excision.

For example, let $x, y > 2$, $n = 5$, and consider the calculation of $\mathfrak{A}_4(x, Y)$, where $Y = \mathfrak{A}_5(x, y)$.

$$\mathfrak{A}_5(x, y+1) = x \star \mathfrak{A}_5(x, y)$$

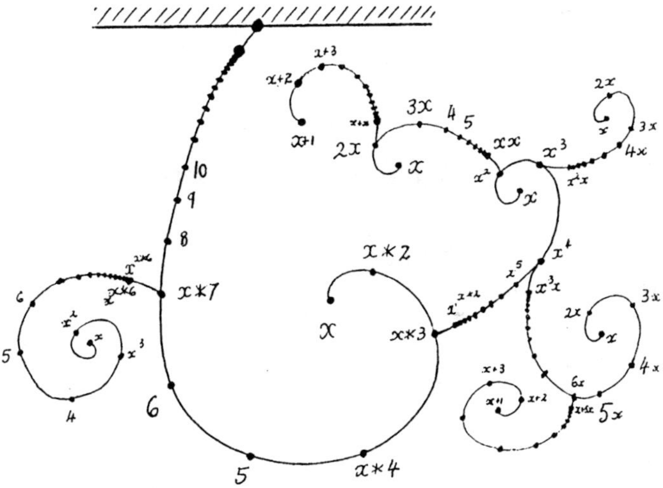

Remarks

1. In some cases '$[\mathbf{m}, i]$' is abbreviated to 'i,' $m \geq 4$.
2. The marked vertices are not intended to indicate convergence; however, generally very many steps occur, before the final step

$$[\mathbf{3}, [\mathbf{4}, i]] = [\mathbf{4}, i+1]$$

from leaf to spine, as indicated by a longer edge.

3. As x, y become larger, so the resemblance to the dual **XP** - fd becomes more evident.
4. This figure is much different from the (finite) $\mathbf{EP}[\varepsilon, r]$ - feather, on account of 2. above.
5. The figure is 'the explication of $\mathfrak{A}_n(x, y)$ by brackets'.

Comparison of Structures

Of course the cfs assignment of defn. 21, applied to the defn. before thm. 35 [28] gives for $\mathfrak{A}$ an (infinite) fd, more complex than for any polynomial $\mu \prec \mathfrak{A}$ in **E2**, in particular for $\mu = \psi$. There is therefore the most remarkable ascent in the hierarchy, in the passage from the (superstructure of the) calculation fd of a polynomial as obtained above, to the diagonalization fd.

Larger Classes

The feather-representation is unlimited: diagrams can be defined for all of the classes $\mathbf{XP}, \mathbf{EP}, \mathbf{EP}^\star, \mathbf{E2}, \mathbf{E}$ and beyond. The 'gradually ascending' property of **EP** (II, §2.10) continues into $\mathbf{EP}^\star$ and its extensions, as precisely described by traversal of the feather diagram.

8.8 The BACHMANN Diagonalization

The class **E** marks the closure of the construction, based on **EP**, using the 4 procedures of normal form, assignment, diagonalization and induction. Two points may be made here. Even thus far is to branch from the scope of the present monograph. The diagonalization of ACKERMANN's function based on **EP** is quite sufficient a demonstration of the potent rates of growth attainable by this means. The main task is to secure the foundations of **EP**, particularly the tree-theoretic foundations, especially with reference to the DCC.

The view advanced here is that the considerable work done on 'ordinal notations' in the 20th century represents at least potential advance, but the flaw is the axiom of choice. In a bin of silicon chips, a flaw here or there is of no consequence, provided they are tested; but a flaw, even the possibility of a flaw, in a hierarchy that ascends far beyond the ACKERMANN function is surely unacceptable. Somehow this work must be transcribed into the finite domain, and a fairly close study of $\mathbf{EP}^\star$ and its enlargements **E2** and **E** are given here as evidence that this project is not an 'engineering impossibility'.

Following HARDY's series of papers [57], [58] early in the century, the construction [29] was carried much further, by VEBLEN [155] in 1908. The method of HARDY of obtaining a hierarchy $(\Delta_\alpha)_{\alpha \prec \varepsilon}$ depends on assigning fundamental sequences to

[28] for which the basic notion is diagonalization, rather than calculation.

[29] Ref. §1.

$\alpha \prec \varepsilon$. This requires the 3 operations of $+, \cdot, \exp$ to create the normal form, so that canonical reduction can act on the resulting polynomials. In the first instance, HARDY assumed the CANTOR normal form, based on the similarity operations $(+), (\cdot), (\exp)$ (II, §18, defns. 39, 42).

In order to extend the Δ_α hierarchy further, VEBLEN extended the series $(+), (\cdot), (\exp), \ldots$ by the method of fixed points, [30] the first step of which is outlined at the end of the above-mentioned section. The device used involved functions of infinitely many variables, and the question of further extension remained untouched for many years. Then in 1950 BACHMANN, at one stroke, re-constructed and vastly extended VEBLEN's hierarchy, by using the most powerful method of diagonalization yet discovered. The long 1950 paper was a mere trial run of the method of downwards reflection from CANTOR's third 'number class', to the second. To every $e \in \mathbf{E}$ there is a hierarchy of functionals corresponding to a particular 'number class'; and the downwards reflection has been effected, by those who have built on BACHMANN's idea, for more 'number classes' than these.

Remark

KLEENE's theory of recursive notations (III·5 ·3) provides a means of determining whether a given assignment of fundamental sequences is recursive. In no way does this theory actually contribute to the work of assigning fundamental sequences.

The FD in Diagonalizations via Non-Finitary Objects

It has proved possible to realize CANTOR's 'ordinal' ε_0 and its theory of fundamental sequences via the finitary hierarchy **EP**. This serves to a) justify the original definition, and b) reveal finitary (e.g. combinatorial) structure (and the construction has been carried beyond ε_0 in Ch. VIII). However, assignment of fundamental sequences to the non-finitary objects in order to probe $\mathcal{N}$-GH (VII, §3·9) has in common with the finitary construction the diagonalization of II, §5·2. Thus, suppose corresponding to each object α_n, $f_n \in \mathcal{N}$ has been defined, where $\{\alpha_n\}_{n=1,2,\ldots}$ is a cfs for α. Then $f = \lambda x \cdot f_x(x)$ is defined to correspond to α. Therefore it is immediate by the fundamental defintion of §7·1, that *all ascending arithmetic growth hierarchies obtained by assigning cfs's to 'ordinal notations', define a* fd.

MILLER [110], 1976 contains explicit equivalencing of work on 'ordinal' notations by ISLES [66], 1968; ACZEL [4], 1972; BRIDGE [16], 1975; TAKEUTI [146], 1960; 1975; using BACHMANN's construction. This is a very interesting field, and a survey volume is much needed.

---

[30]It should be noted that the formalization of set theory, ZF, is no more than an inductively defined system of $\mathfrak{L}$-sentences, which under the standard interpretation of the predicate '$\in$', may be expected to include 'most of', and only those sentences true of what are intuitively accepted to be sets. The theory in no way provides the points of the domain of the 'model'. In particular, if S is 'transitive' (viz. $y \in S \;\&\; x \in y \Rightarrow x \in S$) and 'well-ordered' (VI, §3·2), S is an 'ordinal'. ZF does not construct the ordinals. E.g. the successor $\{\{\ \}, \{\{\ \}\}, \ldots, \{\{\ \}, \{\{\}\}, \ldots\}\}$ of the least 'transfinite ordinal'. The infinity existence axiom of ZF asserts $(\exists S)$ with this property, without actually giving any single instance. Thus in practice 'ordinals' are constructed via repeated applications of the existence schemata to fixed-point theory; and finally (as in §4·2) $\alpha = \{S \mid S$ is similar to the 'α-object' $\}$.

8.9 Various extensions

8.9.1 The SKOLEM-LEVITZ Class

As indicated at the start of the chapter, enlargements of **EP** may be obtained by modification of the operations of closure. If, in place of **EP**.3, the operation of closure

$$\mathbf{EP}.3' \; e, f \in \mathbf{EP}' \Rightarrow \lambda x.\{e(x)\}^{f(x)} \in \mathbf{EP}'$$

is taken, then a considerably enlarged class **EP**′ (as in II, §2·16) is obtained. Firstly, it is clear that for all $e \in \mathbf{EP}'$, there is an $f \in \mathbf{EP}$ such that $e \prec f$. Therefore **EP**′ is 'wider' than **EP** with respect to eventual domination. Nonetheless, RICHARDSON has proved that the system $(\mathbf{EP}, \prec)$ is linear. And by using KRUSKAL's i-j embedding result for finite oriented trees, EHRENFEUCHT [33] has shown that the system satisfies the DCC. The proof of KRUSKAL's result, however, is highly non-constructive, giving little or no information on the 'length' of the ordering. A rendering of EHRENFEUCHT's result, using an idea of NASH-WILLIAMS, is given in Ch. IX, where every attempt is made to eliminate any unnecessary generalities, in order to show clearly that no essentially new proof of the DCC for **EP** is obtained.

The problem of establishing the actual order-type of $(\mathbf{EP}', \prec)$ has not been solved, and appears to be hopelessly difficult. However numerous partial results have been obtained by LEVITZ [85]. For example, the initial segment determined by the arithmetic function 2^{ω} is similar to **XP**. The problem then arises; what function correlates the bound with the similarity type? This may be so complicated that it cannot even be specified by a general recursive algorithm (i.e. a TURING machine). The central problem with **EP**′ is that its new rule of closure destroys the (normal) tree structure of the exponentials, from which the linearity and well-foundedness naturally unfold. Furthermore, the key property of **EP** is finiteness. The recurrent polynomials of **EP**′ are defined finitely, of course, but in the view of LEVITZ [86] 'non-standard analysis' is the natural treatment of **EP**′.

8.9.2 Many Variables

LAUGHLI has extended the notion of eventual domination to functions of several natural number variables. E.g. for 2 variables,

$$e \prec f \Leftrightarrow (\exists N)(\forall x)_{x>N}[\lambda y \cdot e(x, y) \prec \lambda y \cdot f(x, y)];$$

leading to extensions of **EP**′.

8.10 Notes on Ch. VIII

1. 8.10.1 Extensions of Domain

The distinguished polynomial $\psi \in \mathbf{EP}$ (II, §2.1) occurs very commonly, as a function of a continuous variable, often in the logarithmic form $\ln\psi = \omega\ln\omega$.

(a) By the prime number thm., the fraction of numbers which do not exceed n, and which are prime, is asymptotic to $\frac{1}{\ln n}$. Therefore the nth prime p_n has $p_n \sim \ln\psi(n)$.

(b) ψ occurs in the definite integral

$$\int_0^1 \frac{dx}{\psi(x)} = \sum_{n\in\mathbf{N}} \frac{1}{\psi(n)} = 1.29128\ldots.$$

(c) LINDELÖF's mean. This form of summability is of importance in the theory of analytic continuation, and makes use of ψ. The applicability of the method follows from its success in the following special case. The series

$$1 + z + z^2 + z^3 + \ldots$$

converges to $\frac{1}{1-z}$ within the disc $z \leq 1$, and diverges outside the disc. However the series may be transformed using ψ, to obtain an analytic continuation:

THEOREM 8.39

$$\lim_{\delta\to 0}\sum_{r=0}^{\infty}\psi^{-\delta}(r)\cdot z^r = \frac{1}{1-z},$$

for all $z \in \mathbf{C} - (1,\infty)$.

Proof, HARDY, *'Divergent Series'*, §4.11, p77.

□

So taking $i = \sqrt{-1}, z = 2i$, the rapidly diverging series $1 + 2i - 4 + 8i - \ldots$ is transformed to $\lim\sum(r^r)(2i)^r = 1/5 + 4/5i$, as may easily be verified. Now let $\sum a_r z^r$ be any complex power series, convergent for small $\mid z \mid$. The series defines an analytic function with a branch, $f(z)$ say, which is regular at the origin. The MITTAG-LEFFLER star of $f(z)$ is obtained by tracing rays from the origin to each singularity of f; and by removing from the plane all points on rays at and beyond each singularity. Thus the domain of convergence in the above thm. is the M-L star of $\frac{1}{1-z}$.

COROLLARY 8.40 *For all z in the M-L star of f, $\lim\sum a_r\psi^{-\delta}(r)z^r$ converges.*

Proof, uses the above thm., ibid., §8.10, p190.

□

Note that the matrix form of the LINDELÖF transform of a power series is $\lim_{\delta \to 0} A$, where A is

$$\begin{bmatrix} 1 & 0 & 0 & \\ 0 & 4^{-\delta} & 0 & \cdots \\ 0 & 0 & 27^{-\delta} & \\ & \vdots & & \end{bmatrix}$$

8.10.2 Passing Values and STIRLING Coefficients

In Ch. II, §2, the *passing value* for $e \prec f$ in **EP** is defined to be the least $M \in \mathbf{N}$ such that $(\forall x \geq M)\ e(x) < f(x)$. Given any $n \in \mathbf{N}$, can $e, f \in \mathbf{EP}$ be found such that the difference $e(x) - f(x)$ changes sign n times, before x reaches the passing value? This problem is readily solved in the affirmative, by taking an extension of the domain from $\mathbf{N}$ to $\mathbf{R}$. Indeed, such $e, f \in \mathbf{XP}$ may be found. Given $k \in \mathbf{N}$, the polynomial

$$P(x) = \prod_{i=1}^{n} (x - ki)$$

of the continuous variable x has the n zeroes $k, 2k, 3k, \ldots, nk$. And P may be resolved into a difference $P = f - e$; the polynomials $f, -e$ respectively contain exactly the positive and negative terms of P. It remains is to choose a suitable value of k, and show that e, f exhibit the required differences between consecutive zeroes. Let $\left[\begin{matrix} x \\ y \end{matrix} \right]; i, n \in$ $\mathbf{N}$; denote the STIRLING numbers (for values, see Appendix C) of the first kind, in the notation of KNUTH ([77], §1.2.6, p66). Via the standard identity

$$\left[\begin{matrix} n+1 \\ i \end{matrix} \right] = n \left[\begin{matrix} n \\ i \end{matrix} \right] + \left[\begin{matrix} n \\ i-1 \end{matrix} \right],$$

a proof by induction gives an explicit expression for the real polynomial P :

$$P(x) = \sum (-1)^{n+1-i} \left[\begin{matrix} n+1 \\ i \end{matrix} \right] k^{n-1+i} x^{i-1}. \tag{I}$$

Thus

$$f(x) = \left[\begin{matrix} n+1 \\ n+1 \end{matrix} \right] x^n + \left[\begin{matrix} n+1 \\ n-1 \end{matrix} \right] k^2 x^{n-2} + \ldots$$

and

$$e(x) = \left[\begin{matrix} n+1 \\ n \end{matrix} \right] k^{k-1} + \left[\begin{matrix} n+1 \\ n-2 \end{matrix} \right] k^3 x^{n-3} + \ldots.$$

The derivative $P' = \sum P/(x - ki)$ of P is non-zero at the zeroes of P. Therefore P alternates in sign as x passes through each zero. And, since the above are consecutive zeroes, by BOLZANO's theorem, in each open interval

$$(-\infty, k),\ (k, 2k), \ldots,\ (nk - k, nk),\ (nk, \infty)$$

the continuous function $P \neq 0$ remains of fixed sign. The coefficients of e, f range over $\mathbf{N}$; and as discrete functions, $e \prec f$, since the leading term of P, x^n, has positive coefficient. Now let $k = 2$. Then for $N > 2n, e(N) < f(N)$; and $e(2N) = f(2N)$. And the above observations on P in the region of its zeroes give

$$e(2n-1) > f(2n-1),\ e(2n-2) = f(2n-2),\ e(2n-3) = f(2n-3), \ldots,$$

with $e(1) < f(1)$ or $e(1) > f(1)$ respectively as n is even or odd. Thus $e, f \in \mathbf{EP}$ have $n-1$ 'pre-passes' before the last after $x = 2n$.

Example. Let $k = 2, n = 6$. Then the expression (I) gives

$$P(x) = x^6 - 2\left[{7 \atop 6}\right] x^5 + 4\left[{7 \atop 5}\right] x^4 - 8\left[{7 \atop 4}\right] x^3 + 16\left[{7 \atop 3}\right] x^2 - 32\left[{7 \atop 2}\right] x + 64\left[{7 \atop 1}\right],$$

therefore

$$e(x) = 42x^5 + 5880x^3 + 56448x, \qquad f(x) = x^6 + 700x^4 + 25984x^2 + 46080,$$

giving

$$e(12) = 21288960 = f(12), \quad e(11) - f(11) = 945, \quad e(10) = f(10),$$
$$f(9) - e(9) = 315, \ldots\ldots, e(2) = f(2), \quad e(1) < f(1).$$

Remarks.

(a) Given any $n \geq 1$, taking $k = 1$, the functions e, f derived from P satisfy $e \prec f@n+1$, and $e(x) = f(x)$ for $x = 1, 2, \ldots, n$.

(b) Consider e.g. $d = f(2n-1) - e(2n-1)$. The foregoing analytic argument establishes that $d > 0$, but without giving any further information on the magnitude of d. But d is an integer, so $d \geq 1$. This result has been obtained by an application of the law of the excluded middle, rather than by calculation.

(c) Note that the STIRLING numbers of the second kind arise as coefficients in an additive 'triangular' table (Appx. C).

I am most grateful to DR. DAVID SINGMASTER for pointing out this application of STIRLING numbers.

2. An inverse of the operation $\star$ of §2.2 may be obtained by writing $\sqrt[\star]{n}$ as the smallest real number x for which $x \star 2 = n$. Thus, e.g.

$$\sqrt[\star]{2} = 1.5596104\ldots.$$

This also raises the problem of evaluating expressions such as

$$F(x) = \int_1^x \sqrt[\star]{y}\, dy.$$

3. The $\star$ operation also leads [31] to 'super-diophantine' equations. E.g. given natural numbers m, n, what are the least solutions in integers, if any, to the equation

$$1 + x \star m = y \star n?$$

The possibilities may be greatly widened, by admitting binary operations of the sequence VII·15 in the production of the equations.

[31] Ref. MORDELL, *Diophantine Equations* , Academic Press, London; and LANDAU, *Theorems of Arithmetic in Their Historical Order*. MORDELL cites the work *Diophantische Gleichungen* , Publ. Springer, Berlin (1938), by Th. SKOLEM.

4. §4.11 In the introduction to [131], ROSE asserts that the theory of **E2** is 'finitist' on account of 'powerful' arguments advanced by KREISEL. This is beyond doubt, as is demonstrated by the inductive defn. 10 of **E2** from the finitary sequential hierarchy (Δ_e).
5. §2.4. e has the 3rd canonical form $e = \sum a_i\omega \star \alpha_i + \sum b_i\omega^{\beta_i}$ in $\mathbf{EP}^\star$.
6. §7.2 The $\mathfrak{A}$- type RT -fd of $\varepsilon = \lambda x \cdot x \star\star 2$ may be expressed:

$$\varepsilon(x) = x^{x^{x^{\cdots^{x^x}}}} = x^{x^{x^{\cdots^{x \cdot x \cdot \ldots \cdot (x+x+\ldots+x+1+1+\ldots+1+1)}}}}.$$

7. **Irreducibility in $\mathbf{EP}^\star$**

The object is to show $\lambda x \cdot x \star e = \lambda x \cdot x^\alpha$ is impossible in $\mathbf{EP}^\star$, with Diag(e). The approach is to show the relation is impossible for certain fixed x. Since, regardless of the linearity of $\mathbf{EP}^\star$, $\omega^\alpha = \omega \star e \Rightarrow \alpha \prec \omega \star e$, this implies the 'VIII·16' form

$$\alpha \prec \omega \star e \Rightarrow \omega^\alpha \prec \omega \star e.$$

Notation. As in II, §3 ·5, $x, y \in \mathbf{N}$, $x \mid y \Leftrightarrow_{df} x$ divides y without remainder.

LEMMA 8.41 *(a) Suppose* Suc(e) *in* $\mathbf{EP}^\star$. *Then*

$$(\exists\alpha \in \mathbf{EP}^\star)(\exists K \in \mathbf{N})\ \{\alpha = \mathbf{1} \vee \mathrm{Diag}(\alpha)\}\ \&\ e = \alpha + \mathbf{K}.$$

(b) For each diagonal e in $\mathbf{EP}^\star$, $x \mid e(x)$.

(c) Suppose $f \in \mathcal{N}$ is si. Then $x \mid x \star f(x)$.

(d) $\omega \star e = \omega^u$ *is impossible in* $\mathbf{EP}^\star$, *for* Suc(u).

Proof.

(a) Suppose otherwise. Then $e^-, e^{--}, \ldots \in \mathbf{EP}^\star$. Now let $x = a$ be fixed, and consider $b = e(a)$. Then $\beta = e^{-\cdots-}$ to b terms would be a successor in $\mathbf{EP}^\star$, with $\beta(a) = 0$, which is impossible.

(b) True for $\mathbf{EP}^\star$'s initial functions. And if true for e, α, then true also for $e + \alpha$, $\omega^e; \omega \star e$. Therefore by induction, also for all $\mathbf{EP}^\star$.

(c) Taking $1 < x$, x is a divisor of x^F, $F = x \star \{-1 + f(x)\}$.

(d) Suppose Suc(u). Firstly, $\omega \star \mathbf{1} \prec \omega^u$, so $e \neq \mathbf{1}$. Secondly by lemma 41·1, $(\exists v)\ u = v + \mathbf{K}$, v an $\mathbf{EP}^\star$–diagonal, for some natural number K. Well,

$$\{(\forall x)\ x \star e(x) = x^{u(x)}\} \Rightarrow u(x) = x \star \{-1 + e(x)\}.$$

Now let $x = K + 1$. Then $x \mid u(x)$, by lemma 41·3. But by lemma 41·2, $x \mid v(x)$, since Diag(v). Therefore $x \not\mid u(x) = v(x) + x^-$, a contradiction.

□

Remark If $e = \mathbf{1}$, $\omega \star e = x^u$ is satisfied by $u = \mathbf{1}$. If Suc(e), $\omega \star e = \omega^u$ is satisfied by $u = \omega \star e^-$.

Proof of Irreducibility

THEOREM 8.42

$$\omega \star e = \omega^u \tag{I}$$

is impossible in **EP*** *for* Diag(e).

Proof, by contradiction. Suppose (I) holds. The case for Suc(u) was proved in lemma 41·4, so it may be assumed that Diag(u). $u \neq \omega \star v$ in **EP***, since

$$\omega^u = \omega \star v^+ \neq \omega \star e, \text{ for Diag}(e).$$

The remaining possibility is that u is of the form

$$u(x) = x \star \{K; v(x)\},$$

for some $K > 0$, and $v = e_1 + e_2$ in **EP***. Thus (I) gives

$$\omega \star e = \omega^{\omega \star \{K;v\}} = \omega \star \{K+1; v\}. \tag{II}$$

But for $K + 1 \leq x$,

$$x \star e(x) = x \star \{K+1, x \star (e(x) - K - 1)\} \tag{III}$$

The expressions (II), (III) may therefore be equated, giving

$$v(x) = x \star \{e(x) - K - 1\},$$

and for $K + 2 \leq x$,

$$v(x) = x^{E(x)}, \text{ where } E(x) = x \star \{e(x) - K - 2\}.$$

Now suppose

$$v = \sum_{i=1}^{R} \omega \star_{e_{(i)}} v_i,\ 1 < R,\ e_{(i)} = \mathbf{3} \text{ or } \mathbf{4},\ \ v_i \in \mathbf{EP}^\star$$

(it was supposed $v = e_1 + e_2$ is diagonal in **EP***). Suppose $i \leq R$ is such that $e_{(i)} = \mathbf{4}$. Then

$$\omega \star v_i = \omega^{w_i}, \text{ where } w_i(x) =_{df} x \star \{-1 + v(x)\}$$

(i.e. $w_i \notin$ **EP***). Then

$$v(x) = \sum_{i=1}^{R} x^{v_i(x)},$$

where for each i s.t. $e_{(i)} = \mathbf{4}$, v_i has been replaced by w_i, and then 'w_i' has been re-labelled 'v_i'. Now consider $m = \max(K+2, R+1)$. Let $k = (\mu i \leq R)(\forall j \leq R)\ v_j(m) \leq v_i(m)$. Well, $v(m) = m^{E(m)}$, where

$$E(m) = m \star \{e(m) - K - 2\}.$$

Therefore

$$\begin{aligned} m^{v_k(m)} &< \sum_{i=1}^{R} m^{v_i(m)}, \quad (1 < R) \\ &= v(m) = m^{E(m)}. \end{aligned}$$

I.e. $v_k(m) < E(m)$, so

$$\begin{aligned} m^{E(m)} &\geq m^{v_k(m)+1} = m \cdot m^{v_k(m)} \\ &\geq (R+1) \cdot m^{v_k(m)} \text{ by definition of } m \\ &> R \cdot m^{v_k(m)} \geq \sum_i m^{v_i(m)} = v(m), \end{aligned}$$

a contradiction. (I) is therefore impossible for Diag(e).

□

Remark. The property Diag(e) is used only in the 1st part of thm. 42, viz. the proof that Diag(e) $\Rightarrow u$ is not of the form $\omega \star v$. The second part of the thm. establishes $\omega \star e \neq \omega^u$ in $\mathbf{EP}^\star$, where u 'contains a sum'.

Chapter IX

FURTHER FINITENESS PROPERTIES OF EP

Section Headings

9.1 Introduction

Finiteness properties have played a key rôle in preceding chapters, e.g. with respect to $\prec$: in II·25 ('f dominates its exponents'); and II·47 ('if $e \prec f$, then f passes e by 1+ the greatest explicit coefficient of e'). Also with respect to the DCC, Ch. IV, corol. 48 (and to its further corollary, VIII·19), with respect to images, VII·49, and to finiteness of number of regular assignments, III·5·3. It may also be said that gen.

ind. is used (VI·23) to prove the finiteness of definition by gen. rec. And all these properties find some degree of generalization in the VIII-extensions of **EP**.

Here, further finitenesses are considered, especially those arising from the relation of immediate domination, Ch. II, defn. 8.

9.2 Eventual Domination in EP

9.2.1 Primitive Finiteness Property

The following basic lemma, which lies on the route to the main thm. of this §, relates the value $e(x)$ to the order $\mid T \mid$ of the normal tree $T(e)$. Let $\mathfrak{T}$ be as in IV·19; $\Phi : (\mathfrak{T}, \prec) \to \mathbf{EP_0}$ as in IV·47; and let $T \in \mathfrak{T}$ have $(n+1)$ vertices, where $e = \Phi(T)$.

LEMMA 9.1 $x > 1 \Rightarrow e(x) \geq n$.

Proof, by induction on edge-height.

□

As noted after II·35, **EP** contains a spanning subsystem of exponentials e, for which $e(1) = 1$. In contrast, given $y > 1$,

THEOREM 9.2 **EP** *contains only finitely many e for which*

$$(\exists x > 1)\, e(x) = y.$$

Proof. Suppose $(\exists x > 1)\, e(x) = y$. Consider the normal tree $T(e)$. Since $x > 1$, by lemma 1, $T(e)$ has no more than $y + 1$ vertices. But, as is beyond dispute, the number of of trees which can be formed from a finite number of vertices is *finite.* By IV·47, $e = \Phi(T)$ is unique. Thus the number of e with $\mid T(e) \mid \leq y + 1$ is finite.

□

Example. Given the integer M, $e \prec f @ M \Rightarrow f(M^-) \leq e(M^-)$. There are therefore only fm such $f \in \mathcal{R}$, the remainder of **EP** defined by e. Conversely, $e(M) < f(M)$; therefore for each f, there are only fm such $e \in \mathcal{S}$, the segment defined by f.

Consider any diagonal e, and let $e_n = (e, n)$. As e increases with $\prec$, by thm. 2, after an ascent through the numbers $e_n(2)$, $n = 1, 2, 3, \ldots$, and after encountering a descent to the number $e(2)$, it is found that the smaller values have been 'used up', imposing a lower bound on $e(2)$. Therefore

THEOREM 9.3

$$\lim_{\alpha \in (\mathbf{EP}, \prec)} \{\inf_{\alpha \prec e} e(2)\} = \infty.$$

□

9.2.2 Vertex Order and Passing Number

Multiplicity

Suppose $e = \sum_{r=0}^{0} a_i \omega^{e_i}$. Then writing as usual $T(\mathbf{0}) = [\,]$,

$$| T(e) |= 1 + \sum a_i \mid T(e_i) \mid .$$

Thus by defn. II·44,

LEMMA 9.4 $N(e) \leq\mid T(e) \mid .$

□

So by lemma 1, $N(e) \leq e(2)^{+}$; by II·47, $e \prec f@M$ in **EP** implies $M \leq N(e)$. Therefore lemma 4 gives $M \leq\mid T(e) \mid$, and another proof of thm. II·48:

LEMMA 9.5 $M \leq e(2) + 1$.

□

Notation. It may be convenient to replace '$e \prec f@M$' by '$e \prec_M f$.' **Remark.** The notion of diagonal $e = \text{diag}_n\ e_n$, (viz. for all x, $e(x) = e_x(x)$; III, defn. 3), does not in itself rule out the possibility that

$$e \prec e_n,\ n = 1, 2, 3, \ldots . \tag{I}$$

But this gives $e \prec e_n@M = M(n) > n$ for each n. And as in the remark before the lemma, (I) implies $(\forall n) \mid T(e) \mid> M(n)$, impossible since $\mid T \mid$ is finite. Therefore no $e \in \mathbf{EP}$ is the diagonal of a sequence of terms $\succ e$. It may be remarked that this observation in no way uses IV·48, which rules out

$$e \prec \ldots \prec e_n \prec \ldots \prec e_2 \prec e_1 .$$

9.2.3 A Second Proof of II·47

The uniformity proof given in II §3 may be re-cast, by applying specialized definitions to the notion of separating chain, IV, §3·2.

Complete Polynomials

Let $\zeta \in \mathbf{EP}$ be of x-ht h, and let a =mult(ζ) be the largest ec of ζ. The *complete polynomial* $\xi_{a,h}$ *on* a *of height* h is defined by:

$$\xi_{a,o} = \mathbf{a};\ \text{ and for } h > 0:\ \xi_{a,h} = \sum_{0 \leq i < (a+1) \star h} a \cdot \omega^{e_i},$$

where $\mathbf{0} = e_0 \prec e_1 \prec e_2 \prec \ldots$ is the (ascending) enumeration of the (complete) class $\mathcal{X}_{a,h} \cup \{\mathbf{0}\}$ of all $e \in \mathbf{EP}_0$ for which each ec $\leq a$; and x-ht$(e) < h$ (see II·45; also corol. after III·22, and V·15). Thus, if $h = 1$, $\xi = a \cdot \omega^a + a \cdot \omega^{a-1} + \ldots + a \cdot \omega + \mathbf{a}$. **A property** of $\mathcal{X}_{a,h}$ follows imediately by induction on x-ht: for every $1 \leq N < (a+1) \star h$, $(\exists \alpha \in \mathcal{X}_{a,h})\ N = \alpha(a+1)$. Now suppose the multiplicity $a > 2$, and let x take any value $a+1, a+2, \ldots$.

LEMMA A 1. $\zeta(x) \leq \xi_{a,h}(x)$; 2. $\xi_{a,h}(x) < x \star (h+1)$.

Proof. 1. Since a is the max ec of ζ, by defn. ζ is a subpolynomial of $\xi_{a,h}$.
2. Follows from the case $x = a+1$; for which it may be shown

$$\xi_{a,h}(a+1) = -1 + (a+1) \star (h+1).$$

Thus initially, $\xi_{a,0}(a+1) = x - 1$; and given the relation for x-ht h, the inductive step is but an application of the above property to II·46:

$$\begin{aligned} x \star (h+2) &= x \cdot x^{-1+x\star(h+1)} = a \cdot x^{\xi_{a,h}(x)} + x^{\xi_{a,h}(x)} \text{ by i.h.} \\ &= \sum_{k=1}^{\xi_{a,h}(x)} a \cdot (a+1)^k + a + 1 = \sum_{e_i \in \mathcal{X}_{a,h+1}} a \cdot x^{e_i(x)} + a + 1 \text{ by II} \cdot 46 \\ &= \xi_{a,h+1}(a+1) + 1. \end{aligned}$$

□

Remark. The form $\xi_{a,h}$ contains $a \star (h+1) + 1$ terms.

Comparison of Lower Parts

Let $\alpha \prec e$ @ M be an eventual domination in **EP**. Let $T\alpha = S$, $Te = T$. Let $C = C(S,T)$ be the separating chain for $S \prec T$; and consider the upper part $T_C (= S_C)$ and the lower parts S^C, T^C. Let the subpolynomials $\hat{\alpha}, \hat{e}$ of α, e respectively be defined to be those corresponding (via IV, §4) to S^C, T^C; and the subpolynomial β of both α, e to be that corresponding to T^C (ref. IV, §4 ·3, ex. 5). The combining of the lower part $\hat{\alpha}$ (or $\hat{e}$) with the part β to give α (or e) is not always a process of addition. However, since the terms of the common part β contribute equal values:

LEMMA B $(\forall x)\ \{\alpha(x) < e(x) \text{ iff } \hat{\alpha}(x) < \hat{e}(x)\}$.

□

Turning now to the main result, by defn. of C above, eh$S^C = h+1 \Rightarrow$ eh$T^C =$ eh$C = h+2$ (where $h \geq 0$). Now suppose mult$(\hat{\alpha}) = a$.

THEOREM [II·47] The greatest value for which $e(x) \leq \alpha(x)$ is $x = a$.

Proof. Let $x > a$. Then by lemma B, it suffices to prove

$$\begin{aligned} \hat{e}(x) &\geq x \star (h+1) \text{ since } x\text{–ht}(\hat{e}) \geq h+1 \\ &> \xi_{a,h}(x) \text{ by part 2. of lemma A} \\ &\geq \hat{\alpha}(x) \text{ by part 1.} \end{aligned}$$

□

9.2.4 Fine Classification of ec's

Let e be as in §2·2. Explicit coefficients a of e may be individually ranked within e in terms of $\prec$ –influence, via R-to-L pre-order.

Tree of Coefficients

The tree τ of V, §5·2 has vertices (indexed by) the ec's of e; the incidence relations are defined by: if α is an explicit exponent of e, occurring as $\ldots^{a\cdot\omega^{\alpha}}$ in e, and b is an ec of α, then $b \;\widetilde{\mathrm{Sc}}\; a$. Now let a, b be ec's of e. Let $\alpha, \beta \in \mathbf{EP}$ be obtained respectively from e by adding a unit to the values of the ec's a, b. Thus, e.g. the ec of value a in e occurs in α with value $a+1$. Other ec's of e which happen to have the value a remain invariant.

THEOREM $a <_{\text{pre}}^{\text{R-to-L}} b$ in $\tau(e)$.

If $e = \Sigma a_i\omega^{e_i}$, then evidently, whether or not $e \in \mathbf{XP}$, an increment upon a_{i+1} causes a greater $\prec$-increase than one upon a_i. Assuming now, if necessary, the pre-order result in $\tau(e_i)$; by induction the result is obtained generally for all ec's of e.

□

Remarks

1. Taking e.g the basic ec's of $e = \psi + \omega^2$ establishes that '$\alpha \ll \beta$' may not be substituted for the conclusion of the thm.
2. By the rem. at the end of IV, §2·5, the reverse of the pre-ordering of coefficients is the L-to-R post-order. Therefore by the thm. and IV·17, the $\prec$-influence of the coefficient corresponding to n_i in $e = \varepsilon_{n_1 n_2 \ldots n_R}$ diminishes with increase in i; as follows from V, §1.

9.2.5 $\prec$-Transitivity and Passing Value

Consider an ascending scale

$$e_1 \prec_{M_1} e_2 \prec_{M_2} e_3 \prec_{M_3} \cdots \qquad \text{(I)}$$

in **EP**. Is $2 < M_1 \le M_2 \le M_3 \le \ldots$ possible for (I) infinite? Well, were it the case that $\alpha \prec_M e \Rightarrow e(x) \le \alpha(x)$ for *all* $x \le M$, then such a scale would be ruled out: $e_1(M_1^-) \ge e_2(M_1^-) \ge e_3(M_1^-) \ge \ldots$, for which the e_i are finite in number, by thm. 2. Note that this argument also proves that for no constant $M > 2$ can (I) be infinite, while $M_1 = M_2 = \ldots = M$. It is easy to find a scale (in **XP**) for which the M_i are *unbounded above*. If the terms are restricted to **XP**, such that $e_i = a_i \cdot \omega^{n_i}$, then $n_i < n_{i+1}$ and $M_{i-1}^{n_{i+1}-n_i} \le \frac{a_i}{a_{i+1}}$, and (I) is finite in all cases for which $M_i \ge 3$.

The special case $e_1 \lhd_{M_1} e_2 \lhd_{M_2} e_3 \lhd_{M_3} \ldots$ of (I) is finite. By the property of the ei, III, §3·3, e_{i+1} is uniquely determined by e_i, $i = 1, 2, 3, \ldots$. But by V, §2·2, rem. 3, $e_{i+1} = \varepsilon_{\underline{n}} \Rightarrow e_i = \varepsilon_{\underline{n}, M_i}$. Therefore, given the V·12 normal form $e_1 = \varepsilon_{n_1 n_2 \ldots n_R}$, it follows $e_1 = \varepsilon_{M_R M_{R-1} \ldots M_1}$, and the given ascending canonical scale contains $R < \infty$ terms. The ascent was only made possible by e_1 'containing' (via V·14) all the numbers M_i.

Remark

The last term e_1 of the (exponential sequence corresponding to the) LOEBL-MATOUĈEK reduction (VII, §5·1) gives such an ascent, for which $e_R = i \star (1 + x\text{-hte}_1)$.

Returning to the general problem, it appears necessary to show:

1. that e_i must contain an ec whose value depends on M_i (as in II·47); and

2. $e_{i+1}(M_i^-) \leq e_i(M_i^-)$ implies that each ec a of e_{i+1} must be 'covered' by an ec of e_i, for which there is some increase beyond a, either in terms of value, or of 'rank'.

However, it seems unlikely that this can be done, without first overcoming the problems highlighted in II·47, ex. 4 and IV ·50, ex. 6.

9.3 Main Sequence Structure

Some specialized finiteness properties of **EP** are considered here, including STACEY's main sequence lemma, as in Mc BETH [103], (1987).

9.3.1 Main Sequences in EP

Let $\mathcal{S} = e_1, e_2, \ldots$ be an arbitrary [1] endless sequence of pairwise distinct points in a sequentially ordered system such as **EP**.

THEOREM 9.6 $(\exists i)\ e_i \prec e_{i+1}$.

Proof. By II, thm. 18, '$\prec$' is connected, so the falsehood of the thm. would imply $e_1 \succ e_2 \succ \ldots$, contrary to the DCC IV, §4.

□

With $\mathcal{S}$ as above,

THEOREM 9.7 (STACEY) [2] *$\mathcal{S}$ contains an infinite sub-sequence, the* main sequence *of $\mathcal{S}$, which comprises a finite* head $(e_{m(i)})_{i=1,2,3,\ldots}$, *whose terms are followed by a recursively definable* tail $(e_{n(i)})_{i=1,2,\ldots}$, *where*

$$e_{n(1)} \prec e_{n(2)} \prec \ldots .$$

There are only fm natural numbers $m(1) < m(2) < \ldots$ for which

$$(\forall j)\ j > m(i) \Rightarrow e_j \prec e_{m(i)},\ i = 1, 2, 3, \ldots,$$

since otherwise, as in thm. 6 the endless descent $e_{m(1)} \succ e_{m(2)} \succ \ldots$ contradicts the DCC. $m(1), m(2), \ldots$ thus ends at a greatest term, $m(K)$, say: $m(1) < \ldots < m(K)$. So $(e_{m(i)})_{1 \leq i \leq K}$ may be taken as the head of $\mathcal{S}$. Now let

$$n = \begin{cases} 0 & \text{if } (\forall i)(\exists j)\ e_i \prec e_j \\ m(K) & \text{otherwise.} \end{cases}$$

The infinite sequence $n(1), n(2), \ldots$ is defined by: $n(1) = 1 + n$; and for $i \geq 1$, given $n(i) \geq n$, $n(i+1) =$ the smallest $j > n(i)$ such that $e_{n(i)} \prec e_{n(j)}$ (j is always found, because $e_{n(i)}$ is not in the head of $\mathcal{S}$). Then $e_{n(1)} \prec e_{n(2)} \prec \ldots$ may be taken as the tail of $\mathcal{S}$.

□

Remarks.

[1] It may be assumed that $\mathcal{S}$ is predicatively given.

[2] This construction is based on NASH-WILLIAMS [17], 1967, and follows that of IV·77.

1. In the proof, $n(1) = 1 + n$, but any value $n(1) > n$ gives $e_{n(1)}$ not in the head of $\mathcal{S}$.

   Suppose $k > n + 1$ is the smallest index such that $e_{n+1} \prec e_k$. Now suppose $n + 1 < n(1) < k$ is chosen, as may be possible for certain $\mathcal{S}$. It is also possible that

   $$e_{n+1} \succ e_{n(1)}$$

   so that there may be an index $< k$, to be taken as $n(2)$, such that

   $$e_{n(1)} \prec e_{n(2)} \prec e_k.$$

   (Note that $e_{n(2)} \prec e_{n+1}$, with equality ruled out). Again it is possible that for some $n(3) < k$, $e_{n(2)} \prec e_{n(3)} \prec e_k$, and the process may be repeated. But, after fewer than k ascents, e_k is reached. Thus, aside from an initial segment, the tail is a fixed subsequence of $\mathcal{S}$. Since $e_{n(i)} \succeq e_r$, $n(i) \leq r < n(i+1)$, the tail spans the remainder $e_{n+1}, e_{n+2}, e_{n+3}, \ldots$ of $\mathcal{S}$.

2. If $\mathcal{S}$ is an enumeration of **EP**, then $n = 0$; and $\sup_i e_{n(i)} = \varepsilon$.

3. The numbers $m(K), n$; required in the proof are not obtained by constructive means. Thus, given the notion of an arbitrary sequence $\mathcal{S}$ in **EP**, the theorem assures that after a finite number of steps, the number $m(K)$ will be reached; but there is no way generally of knowing when this has happened. For example, take 2 sequences $e_1, e_2, e_3, \ldots,\ \ \alpha_1, \alpha_2, \alpha_3, \ldots$; where every term of the first is $\prec$ −greater than every term of the second sequence, as with $\alpha_n = \omega \star (k; n)$; $e_n = \omega \star (k+1; n)$, for some fixed $k \geq 1$. Now let $GOLD(n)$ be as in VII·26. Then $\mathcal{S} = E_1, E_2, E_3, \ldots$ defined by

   $$E_n = \begin{cases} e_n & \text{if } GOLD(n) = 1, \\ \alpha_n & \text{otherwise.} \end{cases}$$

   would have a very large $m(K)$ if the classic conjecture turns out to be false.

9.4 Immediate Domination in EP.

Recall the defn. II·35 of immediate domination $e \ll f$ of e by f (in **EP**): $e(1) \leq f(1)$ & $(\forall x > 1)\ e(x) < f(x)$. Of course: $e \ll f \Rightarrow e \prec f$.

THEOREM 9.8 *The system* $(\mathbf{EP}, \ll)$ *satisfies the DCC.*

Proof. By the remark this is a special case of IV, corol. 48. But the corollary rests on the argument of IV, §3. IX·8 can be proved so easily, compared with the argument for $\prec$, by: if $\ldots \ll e_2 \ll e_1$, then $\ldots < e_2(2) < e_1(2)$. Therefore the given $\ll$ −descending sequence contains at most $e_1(2)$ terms.

□

Immediate domination occurs in every relation $e \prec f$ in **EP**. Either f contains every term of e, in which case $e \ll f$, or for some $h \geq 1$, $f_{(h)}$ is the dominating explicit exponent of f for e, as in IV·50·2. Then the corresponding ee's of e give

$$e_{(1)} \prec f_{(1)},\ e_{(2)} \prec f_{(2)}, \ldots,\ e_{(h-1)} \prec f_{(h-1)},\ e_{(h)} \ll f_{(h)},$$

by rem. 2 after IV·49.

Now consider $e \ll f$ so that $e(2) < f(2)$. By thm. 2, for each of the values $y = 1, 2, \ldots, f(2) - 2, f(2) - 1$, there are only fm e such that $e(2) = y$. Therefore

THEOREM 9.9 $(\forall f)$ *there are only fm e such that $e \ll f$, in* **EP**.

□

Remark. $\alpha \in$ **EP** is associated with a finite tree T, whose main vertex is α; and such that the successors of each vertex β are the $\gamma \ll \beta$. By thm. 9, the vertices are of finite valency. By thm. 8, the chains are of finite length. Therefore by IV·15, KÖNIG's lemma, T is *finite*.

9.4.1 EHRENFEUCHT's Construction

The notion of well-partial-ordering (or well-quasi-ordering, wqo), was introduced in IV, §8 (defn. 71). As with linear well-foundedness, there are different characterizations, depending on whether or not the AC is used. Upholding the principle of AC-excision, the defn. of IV, §8 will be used. The logically stronger notion, founded on the AC, is: $(\mathcal{X}, <)$ is wqo iff po and every $\mathcal{Y} \subseteq \mathcal{X}$ has fm minimal elements. The following is a result of EHRENFEUCHT [33], 1973, whose proof uses KRUSKAL's thm. IV·77, [81] (1960). The proof [3] given here is a derivation from first principles, which avoids the AC. Let $\mathcal{S} = e_1, e_2, \ldots$ be any given infinite sequence of pairwise distinct functions in **EP**.

THEOREM 9.10 [4] $(\exists i, j)\, i < j \;\&\; e_i \ll e_j$.

Proof. By thm. 7, $\mathcal{S}$ contains a main sequence with an infinite tail $e_{n(1)} \prec e_{n(2)} \prec \ldots$. By rem. 1 following the thm., $n(1)$ may be any value greater than a fixed n determined by $\mathcal{S}$. Thus, let $y = \min_{n<i} e_i(1)$. Let $n(1)$ be the smallest integer $> n$ such that $e_{n(1)}(1) = y$. Now suppose

$$\neg e_{n(1)} \ll e_{n(2)} \;\&\; \neg e_{n(1)} \ll e_{n(3)} \;\&\; \ldots . \tag{I}$$

Then by II, defn. 35, there are $q(1), q(2), \ldots$, such that

$$e_{n(2)}\{q(1)\} \leq e_{n(1)}\{q(1)\} \;\;\&\;\; e_{n(3)}\{q(2)\} \leq e_{n(1)}\{q(2)\} \;\;\&\;\; \ldots, \tag{II}$$

[3] McBETH [103] (1987), preprint.

[4] [103], thm. 4·4.

where by the choice of $n(1)$, and by the above defn., it may be assumed that $q(i) > 1, i = 1,2,3,\ldots$. The uniformity result II·47 states:

$$(\exists N)(\forall e \in \mathbf{EP})\ e_{n(1)} \prec e@M \Rightarrow M \leq N,$$

viz. N is an upper bound for the sequence $q(1), q(2), \ldots$. Thus, let $\sup_{1\leq i} q(i) = q$, and let $Y = e_{n(1)}(q)$. Then by thm. 2 there are only fm e such that $e(\bar{x}) = y$ for some $x > 1$ and $y \leq Y$. Therefore the inequalities (II) are finite in number. Viz. the conjunction (I) is finite, ending in a last term, $\neg e_{n(1)} \ll e_{n(K)}$, say. Therefore $e_{n(1)} \ll e_{n(K+1)}$.

□

Since $(\mathbf{EP}, \ll)$ is partially ordered, by IV·71,

COROLLARY 9.11 $(\mathbf{EP}, \ll)$ *is well-quasi-ordered.*

□

Remark. Thm. 10 is of the weaker i, j form, rather than the $i, i+1$ form of thm. 6. So e.g. the sequence

$$\omega^+, \omega^3, \omega^{2+}, \omega^4, \omega^{3+}, \omega^5, \ldots$$

has no $i, i+1- \ll -$inclusions. Using rem. 1 following thm. 7, for $\mathcal{S}$ as in thm. 10,

COROLLARY 9.12 *The tail of $\mathcal{S}$ contains an infinite subsequence $(e_{n\{i(j)\}})$ for which*

$$e_{n\{i(1)\}} \ll e_{n\{i(2)\}} \ll \ldots .$$

□

Thus, e.g. every given enumeration of **EP** contains such a sequence.

Immediate domination sequences determine si functions $\lambda x \cdot e_{n(x)}(y)$ for every $y = 2,3,4,\ldots$. Thus in **EP**, given $\alpha_1 \prec \alpha_2 \prec \ldots$,

COROLLARY 9.13 *For no α is it true that $(\forall i)\ \alpha_i \ll \alpha$.*

□

Note that for the above sequence

COROLLARY 9.14 $(\forall n)(\exists I)(\forall i)$

$$i \geq I \Rightarrow \mid T(\alpha_i) \mid > n.$$

□

IV, thm. 77 is not sufficient to establish that $(\mathbf{EP}, \ll)$ is wqo, owing to the incomparability of the relations $T(e) \subseteq_{homeo} T(f)$ and $e \ll f$.

9.5 Lattice Ordering of EP

By Ch. IV, corol. 81, the lattice $(\mathfrak{T}, \angle)$ satisfies the DCC. The bijection of IV·47 thus proves the DCC for $(\mathbf{EP}, \angle)$, where $\alpha \angle \beta$ in **EP** iff $T\alpha \angle T\beta$ in $\mathfrak{T}$. A general result of HIGMAN [62], §1, pp326-7, 332 establishes the stronger result, of the previous section, for $(\mathbf{EP}, \angle)$. Let $\mathcal{W}$ be the system of finite words [5] over the class T. In particular, $\mathcal{W}$ contains the single-letter words (viz. trees of $\mathfrak{T}$). And $\mathcal{W}$ is closed under juxtaposition: [6] if $u, v \in \mathcal{W}$, then $uv \in \mathcal{W}$. This operation is associative. Taking the null word as identity, $\mathcal{W}$ is a semigroup. Now define $\subseteq$ on $\mathcal{W}^2$ as in [62], p332:

$$u \subseteq v \Leftrightarrow u = S_1S_2\ldots S_q, \quad v = T_1T_2\ldots T_r; \quad q \leq r$$

and

$$S_1 \preceq T_{k(1)},\ S_2 \preceq T_{k(2)},\ \ldots, S_q \preceq T_{k(q)},$$

for some numbers $1 \leq k(1) < k(2) < \ldots < k(q) \leq r$. Thus $u \subseteq v$ provided the letters of v are sufficiently numerous and $\prec$-large. Now

$$(\forall u, v, a, b \in \mathcal{W})\ u \subseteq v \Rightarrow [aub \subseteq avb]\ \&\ u \subseteq aub.$$

$\mathcal{W}$ is therefore a divisibility ordered semigroup. The alphabet $(\mathfrak{T}, \prec)$ has HIGMAN's finite basis property (fbp), viz. the closed [7] subsystems are the remainders (in the sense of Ch. VI, §2.2), the closure of a single letter. Note it is here that the DCC property of the alphabet is required. HIGMAN's theorem states that if A is a divisibility-ordered algebra, and if any set of generators for A has the fbp, then A itself has the fbp. Thus, as $(\mathfrak{T}, \subseteq) \simeq (\mathfrak{T}, \preceq)$ generates $\mathcal{W}$,

LEMMA 9.15 *The system $\mathcal{W}$ has the fbp, viz. the closed $\mathcal{X} \subseteq \mathcal{W}$ are each the closure of only fm words.*

□

It is well-known that the fbp is equivalent to the ERDÖS-RADO property: for any sequence $u_1, u_2, \ldots$ in $\mathcal{W}$,

$$(\exists i, j)_{i<j}\ u_i \subseteq u_j.$$

But any (diagonal) **EP**−sequence $e_1, e_2, e_3, \ldots$ defines a $\mathcal{W}$−sequence $u_1, u_2, u_3, \ldots$,

$$u_i = T(e_r^{(i)})T(e_{r-1}^{(i)})\ldots T(e_1^{(i)}), \quad r = r(i),$$

where $e_i = \omega^{e_r^{(i)}} + \ldots + \omega^{e_1^{(i)}}$ in the II·22-normal form: $e_r^{(i)} \succeq e_{r-1}^{(i)} \succeq \ldots \succeq e_1^{(i)}$. Therefore, for such a sequence, in the relation $u_i \subseteq u_j$, $k(r), k(r-1), \ldots, k(1)$ may

[5] In IX, §5, u,v,w etc. will denote words, rather than as in IV, vertices.

[6] This operation is not quite that of ex. 5 foll. IV·1.

[7] Viz. $\mathcal{X} \subseteq \mathcal{W}$ is *closed* iff $(\forall u \in \mathcal{X})(\forall v \in \mathcal{W})\ u \subseteq v \Rightarrow v \in \mathcal{X}$ (cf. IV, §9·2).

be taken to be $r(j), r(j)-1, \ldots, r(j)-r(i)+1$. So by the defn. given before IV, corol. 81,

$$e_i \quad e_j \Leftrightarrow u_i \subseteq u_j.$$

Thus for an arbitrary **EP**- sequence $e_1, e_2, e_3, \ldots$,

COROLLARY 9.16 $(\exists i,j)_{i<j}\ e_i \trianglelefteq e_j$.

□

Therefore, by IV, defn. 71,

COROLLARY 9.17 *The lattice* $(\mathbf{EP}, \trianglelefteq)$ *is well-quasi-ordered.*

□

As noted after corol. 11, using the correspondence between normal trees and exponential polynomials, the example

$$\mathbf{22} \not\subseteq \mathbf{311} \not\subseteq \mathbf{42} \not\subseteq \mathbf{511} \not\subseteq \ldots$$

shows that the stronger $i, i+1$ form of the corollary does not hold for $\mathcal{W}$. In place of the observation about remainders, used to obtain lemma 15, the equivalent ERDÖS-RADO property for $(\mathfrak{T}, \prec)$, a special case of thm. 6, might have been used.

CATALAN Numbers

THEOREM 9.18 *The number of vectors* $x_r \geq x_{r-1} \geq \ldots \geq x_1$ *of non-increasing integer components for which*

$$x_i \leq i, \quad i = 1, 2, \ldots, r$$

is c_{r+1}, *the* $(r+1)$*st CATALAN number.*

□

(Ref. IV, §2.7; Appx. C, §3; TAMARI [150], Introdn).

9.6 Further Properties of $(\mathbf{EP}, \Omega, \prec, \ll)$

9.6.1 Properties Relating Directly to EP

In this §, [8] some finiteness properties relating to the standard assignment Ω are derived: let

$$(e,1) \prec (e,2) \prec \ldots$$

be the canonical fundamental sequence for $e \succeq \omega$ (in **EP**).

[8] See [103], §5.

Ω and Immediate Domination

By II, defn. 35, the basic property III·7·3 gives

THEOREM 9.19 $(\forall n \geq 2)\, \neg(e, n) \ll e$.

DEFINITION 9.20 *The* level *of* $l = l(e)$, *viz. the* $x-$*height of* rp e, *is*

1. 1 *if* e *is simple, or of the first kind; or*

2. $k+1$ *if* $e = \sum +\omega^{\alpha}$, *where* α *is a diagonal of level* k.

□

The following is III·7·2:

THEOREM 9.21 $(\forall n \geq 1)\, (e, n) \ll (e, n+1)$.

Proof, by induction on l, using the fact that $\alpha \ll \beta \Rightarrow \omega^{\alpha} \ll \omega^{\beta}$.

□

The tail of a fundamental sequence, guaranteed by thm. 7, is thus the sequence itself:

$$(e, 1) \ll (e, 2) \ll \ldots.$$

There now follow a few results on diagonals e and their ct's. In contrast to thm. 19,

THEOREM 9.22 $(e, 1) \ll e$.

Proof. $(e, 1)(1) = e(1)$; and $(e, 1)(x) < (e, x)(x) = e(x)$, $x > 1$.

□

Thus the $\prec$ −regressive operator $F(e) = (e, 1)$, e nc, has also the stronger property of being $\ll$ −regressive: $F(e) \ll e$. Now let f be diagonal, and $n \geq 1$.

COROLLARY 9.23

$$(f, n) \prec e \preceq (f, n+1) \Rightarrow (f, n) \ll e.$$

Proof. In the sense of V, §3, by III·14: $(f, n) \mathrel{\tilde{\lhd}} e$. But by II·35, ex. 1; and the previous result, $\alpha \lhd \beta \Rightarrow \alpha \ll \beta$. Therefore by the transitivity of $\ll$, $(f, n) \ll e$.

□

The similarity may be noted between the present thm. and VII·41.

For e diagonal,

THEOREM 9.24 $(\forall n)\,(e,n)(2) \geq n$.

Proof. By III, thm. 7·4, for $x \geq 2$, $\lambda n \cdot (e,n)(x)$ is si. But $(e,1)(x) \geq 1$, so $(e,n)(x) \geq n$. Alternatively, lemma 1 may be applied to: $\mid T\{(e,n)\} \mid \geq n$, which follows from lemma 4, and e.g. III, §3·3.

□

Now suppose e, f are diagonals for which $(\forall n)(e,n) \ll (f,n)$. Then applying again the above,

$$(\forall x)\; e(x) = (e,x)(x) \leq (f,x)(x) = f(x),$$

with strict inequality for $x > 1$, so under the above conditions

THEOREM 9.25 $e \ll f$.

□

The converse does not hold, either for $\ll$, or $\ll$ replaced by $\prec$.

Ω and Eventual Domination

For each n,

$$(\forall x > n)\; e(x) = (e,x)(x) > (e,n)(x).$$

$(e,n)(n) = e(n)$ now gives the following most basic relation between a diagonal and its ct's:

THEOREM 9.26 $(\forall n)\,(e,n) \prec e \;@\; n+1$

□

Note the consistency of thm. 26 with II·47: by III·7·5, n is an ec of (e,n).

Let $e \vartriangleleft \ldots \vartriangleleft f$ be the reduction of V, thm. 3. Each predecession corresponds to '$\ll$;' each reduction to the n-th ct to '$\prec @n+1$' by IX·26. Therefore (by II, rem. foll. defn. 13), $M \leq 1 + \max n$. This sharpens the upper bound of II·47 in certain cases (since e may contain ec's other than those determined by cr from f). However $M = 1 + \max n$ only when the values n are nd in the course of the reduction.

Suppose $\alpha \prec e$, when Diag(α), as e.g. $\omega \star (4;4) \prec \varepsilon_5\ @5$. Via thm. 24, a kind of converse to II·47 arises, in that M in $(\alpha,n) \prec e@M$ may be made arbitrarily large, by taking n sufficiently large.

Now suppose Diag(e), and α is such that

$$(e,1) \preceq \alpha \prec e.$$

Then the dee (IV, §§3·1, 4·1; see also V, §3, rem. 3 and V, §6·1) of e for α is rp(e). Furthermore, under the condition stated in the last case of footnote 4, III, §3·3 $\mathrm{rp}\{(e,n)\} = \{\mathrm{rp}(e)\}^-$, while $\mathrm{rp}(e,n)$ is a subpolynomial of a diagonal $\{\mathrm{rp}\ e\}^-$. Therefore using ex. 2 after II·35,

$$\mathrm{rp}\{(e,n)\} \ll \mathrm{rp}(e), \quad n = 1,2,3,\ldots$$

iff e is imperfect.

9.6.2 Properties Relating to Finite Types

A result of II, §4 may be used to give a more formal proof [9] of the characterization $\mathbf{EP} = \mathcal{X}$ of II, thm. 52. This leads to an immediate derivation of the $E-$string of (e, n), given that of the diagonal e.

Incremented String

Recalling the notation of II, §4, for any $e \in \mathcal{X}$, $\underline{E}(e)$ denotes any $E-$ string which determines e. $\underline{E}$ is *final* if $\underline{E}$ contains only the subscript 1. Now let $\underline{E}_{(+)}(e)$ denote the *incremented* $E-$string (of e): every subscript occurring in $\underline{E}(e)$ is incremented by unity, and the result is placed to the *left* of E_1. Note that the incremented string is well-formed.

Examples

1. For $e = \omega + \mathbf{2}$,

$$\underline{E}(e) = (E_2E_1)E_1^{(2)} = \underline{E}(\omega)\underline{E}(\mathbf{2}); \ \ \underline{E}_{(+)}(e) = (E_3E_2)E_2^{(2)}E_1.$$

   Therefore $\underline{E}_{(+)}(e) = \omega^{\omega+\mathbf{2}} = \omega^e$.

2.

$$\begin{aligned} \underline{E}\{\omega^{\omega^2} + \omega^{\omega+\mathbf{2}}\} &= \{(E_3^{(2)}E_2)E_1\}\{(E_3E_2)E_2^{(2)}E_1\} \text{ by II, lemma 50} \\ &= \underline{E}_{(+)}(\omega^2)\underline{E}_{(+)}(\omega + \mathbf{2}). \end{aligned}$$

Calculation of the Form $E_{(+)}(e)(x)$

As seen in II·52, $\mathcal{X} \subseteq \mathbf{EP}$; the converse has to be proved. Thus, let $\underline{E}(x)$ be a well-formed string, with $\underline{E}(x) = e(x)$. Then (ref. II, §4·4) the calculation of $\underline{E}$ gives

$$\underline{E}(x) = E_1^{\{e(x)\}}(x).$$

But every step $E_hF_{h-} = F_{h-}^{(x)}$ in the calculation becomes $E_{h+}F_h = F_h^{(x)}$ in that for $\underline{E}_{(+)}(x)$. Therefore

$$\underline{E}_{(+)}(x) = E_2^{\{e(x)\}}E_1(x) = x^{e(x)},$$

the second equality following by II·49. Therefore:

THEOREM 9.27

$$\underline{E}_{(+)}(e) = \underline{E}\{\omega^e\}.$$

From this may be obtained a proof of II·52:

---

[9] Mc BETH 'Finite Types, Exponential Forms and Complete Trees', *SIGACT News* **82** (1991).

COROLLARY 9.28

$$\mathcal{X} = \mathbf{EP}.$$

Proof. $1 = E_1 \in \mathcal{X}$. And if $\underline{E}(e), \underline{E}(f)$ are strings in $\mathcal{X}$, then so is $\underline{E}(e)\underline{E}(f) = \underline{E}(e+f)$. Therefore by the previous result, $\mathcal{X}$ has the closure operations of **EP**, and $\mathbf{EP} \subseteq \mathcal{X}$. Conversely, by II, lemma 50, no E_h, $h = 1, 2, 3, \ldots$ leads out of **EP**, so $\mathcal{X} \subseteq \mathbf{EP}$.

□

And furthermore, using the normal form thm. for exponential polynomials:

COROLLARY 9.29 *Suppose* $e = \sum_{i=r}^{1} a_i\omega^{e_i} + \mathbf{a}$. *Then*

$$\underline{E}(e) = \prod_{i=r}^{1} \{\underline{E}_{(+)}(e_i)\}^{(a_i)} \cdot E_1^{(a)},$$

where $\prod$ *denotes composition.*

□

Remark. The form $E_{h+1}\{E_h^{(a)}\}\ldots$, $a > 1$ never occurs in the normal string.

Therefore if e is diagonal, for $n = 1, 2, 3, \ldots$, the E–strings of the ct's are given by:

COROLLARY 9.30

$$\underline{E}\{(e,n)\} = \prod_{i=r}^{2} \cdot \{\underline{E}_{(+)}(e_1)\}^{a_1^-} \cdot \begin{cases} \{\underline{E}_{(+)}(e_1^-)\}^{(n)} & \text{if } e_1^- \in \mathbf{EP}; \\ \underline{E}_{(+)}\{(e_1, n)\} & \text{otherwise.} \end{cases}$$

□

Corollary 29 thus marks a 3rd appearence of the basic coefficients of e, after those in the normal form and within the iterated canonical reduction (V, §5).

9.6.3 Properties Relating to the Operators Δ_e

For diagonal e,

THEOREM 9.31 *1.* $\Delta_{(e,n)} \ll \Delta_{(e,n+1)}$.

2. $\Delta_{(e,1)} \ll \Delta_e$.

3. $(\alpha, n) \prec e \preceq (\alpha, n+1) \Rightarrow \Delta_{(\alpha,n)} \preceq \Delta_{(e,1)}$.

4. $\Delta_e \succ \Delta_{(e,n)} @ n+1$.

Proof.

1. By Ch VII, lemma 19, (e, n) occurs in the index sequence in the calculation of $\Delta_{(e,n+1)}$, so $\Delta_{(e,n+1)}(y) = \Delta_{(e,n)}(Y) > \Delta_{(e,n)}(y)$, since $Y > y$.
2. As part 1.
3. By the BACHMANN property, and by VII, corol. 23.
4. By VII, corol. 23.

□

Note it is possible that $e \not\ll f$ in **EP**, and yet, taking $x = 3$, $\Delta_e \ll \Delta_f$, since $\Delta_{\omega^+}(3,2) = \Delta_\omega(3,3) = 27 < \Delta_\omega(3,27) = \Delta_{\omega^+}(3,3) = \Delta_{\omega+\mathbf{2}}(3,2) = \Delta_{\omega^2}(3,2)$. On the other hand, it seems a fair assumption that $e \ll f$ in **E** implies $\Delta_e \ll \Delta_f$.

Second Proof of Finite Persistence

For $\mid T(e) \mid = n+1$, lemma 1 not only gives $e(2) \geq n$, but using the relation $(\forall x \geq 3)\ e(x) < \Delta_e(x)$ gives as well $\Delta_e(x) \geq n$. So: for every y, and $x > 2$, there are only fm e for which $\Delta_e(x) \leq y$, giving a very much shorter proof of VII·45. But the longer proof exposes other details of values taken by the Δ_e. Note that $y \in \text{im}\ \Delta_e$ does not necessarily imply $y \in \text{im}\ e$.

Domination in Extensions of EP

Suppose, in the notation of corol. 18 of the previous Ch., that

$$e \;=\; \sum_{i \leq r}[e_i, f_i] \;\prec\; \sum_{i \leq s}[E_i, F_i] \;=\; E$$

is an eventual domination in **E2**. It is possible (i) that $r < s$, with $[\ldots]_e = [\ldots]_E$ for $i \leq r$. Otherwise (ii) there is a least i such that

$$[e_{r-i}, f_{r-i}] \prec [E_{s-i}, F_{s-i}].$$

Now *if* $e_{r-i} = E_{s-i}$ the analysis may be continued, as in IV, §§3·1, 4·1. By VIII, ·10, after a finite # of steps, either unequal indices are encountered, or an instance of (i). To summarize: in a limited number of cases, a domination $e \prec E$ in **E2** may determine a dee α of E for e. In particular, this is true of $\mathbf{EP}^\star$, VII, §2. But it is a distinction of the very much smaller hierarchy **EP** that here $\prec$ is 'dee-complete'.

Remarks

1. As in **EP**, the ec α of E contains an 'excess of terms' over its corresponding ec in e.
2. Conversely, suppose e, E are arbitrary normal forms in **E2**. Then if the dee of E for e is found, by the much simpler argument of II·18, it is proved: $e \prec E$.

For α, β as in VII, corol. 23, let $\beta = \varepsilon_{n_1,\ldots,n_R}$. Then $\alpha = (\varepsilon_{n_1,\ldots,n_R}, y, y_2, \ldots, y_t)$ denotes the most rapid possible canonical descent from ε to β, followed by descent from β to α subject to the braking effect of taking ct's of rapidly increasing index.

Chapter X

GOODSTEIN SEQUENCES

Section Headings

10.1 Introduction

The purpose of the following is to give a new and finitary proof of the theorem that all GOODSTEIN sequences are finite. The theorem, first proved by GOODSTEIN [45](1944) pp35-39, is equivalent to the form of the descending chain condition (DCC) required for GENTZEN's consistency proof for $\underline{N}$ (ref. [40] §4; 15.4, p191). KIRBY and PARIS [71] proved a first-order case of the theorem, at the same time proving that it is undecidable in the system $\underline{N}$. CICHON [22] obtained a shorter proof of this meta-result, using that of VIII, §4·12.

GOODSTEIN introduced the sequences in order to give a non-infinitary proof of the DCC. But simpler proofs have appeared, and it becomes reasonable to facilitate the study of GOODSTEIN sequences by proving the finiteness property from the DCC.

The well-foundedness of the 'ordinal numbers below ε_0' is usually derived from the 'theorem of transfinite induction'. GENTZEN's proof requires the DCC. It is therefore [1] commonly claimed that the proof requires 'transfinite induction'. But this claim is misleading [2] if not false: infinitary well-order types may be replaced e.g. by arithmetic exponentials of the well-founded system $(\mathbf{EP}, \prec)$. The scheme of generalized induction of Ch. VI, which is based on the standard assignment Ω, is proved from the DCC for **EP**, IV, corol. 49.

On the basis of the canonical Ω, gen. ind. and the DCC are established as logically equivalent (Ch. VI, corol. 17). In other words, GENTZEN's consistency proof for $\underline{\mathbf{N}}$ does *not* require the axiom of choice. Since Ω is not used in the consistency proof, the DCC is in this sense more straightforward than gen. ind. Noting again that gen. ind. is non-infinitary, it follows that the above claim is doubly misleading. GENTZEN himself chose certain positive rationals under the natural ordering to effect his consistency proof. But what is at issue is the DCC for the chosen system.

The paper of KIRBY and PARIS contains, as indicated above, an independence proof for the finiteness of GOODSTEIN sequences. The proof depends on a certain proposition ([71], prop. 10.3.1, p301). As a demonstration of the utility of the proof-theoretic exponential polynomials, a finitary proof of this proposition is given in §3.

It may be remarked that the proof of the GOODSTEIN result, of the present chapter, relies on the uniformity property, II, thm. 47. Thus the uniformity property finds another general area of application, besides the proof of the ERDÖS-RADO property for $(\mathbf{EP}, \ll)$, IX·10. The following treatment is based on McBETH [105] [3] (1988):

10.1.1 Exponential Polynomials

GOODSTEIN's representation of numbers n to a scale m, where $m, n \geq 1$, is greatly simplified by use of SKOLEM's system $\mathbf{S} = \mathbf{EP_0}$. Recall Ch. II, defn. 23·3 of the explicit coefficients of e :

DEFINITION 10.1 *The integer-valued function* 'mult' *on* **EP** *is defined by:* $\mathrm{mult}(e) =$ *the greatest explicit coefficient of* e.

□

Thus II, defn. 44 gives $N(e) = 1+\mathrm{mult}(e)$; and II·47 gives

THEOREM 10.2 $\{e \prec f \ \& \ \mathrm{mult}(e) < x\} \Rightarrow e(x) < f(x)$.

□

[1] See e.g. KLEENE, [74] §79, 476-479.

[2] In respect of his own 'purely finitist' proof of 'accessibility', TAKEUTI ([147], Ch. II, §12.10), makes precisely the same assertion.

[3] 'GOODSTEIN sequences', manuscript.

This thm. affords a clear rendering of GOODSTEIN sequences. Note that $m > M = \max\{\text{mult}(e),\text{mult}(f)\}$ in **EP** and thm. 2 give: $e(m) = g(m) \Rightarrow e = f$. So for $e \neq f$,

COROLLARY 10.3 *The natural number solutions to the exponential equation* $e(x) = f(x)$ *lie in the range* $1 \leq x \leq M$.

□

10.2 GOODSTEIN Sequences

Let $m, N \in \mathbf{N}$.

DEFINITION 10.4 *If* $(\exists e \in \mathbf{EP})\ N = e(m)\ \&\ \text{mult}(e) < m$, *then* e represents N *to* scale m.

It follows by induction that all $N \geq 1$ can be m–represented, $m \geq 2$. For $1 = \mathbf{1}(m)$ and $\mathbf{1} \in \mathbf{EP}$; and if $n < N$ can be represented, where $N > 1$, then

$$N = \sum_{i=r}^{1} a_i m^{N_i} + a, \qquad \text{(I)}$$

where $a_i < m$; $N > N_r > N_{r-1} > \ldots > N_1 \geq 1$, and where, otherwise r, $\sum$, a_i and a are as in II, thm. 22; as a sum of powers of m. Thus $N = e(m)$, where $e = \sum_i a_i \omega^{e_i} + \mathbf{a}$, $N_i = e_i(m)$, $e_i \in \mathbf{EP}$ by i.h.; and $e \in \mathbf{EP}$, with $\text{mult}(e) < m$.

Now suppose e, f represent N to scale m :

$$e(m) = N = f(m).$$

An inequality, e.g. $e \prec f$, is impossible, since by thm. 2 this would imply $e(m) < f(m)$. Therefore for all $m > 1, N$:

THEOREM 10.5 (GOODSTEIN) *N is represented to scale m by a unique* **EP**-*function* e.

□

Note that $m \leq N \Rightarrow e$ is nc.

Example.

The unique e representing 7 625 597 504 677 to scale 3 is $\omega \star 3 + \omega \star (2; \mathbf{2}) + 2\omega + \mathbf{1}$.

Remarks.

1. For e an arbitrary polynomial in **EP**, if $m >$mult(e), then e is the *unique representing function (urf)* of $N = e(m)$. The remark after II, defn. 44 applies equally to mult; so given the representation $N = e(m)$ and the **EP** factorization $e = \alpha \cdot \beta$, α, β are the urf's of $\alpha(m), \beta(m)$.

2. It is a remarkable and most significant fact that any integer in the range say of 1 to 10^{10} can be written down in the usual decimal notation, with hardly more than a few strokes of the pen (while any number in the range 1 to, say, $10^{60\,000}$, may be flashed in an instant on a television monitor). GOODSTEIN's **EP** -representation of a number extends the decimal notation, but not so as to enable the specification of any number in the range of, say, 1 to $10 \star 10$, in less than 'overflow' time. Some numbers in the given range have an easy **EP** -expression. But in general there is little difference; and in cases where '0' does not occur in the standard decimal expression, the extra information given involves more digits.

3. Given the scale $m \geq 2$, GOODSTEIN's theorem has a proof by induction: 1 is its own representation. Now let $N > 1$, and suppose $n = 1, 2, \ldots, N-1$ are well-represented, to scale m. Then there are *unique* $c \in \mathbf{N}_0, c < N$, such that

$$m^{c+1} > N \geq m^c > 0 \text{ and } (a+1)m^c > N \geq am^c,$$

for a a unique integer in the range $1 \leq a < m$. Then $N = am^c + N'$, where $0 \leq N' < m^c \leq N$. Since $c < N$, by i.h. $(\exists! \alpha, \beta \in \mathbf{EP_0})\ N = am^{\alpha(m)} + \beta(m)$, giving the unique representing function

$$e(x) = ax^{\alpha(x)} + \beta(x); \quad e = a\omega^\alpha + \beta \in \mathbf{EP}$$

of N to scale m.

Theorem 2 is thus not required for this result, although it does provide a most direct proof. On the other hand, given a representation

$$e(m) = M < N = f(m),$$

it is not quite clear from the result just proved that $e \prec f$. In this sense thm. 2 is stronger than the unique representation result.

10.2.1 Ordering of Representations

Let m be any number greater than each of mult(e),mult(f) in **EP**. Then by thm. 2, $e \prec f$ implies $e(m) < f(m)$. On the other hand, $e(m) < f(m) \Rightarrow e \neq f$; and again, $e(m) < f(m) \Rightarrow \neg f \prec e$. So II·18 gives

$$e \prec f \Leftrightarrow e(m) < f(m),$$

in accordance with the definition of ordering of representations given by GOODSTEIN ([45], p34).

In particular, if the representations of N to scales m, n, are respectively $f(m), e(n),\ \ 2 \leq m < n \leq N$, then $e(m) < e(n) = f(m)$, viz.

THEOREM 10.6 $e \prec f$.

□

Given $m \geq 2$, the normal form of a function e representing a number to scale m may contain arbitrarily small terms. However

THEOREM 10.7 ($\forall m \geq 2$) *the functions representing some number to scale m have, under eventual domination, the order-type of the natural numbers.*

Proof, by the result given before thm. 6.

□

10.2.2 GOODSTEIN Sequences

Let (m_r) be a non-decreasing $\mathbf{N}-$sequence, with $m = m_1 \geq 2$ (i.e. $m_r \leq m_{r+1}, 1 \leq r$). Given $N \geq 1$, the sequence (n_r) is defined by recurrence:

$$e_1(m_1) = n_1 = N; \ \text{ if } \ r \geq 1 \ \text{ and } n_r > 1,$$

$$n_{r+1} = e(m_{r+1}) - 1, \ (= e_{r+1}(m_{r+1})),$$

where $e = e_r$ is the representation of n_r to scale m_r. The functional $\mathcal{G}[N, (m_r)] = (n_r)$ transforms non-decreasing arithmetic sequences to *GOODSTEIN sequences.* In the KIRBY- PARIS case, $(m_r) = (2, 3, 4, \ldots)$.

Now suppose $f(m)$ is the $m-$representation of $N:\ N = f(m)$. Let $e(n)$ be the $n-$representation of $f(n) - 1, n \geq m:\ f(n) - 1 = e(n)$.

THEOREM 10.8 $e \prec f$

Proof. Suppose otherwise, so that $f = e$ or $f \prec e$. Well $e(n) = f(n) - 1$, so $e \neq f$. Thus, suppose $f \prec e@i$, i.e. such that $i \geq 1$ is the smallest number such that $(\forall x)\ i \leq x \Rightarrow f(x) < e(x)$. Then by thm. 2, $i \leq 1+\text{mult}(f)$. By assumption, $n \geq m >\text{mult}(f)$. Therefore $f(n) < e(n)$; but $f(n) - 1 = e(n)$, a contradiction, and the thm. is proved.

□

Remark It is actually possible to construct $e \prec f$ as above, the construction being simpler if (n_r) is an 'easy' sequence. Thus respective representing functions f, e of consecutive terms in an arbitrary GOODSTEIN sequence are of the form of the thm., giving

COROLLARY 10.9 [4] *Every GOODSTEIN sequence is finite.*

Proof. For $r = 1, 2, 3, \ldots$, if $m_r \leq m_{r+1}$, then $e_{r+1} \prec e_r$ by the theorem, and by the DCC for **EP**, the corol. is proved.

□

Now suppose $e \prec f$ in **EP**. Let $m > \max\{\text{mult}(e), \text{mult}(f)\}$, with $d = e(m) - f(m)$. Note that $d > 0$ by thm. 2. Let $m_1 = m_2 = \ldots m_{d+1} = m$; $m_r \leq m_{r+1}, r \geq d$. Consider the GOODSTEIN sequence $\mathcal{G}[e(m), (m_r)]$:

$$n_1 = f_1(m) = f(m)$$

$$n_2 = f_1(m) - 1 = f_2(m) = n_1 - 1$$

$$\vdots$$

$$n_{d+1} = f_d(m) - 1 = f_{d+1}(m) = n_d - 1 = n_1 - d = e(m),$$

where by thm. 5, e is the uniquely determined scale$-m$ representing function of n_{d+1}. Thus, by a suitable choice of (m_r), e can be made to occur in the GOODSTEIN sequence starting with f. This procedure may be repeated, so

THEOREM 10.10 *The number N and the sequence (m_r) may be chosen so that the sequence (e_r) of representing functions occurring in $\mathcal{G}[N, (m_r)]$ contains any given descending sequence in* $(\mathbf{EP}, \prec)$.

□

Therefore, since the proof of corol. 9 rests on the DCC theorem,

COROLLARY 10.11 *GOODSTEIN's theorem and the DCC thm. for* $(\mathbf{EP}, \prec)$ *are logically equivalent.*

□

Another formulation of this equivalence is included in §4.

10.2.3 GOODSTEIN Normal Form is CANTOR Normal Form

Suppose $N = e(m)$ is the $m-$representation of N. Then by §2.1, the representation $M = f(m)$, $M < N$, gives $f \prec e$. Now suppose $N = \sum a_i m^{N_i} + a$, $a, a_i < m$ as in the argument following definition 4. Then the representation $N_i = e_i(m)$ gives

$$e_i \prec e_{i+1}, \quad i = 1, 2, \ldots, r - 1.$$

It therefore follows that the normal form induced by the expression (I) following defn. 4 coincides with the II·22-nf.

[4] See [105], corol. 2·6.

Remarks.

1. Suppose $m_r = m \geq 2$, $r = 1, 2, 3, \ldots$ in the sequence (m_r). Then evidently

$$\mathcal{G}[N, (m_r)] = (N, N-1, \ldots, 1).$$

2. Suppose now that (m_r) is si. What form does the initially (usually) si but finite (n_r) take? Clearly, as long as e_r is nc $n_{r+1} = e_r(m_{r+1}) - 1 \geq e_r(m_r) = n_r$. In particular, if $(\exists n)\ e_r = \lambda x \cdot (x + n)$, then $n_{r+1} \geq n_r$; and if $e_r \succ 2\omega$ then $n_{r+1} > n_r$.

3. As to the final terms of (n_r), let (m_r) be si. If $n_r > 1$, n_{r+1} is defined. So by corol. 9, there is a k such that $n_k = 1$. Now $e_k \succ \mathbf{1} \Rightarrow e_k(m_k) \geq 2$. Thus $e_k = \mathbf{1}$.

10.3 Normal Form of Representations

10.3.1 Augmented GOODSTEIN Reduction is Slower Than CR

Consider the class of GOODSTEIN sequences $(n_r) = \mathcal{G}[N, (m_r)]$ for which

$$(m_r) = (m+1, m+2, m+3, \ldots),\ m \geq 1.$$

(n_r) is of the form

$$n_1 = e_1(m+1), \text{ and for } r \geq 1,$$

$$n_{r+1} = e_r(m + r + 1) - 1 = e_{r+1}(m + r + 1).$$

By corollary 9, (n_r) is finite. Thus let n_k be the final term of (n_r). From the representing function e_1 and the sequence (m_r) is obtained another sequence $(e_{(r)})$ of functions in **EP**, namely the rth (m_r)−canonical iterate, introduced in Ch. III, defn. 24. The proposition used by KIRBY and PARIS, cited in §1, follows from the result $e_{(k)} = \mathbf{1}$.

NOTATION 10.3.1 *1. For $m \geq 1$, let the notation (e, m) of Ch. III, defn. 6, be amended by:*

(a) $(\mathbf{1}, m) = \mathbf{1}$.

(b) $(e^+, m) = e$, in **EP**.

2. Now let $e \prec f$ in **EP**. *Then modulo a given m, $e \tilde{\lhd} f$ is the transitive closure of the relation $e^+ = f$, or $e = (f, m)$. Note that $\lhd$ implies $\prec$.*

3. Let $e \in$ **EP** *be diagonal, and let m be given. Then $[e, 1] = (e, m)$; and for $i > 1$, if $[e, i]$ is diagonal, then $[e, i+1] = ([e, i], m)$.*

□

(Notation 3.1(3) is for Ch. X, §3 only; so there should be no confusion with the notation of Ch. IV, §2 et seq. or of Ch. VII, §4.1). Note that for $n \geq 1, [e,n] \prec e$.

Consider now the class of GOODSTEIN sequences introduced at the start of §3. Let e be diagonal, and let $m \geq 1$.

LEMMA 10.12 *There is an $n \geq 1$ and a successor $f \in \mathbf{EP}$ such that*

$$f = [e,n] \; \& \; (e,m) \tilde{\triangleleft} f,$$

where the expression $[e,n]$ and the relation $\tilde{\triangleleft}$ are modulo $m+1$.

Proof, by induction. $(e,m) \tilde{\triangleleft} [e,1] = (e,m+1)$ by BACHMANN's thm., III·14, and by IV·48, the DCC. Note that $(e,m) \prec [e,1]$. Now given $r \geq 1$, and $(e,m) \tilde{\triangleleft} [e,r]$, if $[e,r]$ is diagonal, by the above 2 results,

$$(e,m) \tilde{\triangleleft} ([e,r],1) \tilde{\triangleleft} ([e,r],m+1) = [e,r+1].$$

but $[e,r+1] \tilde{\triangleleft} [e,r]$, i.e. $[e,r+1] \prec [e,r]$, so the DCC gives $(\exists n \geq 1)\; f = [e,n] \; \widetilde{\trianglerighteq} \; (e,m)$ is a successor.

□

Remark. By III·22, the DCC need not be invoked, in the proof of lemma 12.

The next lemma is proved by following the intricate process of descent, precipitated by the subtraction of a unit, in the formation of e_{r+1} from e_r.

LEMMA 10.13 $(\forall r < k)\; (e_r, m+r) \; \tilde{\trianglelefteq} \; e_{r+1}$, *where the relation $\tilde{\triangleleft}$ is modulo* $(m+r)$.

Proof. It may be assumed that e_r is nc. Thus, let the normal form of e_r be:

$$e_r(x) = \sum(x) + x^{e(x)}, \; e \in \mathbf{EP},$$

$e \prec e_r$ and $\mathrm{mult}(e) < m+r$. Define $s = m+r+1$.

Case (i), $e = \mathbf{1}$.

Then $e_r(s) - 1 = \sum(s) + s - 1$, and the unique representing function (urf) is

$$e_{r+1} = (e_r, m+r).$$

Case (ii), e is a successor.

Then $e = f + \lambda x \cdot n$ for some diagonal $f \in \mathbf{EP}$, or $f = \mathbf{0}$; and some $n \in \mathbf{N}$ (where $n > 1$ if $f = \mathbf{0}$) by Ch. II, thm. 18. Therefore

$$\begin{aligned} e_r(s) - 1 &= \sum(s) + s \cdot s^{e^-(s)} - 1 \\ &= \sum(s) + s^- \cdot s^{e^-(s)} + \ldots + s^- \cdot s^{f^+(s)} + s \cdot s^{f(s)} - 1. \end{aligned}$$

Therefore the urf is

$$\begin{aligned} e_{r+1} &= \sum(s) + s^- \cdot \omega^{e^-} + \ldots + s^- \cdot \omega^{f^+} + s \cdot \omega^f - \mathbf{1} \\ &\mathrel{\widetilde{\triangleright}} \sum(s) + (m+r) \cdot \omega^{e^-}, \\ &= (e_r, m+r). \end{aligned}$$

For case (iii), suppose e is diagonal. Then (mod $s = m_{r+1}$):

$$\begin{aligned} e(s) &= (e,s)(s) = [e,1](s) \\ &= [e,2](s) = \ldots = [e,n](s), \end{aligned}$$

for some $n \geq 1$, where $[e,n] \succ \mathbf{1}$ is a successor. If $[e,n]$ is nc, by V, thm. 14, $[e,n] = \alpha + \lambda x.s$, where the diagonal α satisfies mult$(\alpha) < s$. Therefore

$$\begin{aligned} e_r(s) - 1 &= \sum(s) + s^{\alpha(s)+s} - 1 \\ &= \sum(s) + s^- \cdot s^{\alpha(s)+s^-} + \ldots + s^- \cdot s^{\alpha(s)^+} + s \cdot s^{\alpha(s)} - 1. \end{aligned}$$

Therefore the urf is

$$\begin{aligned} e_{r+1} &= \sum + s^- \cdot \omega^{\alpha+s^-} + \ldots + s^- \cdot \omega^{\alpha^+} + s \cdot \omega^{\alpha} - \mathbf{1} \\ &= \sum + s^- \cdot \omega^{[e,n]^-} + \ldots + s^- \cdot \omega^{\alpha^+} + s \cdot \omega^{\alpha} - \mathbf{1}. \end{aligned}$$

But $(e, m+r) \mathrel{\widetilde{\trianglelefteq}} [e,n]^-$ by lemma 12. So considering the trailing terms in the e_{r+1}−expression,

$$e_{r+1} \mathrel{\widetilde{\triangleright}} \sum + \omega^{(e,m+r)} = (e_r, m+r),$$

and the lemma is proved in all cases.

□

From the representing function e_1, for $r = 1, 2, \ldots, k$, recall the definition

$$e_{(r+1)} = (e_{(r)}, m+r)$$

(vide Ch. III, defn. 24). Then

PROPOSITION 10.3.1

$$e_{(k)} = \mathbf{1}.$$

Proof [5], by induction on r. Firstly, $e_{(1)} = e_1$. Now suppose $r \geq 1$, and $e_{(r)} \mathrel{\tilde{\lhd}} e_r$, where the relation $\tilde{\lhd}$ is modulo $m + r + 1$. Then

$$\begin{array}{rcl} e_{(r+1)} & = & (e_{(r)}, m + r) \\ & \tilde{\lhd} & (e_r, m + r) \text{ since } e_{(r)} \tilde{\lhd} e_r \\ & \tilde{\lhd} & e_{r+1} \text{ by lemma 13.} \end{array}$$

Therefore by induction $(\forall r \leq k)\ e_{(r)} \mathrel{\tilde{\lhd}} e_r$. Thus $(\forall r \leq k)\ e_{(r)} \preceq e_r$. But as noted in rem. 3 following corol. 11, $e_k = \mathbf{1}$. Therefore $e_{(k)} = \mathbf{1}$.

□

Remark. In the GOODSTEIN reduction given at the head of §3, let $m = 1$. Let k be the smallest index for which $n_k = 1$. Then it follows from the proposition and from the SOLOVAY-KETONEN result VII·37 that

$$H_{e_1}(1) - 1 \leq k,$$

in the notation of Ch. VII.

10.3.2 CICHON's Characterization

Taking now the basic relation $\lhd$ rather than the transitive closure, let the diagonal $e = e_r$ be as in lemma 13, and let the number $s \geq 3$. Then the unique $f \lhd e(\text{mod } s)$ is either a diagonal or a successor, but $f \neq \mathbf{1}$, as $s > 1$. So by the DCC, the descent

$$\ldots \lhd f \lhd e$$

reaches a successor, β say, in a finite number of steps. Take β as the first such successor to be encountered. CICHON ([21], p705, following rem. 2) states, without proof, that

$$e_{r+1} = \beta^{-}.$$

The proof of this is contained in lemma 13 above. It may be noted that the result of Ch. V cited in lemma 13 depends on discussion of the register polynomial of a cd of **EP**, of the 2nd kind.

10.4 Notes on Chapter X

1. §1 The Science Museum Library, Kensington contains a volume comprising a number of works by GOODSTEIN, [46] included. The collection is stamped 29 Sep 50, and titled *'Function theory in an axiom-free equation calculus, etc.'*

[5] See [105], proposition 3.3.

2. §1.1 Although corol. 3 provides an easy algorithm to decide the roots, while M may take an 'everyday' value, the calulation of $e(x), f(x),\ x \le M$, in the search for roots, may be expected to take 'overflow' time.

3. **10.4.1 Recovery of a Given Descending Sequence**

As noted in §1, the theory originates with the estimation of a given descending sequence.

Via Iterated CR

Suppose $\mathcal{S}$: $\quad \ldots \prec e^{(3)} \prec e^{(2)} \prec e^{(1)} = e$ to be a descent in **EP**. Let (m_r) be si, and consider the sequence

$$\mathcal{S}' : \qquad \ldots \prec e_{(2)} = (e, m_1) \prec e_{(1)} = e$$

in the notation recalled after lemma 13. By rem. after III·23, taking e.g. $m_1 = m_2 = \ldots = N\{e^{(2)}\} + 1$ gives $e^{(2)} = e_{(s)}$ for some $s > 1$. And generally, (m_r) can be chosen so that for every $i = 1, 2, 3, \ldots$, provided $\mathcal{S}$ contains an i-th term, there is an $r = r(i)$ for which

$$e^{(i)} = e_{(r)}.$$

A **proof of the DCC** now follows: first note that the sequence $m + r,\ r = 1, 2, 3, \ldots$ may be replaced by any si sequence $(m_r),\ m = m_1 \ge 2$, without affecting the validity of lemmas 12, 13; or of the rem. after lemma 12. By the proof of proposition 3·1, for the GOODSTEIN sequence $\mathcal{G}[e(m), (m_r)]$, the urf's $e = e_1, e_2, \ldots$ satisfy $e_{(r)} \preceq e_r,\ r = 1, 2, 3, \ldots$. Therefore by GOODSTEIN's theorem 9, $\mathcal{S}'$ is *finite*; and the result follows by the embedding $\mathcal{S} \subseteq \mathcal{S}'$. However, see rem. 1 after IV·49.

Via GOODSTEIN Sequences

§2·2. As remarked by GOODSTEIN, a given (**EP**-)descending sequence $\mathcal{S}$: $\ldots \prec \xi \prec e$ can be obtained as $e_1 =$ urf of $e(m)$ to base $m >$ mult(e); $\xi =$ urf $\xi(M)$ to base $M >$mult(e), mult(ξ);... : where $\ldots < \xi(M) < e(M)$. On the other hand, as seen before thm. 10 a suitably increasing base sequence will recover $\mathcal{S}$ via the unit reduction of §2·2. The sketched proof of the thm. may be considered in a little further detail, as follows: suppose in fact mult(e) >mult$(f) = k$, say. Thus $k + 1 < m$. Then $m_1 = k + 1$, and for $i > 1$, $m_i = m$ as above may be taken as the base sequence since

$$n_1' = f_1(k+1) = f(k+1) \qquad ^{6}$$

$$n_2 = f_1(m) - 1 = f_2(m) = n_1 - 1$$

$$\vdots$$

giving the same reduction $f = f_1, f_2, \ldots, f_{d+1} = e$. (Diminution of $m_2, m_3, \ldots$ may also be possible; however e.g. m_2 must be large enough to ensure that the urf f_2 of $f_1(m_2) - 1$ to base m_2 satisfies $e \preceq f_2$). Now consider the repetition of the above process, so that the reduction is to include say $\xi \prec e \prec f$. Clearly, if mult$(\xi) \le m$, taking $m_{d+2} = m_{d+3} = \ldots = m$ will reach ξ. Thus suppose mult$(\xi) = M > m$. Let

[6] Where $n_1' < n_1$ if f is nc.

$e(M) - \xi(M) = D \geq 1$. Then as above, the base sequence $\ldots m_{d+1} = m, m_{d+2} = M = m_{d+3} = \ldots = m_{d+D+1}$ results in a GOODSTEIN descent from e to ξ, and this construction may be applied to any later terms of $\mathcal{S}$. In other words, the choice of (m_r) for the f to e stage need in no way be affected by that for the e to ξ stage. Thus it is *not* necessary to know in advance that the integer sequence

$$\ldots, \mathrm{mult}(e^{(2)}), \mathrm{mult}(e^{(1)})$$

is bounded, in order to deduce thm. 10 from corol. 9.

4. §2·2. Corol. 9. It is incorrect to speak of the 'convergence of GOODSTEIN sequences'; convergence being an infinitary process, and the termination of a gs a finiteness property.

5. §3·1. The relation $e_{(r)} \preceq e_r$ of proposition 3.1 asserts: 'A GOODSTEIN sequence descends more slowly than the corresponding canonical reduction sequence'. Thus defn. VII·15 may be improved e.g. by taking $\Delta_e(x,y) = \Delta_f(x,y)$, in the case Diag$(e)$, where $f(y) = e(y) - 1$; $f \in \mathbf{EP}$, $\mathrm{mult}(f) < y$, $(e,y) \prec f$. However, the proof of thm. 13 shows that the 'improvement' is less than would be obtained by taking yet instead $\Delta_e(x,y) = \Delta_{(e,y+1)}(x,y)$. Therefore for the sequence (n_r) of §3, starting at $n_1 = e_1(m+1)$, the index k of the remark after prop. 3.1 may be bounded above by

$$k \leq \Delta'_{e_1}(2, n_1),$$

in the notation of Ch. VII, §5. Finally, of the 'descent functional' $\mathcal{D}$ (foll. VII·37), it may be remarked that $\mathcal{D} \prec \mathcal{G}$.

6. §3·1. For $m \geq 2$, let $(e \mid m)$ be the urf of $e(m) - 1$ to base m. Thus $(e \mid m)(m) = e(m) - 1$; and the defn. of §2·2 may be re-stated (using the si (m_i)) as:

$$e_1 = \text{urf of } N \text{ to base } m_1; \quad e_{r+1} = (e_r \mid m_{r+1}).$$

For this, analogues of the results III·7 hold. E.g. $m < M \Rightarrow (e \mid m) \prec (e \mid M)$. The general result of §3·1 may now be succinctly expressed: $(e \mid m) \prec (e \mid m+1)$. (i) Thus, given $e_{(r)} \prec e_r$, the si seq. (m_r) is used in the proof of prop. 3.1 to establish that $(e_{(r)}, m_r) \prec (e_r \mid m_{r+1})$. (ii) In terms of trees, the analogue of the observation of IV, §7·6 on IV·59 follows from the proof of lemma 13. Thus the normal tree of $(e \mid m+1)$ contains m $\prec$-reduced branches + lower terms, resulting in the inequality. On the other hand, since given # of lower terms contribute less than 1 reduced base, $(e \mid m) \prec (e, m)$.

7. Further references to the meta-theory of GOODSTEIN sequences include (i) Van de WIELE, J. In: Seminaire General de Logique, Paris 1983-4; Publ. Math. Univ. Paris VII, **27** (1986). (ii) ABRUSCI, M. In: Logic and Combinatorics Proc., 1985, Arcata, Calif., Publ. A.M.S. Contemporary Math. Ser. **65** (1987). (iii) DELON, F., LASCAR, D., PARIGOT, M. In: Seminaire General, Paris, 1983-84. (iv) BECKMAN, F., McALOON, K. Lecture notes, A.M.S. Summer Sch., Ithaca, N.Y. 1982.

Chapter XI

EXPONENTIAL FORMS AND CONTINUED FRACTIONS

Section Headings

11.1 Introduction

Recalling the comments on GENTZEN's $\underline{\mathrm{N}}$–consistency proof in the introduction to the previous Ch., [1] consider now the objects used in the proof. The mapping of $(\mathbf{EP}, \prec)$ into the rational system under the natural order, $(\mathcal{Q}, <)$ is illuminating in itself. One facile way [2] to obtain an appropriate proper subclass [3] $\mathcal{Q}_\varepsilon$ of $\mathcal{Q}$ is by application of (terminating) continued fractions. It is hoped that an AC-free construction will strengthen the foundations of the consistency proof, and at the same time re-awaken interest in GENTZEN's well-founded transformation of formal derivations to points of the real axis.

[1] See also Ch. VIII, end of §4.10.

[2] Mc BETH [104] (1987), 'Proof-theoretic rational numbers obtained as continued fractions', preprint.

[3] Ref. Ch. I, §2.2.

11.2 The Class $\mathcal{Q}_\varepsilon$ of Proof-Theoretic Rationals

To forestall criticism that his proof of consistency for $\underline{\mathbf{N}}$ is not entirely constructivist, GENTZEN defined a subsystem $\mathcal{X}$ of the system $\mathcal{Q}$ of rational numbers (of the form m/n, where m, n are integers, $n \neq 0$), under the natural order, to stand in place of the 'transfinite ordinal numbers below ε_0'.

Each polynomial e of **EP** is to be assigned a unique rational number $q(e)$, such that for all e, f,

$$e \prec f \text{ (the relation of eventual domination in } \mathbf{EP})$$

$$\text{iff } q(e) < q(f) \text{ (in the natural order).}$$

In this way a class $\mathcal{Q}_\varepsilon$ is obtained, distinct from but similar to GENTZEN's $\mathcal{X}$. As stated in §1, the method is based on finite continued fractions:

Suppose $e = \varepsilon_{n_1, n_2, \ldots, n_r}, r \geq 1$, in the ε-normal form, V, §2.

DEFINITION 11.1 *1.* $q(e) = n_1 - \cfrac{1}{n_2^+ - \cfrac{1}{n_3^+ - \cfrac{1}{n_4^+ - \ddots - \cfrac{1}{n_r^+}}}}$

$$= \{n_1, n_2, \ldots, n_r\}.$$ [4]

2. $\mathcal{Q}_\varepsilon(q) \Leftrightarrow (\exists e \in \mathbf{EP})\, q = q(e)$.

□

Thus, taking ex. 1. following V·14,

$$q(e) = 2 - \frac{1\;1\;1\;1}{4 - 6 - 9 - 6} = \frac{2078}{1195}.$$

Note $q(\varepsilon_n) = n,\ n = 1, 2, \ldots$.

Remarks.

1. It is emphasized that $\mathcal{Q}_\varepsilon$ is distinct from GENTZEN's $\mathcal{X}$, which is defined in terms of terminating decimals.

2. For every finite sequence $\underline{n}$ of $\mathbf{N}^k$, $\{\underline{n}\}$ is defined; this is not so for the expressions $\varepsilon_{\underline{n}}$, (V, §3; rem. 3. before V·13); the coefficients n_i occur among the ec's of e in post-order, according to V·12,14; and by ex. 2. after V·12, n_r is the 'constant part' of rp e, iff e is a ct of a perfect form.

[4] There is no clash with the occasional usage of script brackets $\{,\}$ in earlier chapters.

3. For $r > 1$,

$$\begin{aligned}\{n_1, n_2, \ldots, n_r\} &= n_1 - \frac{1}{\{n_2^+, n_3, \ldots, n_r\}},\\ &= n_1 - \frac{1}{1 + \{n_2, n_3, \ldots, n_r\}}.\end{aligned}$$

4. The integers $n_i \geq 1$ occurring in defn. 1 may be replaced by continuous variables $x_i \geq 1$.

11.2.1 Initial Values

The least n_i give the reciprocals:

LEMMA 11.2 $(\forall r)\{1_1, 1_2, \ldots, 1_r\} = \frac{1}{r}$.

□

LEMMA 11.3 $(\forall r)$ $q = \{n_1, n_2, \ldots, n_r\}$ *is si in each variable.*

Proof. Let $\overline{q}_i = \{n_{r-i+1}^+, n_{r-i+2}, \ldots, n_{r-1}, n_r\}$, $1 \leq i < r$. Note that $q = \{n_1, \ldots, n_{r-i} - \frac{1}{\overline{q}_i}\}$. $\overline{q}_1 = n_r$ is si. Thus, let $i \geq 1$, and suppose $\overline{q}_i$ is si in its i variables. Now

$$\overline{q}_{i+1} = n_{r-i}^+ - \frac{1}{\overline{q}_i}.$$

Therefore $\overline{q}_{i+1}$ is si in n_{r-i}; n_{r-i} does not occur as a variable in $\overline{q}_i$, which by hypothesis is si: an increase in any of $n_r, n_{r-1}, \ldots, n_{r-i+1}$ produces an increase in $\overline{q}_i$, a fall in $\overline{q}_i^{-1}$, an increase in $-\overline{q}_i^{-1}$, and so also in $\overline{q}_{i+1}$. Thus by induction it follows that $\overline{q}_{r-1}$ is si. But $q = n_1 - \overline{q}_{r-1}^{-1}$, and is thus also si in every variable.

□

Remark. The above shows that lemma 3 remains true after replacement by continuous variables, as in rem. 4.

Now let $q_i = \{n_1, n_2, \ldots, n_i\}$ be the ith convergent to q, $i \leq r$.

LEMMA 11.4 $q_{i+1} < q_i$, $i < r$.

Proof.

$$\begin{aligned} q_{i+1} &= \{n_1, n_2, \ldots, n_{i-1}, n_i, n_{i+1}\}\\ &= \{n_1, n_2, \ldots, n_{i-1}, m_i\}, \text{ where } m_i = n_i - \frac{1}{n_{i+1}} < n_i\\ &< \{n_1, n_2, \ldots, n_{i-1}, n_i\} \text{ by remark, } = q_i.\end{aligned}$$

□

Note that lemma 2 gives

COROLLARY 11.5 *For all values* $n_i \geq 1, \quad 0 < \{n_1, n_2, \ldots, n_r\}$.

LEMMA 11.6 *For* $n_k > 1, \quad \{n_1, n_2, \ldots, n_k - 1\} < \{n_1, n_2, \ldots, n_k, \ldots, n_r\}$.

Proof. Consider the $\overline{q}_i$ of lemma 3. $\overline{q}_1 = n_r^+ > 1$. Now for $i \geq 1$, suppose $\overline{q}_i > 1$.

Then

$$\overline{q}_{i+1} = n_{r-1}^{+} - \frac{1}{\overline{q}_i} > n_{r-1} \text{ by supposition, } \geq 1.$$

Thus by induction $\overline{q}_i > 1$, for $1 \leq i < r$. But

$$\begin{aligned}\{n_1, n_2, \ldots, n_k - 1\} &< \{n_1, n_2, \ldots, n_k - \frac{1}{\overline{q}_{r-k}}\} \text{ (by lemma 3, since } \overline{q}_{r-k}^{-1} < 1) \\ &= \{n_1, n_2, \ldots, n_k, \ldots, n_r\}.\end{aligned}$$

□

Thus, combining lemmas 4 and 6, for $n_i > 1$,

$$\{n_1, n_2, \ldots, n_i^-\} < \{n_1, n_2, \ldots, n_i, n_{i+1}\} < \{n_1, n_2, \ldots, n_i\} :$$

the 'adjustment' to the fraction brought about by an arbitrary increase in n_{i+1} is less than that resulting from a unit increase in n_i.

COROLLARY 11.7 *The rational $q = \{n_1, n_2, \ldots, n_r\}$ is uniquely determined by the $n_1, n_2, \ldots, n_r$.*

Proof. Suppose

$$\begin{aligned} q &= \{m_1, m_2, \ldots, m_i, m_{i+1}, \ldots, m_r\} \\ &= \{m_1, m_2, \ldots, m_i, n_{i+1}, \ldots, n_s\},\end{aligned}$$

where $m_{i+1} \neq n_{i+1}$, say $m_{i+1} < n_{i+1}$. Then since $i + 1 \leq r$.

$$\begin{aligned} q &\leq \{m_1, m_2, \ldots, m_i, n_{i+1}^-\} \text{ by lemmas 3, 4} \\ &< \{m_1, m_2, \ldots, m_i, n_{i+1}, \ldots, n_s\} \text{ by lemma 6} \\ &= q, \text{ a contradiction.}\end{aligned}$$

□

The preliminaries give the main result of this §, viz.

THEOREM 11.8 [5] *Definition 1 defines an injection from* **EP** *to $\mathcal{Q}$, for which*

$$q(e) < q(f) \text{ iff } e \prec f.$$

Proof. Each e determines unique ε−coefficients (V·11). Since by the previous corollary q is unique, the mapping is an injection. Now suppose $e = \varepsilon_{m_1, m_2, \ldots, m_r}$, $f = \varepsilon_{n_1, n_2, \ldots, n_s}$ have $e \prec f$ in **EP**. Then by V·18, either

[5][104], thm. 3·8.

1. $s < r$ & $m_1 = n_1, \ldots, m_s = n_s$. Thus $q(f)$ is the $s-$convergent $q_s(e)$ to $q(e)$, and $q(e) < q(f)$ by lemma 4. Or

2. $\{\exists i \leq min(r,s)\}\ m_1 = n_1, \ldots, m_{i-1} = n_{i-1}$ & $m_i < n_i$. First, $n_i > 1$. Then

$$\begin{aligned} q(e) &\leq q_i(e) \leq q(\varepsilon_{n_1,n_2,\ldots,n_{i-1},n_i^-}) \text{ as } m_i \leq n_i^- \\ &< q(\varepsilon_{n_1,n_2,\ldots,n_i,\ldots,n_s}) \text{ by lemma 6}, \ = q(f). \end{aligned}$$

Conversely, given $q(e) < q(f)$, lemmas 4 and 6, together with the 'if' part of V·18, give $e \prec f$.

□

By IV·48, $(\mathcal{Q}_\varepsilon, <)$ is well-founded; the least fraction correlated with an exponential is

$$\inf_{e \in \mathbf{EP}} q(e) = q(\mathbf{1}) = q(\varepsilon_{1,1}) = 1/2.$$

11.3 Some Properties of $\{n_1, n_2, \ldots, n_r\}$

Besides the application of $\mathcal{Q}_\varepsilon$ in the context of GENTZEN's proof, a few simple properties are given here, together with a brief comparison with the usual formulation of continued fractions, in which occurrences of '$-$' in defn. 1 are replaced by '$+$'.

First of all, it may be noted that continued fractions [6] are an instance of ill-founded systems of parentheses. E.g. the classical definition gives

$$a + \cfrac{1}{b + \cfrac{1}{c + \ddots}} = a + (b + (c + \ldots)^{-1})^{-1}.$$

(The initial element a is distinguished in not entering into the system of parentheses). Many ill-founded systems of brackets (based on division, or indeed on other operations) are reducible to continued fractions. On the other hand, in other cases this reduction may be difficult, or impossible. For example, (ignoring the initial element) the usual recursive definition

$$F = \frac{1}{n + F}$$

may be replaced by

$$F = \frac{1}{n + F + F}.$$

In the finite case, the two are mutually reducible, but in the infinite case it is not clear how the second form may be re-expressed classically. Furthermore, this is just

[6] C.f. Note 4, VI, §6.

one of a great many possible variations on associations formed with infinitely many brackets, e.g. the number of occurrences of 'F' may be determined as a function of level, or infinitely many occurrences may be allowed. The striking result is that among this range of possibilities, continued fractions are of outstanding practical use in getting theoretical results.

KHINTCHINE's outstanding monograph [69] contains a most clear account of the difference between a best rational approximation to a real number, and an integer approximation to an integer-real product. The first type of approximation requires the introduction of *intermediate fractions*: thus if $\frac{p_n}{q_n}, \frac{p_{n+2}}{q_{n+2}}$ are (classical)consecutive ascending (or descending) convergents, then every such approximation of least denominator is either such a convergent, or

$$\frac{p_n + kp_{n+1}}{q_n + kq_{n+1}},$$

for k in a certain range: indeed, as k increases, the intermediate fraction steadily passes from $\frac{p_n}{q_n}$ until reaching the value $\frac{p_{n+2}}{q_{n+2}}$. Two remarks may be made here: the convergents, of course, 'close up' as the value of the fraction is approached. But with increasing index, the greatest value of k above also increases. This pleasing hierarchy of ever closer approximations to the fraction resembles the BACHMANN property of Ch. III, thm. 14. But whereas the sequences of intermediate BACHMANN terms for the **EP**-hierarchy are finite downwards and infinite upwards, for a finite or infinite continued fraction, the approximant hierarchy is doubly finite. Also, the apparatus of defn. 1, with its slight deviation from that of the classical definition, has *purely ascending convergents.*

Limiting Values

THEOREM 11.9

$$\lim_{n\to\infty} \{n_1, n_2, \ldots, n_r, n\} = \{n_1, n_2, \ldots, n_r\}$$

Proof, by the continuity of the function of defn. 1.

□

Now write e_n for the nth ct of the diagonal e.

COROLLARY 11.10 [7]

$$q(e) = \lim_{n\to\infty} q(e_n).$$

Proof. Let $q(e) = \{n_1, n_2, \ldots, n_r\}$. By V, §4, provided $n > 1$, the ε−expression of e_n is obtained by affixing n to the right of the index string of that of e. Therefore the assumed expression for $q(e)$ gives $q(e_n) = \{n_1, n_2, \ldots, n_r, n\}$, and the result is an instance of thm. 9.

[7] This is [104], corollary 4.2.

□

Remarks

1. The ascending sequence $q(e_1), q(e_2), \ldots$ of rationals is distinguished in that its limit, $q(e)$, is *rational.*

2. By definition, e of corol. 10 is related to its ct's by

$$e = \text{diag}(e_1, e_2, e_3, \ldots) = \text{diag}_n e_n.$$

Thus the result, which may be regarded as a consequence of the $\prec$ −maximal property (start of III, §2.2), links the diagonalization of integer-valued functions and the limiting operation on sequences of reals. In particular, the definition of VI, §2·5, ex. 2, together with $x_{\mathbf{1}} = q_{\mathbf{1}}$; $x_{\alpha+} = q_{\alpha+} - q_\alpha$ and $x_\alpha = 0$ for $\text{Diag}(\alpha)$ give: $\Sigma_{\alpha \prec \beta+} x_\alpha = q_\beta$.

3. The corol. implies the metrization of III·14, as in VIII, §7·2.

By an easy modification of OLDS [118], argument of thm. 1·2

THEOREM 11.11 *For every rational* q, *there is an* n_1, *and unique positive* $r, n_2, \ldots, n_r$ *such that*

$$q = \{n_1, n_2, \ldots, n_r\}.$$

□

Remarks

1. A denominator of the form $\because n_r^+ - \frac{1}{1}$ at the bottom of the fraction does not detract from the uniqueness: the last denominator is $n_r^+ > 1$.

2. $q(\mathbf{1}) = \frac{1}{2}$; $q(\mathbf{2}) = \frac{2}{3}$. Therefore no $e \in \mathbf{EP}$ has $\frac{1}{2} < q(e) < \frac{2}{3}$.

3. In particular, defn. 1 can be used to obtain a subset of $\mathcal{Q}$ which contains a 'gap' (SIERPIŃSKI [138] , XI, §5, p204). Thus, let

$$E(n) = \{1, n\};\ F(n) = \{2, 1_1, 1_2, \ldots, 1_n\}.$$

Then for each n, $E(n) < E(n+1)$ by lemma 3, and $F(n+1) < F(n)$ by lemma 2; the gap in the values of E and F being $\lim E = 1 = \lim F$.

As an alternative to VI, defn. 3, let $l(e)$ = length of fundamental sequence of the diagonal $e \in \mathbf{EP}, = q(e) - q(e_1)$. Then

THEOREM 11.12 $\lim_{\text{ht}(e)\to\infty} l(e) = 0$.

Proof. Uses IX, §5.

□

11.4 Correlation with Irrational Numbers

An alternative correlation is given here, of (higher) polynomials e with irrational numbers $\Xi(e)$. In place of the the ec's of e, the explicit values $e(1), e(2), \ldots$ occur in the definition.

Suppose $e = \varepsilon_{n_1,n_2,\ldots,n_r}$. Define the operator $\tilde{}$ by:

$$\tilde{e}(x) = 1 + \min_{i \le r} e^{(i)}(x), \quad 1 \le x,$$

where $e^{(i)} = \varepsilon_{n_1,n_2,\ldots,n_i}$. Write $\tilde{\varepsilon} = \varepsilon$. Note that $e^{(i+1)} \prec e^{(i)}$, $i < r$, by III·7·3. Thus, taking M as the largest of the passing numbers of $e^{(i)}$ for $e, i < r$,

$$x \ge M \Rightarrow e(x) < e^{(i)}(x)$$

for $i < r$. In other words, $\tilde{e}$ is 'eventually equal' to e. Therefore

$$\alpha \prec e \text{ in } \mathbf{EP} \Rightarrow \tilde{\alpha} \prec \tilde{e}.$$

Examples.

1. $\widetilde{4\omega}$ takes the values $1, 4, 9, 16, 20, \ldots$, as $x = 1, 2, 3, 4, 5, \ldots$.

2. For $e = 3\omega + 4$, $\tilde{e}(x) = 1, 4, 9, 16, 19, \ldots$, as $x = 1, 2, 3, 4, 5, \ldots$.

3. (a) Suppose $e = \alpha^+$. Then $\tilde{\alpha} \ll \tilde{e}$.
   (b) Suppose $\alpha = (e, n)$. Then $\tilde{\alpha}(x) \le \tilde{e}(x)$.

Now suppose $\alpha \prec e$ in **EP**, so that

$$\alpha = e_K \lhd e_{K-1} \lhd \ldots \lhd e_2 \lhd e_1 = e.$$

Then by (a), (b) above, for each argument value x,

$$\tilde{e}_{i+1}(x) \le \tilde{e}_i(x), \; i = 1, 2, \ldots, K-1.$$

Therefore for α, e as above

THEOREM 11.13 *1.* $\tilde{\alpha} \prec \tilde{e}$.

2. $(\forall x)\, \tilde{\alpha}(x) \le \tilde{e}(x)$.

□

In fact, if the $\tilde{e}$ are extended to functions of a real variable, then the curves obtained, convex on the side of greater x, give a feather diagram, (VIII, §7.2).

DEFINITION 11.14 $(\forall e \in \overline{\mathbf{EP}})_{e\neq 1}$

$$\Xi(e) = 1 - \cfrac{1}{\tilde{e}(1) - \cfrac{1}{\tilde{e}(2) - \cfrac{1}{\ddots}}} = \{\tilde{e}(1)^-, \tilde{e}(2)^-, \ldots\}.$$

□

Remarks.

1. $\Xi(\mathbf{n}) = 1 - \frac{2}{n \overset{+}{-} \sqrt{n^2-4}}$, $n \neq 1$; $\Xi(\mathbf{2}) = 0$.

2. For no nc $e \in \mathbf{EP}$ is $\Xi(e)$ rational.

Examples.

1. $\Xi(\omega) = 1 - \frac{1}{2 - \frac{1}{3 - \frac{1}{\ddots}}}$.

2. $\Xi(\varepsilon_4) = \{1, 4, 3^{27}, 4 \star 4, 5 \star 4, \ldots\}$.

3. For $\alpha = \varepsilon_{4,2} = \omega \star (3; 2)$, $\Xi(\alpha) = \{1, 4, 3^{27}, 4^{4^{16}}, \ldots\}$.

According to thm. 3, the fraction is si in every argument. Therefore by thm. 13,

THEOREM 11.15

$$\alpha \prec e \Rightarrow \Xi(\alpha) < \Xi(e).$$

□

Now let Diag(e), $(e, n) = e_n$, $1 \leq n$.

THEOREM 11.16

$$\Xi(e) = \lim_n \Xi(e_n).$$

Proof. By III·7·6, $e(x) < e_n(x)$ for $x < n$. Therefore, as n increases, the coefficients in the fractions defining $\Xi(e)$, $\Xi(e_n)$ are equal, for larger initial segments. The thm. thus follows by the remark after lemma 6.

□

11.5 Notes on Chapter XI

1. Defn. 1 may be extended to any extension of **EP** having an analogue of the V·12-normal form.

2. Continued fractions, a ready choice for sections §2,3, need not be used in the constructions of §4. For example, $e \in \mathbf{EP}$, $\omega^2 \preceq e$ may be correlated with

$$\Xi'(e) = \sum_{x=1}^{\infty} \frac{1}{\tilde{e}(x)}.$$

If Diag(e), since $\frac{1}{\tilde{e}(x)} \to 0$, this would give:

$$\lim_{n\to\infty} \Xi'(e_n) = \Xi'(e).$$

This, however, leads to another question. It is well-known that

$$\sum_{n=\infty}^{1} \frac{1}{n \ln n}$$

diverges. Is it true that, e.g. $\sum \frac{1}{n \ln n \ln \ln n}$ diverges? The general idea is to find as slow-growing as possible an f, such that

$$\sum \frac{1}{nf(n)} < \infty,$$

for which a very small decrement in rate of growth (as measured for example by the functions of classes discussed in II, §6), say from f to ϕ, produces a divergent series: $\sum \frac{1}{n\phi(n)} = \infty$. Note that every such ϕ is associated with an $F \in \mathcal{N}$ of rapid growth: $F(x) =$ the number of terms for which $\frac{1}{1\cdot\phi(1)} + \frac{1}{2\cdot\phi(2)} + \ldots > x$.

3. §4. It may be noted that it is not known whether the number

$$e + \pi = (2 + \frac{1}{1,2,1,1,4,\ldots}) + (3 + \frac{1}{7,15,1,292,1,1,\ldots})$$

is irrational, still less transcendental.

Chapter XII

EP AND GENERATING FUNCTIONS

Section Headings

12.1 Introduction

Aside from properties which find application in logic and arithmetic, such as the provision of finitary foundations, free of reliance on the AC, for GENTZEN's proof of consistency, for diagonalization of ACKERMANN's function and for a clear and finitary proof of GOODSTEIN's theorem, the exponentials of **EP** have a third branch of remarkable, combinatorial, properties.

Arthur CAYLEY [18] , in 1857, published the identity

$$\mathfrak{T}(x) = \frac{1}{(1-x)^{\mathfrak{n}_1}(1-x^2)^{\mathfrak{n}_2}\dots} = \mathfrak{n}_1 + \mathfrak{n}_2 x + \mathfrak{n}_3 x^2 + \dots, \tag{I}$$

which holds for all $x \in \mathbf{C}$, $|x| < r = 0.33832\ 18568\ 992\dots$ [1]; where $\mathfrak{n}_1, \mathfrak{n}_2, \mathfrak{n}_3, \dots$ are

[1] Ref. BALABAN [10] (1970); ROBINSON and SCHWENK [130] (1975).

the numbers obtained by considering all pv-oriented trees [2] which may be formed respectively from $1, 2, 3, \ldots$ vertices. By the normal form thm. IV·26, and its following rem. 1, the identification between pv-oriented and normal trees establishes the CAYLEY result for $\mathfrak{T}$. [3]

Thus, the recursive GOODSTEIN representation of natural numbers defines oriented trees: precisely those whose vertex coefficients are generated by the infinite product. The following includes a proof of the identity; to provide a setting, some discussion of EULER's identity for partitions is included. A proof of the multiple choice lemma is contained in §4. Furthermore, Appx. C contains tables of the numbers n_i, $1 \leq i \leq 100$ in multiple precision, and some remarks on the asymptotic form of this function. The identity (I), in the $\mathfrak{T}$–context, was shown to me by Mr. Rob STACEY. A recurrence for the integers $\mathfrak{n}_i$, together with a proof based on formal derivation is included.

The closing section of the Ch. contains a short account of an application of **EP** to the recursive cycle structure of permutations.

12.2 Generating Functions

12.2.1 Integer Partitions

CAYLEY's identity of §1 is obtained by an elaboration of the technique used by EULER [35] in 1777 to enumerate the number $p(n)$ of partitions (c.f. Ch. IV, §5) of a natural number n. Thus

$$P(x) =_{df} \frac{1}{(1-x)(1-x^2)(1-x^3)\ldots} = 1 + \sum_{n=1}^{\infty} p(n) \cdot x^n .$$

This may be proved, by starting with

$$\frac{1}{1-x^k} = 1 + x^k + x^{2k} + x^{3k} + \ldots, \quad |\, x \,| < 1,$$

so that the product is

$$\begin{array}{c|l} & \text{multiplicity} \rightarrow \\ \text{part} & \{\ 1 + x + x^2 + x^3 + \ldots\ \} \\ \downarrow & \{\ 1 + x^{1(2)} + x^{2(2)} + x^{3(2)} + \ldots\ \} \\ & \{\ 1 + x^{1(3)} + x^{2(3)} + x^{3(3)} + \ldots\ \} \\ & \qquad\qquad \vdots \end{array}$$

[2] For which plane structure is not taken into consideration, ref. Ch. IV, §2; and cf. ex. 4. after IV·19.

[3] Ref. STACEY and Mc BETH[100] manuscript, 1981, now entitled 'CAYLEY's identity for normal trees'.

when the expressions in script brackets are multiplied out, the terms which are products of only fm factors $\neq 1$ [4] will give terms of the form x^k for some k. All other terms $= 0$. The coefficient of x^k is obtained by considering the ways in which exponents, chosen one from each bracket, may add up to k. Thus, if

$$a + 2b + 3c + \ldots = k$$

is a partition of k, a unit is contributed to the coefficient of x^k. Conversely, every contribution determines a partition, therefore the coefficient of x^n is $p(n)$.

The generating function $P(x)$ is related to that of the sum $\sigma(n)$ of divisors of n by

$$\frac{x \cdot P'(x)}{P(x)} = \Sigma\sigma(n) \cdot x^n; \text{ and } p(n) = \frac{1}{2\pi i}\int_C \frac{P(x)}{x^{n+1}}dx,$$

by CAUCHY's closed-contour theorem. Further properties [5] of $p(n)$ include

$$p(n) \sim A_n e^{a\sqrt{n}}; \ A_n = \frac{1}{4n\sqrt{3}}; \ a = \pi\sqrt{\frac{2}{3}}; \quad \text{and} \quad p(5n+4) \equiv 0(\text{mod } 5).$$

ADLER's paper [6], pp733-745 contains an account of different generating functions, for partitions on which stated conditions have been imposed. For example, if the parts are to be taken from the sequence $a_1, a_2, a_3, \ldots$, then the product

$$\frac{1}{(1 - x^{a_1})(1 - x^{a_2})\ldots}$$

enumerates the numbers of such partitions.

Ordered Partitions

Let $T = [T_r, T_{r-1}, \ldots, T_1] \in \mathrm{T}$, $|\, T \,| = N + 1$. Let $q_i = |\, T_i \,|$, $i = 1, 2, \ldots, r$ be the orders of the principal branches of T. If the q_i are written in nd order [6], so that

$$-1 + |\, T \,| = p_k + p_{k-1} + \ldots p_1,$$

then T defines a normal partition P of N (IV, §5.1). T is said to *belong* to P.

Derivation of CAYLEY's Identity

Let $\mathfrak{n}_i$ be the number of $T \in \mathfrak{T}$ for which $|\, T \,| = i$. It may be found by inspection that $\mathfrak{n}_i = 1, 1, 2, 4, 9, 20, \ldots$ for $i = 1, 2, 3, 4, 5, 6, \ldots$.

---

[4]This happens to be the finiteness condition which arises in VEBLEN's construction, cited in VIII, §5.

[5]See M. HALL, *Combinatorial Theory* , Wiley (1986), Ch. 4.

[6]Note that the defining condition $T_i \preceq T_{i+1}$ does not in general imply $|\, T_i \,| \leq |\, T_{i+1} \,|$.

THEOREM 12.1 $(\exists \rho \in \mathbf{R})_{>0}(\forall x \in \mathbf{C}) \mid x \mid < \rho \Rightarrow$

$$\frac{1}{(1-x)^{\mathfrak{n}_1}(1-x^2)^{\mathfrak{n}_2}(1-x^3)^{\mathfrak{n}_3}\cdots} = \mathfrak{n}_1 + \mathfrak{n}_2 x + \mathfrak{n}_3 x^2 + \ldots. \tag{I}$$

Proof. For $n, p \geq 0$, let

$$T(n,p) = \text{ the triangular number of dimension } n \text{ and index } p =$$

$$= \binom{n+p}{p} = \binom{n+p}{n} = \frac{(n+p)!}{n!p!};$$

and for $n \geq 1$, set

$$t(n,p) = T(n-1,p).$$

Thus, $t(n,0) = t(1,p) = 1$; $t(n,1) = n$, and generally $t(n,p) =$ the number of ways of choosing p objects from n, with repetitions. As is well-known,

$$\frac{1}{(1-x)^n} = \sum_{p=0}^{\infty} t(n,p) \cdot x^p.$$

So the LHS of (I) may be re-written

$$\begin{array}{l} (\ 1 + \mathfrak{n}_1 x + t(\mathfrak{n}_1,2)x^2 + t(\mathfrak{n}_1,3)x^3 + \ldots\ ) \cdot \\ (\ 1 + \mathfrak{n}_2 x + t(\mathfrak{n}_2,2)x^2 + t(\mathfrak{n}_2,3)x^3 + \ldots\ ) \cdot \\ (\ 1 + \mathfrak{n}_3 x + t(\mathfrak{n}_3,2)x^2 + t(\mathfrak{n}_3,3)x^3 + \ldots\ ) \cdot \\ (\ 1 + \mathfrak{n}_4 x + t(\mathfrak{n}_4,2)x^2 + t(\mathfrak{n}_4,3)x^3 + \ldots\ ) \cdot \\ \qquad\qquad\vdots \end{array} \tag{II}$$

The proof is by induction on the index i. In the initial case, ${}_1 = \prod_{k=1}^{\infty} 1 = 1$, so for $i \geq 1$, suppose the numbers $\mathfrak{n}_1, \mathfrak{n}_2, \ldots, \mathfrak{n}_i$ are known. Since every $\mathfrak{T}$-tree of $i+1$ vertices belongs to exactly 1 normal P of i, $\mathfrak{n}_{i+1}$ may be found by counting the normal trees which belong to each P, and summing.

Let $T = [T_R, \ldots, T_2, T_1] \in \mathfrak{T}$ belong to P, $\mid T \mid = i+1$. By definition

$$T_R \succeq \ldots \succeq T_2 \succeq T_1. \tag{III}$$

First, take the case in which P has all parts equal, to p say. Then

$$i = pR.$$

Write the $\mathfrak{n}_p$ distinct normal trees on p vertices in $\succ -$ order, and consider a combination C of R of them, allowing repetition. A permutation $\mathcal{P}$ of C may then be chosen so as to satisfy (III). If C contains repetitions, then this can be done in > 1

way. But as these different $\mathcal{P}$ arise because of repeated T_j, they determine the same T. Therefore the T belonging to P is the number of possible C, viz.

$$\binom{\mathfrak{n}_p + R - 1}{\mathfrak{n}_p - 1} = t(\mathfrak{n}_p, R),$$

using the lemma. Now allowing unequal parts, let

$$N = p_{(S)} + p_{(S-1)} + \ldots + p_{(1)}, \quad S \leq R,$$

where the $p_{(j)}$ are blocks

$$p_{j,R(j)} = p_{j,R(j)-1} = \ldots = p_{j,1} = p^{(j)},$$

say, of equal parts of i, with $j < k \Rightarrow p^{(j)} < p^{(k)}$. Now:

1. for $V, W \in \mathrm{T}$, $|V| \neq |W| \Rightarrow V \neq W$, so $V \prec W$, or $W \prec V$; and
2. for various combinations C_j of the principal branches in T of exactly $p^{(j)}$ vertices, permutations $\mathcal{P}_j$ may be chosen, as in the equal-parts case.

Therefore once the $\mathcal{P}_j$ are chosen, they may be merged into *exactly one* permutation which satisfies (III).

But for various C_j, the number of $\mathcal{P}_j$ giving distinct T is $t\{\mathfrak{n}_p^{(j)}, R(j)\}$. Therefore the part contributed by P to $\mathfrak{n}_{i+1}$ is obtained by multiplying all these numbers together. Thus, summing over all normal partitions P of i :

$$\mathfrak{n}_{i+1} = \sum_P \prod_{j=S}^{1} t\{\mathfrak{n}_p^{(j)}, R(j)\},$$

which is precisely the coefficient of x^i in the product (II), and the thm. is proved by induction.

□

CAYLEY's 1857 proof was given for oriented trees. The proof for normal trees (the same result) makes use of plane structure.

12.2.2 A Recurrence Formula for the Numbers of Normal Trees

The following recurrence [7] formula, satisfied by the CAYLEY sequence $\mathfrak{n}_1, \mathfrak{n}_2, \mathfrak{n}_3, \ldots$, may be derived from XII·1:

[7] OTTER, R. [119] 'The number of trees', *Ann. Math.* **49** (1948), 583-599.

COROLLARY 12.2 ($\forall i \geq 1$)

$$\mathfrak{n}_{i+1} = \frac{1}{i} \cdot \sum_{k=1}^{i} k \cdot \mathfrak{n}_k \cdot (\mathfrak{n}_{i-k+1} + \mathfrak{n}_{i-2k+1} + \ldots + \mathfrak{n}_{i-[\frac{i}{k}].k+1})$$ [8]

Proof. $\mathfrak{T}(x) = \prod_{i=1}^{\infty}(1-x^i)^{-\mathfrak{n}_i}, = \frac{1}{x}\sum_{i=1}^{\infty}\mathfrak{n}_i x^i$ by thm. 1. The method is to obtain different expressions [9] for the derivative $\mathfrak{T}'$:

$$\ln\mathfrak{T}(x) = -\sum_{i=1}^{\infty}\mathfrak{n}_i \ln(1-x^i) = \sum_{m,i=1}^{\infty}\frac{\mathfrak{n}_i}{m}x^{mi}.$$

Therefore

$$\frac{\mathfrak{T}'(x)}{\mathfrak{T}(x)} = \frac{d}{dx}\ln\mathfrak{T}(x) = \sum_{m,n=1}^{\infty} i\mathfrak{n}_i x^{mi-1}.$$

So, using $\sum_{a,b} x^{ab} = \sum_{k=1}^{\infty}(\sum_{a|k} 1_a)x^k$,

$$\frac{x\mathfrak{T}'(x)}{\mathfrak{T}(x)} = \sum_{m,i} i\mathfrak{n}_i x^{mi} = \sum_{k=1}^{\infty}(\sum_{i|k} i\mathfrak{n}_i)x^k.$$

Therefore, putting $t(m) = \sum_{i|m} i\mathfrak{n}_i$,

$$x\mathfrak{T}'(x) = \mathfrak{T}(x) \cdot \sum_{m=1}^{\infty} t(m)x^m,$$

i.e.

$$\begin{aligned}\sum_{1}^{\infty} i\mathfrak{n}_{i+1}x^i &= (\sum_{0}^{\infty}\mathfrak{n}_{i+1}x^i) \cdot \sum_{m=1}^{\infty} t(m)x^m \\ &= \sum_{i=1}^{\infty}\{\sum_{m=1}^{i}\mathfrak{n}_{i-m+1}t(m)\}x^i.\end{aligned}$$

Equating coefficients,

$$\begin{aligned}i\mathfrak{n}_{i+1} &= \sum_{m=1}^{i}\mathfrak{n}_{i-m+1}t(m) = \sum_{m=1}^{i}(\sum_{k|m} k\mathfrak{n}_k)\mathfrak{n}_{i-m+1} \\ &= \sum_{k=1}^{i} k\mathfrak{n}_k \cdot (\mathfrak{n}_{i-k+1} + \mathfrak{n}_{i-2k+1} + \ldots + \mathfrak{n}_{i-[\frac{i}{k}]\cdot k+1}),\end{aligned}$$

since the $m \leq i$ for which $k \mid m$ are $k, 2k, \ldots, [\frac{i}{k}] \cdot k$; and the corol. is proved.

□

This recurrence is of particular value in calculation. The table in Appendix C shows values obtained in 1988. These agree with the values quoted in HARARY and PALMER [56], 1976.

[8]'[ ]', '|' have the standard interpretations , viz. 'integral part' and 'divides', respectively.
[9]Throughout §2, " ′ " denotes formal dervative.

12.2.3 Open Problems

Let $\mathfrak{s}_1, \mathfrak{s}_2, \mathfrak{s}_3, \ldots$ be the number of strictly normal trees T (IV, §3.9) for which $\mid T \mid = 1, 2, 3, \ldots$ respectively.

Problem

Can an infinite product or other form be found for the power series $S(x) = \sum_{i=1}^{\infty} \mathfrak{s}_i x^i$? Similar problems arise for each of the many restrictions on $\mathfrak{S}$ or on an already specialized subclass such as $\mathfrak{T}$ above, or $\mathfrak{D}$.

12.2.4 Generating Functions for Various Systems of Trees

CATALAN Numbers

The most basic link was found between the classes $\mathfrak{S}$ and $\mathfrak{D}$ (IV, §2) by HARARY, PRINS and TUTTE [36], 1964. The number of trees in $\mathfrak{S}$ on n vertices and in $\mathfrak{D}$ on $2n-1$ vertices is in each case $\mathfrak{C}_n$, the nth CATALAN number, (above Ch., §2.7). In 1967, de BRUIJN and MORSELT [10] [29] published a note giving 3 separate ways to establish the necessary 1-1 correspondence between the oriented and the oriented trivalent plane structure trees. The first of these methods (with slight alterations of convention) may be given by example: [11]

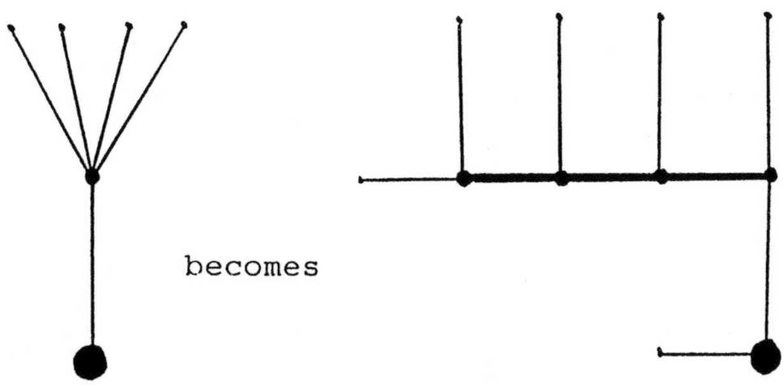

[10]See also N. G. de BRUIJN, in *Applied Combinatorial Mathematics*, Ed. E. F. BECKENBACH (1964), publ. Wiley.

[11]Note that $v^\sharp \in \mathfrak{D}$ by IV·16(2); so $(v^\sharp)' = v^\sharp$ may be taken.

Conversely, e. g.

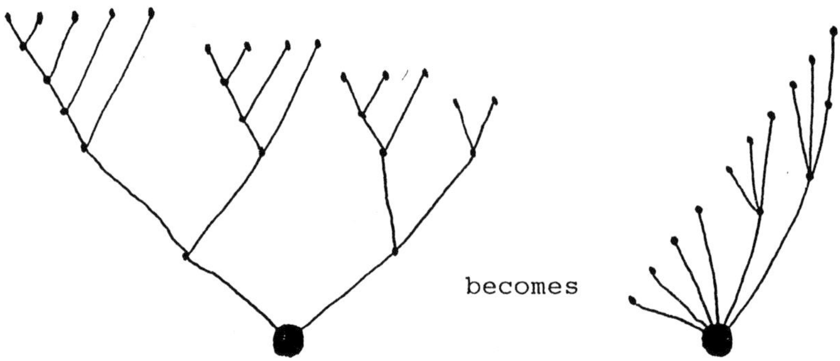

The second method relates the bracket pattern of $T \in \mathfrak{S}$ to that of $T' \in \mathfrak{D}$, via a special characterization. Now let the *transpose* of an $\mathfrak{S}$–bracket pattern be obtained by replacing each occurrence of

1. '[ ]' by '['; and
2. '] [' by ']'.

The 3rd method amounts to the observation that the transpose of the pattern of T, interpreted as an algebraic expression, is that of T', interpreted as a forward polish expression. [12]

The Quadratic Equation Satisfied by the Generating Function of $\mathfrak{S}$

Let $\mathfrak{c}_n$ = the number of distinct $\mathfrak{S}$-trees on n vertices = the nth CATALAN number (see also IX·18). Thus, by IV, §2.7,

$$\mathfrak{c}_n = \binom{2n-2}{n-1}/n,$$

giving the values $1, 1, 2, 5, 14, 42, 132, 429, \ldots$ for small n. Let $(\mathfrak{c})(x) = \sum_{n=1}^{\infty} \mathfrak{c}_n x^n$, and let $\mathfrak{c}_{n+1,k}$, for $n \geq 1, 1 \leq k \leq n$, be the number of $T \in \mathfrak{S}$ on $n+1$ vertices, with pv-valency of k. Then

$$\mathfrak{c}_{n+1} = \sum_{k=1}^{n} \mathfrak{c}_{n+1,k}. \tag{I}$$

For $k = 1$, there are $\mathfrak{c}_n$ possible T, i.e.

$$\mathfrak{c}_{n+1,1} = \mathfrak{c}_n. \tag{II}$$

[12] The mapping here is taken from $\mathfrak{S}$ to $\mathfrak{D}$; and brackets correspond to *vertices*.

Now consider the tree

$$T = [T_k, T_{k-1}, \ldots, T_1]$$

of $n+1$ vertices. Let $|T_i| = r_i$, $i = 1, 2, \ldots, k$. Then for each ordered partition P of n,

$$r_1 + r_2 + \ldots + r_k = n$$

of exactly k parts $r_i \neq 0$, there are $\prod_{i=1}^{k} \mathfrak{c}_{r_i}$ possible T. So

$$\mathfrak{c}_{n+1,k} = \sum_P \prod_{i=1}^{k} \mathfrak{c}_{r_i}. \tag{III}$$

LEMMA 12.3 *The coefficient of* x^n *in* $\{(\mathfrak{c})(x)\}^k$, $k \leq n$, *is* $\mathfrak{c}_{n+1,k}$.

Proof, by multiplying out. When powers of x are interpreted as parts of n, the coefficient of x^n is the RHS of (III).

□

Now

$$(\mathfrak{c}) = (\mathfrak{c})(x) = \mathfrak{c}_1 x + \mathfrak{c}_2 x^2 + \mathfrak{c}_3 x^3 + \ldots = \mathfrak{c}_{2,1} x + \mathfrak{c}_{3,1} x^2 + \mathfrak{c}_{4,1} x^3 + \ldots,$$

by (II). Therefore, applying the lemma,

$$(\mathfrak{c}) + (\mathfrak{c})^2 + (\mathfrak{c})^3 + \ldots = \begin{array}{ccccc} \mathfrak{c}_{2,1}x + & \mathfrak{c}_{3,1}x^2 + & \mathfrak{c}_{4,1}x^3 + & \ldots \\ & \mathfrak{c}_{3,2}x^2 + & \mathfrak{c}_{4,2}x^3 + & \ldots \\ & & \mathfrak{c}_{4,3}x^3 + & \ldots, \\ & & & \vdots \end{array}$$

and summing by columns,

$$\sum_{n=1}^{\infty} (\mathfrak{c})^n = \sum_{n=1}^{\infty} \left(\sum_{k=1}^{n} \mathfrak{c}_{n+1,k}\right) x^n = \sum_{n=1}^{\infty} \mathfrak{c}_{n+1} x^n$$

by (I). Therefore

$$x \sum (\mathfrak{c})^n = \sum \mathfrak{c}_{n+1} x^{n+1} = (\mathfrak{c}) - \mathfrak{c}_1 x = (\mathfrak{c}) - x,$$

giving

$$(\mathfrak{c}) = \frac{x}{1 - (\mathfrak{c})}, \quad x = (\mathfrak{c}) - (\mathfrak{c})^2.$$

Thus $(\mathfrak{c})(-1) = -\cdot 6180339\ldots$ (= –'golden ratio').

Remark

Note that, since

$$\left(\sum_{n=0}^{\infty} x^m\right)^r = \sum_{n=0}^{\infty} t(r, n) \cdot x^n, \qquad |x| < 1,$$

it follows that $t(r, n)$ = the number [13] of ordered $p_1 + p_2 + \ldots + p_r = n$, allowing $p_i = 0$.

[13] C.f. Ch. IV, §5, footnote 10.

A Recurrence Relation Satisfied by the CATALAN Numbers

In lemma 3, in the particular case $k = 2$,

$$\begin{aligned}(\mathfrak{c})^2 &= \sum_{n=2}^{\infty}\left(\sum_{i=1}^{n-1} \mathfrak{c}_i \mathfrak{c}_{n-i}\right)x^n \\ &= (\mathfrak{c}) - x \text{ by the quadratic relation.}\end{aligned}$$

From the above therefore follows the relation (on the right)

$$\mathfrak{c}_{n+1,2} = \mathfrak{c}_{n+1,1} = \mathfrak{c}_n = \sum_{i=1}^{n-1} \mathfrak{c}_{n-i} \cdot \mathfrak{c}_i, \quad n \geq 2,$$

known to EULER and proved by CATALAN (see TAMARI [150] (1962), §1; GOULD [51] (1971), Introduction).

The Exponential Equation Satisfied by the Generating Function of the Vertex-Indexed Oriented Trees

Let $\mathfrak{b}_i$ = the number of vertex-indexed oriented trees on i vertices. Thus, for example, let $i = 4$. The outline may be taken as a normal tree, and the $\mathfrak{n}_i = 4$ trees

$$[[\,]],\ [2],\ [1,[\,]],\ [[\,],[\,],[\,]]$$

respectively contribute 24, 12, 24, 4 indexed trees, giving $\ _4 = 64$. Now let

$$T(x) = \sum_{n=1}^{\infty} \frac{\mathfrak{b}_n x^n}{n!}.$$[14]

Then, as with the series $\{(\mathfrak{c})(x)\}^k$ above, the coefficient of x^n in $\{T(x)\}^k$, $n \geq k \geq 1$, is

$$\mathfrak{b}(k,n) = \sum_P \prod \frac{\mathfrak{b}_{r_i}}{r_i!}, \qquad \text{(I)}$$

where P varies over the k-fold partitions of n.

Now consider the sequences

$$\mathcal{S} = (T_1, T_2, \ldots, T_k)$$

of k trees of $r_1, r_2, \ldots, r_k$ vertices respectively. The $n = \sum r_i$ different vertices may be distributed among the T_i in

$$\frac{n!}{r_1! r_2! \ldots r_k!}$$

[14]Thus:

$$T(x) \text{ converges for } |x| < \rho, \text{ where } \frac{1}{\rho} = \limsup_n \sqrt[n]{\frac{\mathfrak{b}_n}{n!}}.$$

different ways. For every such distribution, there are $\mathfrak{b}_{r_1}\mathfrak{b}_{r_2}\ldots\mathfrak{b}_{r_k}$ sequences $\mathcal{S}$. Now the pv of each T_i of $\mathcal{S}$ may be connected by an edge to a vertex v, to form a tree T of $n+1$ vertices, for which, therefore, there are exactly k edges incident on the pv. The order of connection to v is ignored, therefore the $k!$ permutations of $\mathcal{S}$ give the same T. Any 1 of the n vertices may have been chosen in place of v, as pv, making $n+1$ choices of pv. Therefore by (I) the k–fold P give rise to

$$\sum_P \frac{n!}{r_1!r_2!\ldots r_k!}\cdot \mathfrak{b}_{r_1}\mathfrak{b}_{r_2}\ldots\mathfrak{b}_{r_k}\cdot\frac{1}{k!}\cdot(n+1) = \frac{n+1!}{k!}\cdot\mathfrak{b}(k,n)$$

different T. So there are

$$n+1!\cdot\sum_{k=1}^{n}\frac{\mathfrak{b}(k,n)}{k!} = \mathfrak{b}_{n+1} \qquad \text{(II)}$$

trees of $n+1$ vertices. Therefore

$$\begin{aligned} e^{T(x)} &= 1 + T(x) + \frac{\{T(x)\}^2}{2} + \ldots + \frac{\{T(x)\}^k}{k!} + \ldots \\ &= 1 + \sum_{n=1}^{\infty}\sum_{k=1}^{n}\frac{\mathfrak{b}(k,n)}{k!}x^n = 1 + \sum_{n=1}^{\infty}\frac{\mathfrak{b}_{n+1}x^n}{n+1!} \quad \text{by (II)} \\ &= \frac{T(x)}{x}, \end{aligned}$$

since $\mathrm{b}_1 = 1$. So for x ranging over the complex plane,

$$xe^{T(x)} = T(x).$$ [15]

The Formula $\mathfrak{b}_{n+1} = (n+1)^n$ Derived Using the Inversion Formula of LAGRANGE

The LAGRANGE formula states that if $y = F(x)$ satisfies $y = x\phi(y)$ for a certain ϕ, then

$$\psi(y) = \psi(0) + \sum_{n=1}^{\infty}\frac{x^n}{n!}D^{n-1}[\psi'(y)\{\phi(y)\}^n]_{y=0},$$

for certain ψ, where D^i denotes i–fold differentiation. [Ref. e.g. WHITTAKER and WATSON [158], 'Mathematical Analysis', pp132-3]. Putting $y = T(x)$, $\phi(y) = e^y$,

[15] MOON, J. W., [112] *Counting Labelled Trees* , (1970) cites, as the source of this result, RIDDEL, R. J. and UHLENBECK, G. E. 'On the theory of the virial development of the equation of state of mono-atomic gases', J. Chem. Physics, **21** (1953), 2056-2064. See also GILBERT, F. N., 'Enumeration of Labelled Graphs', *Canadian J. Math.* , **8** (1956), 405-411. However de BRUIJN 'Enumeration of tree-shaped molecules', pp59-68, in *Recent Progress in Combinatorics* Ed. TUTTE, Academic Press (1969), attributes it to CAYLEY.

and $\psi(y) = y^k$, for successive values $k = 1, 2, 3, \ldots$,

$$\begin{aligned} D^{n-1}\psi'(y)\phi^n(y)\mid_{y=0} &= D^{n-1}ky^{k-1}e^{ny}\mid_{y=0} \\ &= k\sum_{i=0}^{n-1}\binom{n-1}{i}D^i(y^{k-1})D^{n-1-i}(e^{ny})\mid_{y=0} \\ &= k\sum_{i=0}^{n-1}\binom{n-1}{i}(k-1)\ldots(k-i)y^{k-1-i}n^{n-1-i}e^{ny}\mid_{y=0} \\ &= k\binom{n-1}{k-1}(k-1)!n^{n-k}, \end{aligned}$$

since at $y = 0$, $y^{k-1-i} = 0$ for $i < k-1$ but $y^{k-1-i} = 0^0 = 1$ for $i = k-1$; and $(k-1)(k-2)\ldots(k-i) = 0$ for $i > k-1$. Now $\psi(y) = \{T(x)\}^k$, therefore the coefficient of x^n is

$$\frac{1}{n!}\cdot D^{n-1}\psi'(y)\phi^n(y)\mid_{y=0} = \mathfrak{b}(k,n).$$

But $\mathfrak{b}_{n+1} = (n+1)!\sum_{k=1}^{n}\frac{\mathfrak{b}(k,n)}{k!}$ by (II), therefore

$$\begin{aligned} \mathfrak{b}_{n+1} &= (n+1)\cdot\sum_{k=1}^{n}\frac{(n-1)!}{(n-k)!(k-1)!}n^{n-k} \\ &= (n+1)\cdot\sum_{k=0}^{n-1}\binom{n-1}{k}\cdot n^{n-1-k} = (n+1)^n. \end{aligned}$$

Remarks

1. The formula gives the radius ρ of convergence of the infinite series $T(x)$ above (ref. footnote 11; see APOSTOL [8], Th. 8·26) as

   $$\rho = \{\limsup_n \sqrt[n]{\frac{(n+1)^n}{n!}}\}^{-1}.$$

   So by STIRLING's asymptotic formula for the Γ-function,

   $$\frac{\mathfrak{b}_{n+1}}{(n+1)!} \sim \sqrt[n]{e^n\cdot(n+1)^{-\frac{3}{2}}\cdot(2\pi)^{-\frac{1}{2}}},$$

   giving $\rho = \frac{1}{e}$. It therefore follows that the vertex-indexed oriented tree generator $T(x)$ has the expression

   $$T(x) = xe^{xe^{xe^{\cdots}}},$$

   for $|\, x \,| < \frac{1}{e}$, in accordance with the convergence of the EULER-MAURER exponential (II, §2.4).

2. If T is a vertex-indexed (vi) free tree of n vertices, any 1 of the n may be chosen as the pv of T. Therefore the $\mathfrak{b}_n$ distinct vi oriented trees of n vertices determine $\frac{\mathfrak{b}_n}{n} = n^{n-2}$ distinct vi free trees of n vertices. I.e. the formula of CAYLEY used in IV, §2.4, note 1. for the number of indexed free trees of n different 'knots'.

The Exponential Equation Satisfied by the Generating Function of the Normal Trees

Changing slightly the notation of §1 (so that edges rather than v's form the basis of counting), let

$$\mathfrak{T}(x) = x + 2x^2 + 4x^3 + \ldots + \mathfrak{n}_{i+1}x^i + \ldots .$$

In 1937, POLYA [124], (1937) applied his fundamental theorem [16] on enumeration to obtain another identity of CAYLEY:

$$\mathfrak{T}(x) = xe^{\{\mathfrak{T}(x)+\frac{1}{2}\mathfrak{T}(x^2)+\frac{1}{3}\mathfrak{T}(x^3)+\ldots\}}.$$

ROBINSON and SCHWENK (loc. cit) remark that the recurrence of corol. 2 may be obtained by differentiating this expression and equating coefficients.

12.3 EP and the Recursive Structure of Permutations

This section touches on a fundamental many-one correspondence between permutations on fm letters, and **EP**- exponentials. The group-theoretic ramifications of **EP** are omitted. Thus, let $\sigma \in S_N$ be a permutation on N letters. Let

$$\sigma = \gamma_R\gamma_{R-1}\cdots\gamma_1$$

be the resolution [17] of σ as a finite product of cycles. For some $i \leq R$, suppose the cycle $\gamma = \gamma_i$ acts on the letters $a_1 < a_2 < \ldots < a_n$, $n = n_i \leq N$. That γ itself is a permutation on $n-1$ letters may be seen from the characterization

$$\gamma = \begin{pmatrix} a_n & b_1 & b_2 & \ldots & b_{n-3} & b_{n-2} & b_{n-1} \\ b_1 & b_2 & b_3 & \ldots & b_{n-2} & b_{n-1} & a_n \end{pmatrix},$$

where the $(b_1, b_2, \ldots, b_{n-1})$ is a permutation of $a_1, a_2, \ldots, a_{n-1}$. It may be noted that, as to each cell what is below at the left, is above at the right; excepting the ends, for which the rule is reversed. Viz. there are $(n-1)!$ different cycles on n letters.

DEFINITION 12.4 *The permutation determined by γ_i is $\sigma_i \in S_{n-1}$.*

Therefore σ comprises a system $(\sigma_R, \sigma_{R-1}, \ldots, \sigma_1)$ of respective degrees $n_R^- \geq n_{R-1}^- \geq \ldots \geq n_1^-$. In the case of a degenerate σ_i of unit degree, $0! = 1$ may be used. Now each σ_i may similarly be resolved, so that σ determines a fixed $S \in \mathfrak{S}$. This tree [18]

[16] See N. G. de BRUIJN 'Polya's theory of counting', Ch. 5 in: BECKENBACH [11]; also SLOANE [142], *Handbook of Integer Sequences* , p12.

[17] Not taking order into consideration. However for $1 \leq i < j \leq R$ it may be assumed that deg $\gamma_i \leq$ deg γ_j, deg $\gamma = \#$ letters acted on by γ.

[18] C.f. the $\mathfrak{S}$ -tree $S(x)$, IV,§6.

is arbitrary, in the sense that the cycles were ordered by degree, not reflecting at all the way they are meshed within σ. Of all possibilities therefore, the natural choice is the IV, §4-normal form.

THEOREM 12.5 *Every permutation $\sigma \in S_N$ determines a unique $\mathfrak{T}$-tree $T\sigma$.*

Remark. The first letter, for example, acted on by γ may be embedded within σ in any of n_i locations, giving $(n_i - 1)!n_i = n_i!$ possible cycles.

Mesh Structure

Cycles may be meshed (cf. rem. 2. after IV·26) in

$$\frac{n!}{n_R!n_{R-1}!\dots n_1!}$$

ways. Therefore different permutations (of like degree) can determine the same $\mathfrak{T}$-tree.

Applications

It may be accepted that $\sigma, \rho \in S_N$, $\mathrm{eh}T\sigma < \mathrm{eh}T\rho$ implies that σ is 'simpler' than ρ. Thus in this reckoning, after the identity (which deranges no letter), the simplest σ shift a_i to a_{i+1} mod N; while the least simple is the unique σ with a recursive decomposition into cycles of degrees $N, N-1, N-2, \dots, 1$. (E.g. for $N = 7$, this is 3675142). This may have an application to the analysis of very long permutations, such as occur in chromosomes.

12.4 Notes on Ch. XII

1.§2.2. The following proof of the formula $t(n,p) = \binom{n+p-1}{p}$ is based on KNUTH [77], p488: $t(n,p)$ is the number of integer solutions to

$$1 \le a_1 \le a_2 \le \dots \le a_p \le n \qquad \text{(I)}$$

(by putting each choice in non-decreasing order, and ordering lexicographically). (I) may be re-expressed

$$0 < a_1 < a_2 + 1 < \dots < a_p + p - 1 < n + p. \qquad \text{(II)}$$

The solutions to (II) are in 1−1 correspondence with those of

$$0 < \mathrm{b}_1 < \mathrm{b}_2 < \dots < \mathrm{b}_p < n + p;$$

and the number of these is $\binom{n+p-1}{p}$.

Appendix A

APPENDIX: THE PREDICATE AND SEQUENT CALCULI

Section Headings

A.1 Introduction

The following account is included for the purpose of articulating the finitist view of logic.

Non-constructive or 1st order predicate logic $\mathfrak{L}$ derives from the symbolic calculus or concept-script ('Begriffsschrift') of Gottlob FREGE (1879). [1] The class $\mathfrak{L}_{\text{form}}$ is

[1]FREGE paved the way for the acceptance and formalization of CANTOR's set theory, by founding our grasp of numbers on *concepts*. This at once led him to the first-order predicate calculus, and to the infinite concept (viz. "is finite") or set. In fact it is quite difficult to explain

that of the first order predicate formulae. These are (well-formed) finite strings of primitive symbols, corresponding to generally accepted ideas of logic, such as $\neg$ (negation), and P(a) (the constant a has the property P). A *sentence* is a formula which may be assigned truth or falsehood. The *axioms*, or *initial logical sequents* of 1st order logic are respectively chosen or constructed from $\mathfrak{L}_{\text{form}}$. The rules of inference adopted here are those of GENTZEN's sequent calculus. It is convenient to analyze $\mathfrak{L}_{\text{form}}$ into a subset $\mathfrak{L}_{\text{atom}}$ of *atomic formulae*; the remaining *molecular formulae* are obtained by taking closure of $\mathfrak{L}_{\text{atom}}$ under the symbolic operations of logical combination and prefix by quantifier.

1st order logic has many extensions, besides taking higher-order forms (for which quantifiers apply to predicate symbols). Since these schemata are all reducible to formalism within $\mathfrak{L}$, the first order logic occupies centre place in all studies of formal logic.

A.2 Zero Order Logic $\mathfrak{L}^{(0)}$

This is often called the propositional or sentence calculus, and is defined inductively from a class $\mathfrak{L}^{(0)}_{\text{elem}}$ of elementary sentences or *truth variables*. Each $A \in \mathfrak{L}^{(0)}_{\text{elem}}$ may take, exclusively, either the value T = true, or F = false. The class $\mathfrak{L}^{(0)}_{\text{sen}}$ of all sentences, is obtained from $\mathfrak{L}^{(0)}_{\text{elem}}$, using the connectives $\neg, \&, \vee, \Rightarrow, \Leftrightarrow$. The subclass $\mathfrak{L}^{(0)}$ of *tautologies* contains exactly those sentences, which remain true, irrespective of the truth values assigned to the elementary sentences which they contain. E.g. suppose A, B are elementary sentences. Then from the truth-tables (Tt's)

A	$\neg A$
T	F
F	T

and

A	B	$A \vee B$
T	T	T
T	F	T
F	T	T
F	F	F

follows

A	B	$A \vee (\neg A)$
T	T	T
T	F	T
F	T	T
F	F	T

Note that $\mathfrak{L}^{(0)}_{\text{elem}} \cap \mathfrak{L}^{(0)} = \{\}$.

A.2.1 A Note on the Class $\mathfrak{L}^{(0)}_{\text{elem}}$

Only the number of sentences in $\mathfrak{L}^{(0)}_{\text{elem}}$ is important, since no $A \in \mathfrak{L}^{(0)}_{\text{elem}}$ has any distinguishing feature, other than truth or falsehood. Thus, consider a boolean function $\phi(A_1, A_2, \ldots, A_n)$. Any one of the 2^n subsets of the n arguments defines a possible assignment of truth values. For each assignment, ϕ returns T or F. There are therefore $2^{2^n} = 2\star(2;n)$ distinct ϕ, each determining a class of sentences, equivalent under assignment of truth values. It is an elementary meta-theorem of $\mathfrak{L}^{(0)}$ that every boolean function may be built up from *propositional variables* and

why the brain is so adroit with numbers, but surely the link lies deeper than concepts.

the logical connectives. In fact the usual logical connectives, taken together, are redundant and may be replaced by the remarkable [2] SHEFFER stroke, defined by

$$A \mid B \text{ iff } (\neg A) \vee (\neg B).$$

Normal Form

It may be remarked (see e. g. HILBERT and ACKERMANN *Introduction to Mathematical Logic* [63], Ch. I, §3) that any $A \in \mathfrak{L}^{(0)}_{\text{sen}}$ has the same Tt as a sentence (of $\mathfrak{L}^{(0)}_{\text{sen}}$) containing only conjunctions of sentences themselves free of conjunctions (viz. containing only disjunctions of elementary or negations of elementary sentences).

Duality

According to this principle (ibid. Ch. I, §5), the truth value of $A \in \mathfrak{L}^{(0)}_{\text{sen}}$ is invariant if: every elementary sentence B in A is replaced by $\neg B$; and if all occurrences of '&' and '$\vee$' are interchanged. Application of this to the normal form gives the dual *disjunctive normal form.*

Deduction in $\mathfrak{L}^{(0)}$

The above-cited *Introduction* , Ch. I, §10; and KLEENE, *Introduction to Meta-Mathematics* [74], 1971, p82 contain lists of axioms. [3] From these, using the rules of inference (axioms and rules of inference are taken together as *postulates* in [74]), the theorems of $\mathfrak{L}^{(0)}$ may be inferred. The theorems include the axioms. If the sentence A is a theorem of $\mathfrak{L}^{(0)}$ this is expressed

$$\vdash A.$$

If, given a sentence or set of sentences $\mathcal{A}$, a sentence A may be inferred in $\mathfrak{L}^{(0)}$, this is expressed $\mathcal{A} \vdash A$.

Some properties of the zero order logic are now listed:

1. Correctness (meta-)theorem: $\vdash A$ only if A is a tautology. This follows, as neither the axioms nor the rules lead out of the class of tautologies.

2. Deduction (meta-)theorem: if $\mathcal{A}, A \vdash B$, then $\mathcal{A} \vdash A \Rightarrow B$.

---

[2] See LOWESMITH [90], 1981.

[3] The *axioms* may be taken as

$$1.\ A \vee A \Rightarrow A \quad 2.\ A \Rightarrow A \vee B$$

$$3.\ A \vee B \Rightarrow B \vee A \quad 4.\ (A \Rightarrow B) \Rightarrow \{(A \vee C) \Rightarrow (B \vee C)\},$$

where the letters denote variables over $\mathfrak{L}^{(0)}_{\text{sen}}$. For these axioms, *the rules of inference* are (i) *modus ponens* and (ii) (simultaneous) substitution for sentence letters.

3. $\mathfrak{L}^{(0)}$ is *consistent*: for no A is $\vdash A$ and $\vdash \neg A$. A proof of this is contained in KLEENE [74], p129.

4. The axioms of $\mathfrak{L}^{(0)}$ (as listed above) are *independent* ([63], I, §13): if any one is omitted, then only a proper part of the sentence calculus remains.

5. $\mathfrak{L}^{(0)}$ is *complete* (LUKASIEWCZ): if A is a tautology, then $\vdash A$. By (1) , a corollary to this is:

6. The theorems of $\mathfrak{L}^{(0)}$ and the tautologies coincide. Since the truth-table of a propositional function may be calculated in a finite number of steps, and checked for identical truth, a corollary to this is:

7. $\mathfrak{L}^{(0)}$ is (recursively) *decidable.*

Ch. VI of KLEENE [74], especially §28 contains a thorough discusion of $\mathfrak{L}^{(0)}$.

Intuitionist Zero-Order Logic

The Intuitionistic logic $\mathfrak{L}^{(0)}_{\text{int}}$ of L. E. J. BROUWER, formalized by A. HEYTING, is obtained by

1. replacing the above axioms by a longer list of *basic tautologies*;
2. removing the tautology $A \vee \neg A$; and
3. taking closure under the above rules.

Thus e.g. $\vdash A \Rightarrow \neg\neg A$ in $\mathfrak{L}^{(0)}_{\text{int}}$; but $\nvdash \neg\neg A \Rightarrow A$.

Boolean Algebras

Zero order logic may be interpreted algebraically. Thus, if the lattice (IV, §10) $(\mathcal{X}, \subseteq)$ has $\mathbf{0}, \mathbf{1} \in \mathcal{X}$, for which

$$(\forall A \in \mathcal{X})\ A \cap \mathbf{0} = \mathbf{0};\ A \cup \mathbf{1} = \mathbf{1};\ A \cap \mathbf{1} = A = A \cup \mathbf{0},$$

then $\mathcal{X}$ is a *bound* lattice. $\mathcal{X}$ is *complemented* if

$$(\forall A \in \mathcal{X})(\exists B \in \mathcal{X})\ A \cup B = \mathbf{1}\ \&\ A \cap B = \mathbf{0}$$

(and B is written $\overline{A}$). A *boolean algebra* (ba) is defined as a complemented, distributive lattice.

Now let $B \in [A]$ in $\mathfrak{L}^{(0)}_{\text{sen}}$ iff $B \equiv A$. Then the system $\mathfrak{L}^{[0]} = \{[A] : A \in \mathfrak{L}^{(0)}_{\text{sen}}\}$ forms a ba under the operations

$$[A] \cup [B] = [A \vee B];\ \ [A] \cap [B] = [A\ \&\ B]$$

(SZASZ [144], §46, p134). The upper and lower bound elements are respectively the equivalence classes of the tautologies, and of the identically false sentences. $\mathfrak{L}^{[0]}$ is the ba of subsets of the class of truth value assignments. Therefore for n elementary sentences, $|\mathfrak{L}^{[0]}| = 2 \star (2; n)$.

A *homomorphism* $h : \mathcal{X} \to \mathcal{Y}$ of boolean algebras satisfies

$$h(\mathbf{1}_{\mathcal{X}}) = \mathbf{1}_{\mathcal{Y}}; \quad h(A \cup B) = (hA) \cup (hB).$$

From these follow many identities (see RASIOWA and SIKORSKI, *The Mathematics of Meta-Mathematics* [126], Ch. II). h is an *isomorphism* iff $h(A) = \mathbf{1} \Rightarrow A = \mathbf{1}$.

Now define (SZASZ [144], pf. of thm. 83, p167): [4]

$$\mathcal{F}(\mathcal{X}) = \text{ the class of ultrafilters of the ba } \mathcal{X}$$

$$\iota(A) = \{\mathcal{F} \in \mathcal{F}(\mathcal{X}) : A \in \mathcal{F}\}, \text{ for each } A \in \mathcal{X}$$

$$\mathbf{F}(\mathcal{X}) = \{\iota(A) : A \in \mathcal{X}\}.$$

Then the representation theorem of M. H. STONE [5] states that $\iota : \mathcal{X} \to \mathbf{F}(\mathcal{X})$ is an isomorphism of boolean algebras.

Remarks

1. Let $X = \{x_1, x_2, \ldots, x_n\}$, $n \geq 1$, be a finite set. Let the ba $(\mathcal{X}, \subseteq)$ be defined by

   $$\mathcal{X} = \{A : A \subseteq X\} \text{ (so that } |\mathcal{X}| = 2^n).$$

   Now all sequences $\ldots \subset A_2 \subset A_1$ in $\mathcal{X}$ satisfy the DCC. Therefore by DILWORTH's theorem, every filter is a dual principal ideal, i.e. of the form $\{B \in \mathcal{X} : A \subseteq B\}$, for some $A \in \mathcal{X}$. Therefore every ultrafilter $\mathcal{F}$ is of the form

   $$\mathcal{F}_x = \{B : \{x\} \subseteq B\},$$

   for some $x \in X$, with $|\mathcal{F}_x| = 2^{n-1}$. Now let $A \in \mathcal{X}$ be fixed. By the above, the class of ultrafilters of $\mathcal{X}$ which contain A is

   $$\iota(A) = \{\mathcal{F}_x : x \in A\}.$$

   Viz. in the important special case of $\mathcal{X}$ = the boolean algebra of subsets of a finite set X, the STONE isomorphism collapses to the identity.

2. It may be noted that, for $\mathcal{X}$ as in 1., X contains n points; $\mathcal{X}$ contains every one of 2^n subsets. Therefore $\mathcal{F}(\mathcal{X})$ is a combination, taken from $2 \star (2; n)$ subsets of these subsets; while $\mathbf{F}(\mathcal{X})$ is a (2^n-)combination taken from $2 \star (3; n)$ subsets of these.

---

[4] A filter (or *dual ideal*; see Ch. IV, §10.2, footnote 19) $\mathcal{F} \subseteq \mathcal{X}$ is an *ultrafilter*, if $\mathcal{F}$ is maximal, viz. properly contained in no other filter. This further condition has various equivalences. Thus

1. $\mathcal{F}$ is the dual of an ideal $\mathcal{I}$ for which

   $$A, B \in \mathcal{X} \Rightarrow \{A \cap B \in \mathcal{I} \Rightarrow (A \in \mathcal{I} \vee B \in \mathcal{I})\}.$$

2. $(\forall A \in \mathcal{X})\ A \in \mathcal{F} \Leftrightarrow \overline{A} \notin \mathcal{F}$.
3. $\mathcal{X} - \mathcal{F}$ is a prime (or maximal, since $\mathcal{X}$ is a ba) ideal.

[5] 'Boolean algebras and their relation to topology', *Proc. Nat. Acad. Sci. U. S. A.* , **20** (1934), 197-202.

A.3 First Order Predicate Logic, $\mathfrak{L}$

Formulae

The class $\mathfrak{L}_{\text{form}}$ of predicate formulae is built up inductively, via *terms*, *atomic formulae*, and closure under symbolic operations. Terms are built up from constant, variable and function symbols, as in: $\mathrm{F}\{x,\mathrm{G}(a,b,y)\}$. *Atomic* formulae are built up by substituting terms s,t, etc. in the argument places of predicate [6] symbols, as in $\mathrm{P}(t)$; $\mathrm{Q}(s,t)$, etc. *(Molecular) formulae* are built up from atomic formulae A, B etc. by closure under

1. linking by logical connectives; and
2. prefixing signs of quantification, as in $(\forall x)$ A. Here,'x' is a *bound* or *apparent* variable.

A formula may contain *free* (unquantified) variables. A formula without free variables is a *sentence.* [7] $\mathfrak{L}_{\text{form}}$ is closed under substitution of terms for occurrences of free variables. [8] $\mathfrak{L}$ is itself the subclass of $\mathfrak{L}_{\text{form}}$ obtained by taking the closure of the axioms under the procedures of inference.

Axioms

The (logical) *axioms* of $\mathfrak{L}$ include those of $\mathfrak{L}^{(0)}$, [9] together with

1. $(\forall x)\ \mathrm{A}(x) \Rightarrow \mathrm{A}(t)$
2. $\mathrm{A}(t) \Rightarrow (\exists x)\ \mathrm{A}(x)$

Inference

Inference in $\mathfrak{L}$ is e.g. via the sequent calculus discussed in the next section. The '$\vdash$' notation is adopted for (theories expressible in) first order logic, as for $\mathfrak{L}^{(0)}$.

A.4 Sequent Calculus

Deduction of formal theorems, within $\mathfrak{L}$, may be implemented by GENTZEN's sequent calculus. Let $\mathcal{A}$ be a finite string $\mathrm{A}_1,\mathrm{A}_2,\ldots,\mathrm{A}_m$, of (predicate) formulae, and let A be a formula. A *sequent* $\mathcal{S}$ is a symbolic string of the form

$$\mathcal{A} \rightarrow \mathrm{A},$$

[6]Including a predicate symbol of 2 arguments, to be interpreted as '='.

[7]A formula without free variables, and which does not contain logical connectives, is an *elementary sentence.*

[8]Some care is necessary, to except such cases as e.g. substitution of t for y in $(\exists x)\ \mathrm{P}(x,y)$ to give $(\exists x)\ \mathrm{P}(x,t)$, where $t = x$ (see H. D. EBBINGHAUS et al, [54], Ch. III, §8).

[9]Or of $\mathfrak{L}^{(o)}_{\text{int}}$ in the case of the intuitionistic sub-system $\mathfrak{L}_{\text{int}}$ of $\mathfrak{L}$.

to be read 'from $\mathcal{A}$ follows A'. The formulae of $\mathcal{A}$ are the *antecedents* of $\mathcal{S}$; A the *succeedent.*

A.4.1 Initial Sequents

1. The *initial logical sequents* are those of the form A→ A, and

   (a) A & B→ A;

   (b) A & B→B;

   (c) A,B→A & B

   (d) A, ¬A→ $1 \neq 1$; [10]

   (e) ¬¬A→A;

   (f) $(\forall x)$ F(x) →F(t), where t is a term.

   These sequents alone are sufficient for the inference of the logical theorems of $\mathfrak{L}$.

2. The *initial mathematical sequents* are those of the form

$$\mathcal{A} \rightarrow \text{A},$$

   such that substitution of free variables by numerals results in the succeedent being true; or in an antecedent being false.

A.4.2 Inference

An *inference* is a symbolic structure

$$\frac{\mathcal{S}_1, \mathcal{S}_2, \ldots, \mathcal{S}_n}{\mathcal{S}}$$

where the sequents $\mathcal{S}_1, \ldots, \mathcal{S}_n$, the *premisses*, and the sequent $\mathcal{S}$, the *conclusion* of the inference, are separated by a *line of inference*.

A *derivation* is a tree (IV·1) $T \in \mathfrak{S}$, whose vertices v correspond to sequents $\mathcal{S}(v)$. If v is an endv, $\mathcal{S}(v)$ is an initial sequent. And if $v_1, v_2, \ldots v_n$ Sc v is an exhaustive list of the successors of v in T, then $\mathcal{S}(v_i)$, $i = 1, 2, \ldots, n$ are the premisses, and $\mathcal{S}(v)$ the conclusion of an inference. Taking, as in Ch. IV et seq., $v^\sharp$ to be the pv of T, $\mathcal{S}(v^\sharp)$, the last sequent of the derivation is the *endsequent.*

New sequents may be derived from given sequents, using the following groupings of rules; these were used by GENTZEN in the proof of consistency later published by BERNAYS, mentioned in I, §4:

[10]Thus $\mathfrak{L}$ is taken to be *predicate calculus with equality,* in which the constant symbol (numeral) '1' is to be included (under the convention of Ch. 1, §3.1.)

1. Structural changes in a sequent:
   (a) Derangement of the order of antecedents
   (b) Addition of any predicate formula to the antecedents
   (c) Delete a repetition in the antecedents
   (d) re-name a variable, bound by a universal quantifier.
2. Inference schemata
   (a) [11]
   $$\frac{\mathcal{A}\rightarrow \mathrm{A},\quad \mathcal{B},\mathrm{A}\rightarrow \mathrm{B}}{\mathcal{A},\mathcal{B}\rightarrow \mathrm{B}} \qquad \text{(cut)}$$
   (b)
   $$\frac{\mathcal{A},\mathrm{A}\rightarrow 1\neq 1}{\mathcal{A}\rightarrow \neg \mathrm{A}} \qquad \text{(negation introduction)}$$
   (c)
   $$\frac{\mathcal{A}\rightarrow \mathrm{A}(a)}{\mathcal{A}\rightarrow (\forall x)\mathrm{A}(x)} \qquad (\forall - \text{introduction}).$$

A.5 Syntax and Interpretation of $\mathfrak{L}$

A.5.1 Normal forms

Formulae A, B are said to be *convertible* if $\vdash$A$\Leftrightarrow$ B. It is a meta-theorem of logic that if A is any formula, there is a formula B, such that A,B are convertible, and B is of the form QC, where Q represents a string of quantifier expressions, and C a string of symbols of $\mathfrak{L}$, not including any quantifier symbols.

A stronger result may be obtained, by weakening the relation between A, B. The formulae are said to be *deductively equivalent* iff A$\vdash$ B when and only when B $\vdash$ A. SKOLEM 1934, showed that, given any A, there is a deductively equivalent B for which the quantifier string Q has all the existential quantifiers ($\exists$) to the *left* of the universal quantifiers.

A.5.2 Theories

The purpose of the logical formalism is to confine within formal inference the deduction, of axiomatic theories. Thus, the *mathematical axioms* of a (1st order, consistent) formal theory $\underline{\mathrm{T}}$ constitute a class $\mathcal{A}\subset\mathfrak{L}_{\text{form}}$, [12] other than the logical

[11] Note that *cut* is a generalization of *modus ponens*.

[12] If $\underline{x}$ is an exhaustive list of the free variables of a formula A, then (ref. HAMILTON, A. G. *Logic for Mathematicians* , Ch. III; Cambridge (1978), ISBN 0 521 29291 3) it may be shown that $\vdash$ A iff $\vdash$ $(\forall\underline{x})$ A. It is therefore possible to take as axioms of $\underline{\mathrm{T}}$ only *sentences*.

axioms of $\mathfrak{L}$ or their negations. $\underline{\mathrm{T}}$ is the closure of $\mathcal{A}$ ùnder logical consequence. If A,B are any sentences, and A & $\neg$A a theorem of $\underline{\mathrm{T}}$; then $\underline{\mathrm{T}} \vdash$ B (I, §3.5).

A *model* or correct interpretation for $\underline{\mathrm{T}}$ is any structure whose description by $\underline{\mathrm{T}}$ is true. Thus, the usage 'model' in contemporary logic is quite opposite to the usage in e.g. 'mathematical model'. That a sentence, set of sentences or whole theory has a model $\mathcal{I}$ is denoted by the 'satisfaction predicate', as in

$$\mathcal{I} \models \underline{\mathrm{T}}.$$

$\underline{\mathrm{T}}$ is *complete* if for any A $\in \mathfrak{L}_{\mathrm{form}}$, either $\vdash$ A or $\vdash \neg$ A. [13] Thus by I, 1·3·1, $\underline{\mathrm{N}}$ is incomplete. *Complete arithmetic* is defined to be the subclass of $\mathfrak{L}_{\mathrm{form}}$ for which N is a correct interpretation.

A.5.3 Completeness

Interpretation and Validity

Validity [14] is a central notion of logic, and it is opportune to reduce it, to the extent possible, to finitistic terms. Validity is usually introduced via interpretation of constants, variables, and predicate and function symbols, over an arbitrary domain D. Any A$\in \mathfrak{L}_{\mathrm{form}}$ is *(universally) valid*, written $\models$ A, if there is no interpretation in which A is false. The notion of arbitrary interpretation is perhaps less acceptable finitistically, than is the axiom of choice.

1. Suppose that A is true in all interpretations for which D is countable (viz. finite, or enumerable by the natural numbers); and that A is *false* under an interpretation $\mathcal{I}$ for which D is uncountable. Then $\neg$ A is true in $\mathcal{I}$. By a well-known result of LOWENHEIM-SKOLEM, $\neg$ A is therefore true under some interpretation for which D is countable, contrary to hypothesis. From this it follows that, for the purposes of establishing validity, only interpretations with *countable* domains need be considered.

2. Now consider e.g. a predicate symbol P=P(x, y) of the language $\mathfrak{L}$. Let $\mathcal{I}$ be an interpretation whose domain D is countable. D can thus be enumerated:

   $$D = (x_1, x_2, x_3, \ldots).$$

   The effect of $\mathcal{I}$ is to assign a truth value Θ to P at x, y :

   $$\mathrm{P}(x, y) = \Theta, \quad \text{where } \Theta = T \text{ or } F.$$

   But there are m, n such that $\mathrm{P}(x_m, x_n) = \Theta$. Therefore the effect of $\mathcal{I}$ is that of the arithmetic interpretation, under which x is interpreted as m, y as n,

[13] Of course by the LM (I, §2): for any $\underline{\mathrm{T}}$, either $\vdash$A or $\nvdash$A.

[14] 'Valid' is used only in the formal sense in the Appendix; rather than in the empirical sense, as throughout the text.

and in which $P(m, n)$ is assigned the truth value Θ. It follows therefore that *the nature of the objects (e.g polynomials, latin squares) of the domain is of no importance at all, in the determination of the truth value of an arbitrary* A $\in \mathfrak{L}_{\text{form}}$. Indeed, 'interpretation' is not the correct notion, since $\mathcal{I}$ is reduced to a truth-value assignment (in the sense e.g of 'assignment of fundamental sequences' of Ch. III, §3). Note that $\mathcal{I}$ applies only to the *mathematical* symbols of $\mathfrak{L}$; the logical symbols $\neg, \&, \forall, \ldots$ are then applied to the result, bearing their fixed interpretations.

3. The validity of A $\in \mathfrak{L}_{\text{form}}$ may now be understood arithmetically, in the light of 1. and 2., as 'infinite tautologies' (c.f. Ch. I, §2.4).

Examples

(a) Let $A \equiv (\forall x)\,[P(x) \vee \neg P(x)]$. Then A is

$$\{P(1) \vee \neg P(1)\} \; \& \; \{P(2) \vee \neg P(2)\} \; \& \; \ldots,$$

true irrespective of the truth values assigned to P, so $\models$ A.

(b) Let $A \equiv \{(\forall x)\ P(x)\} \Rightarrow (\exists y)\ P(y)$. Writing in convertible (or *prenex*) normal form,

$$\begin{aligned} A &\equiv (\exists y)[(\forall x)\,P(x) \Rightarrow P(y)] \\ &\equiv (\exists y)(\exists x)[P(x) \Rightarrow P(y)]. \end{aligned}$$

Now, independently of the assignment, $P(1) \Rightarrow P(1)$, so $\models$ A.

Notes

(a) Suppose A is the sentence $\underline{Q}\ P(\underline{x})$, where $\underline{Q}$ is a quantifier string, and P is a *predicate symbol* of $\mathfrak{L}$. Since for A to be valid, it is necessary (and sufficient) that $P(x)$ holds for *every* assignment for P, $\not\models$ A. This does not violate the previous example, the predicate

$$Q(x, y) \equiv P(x) \Rightarrow P(y)$$

necessarily taking $\Theta = T$ at $Q(x, x)$, whereas a predicate of $\mathfrak{L}$ is necessarily free to take arbitrary truth values at every argument vector.

(b) The formulae A true in no interpretation, viz. the *identically false formulae*, are exactly the A for which $\models \neg$ A, viz. the negations of valid formulae.

(c) An interpretation of a variable x is a (fixed) value. Suppose x occurs free in $A=A(x)$. Then, by the above definition, provided A turns out true at all values of x, *the formula* $A(x)$ *is valid.*

A.6 Meta-Theory of $\mathfrak{L}$

Some properties of $\mathfrak{L}$ are now listed.

Completeness

The SKOLEM normal form may be used to show that $\mathfrak{L}$ is *complete*: [15] the theorems A of $\mathfrak{L}$ coincide with the valid formulae:

$$\vdash \mathrm{A} \text{ iff } \models \mathrm{A}.$$

The proof, which is by contradiction, of $\models$ A implies $\vdash$ A, essentially turns on: any consistent extension of $\mathfrak{L}$ (as in the inclusion of $\neg$A as an axiom, where $\models$ A) *has an interpretation*. This reasoning is surely outstandingly non-constructive.

With A as in footnote 9, note that

$$\models \mathrm{A}(x) \text{ iff } \models (\forall x)\,\mathrm{A}(x).$$

Also that the completeness of $\mathfrak{L}$ may be given in the 'contrapositive' form: if the negation of A is not a theorem of $\mathfrak{L}$, then A turns out to be true, in some interpretation. A formula A is said to be *satisfiable*, Sat(A), iff A is true [16] in some interpretation. A subset $\mathcal{A} \subseteq \mathfrak{L}_{\mathrm{form}}$ is said to be (simultaneously) satisfiable, Sat($\mathcal{A}$), if

1. $\mathrm{A} \in \mathcal{A} \Rightarrow \mathrm{Sat(A)}$, and

2. the theory determined by $\mathcal{A}$ is consistent.

The relation $\mathcal{A} \models A$ is defined by: $\mathrm{Sat}(\mathcal{A}) \Rightarrow \mathrm{Sat}(A)$. According to the *extended completeness meta-theorem*,

$$\mathcal{A} \models \mathrm{A} \text{ iff } \mathcal{A} \vdash \mathrm{A}.$$

Other Properties

1. $\mathfrak{L}$ is *consistent*: for no A is both $\vdash$ A and $\vdash \neg$ A. (Ref. KLEENE [74] (1952), §36, corol. 2, and HILBERT and ACKERMANN [63] (1927), page 65. This may be used to show that the axioms of $\mathfrak{L}$ are *independent* (ibid., III, §9).

2. The *compactness* meta-theorem states that if for every finite subset $\mathcal{A}'$ of $\mathcal{A}$, Sat($\mathcal{A}'$), then Sat($\mathcal{A}$). Syntactically this may be stated: every finite $\mathcal{A}' \subseteq \mathcal{A}$, is consistent, iff $\mathcal{A}$ is itself consistent.

[15] Note that the sense differs from that given after Ch. I, meta-theorem 3.1. Certainly neither the sentence $(\forall x)\ \neg\mathrm{P}(x, y, z)$ of I,3.1, nor its negation is provable in $\mathfrak{L}$ alone. In this sense, $\mathfrak{L}$ is incomplete.

[16] $\mathfrak{L}_{\mathrm{form}}$ is thus partitioned into 3 subclasses: the valid, the (properly) satisfiable and the identically false formulae.

Undecidability of $\mathfrak{L}$

$\mathfrak{L}$ is *undecidable* (CHURCH, 1936; [20] TURING 1936-7; BÜCHI [17] 1962; see also KLEENE [74] §76, 432; and the exposition of this result, ROSSER [135] (1939), in M. DAVIS, 'The Undecidable', pp 140-145): there is no recursive procedure for deciding whether, for any given A $\in \mathfrak{L}_{\text{form}}$ either $\vdash$ A, or $\nvdash$ A (and by the completeness result, nor is there for $\models$ A or $\not\models$ A). Examples of *theories* which are decidable are those of the arithmetic of the real and complex fields; and of the commutative groups. For a more complete list and bibliographic data, see R. L. WILDER [159], §5.3 p277. It is remarkable that by adding certain mathematical formulae to the basic logical formulae of $\mathfrak{L}$, a decidable theory may be obtained. Since $\mathfrak{L}$ itself is undecidable, this should be regarded as exceptional.

Remarks

1. Much as in VIII, §3, it is possible to trace back interdependences in the metatheory of $\underline{\text{N}}$. Thus $\varphi \notin \mathcal{R}$ is the key to the proof of the negative decision result for logic. The description of a TURING machine in $\mathfrak{L}$, used to obtain this result, is the same used to obtain the representability result, Ch. I, §3·4. Finally, the representability of $\mathcal{R}$ establishes the incompleteness of first-order arithmetic, $\underline{\text{N}}$, I,3·2.

2. It is with regret that it must be pointed out that this radically important result of mathematical logic is not treated in the volume [18] edited by Jon BARWISE (with H. J. KEISLER, K. KUNEN, Y. N. MOSCHAVAKIS and A. S. TROELSTRA), a book of some 1165 pages. In Ch. A.1 (of Part A: Model Theory), 'First Order Logic' by BARWISE, the result is not even mentioned. Nor in §7 'Logic and Recursion Theory', of Ch. C.1 (of Part C: Recursion Theory) by H. B. ENDERTON (although CHURCH [20], 1936 is cited). And in Ch. C.3, 'Decidable Theories' by M. O. RABIN, the result is dismissed with the words 'CHURCH proved that the predicate calculus, as well as ... are undecidable'. Not one of the numbered theorems in Ch. C.3 is devoted to the CHURCH-TURING theorem. Nowhere else [19] in the *Handbook* is the result discussed.

3. The result holds as well for intuitionistic logic: $\mathfrak{L}_{\text{int}}$ is undecidable. However the restriction of $\mathfrak{L}$ to predicate symbols of one argument (the *monadic* predicate calculus) is decidable.

---

[17] 'TURING machines and the entscheidungsproblem', *Math. Ann.* **148**, 201-213.

[18] *Handbook of Mathematical Logic* , publ. North-Holland, Stud. Logic # 90

[19] M. DAVIS, pp587-8 of the *Handbook* solves in the negative, via $\mathcal{R}-$unsolvability of a problem in Diophantine equations, HILBERT's 'Entscheidungsproblem' for a particular formula for a particular formal system, which contains non-logical axioms.

Appendix B

APPENDIX: A SUMMARY OF FIRST ORDER ARITHMETIC

Section Headings

B.1 Axioms and Language

The *language* [1] of 1st order arithmetic, $\underline{N}$ is the infinite sentence calculus $\mathfrak{L}^{(1)}$ of Appx. A, containing the following special symbols as required to express the axioms:

1. Numerals: $\underline{1}, \underline{1}', \underline{1}'', \ldots$

2. Predicate symbols: $=$ (including $<$, $\leq$, etc. as auxiliary symbols).

3. Function symbols: $S, +, \cdot$.

Added to the axioms of the language are the non-logical axioms

1. $x = y \Rightarrow (x = z \Rightarrow y = z)$.

2. $x = y \Rightarrow Sx = Sy$.

3. $\neg 1 = Sx$.

[1]The language of (first-order) intuitionistic arithmetic is $\mathfrak{L}_{\text{int}}$ (Appx. A, §3).

4. $Sx = Sy \Rightarrow x = y$.

5. **Induction:** for any formula $\mathrm{A}(x)$,

$$\mathrm{A}(\underline{1}) \Rightarrow [\{(\forall x)\,\mathrm{A}(x) \Rightarrow \mathrm{A}(Sx)\} \Rightarrow (\forall x)\,\mathrm{A}(x)].$$

These are the four axioms plus induction schema originally chosen by Giuseppe PEANO 1897, in English translation in [149].

In case of inference via the sequent calculus (Appx. A, §3.2), the following inductive rule is included:

$$\frac{\mathcal{A} \to A(1) \quad A(a), \mathcal{B} \to A(a+1)}{\mathcal{A}, \mathcal{B} \to A(t)} \quad ,$$

t a term.

B.1.1 Inference in N

The following schemata of proofs of definedness is of foundational importance for **EP**. It has been used implicitly in many places in the text.

1. $\mathbf{x+y}$. $x + 1 = Sx$ by defn. Now suppose $y \geq 1$, and that $x + y$ has already been defined. Then

$$x + Sy = S(x + y)$$

by defn. Therefore by induction on y, for all x, y, $x + y$ is defined.

2. $\mathbf{x \cdot y}$. $x \cdot 1 = x$ by defn. Now suppose $y \geq 1$, and that $x \cdot y$ has already been defined. Then

$$x \cdot Sy = x \cdot y + x$$

by defn. Therefore by induction, for all x, y, $x \cdot y$ is defined.

These, of course, are the first two [2] steps in the definition of the GOODSTEIN-ACKERMANN sequence

$$x + y,\ x \cdot y,\ x^y,\ {}^y x,\ {}_y x,\ x_y, \ldots$$

(GOODSTEIN [47], 1957, *Recursive Arithmetic*).

As to formal $\underline{\mathbf{N}}$–proofs, it may be remarked that the proof of the *associative law* $x + (y + z) = (x + y) + z$ is one of the 'earliest', in the sense that from it flow naturally the commutative, distributive, etc. laws.

As another example, the 'least number principle' (II, §2.18, ex. 2; VI, proof of thm. 5) is a theorem of $\underline{\mathbf{N}}$: for $\mathrm{A} \in \mathfrak{L}_{\text{form}}$,

$$\vdash (\exists x)\mathrm{A}(x) \Rightarrow (\exists a)\{\mathrm{A}(a)\ \&\ (\forall x)\mathrm{A}(x) \Rightarrow a \leq x\}.$$

[2] It is the subsequent step, together with BERTRAND's postulate, which enables WRIGHT to obtain the result of ex. 4 after II·7.

Lastly, as an equivalent the mathematical induction schemata, the 'method of infinite descent' (VI, §1) is $\underline{\mathrm{N}}$–provable:

$$\vdash \{(\forall x)(\neg \mathrm{A}(x+1) \Rightarrow \neg \mathrm{A}(x))\&\mathrm{A}(1)\} \Rightarrow (\forall x)\mathrm{A}(x).$$

For further references, see JEFFREY and BOOLOS [67] (1980); and GOODSTEIN [49] Ch. 6, 'Formal arithmetic'.

B.2 Meta-Theorems

1. CHURCH [20], (1936) has shown that $\underline{\mathrm{N}}$ is undecidable. This result is not to be confused with the incompleteness of $\underline{\mathrm{N}}$, Ch. I, meta-theorem 3.1, which states that there is a sentence of $\mathfrak{L}_{\mathrm{form}}$ *undecidable in* $\underline{\mathrm{N}}$.
2. The $\underline{\mathrm{N}}$–representable functions coincide with $\mathcal{R}$ (I, §3.2).
3. TARSKI's indefinability meta-theorem: first-order complete arithmetic is not $\underline{\mathrm{N}}$-definable: in the primitive recursive enumeration of $\mathfrak{L}_{\mathrm{form}}$, $\{n \mid \underline{\mathrm{N}} \vdash_{\mathrm{compl}} \mathrm{A}_n\}$ is not 'arithmetically' (see I, §4·1) definable.

B.3 Proofs of Consistency

The sequent calculus (Appx. A, §3) is tailored for the GENTZEN proof, as outlined in I, §4. In all versions [3] of the consistency proof, an essential notion is that of *reduction.* Thus, a *reduction step* on a sequent $\mathcal{S}$ eliminates from $\mathcal{S}$ some if its logical structure. $\mathcal{S}$ is in *endform* if

1. the succeedent is a true minimal formula, or
2. an antecedent is a false minimal formula.

GENTZEN proceeds by proving that

1. every initial sequent is reducible (by reduction steps) to endform; and
2. given reduction to endform of the premisses of an inference, such a reduction can be given for the conclusion.

Remark

In 1940, W. ACKERMANN published a consistency proof [2] for $\underline{\mathrm{N}}$. As with GENTZEN's argument, this also depends on gen. ind. (VI·12). It uses HILBERT's 'selection operator' (ref. [84]).

[3] GENTZEN published 2 DCC-based proofs, English translations of which may be found in SZABO's *'Collected Papers'* [143].

Appendix C

APPENDIX: TABLES

Section Headings

C.1 Summary

The following contains 2 classes of numerical and other tables:

1. information on the orders of infinity (eventual rates of growth) of functions of classical arithmetic and combinatorics.

2. the early terms of arithmetic sequences, whose terms arise from enumeration of trees directly or otherwise related to the structure of **EP**.

C.2 Standard Orders of Infinity

C.2.1 Classical Arithmetic

Few tables are available, which contain this interesting data. A list of orders of infinity, as a companion handbook to that of SLOANE [142] would be a demanding undertaking, and here only a few samples can be offered. However, the intention is negative rather than affirmative: some of these functions are quite fast-growing; but all are fiercely dominated by ROBINSON's $\star$- exponential (VIII, §2).

1. The average order of the number $d(n)$ of divisors of n is $\ln n$; of $\sigma(n)$ (XII, §2.1) is $\frac{1}{6}\pi^2 n$. For all $\delta > 0$, $d(n) = O(n^{\delta})$.

2. The nth prime $p_n \sim n \ln n$ (VIII, §10.1). For EULER's totient, $\Sigma_{i=1}^{n}\phi(i) \sim \frac{3n^2}{\pi^2}$.

C.2.2 Combinatorics

Theory of Partitions

1. Van der WAERDEN's function is defined by: $W(n)$ = the least N such that every 2−partition of $1, 2, 3, \ldots, N$ contains in one part at least an AP of length n. Present upper and lower bounds for W are wide apart:

$$2^{0.5n} < W(n) < \mathfrak{A}(n).$$

The known values are $W(1) = 1, W(2) = 3, W(3) = 9, W(4) = 35, W(5) = 178$ (GRAHAM and RÖDL, in *Surveys in Combinatorics* , C. U. P. (1987)).

2. Let $f(n, k, r)$ = the smallest N such that for every k−fold partition of the r−tuples (without repetition) of the integers $1, 2, 3, \ldots, N$ there is always a subset S of the integers, $\mid S \mid \geq n$, for which the r−tuples lie in one single part. ERDÖS and RADO used trees to show $f^{\frac{1}{n}} < k \star (r-1)$; and with HAJNAL ('Partition relations for cardinal numbers, *Acta. Math. Hungarica* **16** (1965) 93-196) that

$$k \star (r-2) < f.$$

No constructive proof has been given that, if the extra condition $\min S \geq n$ is imposed, then $\Delta \approx f$.

C.3 Arithmetic Sequences of Combinatorics

C.3.1 Standard Sequences and Arrays

Binomial Coeficients

As cited in VI, §1, the table of 2-colour RAMSEY numbers $r(m, n)$ dominates that of the binomial coefficients, attributed to PASCAL but known in ancient China:

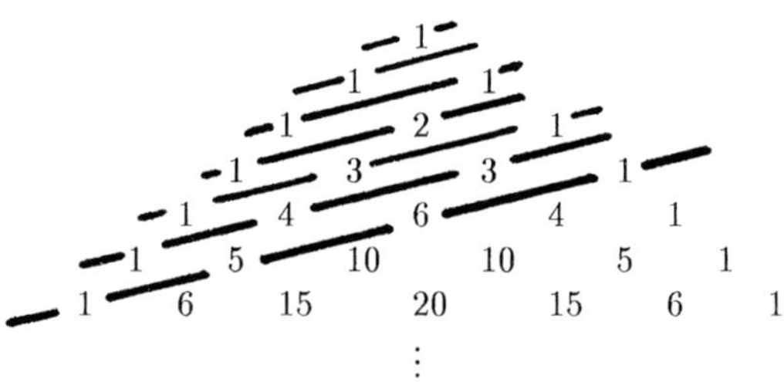

The 'rising diagonals' indicated give the FIBONNACI numbers (IV, §3), as may be seen at once, from the construction of the table.

STIRLING Numbers

STIRLING numbers, of the first and second kinds, may be obtained, as may be the binomial coefficients, by successive addition of rows. The notation (already used in Ch. VIII, section 2) is that of KNUTH [77], pp65-70. This source contains further details and references; including KAROLY (CHARLES) JORDAN, 1947, *Calculus of Finite Differences*, Ch. 4, publ. Chelsea, New York.

The STIRLING numbers of the first kind, $\left[{n \atop m}\right]$, are defined to be the moduli of the coefficients in the expansion of $\Pi_{i=n}^{1}(x-i)$. From this definition follows the recurrence

$$\left[{n \atop m}\right] = (n-1)\left[{n-1 \atop m}\right] + \left[{n-1 \atop m-1}\right].$$

This may be interpreted as the rule which generates the table

$$\begin{array}{rrrrr}
1 & & & & \\
1 & 1 & & & \\
 & 2 & 2 & & \\
\hline
1 & 3 & 2 & & \\
 & 3 & 9 & 6 & \\
\hline
1 & 6 & 11 & 6 & \\
 & 4 & 24 & 44 & 24 \\
\hline
1 & 10 & 35 & 50 & 24 \\
 & & \vdots & &
\end{array}$$

as with the binomials coefficients, except that the added rows must be successively multiplied by $1, 2, 3, \ldots$. It may be noted that the $\left[{n \atop m}\right]$ are the coefficients of x^n in the expansion of $n!\binom{x}{n}$. In the context of XII, §4, it may be noted that from the above identity follow

THEOREM C.1 *1.* $\left[{n \atop m}\right] =$ *the number of* $\sigma \in S_n$ *for which* σ *has* m *cycles.*

2. $(x)_n = \Sigma_{m=0}^{n}\left[{n \atop m}\right] \cdot x^m$. $((x)_n =_{df} x \cdot (x-1) \cdot \ldots \cdot (x-n+1))$.

□

STIRLING numbers of the *second* kind $\left\{{n \atop m}\right\}$ may be defined to be the number of ways to partition a set of n elements into m non-empty subsets. From this follows

$$\left\{{n \atop m}\right\} = m\left\{{n-1 \atop m}\right\} + \left\{{n-1 \atop m-1}\right\} = \frac{1}{n!} \cdot \sum_{i=0}^{n}(-1)^i\binom{n}{i}(n-i)^m.$$

The left identity again gives a table-generating rule:

$$\begin{array}{ccccc}
1 & & & & \\
1 & 1 & & & \\
 & 2 & 1 & & \\
\hline
1 & 3 & 1 & & \\
 & 3 & 6 & 1 & \\
\hline
1 & 6 & 7 & 1 & \\
 & 4 & 18 & 14 & 1 \\
\hline
1 & 10 & 25 & 15 & 1
\end{array} \;;$$

$$\vdots$$

the entries satisfy: $x^n = \Sigma_{m=0}^{n} \left\{ {n \atop m} \right\} (x)_m$.

Connections Between the Two Kinds of STIRLING Numbers

1. Many of the identities satisfied by the STIRLING numbers have a near-duality. E.g.

$$\sum_k \left[{n \atop k} \right] \binom{k}{m} = \left[{n+1 \atop m+1} \right]; \quad \sum_k \binom{n}{k} \left\{ {k \atop m} \right\} = \left\{ {n+1 \atop m+1} \right\}.$$

2. Let $I = \begin{bmatrix} 1 & 0 & 0 & \dots \\ 0 & 1 & 0 & \\ 0 & 0 & 1 & \\ & & \vdots & \end{bmatrix}$ be the infinite multiplicative identity.

   Now further standard identities are

$$\sum_k \left[{n \atop k} \right] \left\{ {k \atop m} \right\} = \delta_{mn};$$

$$\sum_k \left\{ {n \atop k} \right\} \left[{k \atop m} \right] = \delta_{mn}.$$

   From these, it follows

$$I = ([\ ]) \cdot (\{\ \}).$$

   This may be regarded as a companion analysis of I to that of

$$B = \begin{bmatrix} 1 & 0 & 0 & 0 & \dots \\ 1 & -1 & 0 & 0 & \\ 1 & -2 & 1 & 0 & \\ 1 & -3 & 3 & -1 & \\ \vdots & & & & \end{bmatrix}$$

   giving [1]

$$B^2 = I.$$

[1]Ref. HARDY, *Divergent Series* (1949, 1973) Ch. XI, §1.

Bell Numbers

The BELL or *exponential* numbers (II, §2.1)

$$B(1), B(2), B(3), \ldots = 1, 1, 2, 5, 15, 52, 203, 877, 4140, 21147, 115975, \ldots$$

satisfy the defining relation $B(n) = \Sigma_{m=0}^{n}\{{n \atop m}\}$.

CATALAN Numbers

In the notation following IV·17,

$$c_{n+1} = \binom{2n}{n} / \binom{n+1}{1} = \binom{2n}{n} - \binom{2n}{n+1}.$$

The EULER-SEGNER-CATALAN sequence begins

$$1, 1, 2, 5, 14, 42, 132, 429, 1430, 4862, 16796, \ldots$$

(see H. W. GOULD [51]).

The sequence is one of the most important keys to combinatorics, occurring very often. E.g. the integers are the entries of the unique (infinite) 'staircase' array

$$\begin{array}{cccccc} & & & & & \vdots \\ & & & 1 & 1 & 2 \\ & & 1 & 1 & 2 & 5 \\ & 1 & 1 & 2 & 5 & 14 \\ 1 & 1 & 2 & 5 & 14 & 42 \\ \vdots & & & & & \end{array}$$

such that if A is any square sub-array touching the boundary, the determinant $| A | = 1$ (ref. E. R. BERLEKAMP 'Unsolved Problems' in *Progress in Combinatorics* , Ed. TUTTE).

C.3.2 Enumeration Coefficients

These may be integers, functions or formal power series. Only a few examples can be given, 2 to high precision.

The Numbers of Trees on n Vertices

As n takes the values $1, 2, 3, \ldots$, the numbers of possible trees are as given.

1. Free trees: $1, 1, 1, 2, 3, 6, 11, 23, 47, 106, 235, 551, \ldots$.

2. Indexed free trees: $1, 2, 9, 64, 625, 7\,776, 117\,649, \ldots$.

3. Oriented trees: $1, 1, 2, 4, 9, 20, 48, 115, 286, \ldots$.

4. $\mathfrak{S}$-trees: $1, 1, 2, 5, 14, 42, 132, \ldots$.

5. $\mathfrak{T}$-trees: $1, 1, 2, 4, 9, 20, 48, 115, 286, 719, \ldots$.

The Numbers of Graphs and of Normal Trees

The number of graphs (connected or disconnected) on i vertices is

i	
1	1
2	2
3	4
4	11
5	34
6	156
7	1044
8	12 346
9	274 668
10	12 005 168
11	1 018 997 864
12	165 091 172 592
13	50 502 031 367 952
14	29 054 155 657 235 488
15	31 426 485 969 804 308 768
16	64 001 015 704 527 557 894 928
17	245 935 864 153 532 932 683 719 776
18	1 787 577 725 145 611 700 547 878 190 848
19	24 637 809 253 125 004 524 383 007 491 432 768
20	645 490 122 795 799 841 856 164 638 490 742 749 440
21	32 220 272 899 808 983 433 502 244 253 755 283 616 097 664
22	3 070 846 483 094 144 300 637 568 517 187 105 410 586 657 814 272
23	559 946 939 699 792 080 597 976 380 819 462 179 812 276 348 458 981 632
24	195 704 906 302 078 447 922 174 862 416 726 256 004 122 075 267 063 365 754 368.

The numbers of $\mathfrak{T}$-trees on i vertices given in the following table were obtained in 1988, using the recurrence of XII, corol. 2:

i	$\mathfrak{n}_i$	i	$\mathfrak{n}_i$
1	1	51	12221792623697519145558
2	1	52	35086098027065855916488
3	2	53	100780620321271803234688
4	4	54	289635449384901155876900
5	9	55	832818910243238821889344
6	20	56	2395882519509716300708833
7	48	57	6895866957500277715288588
8	115	58	19856988278141228513895444
9	286	59	57204756954104706980343522
10	719	60	164868857260434652052007788
11	1842	61	475364352982258385137776899
12	4766	62	1371166462998366400135821588
13	12486	63	3956614262001721208931721666
14	32973	64	11421465656125033773672476199
15	87811	65	32982186890253964688079282877
16	235381	66	95277787692773674357621397144
17	634847	67	275330186880666751227042565033
18	1721159	68	795905477379813755300857449855
19	4688676	69	2301496939036097417139000617066
20	12826228	70	6657274784058576516143594129944
21	35221832	71	19262641450296837892019025954811
22	97055181	72	55752550702414587697773379772166
23	268282855	73	161413407536932894468267613429655
24	743724984	74	467451978191925223418036844585399
25	2067174645	75	1354109035031915035033847059705011
26	5759636510	76	3923614623603032122192184580462933
27	16083734329	77	11371875316201287173100467337933777
28	45007066269	78	32967646858094047674113593469502511
29	126186554308	79	95598508336727248658168635893011500
30	354426847597	80	277280217189474477900640720862117288
31	997111512798	81	804430737080653465159079471875487200
32	2809934352700	82	2334306732596572585347883172659183044
33	7929819784355	83	6775233527681248890007585500905370888
34	22409533673568	84	19669139822761818769124816439937144322
35	63411730258053	85	57113526048816783834379895655122954244
36	179655930440464	86	165875757651937084210400003786317639688
37	509588049810620	87	481853608146765865443799477958274990077
38	1447023384581029	88	1400017748175369039429679460080519249911
39	4113254119923150	89	4068518045551941348555996021636630446066
40	11703780079612453	90	11825549924330688472148621269684099481355
41	33333125878283632	91	34378507342303500357684123889142370139600
42	95020085893954917	92	99961207861533350362270147847563729960122
43	271097737169671824	93	290705384914695761033230860449681401441522
44	774088023431472074	94	845571057856562018076489379167898267802944
45	2212039245722726118	95	2459920069628085241492658931713365455638900
46	6325843306177425928	96	7157546645942866499428538313329028764712266
47	18103111141539779470	97	20829468259101246375476715907892577283124322
48	51842285219378800562	98	60626365134393295515505702197187224240702999
49	148558992149369434381	99	176487040327433160478675408481705249763216633
50	425976989835141038353	100	513843283516593268803371363950543276441564711 .

Symmetric Polynomials

The *symmetric polynomials* (of degree k in n independent variables) of combinatorial algebra have been studied since the time of NEWTON. They are important in the theory of equations.

Definitions. The *degree* of a polynomial

$$f(x_1, x_2, \ldots, x_i, \ldots, x_n) = \sum_{j \leq N} a_j \cdot x_1^{k_1(j)} \cdots x_n^{k_n(j)}$$

(over the ring $Z = \ldots -2, -1, 0, 1, 2, \ldots$) is the number $k = \max_{j \leq N} \Sigma_{i \leq n} k_i(j)$. If there is a k such that for $\Sigma k_i(j) = k$ for all i, then f is *homogeneous* or a *form* (of degree k)(see, however II, §2·1). If every (non-zero) coefficient = 1, then f is *simple.* f is *symmetric* if f is unchanged, when 'x_i' is replaced by '$x_{\sigma(i)}$' in the expression for f, where σ is any permutation on n ciphers.

The *elementary* symmetric polynomials ${}^n\sigma_r(x_1, \ldots, x_n) = \sigma_r$ are defined by

$$\sigma_r = \sum_{1 \leq i(1) < \ldots < i(r) \leq n} x_{i(1)} \cdots x_{i(r)},$$

as $r = 1, 2, \ldots, n$. $(-1)^r \cdot \sigma_r$ is therefore the coefficient of x^{n-r} in the polynomial $P(x) = \prod_{i=1}^n (x - x_i)$; and

$$\begin{aligned} \sigma_0 &= 1 \\ \sigma_1 &= x_1 + x_2 + \ldots + x_n \\ \sigma_2 &= x_1x_2 + x_1x_3 + \ldots + x_{n-1}x_n \\ &\vdots \\ \sigma_n &= x_1x_2 \cdots x_n. \end{aligned}$$

Evidently, ${}^n\sigma_r$ is homogeneous of degree r, having $\binom{n}{r}$ terms. The *weight* of a polymonial $f(\sigma_1, \ldots, \sigma_n)$ in the elementary forms is the degree of f in $x_1, \ldots, x_n$.

The thm. at the end of VII, §2 may be proved [2] *without* double induction as follows: A polynomial may be expressed as the sum of homogeneous parts, each symmetric[3] if f is. Thus f may be presumed homogeneous. Let the terms of f be R-to-L lexicographically ordered by exponent. If $x_1^{k_1} \cdots x_n^{k_n}$ is the largest term, then by symmetry of f, all terms $x_{\sigma(1)}^{k_1} \cdots x_{\sigma(n)}^{k_n}$ occur, therefore $k_1 \leq k_2 \leq \ldots \leq k_n$. Now consider the largest term in the symmetric form $\pi = (x_1 + \ldots + x_n)^{d_n} \cdot (x_1x_2 + x_1x_3 + \ldots x_{n-1}x_n)^{d_{n-1}} \cdots ({}^n\sigma_r)^{d_{n-r+1}} \cdots (x_1x_2 \cdots x_n)^{d_1}$. π must contain the highest power in x_n, viz. $d_1 + \ldots + d_n$; then the highest in x_{n-1}, viz. $d_1 + \ldots d_{n-1}$; and generally (the largest term in σ_r being $x_{n-r+1} \cdots x_{n-1} \cdot x_n$) the largest term of π is $x_1^{d_1} \cdot x_2^{d_1+d_2} \cdots x_r^{d_1+\ldots+d_r} \cdots x_n^{d_1+\ldots+d_n}$. Now putting $d_1 = k_1$; $d_{r+1} = k_{r+1} - k_r$, $1 \leq r < n$, the highest term becomes that of f. Thus this representation may be eliminated from f. This gives a new polynomial of lower highest term. Since there are only fm possible terms of degree k, after a finite number of steps, every term of f is well-represented.

---

[2] Ref. Nathan JACOBSON (1974), §2·13 *Basic Algebra* Vol. 1, publ. W. H. Freeman, San Francisco.

[3] Since the mapping $x_i \mapsto x_{\sigma i}$ takes forms to forms.

Author Index

Bibliography

[1] W. ACKERMANN. Zum HILBERTschen Aufbau der reellen Zahlen. *Mathematische Annalen*, 99:118–133, 1928.

[2] W. ACKERMANN. Zur Widerspruchsfreiheit der Zahlentheorie. *Mathematische Annalen*, 117:162–194, 1940.

[3] W. ACKERMANN. Konstructiver Aufbau eines Abschnitts der zweiten CANTORschen Zahlenklasse. *Math. Zeitschr*, 53(5):403–413, 1951.

[4] P. G. H. ACZEL. Describing ordinals using functionals of transfinite type. *J. Symb. Logic*, 37:35–47, 1972.

[5] H. L. ADLER. Partition identities - from EULER to the present. *Amer. Math. Monthly*, 76:733–746, 1969.

[6] ANDREWS and LACHER. Translation-invariant subspaces of linear function space. *Proc. Amer. Math. Soc*, 1970.

[7] ANSELONE and KOREVAAR. Exponential polynomials. *Proc. Amer. Math. Soc*, 15:747–752, 1964.

[8] T. M. APOSTOL. *Mathematical Analysis.* Addison-Wesley, 1974.

[9] H. BACHMANN. Die Normalfunktionen und das Problem der ausgezeichneten Folgen von Ordnungszahlen. *Vierteljahrsschr. Naturf. Ges. Zürich*, 95:115, 1950.

[10] BALABAN. Enumeration of chemical trees. 1970.

[11] E. F. BECKENBACH. *Applied Combinatorial Mathematics.* Wiley, 1964.

[12] E. T. BELL. Exponential polynomials. *Annals of Math.*, 35(2), April 1934.

[13] Garrett BIRKHOFF. *Lattice Theory*, volume XXV. Amer. Math. Soc. Colloquium Publications, New York, 1940. Revised edn. 1948.

[14] B. BOLLOBÁS. *Graph Theory.* Springer-Verlag, 1988.

[15] G. BOOLOS. *Notices Amer. Math. Soc.* (under 'Feature Columns', ed. Jon BARWISE), 36(4):388–390, April 1989.

[16] J. BRIDGE. A simplification of BACHMANN's method of generating large countable ordinals. *J. Symbolic Logic*, 40, 1975.

[17] C. St. J. A. NASH-WILLIAMS. Infinite graphs-a survey. *J. Combinatorial Theory*, 3:286–301, 1967.

[18] A. CAYLEY. On the theory of the analytical forms called trees. *Phil. Mag.*, 13:172–176, 1857.

[19] A. CAYLEY. A theorem on trees. *Quarterly J. of Pure and Applied Math*, 23:376–378, 1889.

[20] A. CHURCH. An unsolvable problem in elementary number theory. *Amer. J. Math.*, 58:345–363, 1936.

[21] E. A. CICHON. A short proof of two recently discovered independence results using recursion-theoretic methods. *Proc. Amer. Math. Soc.*, 87(4), 1983.

[22] L. E. CLARKE. On CAYLEY's formula for counting trees. *J. London Math. Soc*, 33:471–474, 1958.

[23] J. P. CLEAVE and H. E. ROSE. $\mathcal{E}^n$–arithmetic. *In: Sets, Models and Recursion Theory*, 1967. Proc. Coll. Math. Logic, Leicester 1965, ed. J. N. CROSSLEY. Publ. North-Holland, Amsterdam.

[24] P. J. COHEN. *Set Theory and the Continuum Hypothesis.* Benjamin/Cummings, first printing 1966.

[25] H. DAVENPORT. *The Higher Arithmetic.* Hutchinson's University Library, London, first published 1952. 6th edn. 1992.

[26] M. DAVIS. *The Undecidable.* Raven Press, New York, 1970.

[27] N. G. de BRUIJN. A converse of CAUCHY's fundamental sequence theorem. *Simon Stevin, Wis-en Natuurkundig Tijdschrift*, 35(3):110–112, Feb. 1962.

[28] N. G. de BRUIJN. Denumerations of rooted trees and multisets. *Disc. Appl. Math.*, 6:25–33, 1983.

[29] N. G. de BRUIJN and B. J. M. MORSELT. A note on plane trees. *J. Combinatorial Theory*, 2(1):27–34, 1967.

[30] E. C. DENNIS-JONES and S. S. WAINER. *Springer Lecture Notes in Mathematics*, 1104:117–128, 1983. Logic Coll. Vol.

[31] C. WALDO DUNNINGTON. *C. F. GAUSS, Titan of Science.* Chelsea, New York, 1955.

[32] R. HONSBERGER (Ed.). *Mathematical Plums.* Math. Assoc. America, 1979.

[33] A. EHRENFEUCHT. Polynomial functions with exponentiation are well-ordered. *Algebra Universalis*, pages 261–262, 1973.

[34] P. ERDÖS and MILLS. Some bounds for .... *J. Combinatorial Theory*, 30:53–70, 1981.

[35] L. EULER. De formulis exponentialibus replicatis. *Acta Acad. Petropolitanae*, 1:38–60, 1777. In: Opera Omnia, Series prima XV, 1927.

[36] PRINS F. HARARY and TUTTE. The number of plane trees. *Indag. Math.*, 26:319–329, 1964.

[37] F. P. RAMSEY, (edited by R. BRAITHWAITE, with preface by R. G. MOORE). *The Foundations of Mathematics.* Kegan Paul, London, 1931.

[38] S. FEFERMAN. Predicative definitions. *Preprint*, 1970.

[39] S. FEFERMAN. Predicative definitions, ii. *Preprint*, 1975.

[40] G. GENTZEN. Die Widerspruchsfreiheit der reinen Zahlentheorie. *Math. Annalen*, 112:493–565, 1936.

[41] G. GENTZEN. Translated in ref. 140 as 'New version of the consistency proof for elementary number theory'. *Forschungen z. Logik u. Grundlegung d. exakten Wissenschaften, Leipzig.*, New Series(4):19–44, 1938.

[42] G. GENTZEN. Beweisbarkeit und Unbeweisbarkeit von Anfangsfällen der transfiniten Induktion in der reinen Zahlentheorie. *Math. Ann.*, 119(1):140–161, 1943.

[43] J.-Y. GIRARD. *Ann. Pure and Appl. Logic*, 1981.

[44] K. GÖDEL. Über formal unentscheidbare Sätze der Principia Mathematica und verwandter Systeme I. *Monat. für Mathematik und Physik*, 38:173–198, 1931. Part II did not appear.

[45] R. L. GOODSTEIN. On the restricted ordinal theorem. *J. Symb. Logic*, 9(2):33–41, June 1944.

[46] R. L. GOODSTEIN. Transfinite ordinals in recursive number theory. *J. Symbolic Logic*, 12(4):123–129, Dec. 1947.

[47] R. L. GOODSTEIN. *Recursive Number Theory.* North-Holland, Amsterdam, 1957.

[48] R. L. GOODSTEIN. *Essays in the Philosophy of Mathematics.* Leicester Univ. Press, 1965.

[49] R. L. GOODSTEIN. *Development of Mathematical Logic.* Logos Press, London, 1971.

[50] R. L. GOODSTEIN. *Mathematical logic.* Leicester Univ. Press, first published 1957, 2nd edn 1961.

[51] H. W. GOULD. Research bibliography of two special number sequences. *Mathematica Monongaliae*, 12, May 1971.

[52] J. W. GROSSMAN and R. S. ZEITMAN. An inherently iterative computation of ACKERMANN's function. *Theoretical Computer Science*, 57(2-3):327–330, 1988.

[53] A. GRZEGORCZYK. Some classes of recursive functions. *Rozprawy Matematyczne*, 4:1–45, 1953.

[54] R. FLUM H. D. EBBINGHAUS, R. THOMAS. *Mathematical Logic.* Springer, Berlin, 1984.

[55] P. HALL. A partition formula connected with ABELian groups. *Comment. Math. Helv*, 11:126–129, 1938.

[56] F. HARARY and E. M. PALMER. *Graph Theory.* 1976.

[57] G. H. HARDY. A theorem concerning the infinite cardinal numbers. *Quart. J. Pure Appp. Math.*, 35:87–94, 1904.

[58] G. H. HARDY. The continuum and the second number class. *Proc. Lond. Math. Soc.*, 1906.

[59] G. H. HARDY. *Orders of Infinity.* Cambridge Tracts in Mathematics and Mathematical Physics, originally published, Cambridge Univ. Press, 1910; reprinted, Hafner, New York, 1971.

[60] J. HERBRAND. Sur la non-contradiction de l'arithmetique. *J. für die reine und angewandte Mathematik*, 166:1–8, 1931.

[61] H. HERMES. *Enumerability, Decidability, Computability.* Springer, Berlin, 1965. English translation by G. HERMANN and O. PLASSMANN.

[62] G. HIGMAN. Ordering by divisibility in abstract algebras. *Proc. Lond. Math. Soc.*, 3(2):326–336, 1952.

[63] D. HILBERT and W. ACKERMANN. *Principles of Mathematical Logic.* Chelsea, New York, 1950.

[64] D. HILBERT and P. BERNAYS. *Grundlagen der Mathematik.* Springer, Berlin, 1939.

[65] E. W. HOBSON. *Theory of Functions of a Real Variable.* Univ. Press, Cambridge, 3rd edn. 1927.

[66] D. ISLES. Natural well-orderings. *J. Symb. Logic*, 36:288–300, 1971.

[67] JEFFREY and BOOLOS. *Computability and Logic.* 3rd edn. 1980.

[68] L. KALMÁR. On unsolvable mathematical problems. *Proc. 10th Internat. Phil. Congress, Amsterdam*, 1:534–536, 1948.

[69] A. Yu. KHINTCHINE. *Continued Fractions.* 3rd edn, Moscow, 1964, first publ. 1935. English translation 1964, ed. by H. EAGLE, publ. Univ. Chicgo Press; and transl. by P. WYNN.

[70] A. KINO. On provably recursive functions and ordinal recursive functions. *J. Math. Soc. Japan*, 20(3):456–476, 1968.

[71] L. KIRBY and J. B. PARIS. Accessible results for PEANO arithmetic. *Bull. London Math. Soc.*, 14:285–293, 1982.

[72] S. C. KLEENE. On notations for ordinal numbers. *J. Symbolic Logic*, 3:150–155, 1938.

[73] S. C. KLEENE. Extension of an effectively generated class of recursive functions by enumeration. *Colloquium Mathematicum*, VI, 1958.

[74] S. C. KLEENE. *Introduction to Metamathematics.* Wolters-Noordhoff/North-Holland, Gröningen and Amsterdam, 1971. First published, 1952, Bibliotheca Mathematica, Vol. I.

[75] G. T. KNEEBONE. *Mathematical Logic and the Foundations of Mathematics - an Introductory Survey.* Van Nostrand, 1963.

[76] R. A. KNOEBEL. Exponentials reiterated. *Amer. Math. Monthly*, 88, 1981.

[77] D. KNUTH. *Fundamental Algorithms.* Addison-Wesley, Reading, Mass., 1968.

[78] D. KNUTH. Big omicron and big omega and big theta. *SIGACT News*, 18, April-June 1976.

[79] D. KÖNIG. *Theorie der endlichen und unendlichen Graphen. Kombinatorische Topologie der Strekenkomplexe.* Akademische Verlagsgesellschaft, Leipzig, 1936. In English transl. as *Theory of Finite and Infinite Graphs* by R. Mc COART, 1992.

[80] G. KREISEL. On the interpretation of non-finitist proofs II. *J. Symbolic Logic*, 17(1):43–58, March 1952.

[81] J. B. KRUSKAL. Well-quasi-ordering, the tree theorem, and VAZSONYI's conjecture. *Trans. Amer. Math. Soc*, 95:210–225, May 1960.

[82] J. B. KRUSKAL. The theory of well-quasi-ordering: a frequently discovered concept. *J. Combinatorial Theory, Series A*, 13:297–302, 1972.

[83] P. LAIRD. On characterizations of exponential polynomials. *Pacific J. Math.*, 80:503–507, 1979.

[84] A. C. LEISENRING. *Mathematical Logic and HILBERT's $\varepsilon-$symbol.* Macdonald Scientific and Technical, London, 1969. Univ. Math. Series.

[85] H. LEVITZ. An ordinal bound for the set of polynomial functions with exponentiation. *Algebra Universalis*, 8:233–243, 1978.

[86] H. LEVITZ. Calculation of an order type: an application of non-standard methods. *Zeitschr. math Logik Grundl. Math*, 28:219–228, 1982.

[87] H. LEVITZ and W. NICHOLS. A natural variant of ACKERMANN's function. *Zeitschr. math. Logik Grundl. Math*, 34(5):399–401, 1988.

[88] J. E. LITTLEWOOD. Sur la distribution des nombres premiers. *Paris Compt. Rend.*, 158:1869–1892, 1914.

[89] M. H. LÖB and S. S. WAINER. Hierarchies of number-theoretic functions I, II. *Archiv für mathematische Logik und Grundlagenforschung*, 13:I:39–51; II:97–113, 1970.

[90] B. LOWESMITH. 3-valued commutative pseudo-SHEFFER functions. *Zeitschr. Math. Logik Grundl. Math.*, 27:161–180, 1981.

[91] A. MACINTYRE. Convergence of $i \star \infty$. *Proc. Amer. Math. Soc*, 17:67, 1966.

[92] A. MACINTYRE. The laws of exponentiation. *In: Springer Lecture Notes in Mathematics, 'Model Theory and Arithmetic'*, 890, 1981.

[93] K. MAHLER. *Introduction to $p-$adic numbers and their functions.* Cambridge Univ. Press (Cambridge Tracts in Mathematics # 64), 1973.

[94] H. MAURER. Über die funktion $x^{[x[x\ldots]]}$ für ganzzatliges Argument (Abundanzen). *Mittheilungen der Mathematische Gesellschaft in Hamburg*, 4:33–50, 1901.

[95] Moh'd (R.) McBETH. Exponential polynomials of linear height. *Zeitschr. math. Logik Grundlagen Mathematik*, 26(5):399–404, 1980.

[96] Moh'd (R.) McBETH. Fundamental sequences for exponential polynomials. *Zeitschr. math. Logik Grundlagen Mathematik*, 26(2):115–122, 1980.

[97] Moh'd (R.) McBETH. A generalization of ACKERMANN's function. *Zeitschr. math. Logik Grundlagen Mathematik*, 26(6):509–516, 1980.

[98] Moh'd (R.) McBETH. A second normal form for functions of the system **EP**. *Zeitschr. math. Logik Grundl. Mathematik*, 30(5):393–400, 1984.

[99] Moh'd (R.) McBETH. Normal trees. *Eur. J. Combinatorics (under consideration)*, 1986. Contains proof of the DCC.

[100] Moh'd (R.) McBETH. CAYLEY's identity for normal trees. Submitted, *Bull. London Math. Soc.*, 1987. Following collaboration with Rob STACEY, and expanded in Ch. XII of this Volume.

[101] Moh'd (R.) McBETH. Fundamental sequences for normal trees. Manuscript, 1987. Contains treatment of normal trees, and of their infinite sequences.

[102] Moh'd (R.) McBETH. A proof of generalized induction. *SIGACT News*, 19, 1987. This note forms the basis of Ch. VI, §2.

[103] Moh'd (R.) McBETH. Some finiteness properties of $(\mathbf{EP}, \Omega, \prec, \ll)$. Pre-print, 1987. Gives proofs of the uniformity result, II, thm. 47, and of the wqo property of $(\mathbf{EP}, <<)$, IX, corol. 11.

[104] Moh'd (R.) McBETH. Continued fractions and exponential forms. Pre-print, 1988. Contains the realization of the 2nd normal form V, defn. 12; of Ch. 11.

[105] Moh'd (R.) McBETH. GOODSTEIN sequences. Manuscript, 1988. Contains the **EP**-proof of the finiteness of GOODSTEIN sequences of Ch. X, § 2.

[106] Moh'd (R.) McBETH. Partitions and normal trees. *Discrete Math.*, publ. North-Holland, 79:95–101, February 1990.

[107] Moh'd (R.) McBETH. The tree structure of exponential calculations. *SIGACT News*, 17(1), summer 1985.

[108] E. MENDELSON. *Introduction to Mathematical Logic.* D. Van Nostrand, New York, Original edition, January, 1964.

[109] J. C. P. MILLER. *Table of Binomial Coefficients.* Cambridge University Press, 1954. Roy. Soc. Math. Tables, Vol. 3.

[110] L. W. MILLER. Normal functions and constructive ordinal notations. *J. Symbolic Logic*, 41(2):439–459, June 1976.

[111] G. MILLS. A Tree Analysis ... *Proc. Symp. Karpacz*, 1978.

[112] J. W. MOON. *Counting Labelled Trees.* Canadian Mathematical Congress, 1970. Canadian Mathematical Monographs.

[113] A. MOSTOWSKI. *Thirty Years of Foundational Studies.* Blackwell, Oxford, 1966.

[114] A. A. MULLIN. On a final multiplicative formulation of the fundamental theorem of arithmetic. *Zeitschr. math. Logik Grundl. Math.*, 10:159–161, 1964.

[115] A. A. MULLIN. A contribution towards computable number theory. *Zeitschr. math. Logik Grundl. Math.*, 11:117–119, 1965.

[116] D. KNUTH N. G. de BRUIJN and S. RICE. The average height of planted plane trees. *In: Graph Theory and Computing*, pages 15–22, 1972. Ed. R. C. READ. Publ. Academic Press, New York.

[117] NEDERPELT. Solution spaces. *Proc. A.M.S.*, 1960.

[118] C. D. OLDS. *Continued Fractions.* Yale University/Random House New Math. Library, 1963.

[119] R. OTTER. The number of trees. *Annals Math.*, 49:583–599, 1948.

[120] R. RADO P. ERDÖS and E. HAJNAL. Partition relations for cardinal numbers. *Acta Math. Acad. Sci. Hungarica*, 16:93–196, 1965.

[121] R. PÉTER. Konstruction nichtrekursiver Funktionen. *Math. Annalen*, 111:42–60, 1935.

[122] R. PÉTER. *Recursive Functions.* Academic Press, 1967.

[123] R. PÉTER. *Recursive Functions in Computer Theory.* Ellis Horwood, New York, 1981. Transl. by I. JUHÁSZ; ISBN 0-85312-164-8.

[124] G. POLYA. Kombinatorische Anzahlbestimmungen für Gruppen, Graphen und chemische Verbindungen. *Acta Math*, 68:145–254, 1937.

[125] M. PRESBURGER. Über die Vollständigheit .... *Kong. Mat. Warszawa*, 1929.

[126] H. RASIOWA and R. SIKORSKI. *The Mathematics of Meta-Mathematics.* Panstwowe Wydawn. Naukowe (PWN), 1960.

[127] A. RÉNYI and G. SZEKERES. On the height of trees. *a*, 1:497–507, 1975.

[128] J. RIORDAN. *An Introduction to Combinatory Analysis.* Wiley, 1958.

[129] J. ROBINSON. Existential definability in arithmetic. *Trans. Amer. Math. Soc.*, 72:437–449, 1952.

[130] R. W. ROBINSON and A. J. SCHWENK. The distribution of degrees in a large random tree. *Discrete Math.*, 12:359–372, 1975.

[131] H. E. ROSE. *Subrecursion: Functions and Hierarchies.* Oxford Univ. Press, 1984. Oxford Logic Guides, 8.

[132] H. E. ROSE. Obituary of R. L. GOODSTEIN. *Bull. London Math. Soc.*, 20:159–166, 1988.

[133] B. ROSSER. *Logic for Mathematicians.* Mc Graw-Hill / Chelsea, New York, 1953.

[134] J. B. ROSSER. Extension of some theorems of GÖDEL and CHURCH. *J. Symb. Logic*, 1:87–91, 1936.

[135] J. B. ROSSER. An informal exposition of proofs of ... *J. Symb. Logic*, 4:53–60, 1939.

[136] B. A. W. RUSSELL. Mathematical Logic as based on the theory of types. *Amer. J. Math.*, 30:222–262, 1908.

[137] R. D. SCHAFER. *An Introduction to Nonassociative Algebras.* Academic Press, New York, 1966.

[138] W. SIERPINSKI. *Cardinal and Ordinal Numbers.* PWN, 1965. 2nd edn. revised.

[139] D. SINGMASTER. An elementary evaluation of the CATALAN numbers. *Amer. Math. Monthly*, 85(5):366–368, 1978.

[140] Th. SKOLEM. Über die Nicht-charakterisierbarkeit der Zahlenreihe mittels endliche oder abzählbar unendlich vieler Aussagen mit ausschliesslich Zahlenvariablen. *Fundamenta Mathematicae*, 23, 1934.

[141] Th. SKOLEM. An ordered set of arithmetic functions representing the least ε−number. *Det Kongelige Norske Videnskabers Selskabs Forhandlinger*, 29(12), 1956.

[142] N. J. A. SLOANE. *A Handbook of Integer Sequences.* Academic Press, 1973.

[143] M. E. SZABO. *The Collected Papers of Gerhard GENTZEN.* North-Holland, Amsterdam, 1969.

[144] G. SZASZ. *Introduction to Lattice Theory.* Academic Press, 1960.

[145] L. SZEKELYHIDI. Note on exponential polynomials. *Pacific J. Math*, 103(2):583–587, 1982.

[146] G. TAKEUTI. Ordinal diagrams, II. *J. Math. Soc. Japan*, 12:385–391, 1960.

[147] G. TAKEUTI. *Proof Theory.* North-Holland Studies in Logic # 81, 1975. See also Springer Lecture Notes in Mathematics, #891.

[148] G. TAKEUTI. Fundamental sequences for ordinal diagrams. *Comment. Math. Univ. S. Paul*, 25(1):1–80, 1976.

[149] G. TAKEUTI. Ordinal diagrams. *J. Math. Soc. Japan*, 9(4), October, 1957.

[150] D. TAMARI. The algebra of bracketings and their enumeration. *Nieuw Archief voor Wiskunde*, 3:131–146, 1962.

[151] Alan M. TURING. *Proc. London Math. Soc., Series 2*, 42, 1936. Ref. p230.

[152] Alan M. TURING. *ibid.*, 43, 1937. Ref. p544.

[153] B. L. van der WAERDEN. *Modern Algebra,* Vols I, II. Frederick UNGAR, New York (original edition Julius Springer, Berlin, 1931), 1953. Revised English edition, translated by F. BLUM.

[154] J. van HEIJENOORT. *A Sourcebook in Mathematical Logic.* Harvard University Press (Source Books in the History of the Sciences), first edition 1967, 2nd printing 1971.

[155] O. VEBLEN. Continuous increasing functions of finite and transfinite ordinals. *Trans. Amer. Math. Soc.*, 9:280–292, 1908.

[156] S. S. WAINER. A classification of the ordinal recursive functions. *Archiv math. Logik Grundlagenforschung*, 13:136–153, 1970.

[157] S. S. WAINER. Ordinal recursion and a refinement of the extended GRZEGORCZYK hierarchy. *J. Symb. Logic*, 37(2):281–292, June 1972.

[158] E. T. WHITTAKER and G. N. WATSON. *Modern Analysis.* University Press, Cambridge, 1960.

[159] R. L. WILDER. *Introduction to the Foundations of Mathematics.* John Wiley, New York, 1965. First published, 1952, chapman & Hall.

[160] M. DUMMETT (with R. MINIO). *Elements of Intuitionism.* Clarendon Press, Oxford, 1977.

[161] E. M. WRIGHT. *Amer. Math. Monthly* , **58**, 616-618 (1951).

[162] A. YASUHARA. *Recursive Function Theory and Logic.* New York, 1971.

[163] A. YESSENIN-VOLPIN. In: Ch. A· 1, *Intuitionism and Proof Theory,* publ. North-Holland, 1970. Ed. A. KINO, J. MYHILL, R.E. VESLEY.

Index

— D —

— **E** —

— **F** —

— G —

— **P** —

— **Q** —

— **R** —